PROGRESS IN MODERN HYDROLOGY: PAST, PRESENT AND FUTURE

Teddington Weir: This is the lowest gauging station on the Thames and provides the longest continuous record of flow in the UK, and one of the longest flow records in the world. Systematic flow measurement began in 1883 and there is a wealth of documentary evidence relating to historical floods. Prior to the twentieth century, snowmelt-aggravated flood events were more common; extreme examples include 1809, when a number of bridges across the Thames were destroyed. Urbanisation and agricultural intensification have influenced flood flows to the river system, and river engineering works particularly following the disastrous 1947 flood have increased the capacity of the Thames. Thus, while peak flows exhibit little trend, peak river levels – the primary cause of overbank flooding – decreased appreciably through the twentieth century. (Source: Reproduced by permission of Environment Agency.)

Progress in Modern Hydrology is dedicated to the memory of Dr Jim McCulloch who passed away while the book was being prepared. Jim will be remembered for his distinguished leadership of the Institute of Hydrology and his tireless promotion of the science during a long career.

This volume was prepared to mark the 50th anniversary of the founding of the Natural Environment Research Council in 1965 and to highlight in excess of 50 years of hydrological research at Wallingford; first at the Hydraulics Research Station, then at the Hydrological Research Unit and the Institute of Hydrology and currently at the Centre for Ecology and Hydrology.

Progress in Modern Hydrology: Past, Present and Future

Edited by

John C. Rodda and Mark Robinson

*Honorary Fellows at the Centre for Ecology and Hydrology,
Maclean Building, Crowmarsh Gifford, Wallingford,
Oxfordshire, OX10 8BB, UK*

WILEY Blackwell

Library of Congress Cataloging-in-Publication Data

Progress in modern hydrology : past, present and future / edited by John Rodda and Mark Robinson.
pages cm
Includes bibliographical references and index.
ISBN 978-1-119-07427-4 (cloth)
1. Institute of Hydrology (Great Britain)–Research. 2. Centre for Ecology and Hydrology (Great Britain)–Research. 3. Hydrology–Research. I. Rodda, J. C., editor. II. Robinson, M. (Mark), 1953- editor.
GB658.8.G7P76 2015
551.48–dc23

2015018213

A catalogue record for this book is available from the British Library.

Wiley also publishes its books in a variety of electronic formats. Some content that appears in print may not be available in electronic books.

Cover image: Flooding of the River Thames at Wallingford, Oxfordshire taken by RAF Benson on 7th January 2003, several days after the peak which was the highest since 1947 in many reaches of the Thames. The Centre for Hydrology and Ecology is situated above the flood waters at the top left hand corner of the photograph. (Source: Crown Copyright 2003, Reproduced under Open Government Licence v2).

Set in 9.5/11.5pt Trump Mediaeval by Laserwords Private Limited, Chennai, India.
Printed in Singapore by C.O.S. Printers Pte Ltd

1   2015

Contents

List of Contributors

MIKE C. ACREMAN *Centre for Ecology and Hydrology, Wallingford, Oxfordshire, UK*

NIGEL ARNELL *Walker Institute for Climate Research, University of Reading, UK*

JAMES BATHURST *Dept Civil Engineering & Geoscience, University of Newcastle upon Tyne, UK*

JOHN BELL *Ex-Institute of Hydrology, Wallingford, UK*

MAX BERAN *Ex-Institute of Hydrology, Wallingford, UK*

KEITH BEVEN *Lancaster Environment Centre, University of Lancaster, Lancaster, UK*

JIM BLACKIE *Ex-Institute of Hydrology*

JAMES R. BLAKE *Centre for Ecology and Hydrology, Wallingford, Oxfordshire, UK*

ELEANOR BLYTH *Centre for Ecology and Hydrology, Wallingford, Oxfordshire, UK*

MIKE BOWES *Centre for Ecology and Hydrology, Wallingford, Oxfordshire, UK*

JOHN BROMLEY *School of Geography & Environment, University of Oxford, Oxford, UK*

DAVID J. COOPER *Ex-Centre for Ecology and Hydrology, Wallingford, UK*

LAURENCE R. CARVALHO *Centre for Ecology and Hydrology, Edinburgh, Midlothian, UK*

HARRY DIXON *Centre for Ecology and Hydrology, Wallingford, Oxfordshire, UK*

MIKE J. DUNBAR *Centre for Ecology and Hydrology, Wallingford, Oxfordshire, UK*

JON EVANS *Centre for Ecology and Hydrology, Wallingford, Oxfordshire, UK*

FRANK FARQUHARSON *Water Resource Associates, Henley on Thames, UK*

DUNCAN FAULKNER *JBA Consulting, Skipton, UK*

KEVIN GILMAN *Ex-Institute of Hydrology, Wallingford, UK*

IAIN D. M. GUNN *Centre for Ecology and Hydrology, Edinburgh, Midlothian, UK*

ALAN GUSTARD *Ex-Institute of Hydrology*

CATE GARDNER *IAHS Press, Wallingford, Oxfordshire, UK*

JAMIE HANNAFORD *Centre for Ecology and Hydrology, Wallingford, Oxfordshire, UK*

ATUL HARIA *Ex-Centre for Ecology and Hydrology, Wallingford, UK*

RICHARD HARDING *Centre for Ecology and Hydrology, Wallingford, Oxfordshire, UK*

MARTIN HODNETT *Ex-Centre for Ecology and Hydrology, Wallingford, UK*

HELEN HOUGHTON-CARR *Centre for Ecology and Hydrology, Wallingford, Oxfordshire, UK*

JIM HUDSON *Ex-Centre for Ecology and Hydrology, Wallingford, UK*

HELEN JARVIE *Centre for Ecology and Hydrology, Wallingford, Oxfordshire, UK*

ALAN JENKINS *Centre for Ecology and Hydrology, Wallingford, Oxfordshire, UK*

ANDREW JOHNSON *Centre for Ecology and Hydrology, Wallingford, Oxfordshire, UK*

IAN D. JONES *Centre for Ecology and Hydrology, Bailrigg, Lancaster, UK*

CELIA KIRBY *British Hydrological Society, UK*

THOMAS KJELDSEN *Department of Architecture and Civil Engineering, University of Bath, Bath, UK*

CEDRIC LAIZÉ *Centre for Ecology and Hydrology, Wallingford, Oxfordshire, UK*

GRAHAM LEEKS *Centre for Ecology and Hydrology, Wallingford, Oxfordshire, UK*

MELINDA LEWIS *British Geological Survey, Wallingford, Oxfordshire, UK*

IAN LITTLEWOOD *British Hydrological Society, UK*

DAVID JONES *Ex-Centre for Ecology and Hydrology, Wallingford, UK*

COLIN LLOYD *Centre for Ecology and Hydrology, Wallingford, Oxfordshire, UK*

STEPHEN C. MABERLY *Centre for Ecology and Hydrology, Bailrigg, Lancaster, UK*

ELEANOR B. MACKAY *Centre for Ecology and Hydrology, Bailrigg, Lancaster, UK*

TERRY MARSH *Centre for Ecology and Hydrology, Wallingford, Oxfordshire, UK*

LINDA MAY *Centre for Ecology and Hydrology, Edinburgh, Midlothian, UK*

JIM MCCULLOCH *Ex-Institute of Hydrology, Wallingford, UK*

CHRISTINE MCCULLOCH *School of Geography and Environment, University of Oxford, Oxford, UK*

ROGER MOORE *British Geological Survey, Wallingford, Oxfordshire, UK*

J. OWEN MOUNTFORD *Centre for Ecology and Hydrology, Wallingford, Oxfordshire, UK*

COLIN NEAL *Centre for Ecology and Hydrology, Wallingford, Oxfordshire, UK*

MALCOLM NEWSON *Tyne Rivers Trust, Corbridge, Northumberland, UK*

ENDA O'CONNELL *Dept. Civil Engineering & Geoscience, University of Newcastle upon Tyne, Newcastle upon Tyne, UK*

HOWARD OLIVER *Ex-Institute of Hydrology*

SYLVIA OLIVER *Ex-Institute of Hydrology*

CHRISTEL PRUDHOMME *Centre for Ecology and Hydrology, Wallingford, Oxfordshire, UK*

GWYN REES *Centre for Ecology and Hydrology, Wallingford, Oxfordshire, UK*

NICK REYNARD *Centre for Ecology and Hydrology, Wallingford, Oxfordshire, UK*

MARK ROBINSON *Centre for Ecology and Hydrology, Wallingford, Oxfordshire, UK*

JOHN C. RODDA *Centre for Ecology and Hydrology, Wallingford, Oxfordshire, UK*

ANDRAS SZOLLOSI-NAGY *UNESCO-IHE Institute for Water Education, Delft, The Netherlands*

BRYAN M. SPEARS *Centre for Ecology and Hydrology, Edinburgh, Midlothian, UK*

LISA STEWART *Centre for Ecology and Hydrology, Wallingford, Oxfordshire, UK*

IAN STRANGEWAYS *Terradata, Wallingford, UK*

CHARLIE J. STRATFORD *Centre for Ecology and Hydrology, Wallingford, Oxfordshire, UK*

JOHN SUTCLIFFE *Ex-Institute of Hydrology, Wallingford, UK*

CHRIS TAYLOR *Centre for Ecology and Hydrology, Wallingford, Oxfordshire, UK*

STEPHEN J. THACKERAY *Centre for Ecology and Hydrology, Wallingford, Oxfordshire, UK*

HELEN WARD *Department of Meteorology, University of Reading, Reading, UK*

PAUL WHITEHEAD *School of Geography and the Environment, University of Oxford, Oxford, UK*

RICHARD WILLIAMS *Centre for Ecology and Hydrology, Wallingford, Oxfordshire, UK*

IAN J. WINFIELD *Centre for Ecology and Hydrology, Bailrigg, Lancaster, UK*

ANDY YOUNG *Wallingford HydroSolutions Ltd., Wallingford, Oxfordshire, UK*

Chapter Reviewers

ROBIN CLARKE *Institute of Hydraulic Research, Federal University of Rio Grande do Sul, Porto Alegre, Brazil*

CLAUDE COSANDEY *Centre National de la Recherche Scientifique (CNRS), Paris, France*

CON CUNNANE *Department of Civil Engineering, University College, Galway, Ireland*

ROBERT GURNEY *Department of Earth Observation Science, University of Reading, Reading, UK*

ZBYSZEK KUNDZEWICZ *Institute for Agricultural and Forest Environment, Polish Academy of Sciences, Poznan, Poland*

FRANK LAW *Ex-Institute of Hydrology*

LARS ANDREAS ROALD *Norwegian Water Resources and Energy Directorate (NVE) Oslo, Norway*

JULIAN THOMPSON *Wetland Research Unit, Department of Geography, University College, London, UK*

JIM SHUTTLEWORTH *Department of Hydrology and Water Resources, University of Arizona, Tucson, USA*

ANNE VERHOEF *Department of Geography and Environmental Science, University of Reading, Reading, UK*

ANDREW WADE *Department of Geography and Environmental Science, University of Reading, Reading, UK*

ROY WARD *Ex-Geography, Environment and Earth Sciences Department, University of Hull, Hull, UK*

ERIC WOOD *Department of Civil and Environmental Engineering, Princeton University, Princeton, USA*

Foreword

From my various vantage points, first behind the Iron Curtain, then in Vienna and Paris, and most recently in Delft, the hydrological scene visible across the seas to those islands fringing the west coast of the Continent has always intrigued and attracted me. A number of scientific landmarks loom out of the mists of space and time: Imperial College London and the universities at Newcastle, Dublin and Galway, to name but a few. However, probably the most noticeable feature is at Wallingford, which currently has the name the Centre for Ecology and Hydrology. So to be asked to write a Foreword to a volume recording the achievements of researchers at Wallingford, perhaps the most celebrated of these institutions, is indeed an honour – something I could not have anticipated when my career in hydrology commenced.

In 1965 I was a senior high school student in Hungary. In that hot summer I worked as an observer at the Bakonynána Hydrological Experimental Station that was set up within the framework of the just launched International Hydrological Decade of UNESCO. My task was to measure infiltration in the field by the Münz-Laine apparatus and compare results with a small artificial rainfall generator developed by a Hungarian agrometeorologist, Béla Kazó. I moved from one site to the other accompanied by a dozen spotty cows who happily slurped water from the Münz–Laine apparatus. Therefore, I had to rely on the Kazó-equipment that had a circulating horizontal bar one metre high with holes in it through which I generated rainfall. I measured the outflow through a V-notch in a cylinder hammered into the soil and plotted the time series of the difference between the two flows until the soil became saturated. Then I moved a few metres, set up the equipment and started the rain simulation again. To my great horror the new infiltration curve parameters were totally different from those of the previous measurement! Then I repeated the experiment at several sites across the field. The results scattered a great deal. I could not explain this and felt that brainwise, I must have lost my senses to my faithful companions, the cows. However, the experiment captured my interest in what happens to the rain. I was sixteen and wanted to become a hydrologist.

Eight years passed and an Austrian friend smuggled out to Vienna two of my manuscripts on Kalman-filtering in hydrological forecasting and mailed them to Professor Eamonn Nash as a potential contribution to the Mathematical Modelling in Hydrology Symposium that Nash was organizing in Galway in 1974. A short while later an invitation came from Professor J. C. I. Dooge to visit him in Dublin, en route to Galway, as he had read my papers and wanted clarification. He received me at Dublin airport and declared upon my arrival that my papers were a piece of junk. He invited me to stay in his house because he wanted to understand 'all the stochastic nonsense', as he put it, that I had written. He also said that I should walk him through every mathematical derivation in my papers every morning before he went into his office. I did that diligently so at the end of the week he said fine, let's go to church. Those days Hungary was very much behind the Iron Curtain, and when I submitted my application for an exit visa I had a flat refusal from the communist authorities, even a rebuttal: 'how dare you keep contacts with people from the West'. Again, through clandestine channels, I informed Professor Dooge of what had happened. What he did to get me out of Hungary

I learned only decades later, but that is a different story. At that time he was the Speaker of the Senate in the Irish Parliament and had some sizeable political clout that he used in order to get me out.

Although at the time Nash and Dooge were two of my idols, I still wanted to use the opportunity to visit the world famous Institute of Hydrology at Wallingford. The Institute for me, as a greenhorn idealist hydrologist, was the Mecca of hydrology, and I wanted to meet the big names in life, including Dr McCulloch, Director, and Dr John Rodda whom I considered as the Holy Men of Hydrology. Jim McCulloch received me, a real nobody in my shabby clothes from Central-Europe, in his posh office at the Institute and with his usual charming direct style asked me: 'What do you want?' I responded that I wanted to learn about the fantastic results of British hydrology and meet the Great Men of Hydrology who changed the empirical muddy waters into a science. 'You have come to the right place and to the right man, young lad', declared Jim.

For decades after my visit to Wallingford, I followed with great admiration what IH did in fundamentally changing the landscape of hydrology, covering a wide range of research topics from experimental methods to flood studies.

The past fifty years has indeed fundamentally changed hydrology. A long journey has been covered from the Instantaneous Unit Hydrograph (which does not exist by the way, yet we still find seventy thousand Google hits on it) to complex nonlinear structured-stochastic 2D models that simulate part of the hydrological cycle, or even the whole cycle at different temporal and spatial scales. Thanks to the fantastic developments in computational and remote sensing technology over the past three decades, many made at Wallingford, basically we do not have any more computational barriers and we can measure almost anything we want in the hydrological cycle. Theories have advanced a great deal, techniques to deal with georeferenced data are common tools at present – yet the situation in the developing world changes very slowly. The passage of time has not improved data availability: for example we possess less data on the hydrology of Africa now than decades ago. Most of our global models are calibrated against data from North America, Europe, Japan and the Murray Darling basin. Conclusions from such biased data sets are necessarily biased themselves. Yet we build policies on those shaky models. Inevitably, those policies are shaky as well. It is, therefore, our common responsibility to build hydrological capacities and ground truthing capabilities for and in the developing countries. Otherwise the sustainability of our common systems will never be achieved. It is so simple: If you don't measure it, you can't manage it.

For many decades IH was *the* lead in hydrological sciences, covering all the elements of the hydrological cycle. I pay tribute to all those fantastic people who contributed to that success and made it happen. This volume gives a wonderful historical summary of the golden days of hydrological science as led by Wallingford. Just read it, and you will see what an incredibly huge impact IH made on what we do today as hydrological science.

Andras Szollosi-Nagy
Rector of the UNESCO-IHE Institute
for Water Education
Former Secretary of the International Hydrological
Programme of UNESCO

Preface

This book is about hydrology, the science of water, and about research in that science. It is unique amongst the volumes that address these subjects because it describes the advances achieved by a particular group of researchers over the last 50 years. This group, the present and past members of the Centre for Ecology and Hydrology at Wallingford, in the United Kingdom, record the research they undertook and its possible future directions – a group which grew from a handful of scientists and engineers to become a multidisciplinary team several hundred strong. Their story describes a great variety of national and international projects, many at the cutting edge of the science, together with the changes in emphasis over time. It consists of twelve chapters, each written by those who have together researched that particular topic. In the 1960s, when the story starts, climate change was not the raging issue it is currently, but then as now, floods, droughts and water resources were. The wide range of topics addressed include impacts of land use change; nitrates in drinking water, acid rain; floods and droughts; the degradation of wetlands; instrument development and many others. The half century spans the transformations brought about by the digital revolution and the advent of the space age.

Of course, concepts and theories that are accepted now have origins somewhere in the past. Many achievements and failures, although history never exactly repeats itself, provide guidance for tackling future challenges. Strong vested interests may have to be overcome and funding cycles accommodated, as the attention of policy makers switches to the immediate and pressing problem, while yesterday's are conveniently forgotten. Funding for droughts and floods research follows the pattern of attention to the natural environment, but a credible level of scientific expertise needs to be maintained across the spectrum of science to combat them and other challenges as they arise.

Water, in the world of Plato and Aristotle, was one of the four fundamental elements. Today, along with shelter, food and energy, it is the basis of civilised society. Without freshwater life would stop, not only human life, but virtually every other form of life on the earth. Indeed, water makes up the bulk of most living things, and it is the main medium for transporting energy and materials across interfaces at scales from the molecular to the global. Water, particularly freshwater, may be thought of as the cement that binds the living world together – the supreme integrating substance. This colourless, tasteless, odourless fluid defies many of the laws of physics and chemistry. Its unusual attributes condition life across the globe. However, water engenders risk to humans, in the form of floods, droughts and security of supply. Polluted water harbours disease: moving water and ice are the main agents of erosion and deposition. Taken for granted where plentiful, a prized possession where scarce, the only substance common in all three phases at normal temperatures and pressures, the abundance of water is *the* single most important feature that sets this planet apart from its fellows in the solar system. However, the global distribution of freshwater is irregular in space and time. This renders water prone to disagreements and disputes within and between societies from headwater streams to the largest river basins. Where basins and aquifers are shared between nations, equitable use of water resources is the aim of the protocols and treaties that have been negotiated. But the majority of these problems are eclipsed by those of *water governance*, an often neglected domain: governance deals with the organisational systems that administer water. Water services are generally weak and fragmented, locally, nationally and internationally and, as a consequence, ill-equipped to deal with the growing pressures on water during the twenty-first century. Amongst the responsibilities of these services, hydrology and hydrological research often feature low on the list of priorities. Few nations operate a unified hydrological service dealing with the whole of the hydrological cycle; if there is research, it is usually spread thinly across universities. Although the situation in the water sector in the United Kingdom is not ideal, the strength of the hydrological research undertaken by the Centre for Ecology and Hydrology and its predecessors provides a capability available to few other countries Consequently, Wallingford hydrology has gained an enviable reputation in the United Kingdom and around the world, amongst different nations and across international bodies.

Acknowledgements

The last 50 years have seen tremendous scientific and technological advances, few more so than in our understanding of the science of hydrology. So when we, as co-editors, first put forward the idea of a recent history of hydrology seen through the eyes of researchers and practitioners based at Wallingford, we received much encouragement from past and present colleagues, together with the inevitable one or two doubtful voices.

We are very grateful for all the support from many individuals and organisations in the preparation of *Progress in Modern Hydrology*. These include current staff at Wallingford, colleagues who have gone on to pursue research and teaching interests in other organisations, as well as a surprising number of retired colleagues who were able to provide invaluable guidance to the background of many key developments that we all too often take for granted.

We are grateful to CEH for the help and encouragement we have received, while we are particularly indebted to Alan Jenkins, John Griffin, John Gash, Katie Muchan, Adrian Smith and Dee Galliford. Celia Kirby, Ian Littlewood and Tim Fuller of the British Hydrological Society have supported us in a number of ways for which we thank them. Grateful thanks are also due to Harvey Rodda and Sabine Rodda of HYDRO-GIS Ltd for their assistance in the preparation of this volume.

Unless it is otherwise stated the illustrations in this volume are reproduced with permission from CEH/NERC for which our thanks are due. We also thank those individuals who have provided us with additional illustrations for their permission to reproduce them. We apologise to the owners of any material not covered by the above statements for our unwitting infringement of their copyrights.

All Royalties from the sale of this book will go the British Hydrological Society (BHS), in its work to support the professional development of young UK hydrologists. It provides travel grants to attend scientific meetings and a small number of studentships towards the tuition costs of Masters degrees at UK Higher Education Institutions.

John C. Rodda and Mark Robinson
Centre for Ecology and Hydrology, Wallingford

We thank our wives, Annabel and Mary, for their considerable support and forbearance during the preparation of this volume.

Acronyms

ABACUS *Arctic Biosphere Atmospheric Coupling at Multiple Scales*

ABRACOS *Anglo-Brazilian Amazonian Climate Observation Study*

ADB *Asian Development Bank*

ADC *Analogue-to-Digital Converter*

AERE *Atomic Energy Research Establishment*

AWS *Automatic Weather Station*

AMMA *African Monsoon Multidisciplinary Analysis*

ARME *Amazon Region Meteorological Experiment*

BATS *Biosphere-Atmosphere Transfer Scheme*

BFI *Base Flow Index*

BGS *British Geological Survey*

BHS *British Hydrological Society*

BL *Boundary Layer*

BOREAS *BOReal Ecosystem-Atmosphere Study*

CAMS *Catchment Abstraction Management Strategy*

CCM *Community Climate Model*

CEH *Centre for Ecology and Hydrology*

CERF *Continuous Simulation of River Flow Model*

CESBIO *Centre d'Etudes Spatiales de la BIO-sphere*

CNPq *Conselho Nacional de Desenvolvimento Cientifico e Tecnologico*

CNRM *Centre National de Recherches Météoro-logiques*

CONGAS *biospheric CONtrols on trace GAS fluxes in northern wetlands*

COSMOS *COsmic-ray Soil Moisture Observing System*

CSIRO *Commonwealth Scientific and Industrial Research Organisation*

CWPU *Central Water Planning Unit*

DAS *Data Acquisition System*

DEFRA *Department for the Environment, Food and Rural Affairs*

DFID *Department for International Development*

DOE *Department of the Environment*

EA *Environment Agency*

EEC *European Economic Community*

EFEDA *ECHIVAL Field Experiment in Desertification threatened Areas*

EMBRAPA *Empresa Brasileira de Pesquisa Agropecuária*

EOP *Extensive Observation Period*

ESCOBA *European Study of Carbon in the Oceans, Biosphere and Atmosphere*

EU *European Union*

EUROFLUXNET *European section of the Fluxnet community*

EUSTACH *European Studies on TRace gases and Atmospheric CHemistry*

FBA *Freshwater Biological Association*

FDC *Flow Duration Curve*

FEH *Flood Estimation Handbook*

FIFE *First ISLSCP Field Experiment*

FLUXNET *international Flux Network*

FSR *Flood Studies Report*

FREND *Flow Regimes from Experimental and Network Data*

FRIEND *Flow Regimes from International Experimental and Network Data*

GCM *Global (or General) Climate (or Circulation) Model*

GWAVA *Global Water Availability Assessment model*

HAPEX *Hydrologic Atmospheric Pilot EXperiment*

HBVMOR *Coupled conceptual model*

HKH FRIEND *FRIEND Project in Hindu Kush Himalayas*

HOST *Hydrology of Soil Type*

HRS *Hydraulics Research Station*

HRU *Hydrological Research Unit*

HYCOS *Hydrological Cycle Observation System*

HydrA-HP *Software package for estimating hydro-power potential*

IAHS *International Association of Hydrological Sciences from 1971*

IASH *International Association of Scientific Hydrology to 1971*

ICE *Institution of Civil Engineers*

ICRISAT *International Crops Research Institute for the Semi-Arid Tropics*

IFC *Intensive Field Campaign*

ICSU *International Council for Science*

IFIM *Instream Flow Incremental Methodology*

IGBP *International Geosphere Biosphere Programme*

IH *Institute of Hydrology*

IHD *International Hydrological Decade*

IHP *International Hydrological Programme*

INPA *Instituto Nacional de Pesquisas da Amazônia*

INPE *Instituto Nacional de Pesquisas Espaciais*

IPCC *Intergovernmental Panel on Climate Change*

IPY *International Polar Year*

ISLSCP *International Satellite Land Surface Climatology Project*

LAPP *Land Arctic Physical Processes*

LBA *Large scale Biosphere-atmosphere experiment in Amazonia*

LFHG *Low Flow Host Groups*

LOCAR *Lowland Catchment Research Programme*

Low Flows 2000 *Software system for low flow estimation*

MAFF *Ministry of Agriculture, Fisheries and Food*

MAM (10) *Mean of the 10 day Annual Minima*

MEDIFLUX *Mediterranean focused part of Fluxnet community*

Met Office *Meteorological Office*

Micro LOW FLOWS *Software system for low flow estimation*

MOBILHY *MOdelisation du BILan Hydrique*

MOSES *Meteorological Office Surface Exchange Scheme*

NASA *National Aeronautics and Space Administration*

NCAR *National Center for Atmospheric Research*

NERC *Natural Environment Research Council*

NOPEX *NOrthern hemisphere climate-Processes land-surface Experiment*

NRA *National Rivers Authority*

ODA *Overseas Development Administration*

ORSTOM *Office de la Recherche Scientifique et Technique Outre-Mer*

OHP *Operational Hydrology Programme*

PBDM *Physically Based Distributed Model*

PDM *Probability-Distributed Model*

Q95 *Flow exceeded for 95% of the time*

Q95 (10) *10-day average flow exceeded by 95% of 10-day average flows*

SA FRIEND *FRIEND project in Southern Africa*

SEBEX *Sahelian Energy Balance EXperiment*

SEPA *Scottish Environmental Protection Agency*

SiB *Simple Biosphere*

SOP *Special Observation Period*

STEPPS *Snow in Tundra Environments: Patterns, Processes and Scaling*

SVAT *Soil-Vegetation-Atmosphere Transfer*

SVP *Saturated Vapour Pressure*

TDR *Time Domain Reflectometry*

TFS *Terrestrial and Fresh Water Sciences Directorate of NERC*

TIGER *Terrestrial Initiative in Global Environmental Research*

TIS *Thermometer Interchange System*

UN *United Nations*

UNESCO *United Nations Educational Scientific and Cultural Organisation*

VPD *Vapour Pressure Deficit*

WAM *West African Monsoon*

WCRP *World Climate Research Programme*

WDU *Water Data Unit*

WFD *European Water Framework Directive*

WHS *Wallingford HydroSolutions*

WINTEX *WINTer EXperiment (of NOPEX)*

WMO *World Meteorological Organisation*

WRA *Water Research Association*

WRB *Water Resources Board*

1 Introduction

JOHN C. RODDA[1], MARK ROBINSON[1], JIM MCCULLOCH[2],
CHRISTINE MCCULLOCH[3], ALAN JENKINS[1],
TERRY MARSH[1], CELIA KIRBY[4], IAN LITTLEWOOD[4],
MAX BERAN[2], AND GRAHAM LEEKS[1]

[1]Centre for Ecology and Hydrology, Wallingford, Oxfordshire, UK
[2]Ex-Institute of Hydrology, Wallingford, UK
[3]School of Geography and Environment, University of Oxford, Oxford, UK
[4]British Hydrological Society, UK

1.1 Starting Point

Hydrology can claim to be one of the *oldest*, yet one of the *youngest* of the natural sciences. Since the start of civilisation it has been applied to control and manage water; yet, hydrology only emerged as a distinct scientific discipline during the latter half of the twentieth century.

This book reviews the development of modern hydrology primarily, but not exclusively, through the experiences of the scientists and engineers at Wallingford, near Oxford, who have been at the forefront of many of the developments in hydrological research over last 50 years. Coming from very differing academic backgrounds, they form an effective multidisciplinary team, one which includes a number of foreign members. Together they describe the results of their scientific research which seeks to improve understanding of the vicissitudes of the hydrological cycle in order to apply knowledge for a variety of purposes. These include data collection on the ground and from space, together with the management and application of these data to: prediction and forecasting of extremes, husbanding surface and groundwater water resources and the control of pollution. Initially, this research aimed to answer to just one question – what is the impact of land use on water resources? But as might be expected, seeking this answer raised a host of other questions. Indeed, as time progressed the ambit of Wallingford hydrology has widened immeasurably to the extent that the complexion of research in 2015 includes hues that were not even thought about in 1965. The new knowledge generated by research emanating from Wallingford has acted as a catalyst on the hydrological community in the United Kingdom and internationally, both informally and through specific technology transfer initiatives.

1.2   Setting the Scene

The story of progress in hydrological research at Wallingford (Chapter 1) mirrors that of advances made by similar institutions in several other nations and in the developments achieved at the international level. Essential to tackling practical questions and advancing scientific understanding is the collection of high-quality data on the ground and from space. This necessitated the development of specialised instrumentation, and often their conjunctive use in experimental basins (Chapter 2). An early concern was the study of extreme flows – initially floods – but later droughts – the importance of the variability of the hydrological cycle in space and time was seen as a key issue (Chapter 3). Central to most hydrological processes is the role moisture in the soil, influencing recharge of groundwater, determining the availability of water for plants and generating runoff (Chapter 4). It became apparent that it was not sufficient to observe differences in behaviour between sites and basins, since the reasons 'why' needed to be understood and predicted through process studies (Chapter 5). Although much of the hydrological development work and its application was UK based, there has always been a strong overseas component to the work at Wallingford, especially in water resources (Chapter 6). As hydrological understanding developed and grew, modelling and prediction of the extremes of flow improved (Chapter 7), along with understanding of water

quality issues (Chapter 8). There is widening recognition of the importance of an ecosystem approach (Chapter 9). The application of the experience gained is needed to meet the challenges of climate change, where water has a key role (Chapter 10). The generation of large amounts of data need skills in quality control, handling, storage and retrieval (Chapter 11). Finally, Chapter 12 tries to capture the essence of the gain in hydrological understanding and looks to future developments in hydrology and the pressures which generate them.

The United Kingdom shares most of the hydrological problems that affect other nations, excepting those concerned with glaciers, ice and large rivers. It occupies one relatively large island, part of another and a series of smaller islands covering a total area of nearly $250,000 \, km^2$ on the eastern edge of the Atlantic Ocean. Westerly airflows predominate but weather systems can arrive from all points of the compass. Consequently, the generally temperate climate is inherently very variable. Ancient igneous and metamorphic formations elevate the north and west, whereas more recent sedimentary deposits floor the south and east. The drainage patterns across the United Kingdom are largely a response to regional and more local contrasts in geology. Carved into the landscape are almost 1500 discrete river systems draining to the sea through over 100 estuaries. The rivers and streams are major agents of landscape modification and, in turn, their characteristics are greatly influenced by the catchments through which they flow. The diverse patterns of climate, topography, geology and land use make for a rich variety of watercourses and aquatic environments.

The regular passage of Atlantic low-pressure systems ensures that the United Kingdom is one of the wettest countries in Europe. Average annual rainfall totals grade across the country from over 4000 mm in the higher north and west to less than 600 in the lower south and east. The daily maximum recorded rainfall has once reached 280 mm and totals exceeding 100 mm are not uncommon in the western highlands. However, short period rainfall intensities are relatively low on the world scale, but the number of days with rainfall is high. Absolute droughts extending over 45 days or more are rare, but accumulated rainfall deficiencies over periods of six months or more can have substantial impacts on society, particularly in relation to water resources.

On average, rainfall is evenly distributed throughout the year but with a tendency towards a late autumn and early winter maximum. By contrast, evaporation losses are highly seasonal with around 80% of the average annual evaporation normally occurring over the April–September period. Evaporation is a relatively stable variable with annual losses generally in the 400–560 mm range, accounting for around 40% of the annual rainfall at the national scale but rising to over 85% in the driest parts of the country.

The residual rainfall (rainfall minus evaporation) ranges from less than 100 mm a year over much of South East England to over 3500 mm in parts of Scotland and Wales. It is markedly seasonal with the bulk of the annual runoff occurring in the November–March period when fluvial and groundwater flooding is most frequent; in contrast, urban flash flooding is most common following intense summer storms. Annual minimum flows in western and northern rivers generally occur in June or July but, to the east and south, flows are typically at their lowest during the late summer and early autumn – and later in rivers draining some permeable catchments where groundwater, via springs and seepages, is a major component of low flows.

The spatial trend in average runoff across the country is largely the reverse of the distribution of the population; this creates problems of water supply. Indeed, much of South East England can be classed, according to the World Bank criterion (less than 1000 m^3 per head per year of available water), as suffering from serious water stress. This is likely to be exacerbated as the population rises by 10% from a total of 63.7 million in 2012 to a forecast 70 million by 2027, with much of the increase occurring where water resources are under most pressure. New environmental controls are also limiting abstractions. Groundwater from the Chalk is the chief source in the South East, whereas some parts of the Midlands and north of England rely on groundwater from Permian, Triassic and Jurassic formations. The Thames is the principal source of London's drinking water and the river supplies a number of other towns and cities along its course. This pattern is replicated in many other UK rivers. The pattern of multiple abstractions and discharges of treated effluent causes concern over the long-term effects on human health, particularly during droughts when dilution is low. Long-standing problems include high levels of nitrate and phosphate as well as newer challenges such as oestrogens and nanoparticles. Reservoirs

in the high rainfall areas of the north and west supply by aqueduct the cities and towns in those parts of the United Kingdom. Usually supplies are more than adequate, but occasional droughts have caused shortages. Problems arose in the1960s and 1970s in these areas because of acid rain – the acid buffering capacity of the soils being very limited. Toxic drainage from abandoned mines occurs in a number of basins, particularly in the north and west. Plans were drawn up in the 1970s for the large-scale transfer of water from the wetter parts of England and Wales to the drier ones, but they were not implemented. Increased interconnectivity of water resources at the regional and local levels has bolstered resilience to drought; however, no new reservoirs have been built for more than 20 years: a few in the south have been extended. Many rivers in the north and west have steep short courses and flows generated by moderate amounts of rain falling on saturated soils can overtop flood defences in a matter of hours. Where rivers in flood reach flatter downstream areas, extensive flooding can occur. Storm surges in coastal areas coupled to river floods have produced some of the worst inundations, although the highest recorded floods in many rivers resulted from heavy rain falling on melting snow in March 1947. Flood forecasts are issued by the Environment Agency (EA), whereas local government emergency bodies provide flood relief to affected localities assisted by the Police. Fortunately, there are few fatalities, but frequently there is considerable property damage and loss of income. Farm land may be flooded for weeks as happened in the winter of 2013–2014. Insurance cover is not universal and where floods are repeated premiums are very high. The Flood and Water Management Act (HMSO, 2010) gave powers to English County Councils to develop, maintain, apply and monitor a flood risk strategy.

Continuation of the recent febrile nature of the climate into the future threatens to alter the variables that affect water resources and the hydrological extremes. Drier summers are predicted for the south and east with rain falling in more intense showers; in contrast, winters are expected to become wetter and milder. The north and west is likely to get wetter with longer periods of rain and less snow. The wetter, warmer atmosphere with its procession of frontal systems moving east across the Atlantic, is interrupted from time to time by periods of high pressure emanating from the east and south. Changes in the balance between the different air masses and the battle between persistence and volatility will continue to characterise the UK weather and climate.

Fifty years ago river and water management was predicated on an assumption that hydrological variability was about a relatively stable long-term mean – rainfall averages were based on a standard 30 year period. This can no longer be taken for granted and the challenge for the future is to build on the firm scientific foundations established at Wallingford and elsewhere to accommodate change, whilst reconciling, the often competing demands on rivers, aquifers and wetlands.

1.3   Early Days at Wallingford

The post-World War II years heralded a period of unprecedented social, political and economic change in the United Kingdom. Industrial production slowly recovered and started to expand, whereas the response to the demand for housing promoted a surge in urban growth. The 'space age' was about to dawn and the technological revolution was beginning to blossom. This was the time when hydrological research at Wallingford made its first impact. The late 1950s saw several ground-breaking papers on predicting runoff from rainfall published by Eamonn Nash, at the then Hydraulics Research Station (HRS) (Nash, 1958, 1960). The advances these papers describe attracted the attention of scientists and engineers, nationally and internationally. To complement his mathematical approach, in 1960, Eamonn launched an experiment on the small clay catchment surrounding the headwaters of the River Ray in Buckinghamshire to determine the factors controlling the flow from the basin. He also developed a novel weighing lysimeter which was installed at Howbery Park. A large cylinder filled with soil floated on a trough of mercury, its changes in level registering the evaporation and rainfall on the grass surface of the lysimeter. This strong scientific base at Wallingford led to the formation of the Hydrological Research Unit (HRU) within HRS in 1962.

The formation of HRU came with an upsurge in the 1960s of UK interest in hydrology (Fig. 1.1). Hitherto hydrology had received little recognition (Rodda, 2006), with the exception of the Postgraduate Course at Imperial College where Peter Wolf was appointed Reader in 1955. Later, courses were launched at Newcastle, Birmingham, University College London and several other universities. At undergraduate level, hydrology became part of the syllabus, mainly in Departments of Geography, such as at Reading, Hull and Aberystwyth. This upsurge was also reflected in several other ways. For example, hydrology gained credibility as a separate profession (Law, 2000), a growing number of scientific papers were published dealing with a wide range of hydrological research topics (Royal Society, 1971) and the Hydrological Discussion Group was started in 1963 at the Institution of Civil Engineers.

As a result of the reorganisation of government science in 1965 (Science & Technology Act, 1965), the Unit was transferred to the newly formed Natural Environment Research Council (NERC). In recognition by the Council of the importance of advancing hydrology, in 1968 it metamorphosed into the Institute of Hydrology (IH). Then on April 1, 2000 the Institute became part of the Wallingford Laboratory of the Centre for Ecology and Hydrology (CEH), just at a time when an ecological approach to river basin management was being widely advocated.

Innovative and perceptive programmes pursued by these bodies placed 'Wallingford' hydrology at the forefront of most national and many international initiatives. Over the last 50 years, the spur of fresh problems facing this 'new' science has carried activities forward to the frontiers of research on a wave of excitement and enthusiasm. This book attempts to capture these sentiments. It highlights what has been achieved and indicates future directions for action (Box 1.1).

1.4   NERC's Role in Promoting Hydrological Research

From its inception in 1965, the Council has, in the words of its first chairman Sir Graham Sutton, 'encouraged and supported research in hydrology and the provision of advice and the dissemination of knowledge' through its component bodies and in the universities. It aimed to promote previously neglected environmental sciences by a dedicated budget separate from the support for the physical, agricultural and medical sciences. Within NERC itself, recognition of hydrology as an embryonic science requiring nurture and shelter from competing established disciplines was achieved with

Natural Environment Research Council

Hydrological Research Unit

Record of Research 1966 and Programme of Research

Fig. 1.1 The First HRU Annual Report in 1966. (Source: Reproduced with permission of CEH.)

Box 1.1   Scientific research at Howbery Park

It may be of interest that the site at Wallingford where HR Wallingford and CEH are located, namely Howbery Park, has been in the forefront of research in the past. Jethro Tull (1674–1741), a pioneer in agricultural research, was the first to introduce mechanical devices, such as the seed drill, to replace hand methods of cultivation. He sowed seeds in rows rather than by broadcasting, and he developed ideas on plant nutrition and soil and plant management. This was at the start of the agricultural revolution.

---

a struggle. The new entrant had to compete for funds against existing research organisations; a convincing case had to be made for removing the small research unit from an engineering body to create a dedicated hydrology centre, rather than spreading support thinly in the universities. At the same time, NERC encouraged universities to train scientists and to cooperate in advancing the subject. From a time when there were few, if any, practicing hydrologists in Britain and minimal activity involving them, the last 50 years has seen a remarkable growth in the effort invested in the science and a rising number of career hydrologists, estimated in 2013 at about 2000. Of course, the very nature of hydrology with its many faceted interests, demands the involvement of a wide range of disciplines – civil and electronic engineering, physics, geography, geology, ecology, chemistry, mathematics and statistics, to name but a few.

The transformation of a small research unit (Fig. 1.1) into an Institute with significant funding, an interdisciplinary remit and an agenda in both applied and basic scientific research was influential in defining the identity of hydrology, both in the United Kingdom and abroad. Disciplinary barriers had to be broken down, instrumentation and measurement improved and systems analysis developed whilst responding to urgent solutions to practical problems. From small beginnings in the late 1960s, a series of NERC thematic programmes have been mounted involving CEH, the British Geological Survey (BGS) and a considerable number of universities. Current programmes bring together hydrologists with scientists from a range of other disciplines. But NERC has not been the sole source of funds for work at Wallingford: since the days of HRU, there have been multiple sources of income including government departments, public and private organisations, the European Union and the World Bank. Increasingly, important have been the programmes of collaborative research undertaken through the different tranches of European Union's 'Framework Programme'. In addition, a large number of projects have been engaged in with British consulting engineers, often within technical assistance programmes. A number of developments in the UK water industry have taken place during the last 50 years which have influenced the direction of research at Wallingford.

1.5   Countering Water Problems

Human activities over the last 250 years have quickened the pace of global change, a pace unlikely to lessen in the future. The population explosion, leading to deforestation, desertification and urbanisation are amongst the main agents of this change. These coupled with climate change are today among the most pressing issues threatening the fabric of society – issues which cause concern amongst politicians, the public as well as scientists. Water is pivotal factor across all of them. And while floods claim the largest toll of life and cause the greatest damage of all natural disasters, many extensive areas in Asia, the Middle East and Africa currently suffer severe shortages of water. It must be recognised that parts of the developed world also face water scarcity, South East England being one. Allied to these problems is the pollution of surface and groundwater. As the world population climbs towards 10 billion by about 2050 AD and contemporary climate change *moves the hydrological goal posts* in many regions,

these shortages will intensify and extend. If only the distribution of the world's water resources were regular and well ordered in space and time. But it is not. That many regions of the globe will suffer severe shortages by about the middle of this century has been predicted for a number of years. For example, the UN Secretary General, Mr Ban Ki Moon, stated recently that by 2030 AD about half the world's population will be facing water scarcity.

Most types of research aim to reduce risk and minimise uncertainty. One of the prime responsibilities of governments is to reduce risks to its citizens and protect them from hazards, especially environmental hazards, so it follows that public funds should be directed to these ends. Most sources of risk in the natural environment need sound environmental data for purposes such as prediction, forecasting, planning and management. These data are required to better comprehend the risks concerned. Capturing reliable and consistent long-term data is vital to establishing the severity of extreme events, together with their frequency, extent and like variables. Then to analyse, apply and disseminate the results aims to minimise loss of life, damage and disruption. This is not an easy task. This unglamorous segment of research is often the subject of cuts in funding because reduced budgets do not have an immediately apparent impact on the public.

1.6   The Beginnings of Experimental Hydrology

In the years immediately after World War II, the water problems facing Britain were remarkably similar to those current during the early years of the twenty-first century. The most severe floods on record swept the country in 1947, and there were several droughts, including an extensive one in 1959. With the resurgence of industrial production, the burgeoning population and the strong post-war demand for housing, there was anxiety in government and in the water industry that the rising consumption of water would soon outstrip the available resource. It was an anxiety reinforced by the results of a small-scale experiment on a stand of conifers, conducted by a highly respected water engineer Frank Law at Stocks Reservoir in Lancashire (Law, 1956, 1957 and 1958). His experiment showed that the trees used about 30% of the water that might have run into Stocks Reservoir, representing an economic loss for the water company. This was at a time when large areas of upland Britain were being planted with conifers by the Forestry Commission as part of government policy.

To investigate this problem, a government committee was established, the Committee on Hydrological Research, with members drawn from various departments and agencies of government and from the water industry. On their recommendation, on April 1, 1962, and with the agreement of the Department of Scientific and Industrial Research's Hydraulics Research Board, the HRU was set up at HRS with Eamonn Nash as its Head. The Unit was charged with the task of confirming or disproving Law's results at a basin scale. However, it first had to search for an experimental site suitable for comparing the water balance of basins in order to answer the question 'Do trees use more water than grass?' After an extensive survey of suitable sites across parts of Scotland, England and Wales, the preferred site was three small south-east-facing basins on the Long Mynd in Shropshire. The geology rendered them watertight, one basin was in coniferous forest and the others in sheep pasture. When the protracted negotiations over access failed, it became necessary to find an alternative site. A new site was found in Central Wales; this was the small basins surrounding the heads of the rivers Wye (upland pasture) and Severn (mostly coniferous forest) on the east flank of Plynlimon. The forested basin was owned by the Forestry Commission, whereas the other landowner was very sympathetic to the aims of the study and agreed access. Despite initial criticism that the area was too wet to distinguish hydrologically between the contrasting short and long vegetation, the Plynlimon Project started on January 1, 1967. Subsequently, it has provided a core project for the Unit and its successors (Robinson *et al*, 2013).

In addition to the primary investigation at Plynlimon and the earlier River Ray, several fresh catchment studies were launched by HRU. These included investigation of the effects of urbanisation (Milton Keynes), afforestation (Coalburn) and, to contrast with the clay River Ray catchment, a study of one mainly composed of chalk (River Cam). Because of the importance of the influence of forests on evaporation, a very detailed micrometeorological study was mounted to measure forest evaporation. This was at Thetford where there is a very extensive area planted with conifers. For Plynlimon, Thetford

and other studies, it was soon found that most of the 'off-the-shelf instruments' then available were not 'fit for purpose'. They were not sufficiently robust, they lacked the required accuracy and in some cases, such as for the measurement of soil moisture, new devices needed to be developed, namely the Wallingford neutron probe. Similarly, for a continuous record of rainfall and the meteorological factors governing evaporation, an automatic weather station (AWS) was devised, the data being logged on a multichannel recorder. In the case of the Thetford study where the turbulent structure of the wind flow was investigated, care had to be taken to select and position the sensors correctly. The way these sensors were used, the neutron probe and the AWS were, at the time, in the forefront of instrument development and operation. In 1966, a Digital Equipment PDP 8 computer was acquired by HRU, one of the first computers in the United Kingdom to be dedicated solely to hydrological research. Alongside these fundamental studies, the Unit's interests expanded to include projects conducted abroad in support of British consulting engineers. To cope with this work, staff numbers increased from 26 in 1966 to 54 in 1968 and to 102 in 1972. In recognition of the importance of hydrology, on April 1, 1968 HRU became a full NERC Institute – IH, with Jim McCulloch as Director. Jim had been appointed Head of HRU in 1964 and had successfully led the development of the young Unit. In 1972, a notable, modernist, prize-winning concrete building was constructed to house the rapidly growing staff and the powerful computing facilities. It was formally opened by Earl Jellicoe, Leader of the House of Lords, as the Maclean Building in May 1973, an event marked by a Scientific Symposium 'A View from the Watershed' (IH, 1973).

1.7 Fighting Floods

In the 1960s, the methods of flood prediction dating from the 1930s that were being employed by British consulting engineers were desperately in need of updating. In 1968, the Institution of Civil Engineers (ICE) asked for urgent government action to initiate a study of the factors affecting floods, particularly with respect to reservoir safety and the flooding of land and urban communities. In response, a 15 strong Flood Studies Team was established at IH in 1970 to address this need with

Fig. 1.2　The Flood Studies Team in the 1970s. (Source: Reproduced with permission of CEH.)

John Sutcliffe as its leader (Fig. 1.2). For the first time, the flood records collected by a variety of disparate bodies across the United Kingdom (and Ireland) were brought together at a national level for analysis in a consistent manner. The Team was able to amass sufficient data, analyse it and produce a methodology that was published in the 5 volume Flood Studies Report (FSR) in 1975 (NERC, 1975). This methodology largely met the then needs for flood prediction. When the work of the Floods Team finished, its members were absorbed into the Institute.

1.8　Gaining International Recognition

The 'Cold War' was at its height during the 1960s with threats of a nuclear holocaust peaking at the time of the Cuban missile crisis in the autumn of 1962. In stark contrast to the ongoing confrontation between east and west, there was a willingness to cooperate in hydrology at the intergovernmental level across the Iron Curtain! This was aptly demonstrated by the launch of the International Hydrological Decade (IHD) by UNESCO in 1965, in cooperation with the World Meteorological Organization (WMO), together with several other agencies of the United Nations (UN) and with the support of the International Association of Scientific Hydrology (IASH) and several other non-governmental bodies. The Decade, the first of its kind for encouraging international cooperation in the field of water, had a number of aims: the stimulation of research, improving education and training and facilitating the exchange of information. The UK was a strong advocate of the decade. It participated in the IHD programme and it was represented at the meetings of its Co-ordinating Council, whereas IH provided the secretariat for the British National Committee for the IHD. The Decade gave extra weight to the programme of research being carried out at IH.

Indeed, the International Symposium on 'World Water Balance' held at Reading University in July 1970 gave global exposure to the youthful Institute (IASH/UNESCO/WMO, 1972b). The symposium revealed how much more data and interpretation were needed to understand local and planetary water movements. Three other symposia held at the start of the 1970s helped shaped hydrological research at Wallingford and globally. They were convened in Wellington (IASH/UNESCO/WMO, 1972a), Koblenz (IAHS/UNESCO/WMO, 1970) and Warsaw (IAHS/UNESCO/WMO, 1971) and dealt with results from representative and experimental basins, hydrometry and mathematical modelling.

In September 1974, the International Conference on the Results of the IHD was held at UNESCO with 299 delegates from 90 nations, including the UK, together with a large number of non-governmental organisations (NGOs). The Conference agreed to mount the International Hydrological Programme (IHP) as a major UNESCO Programme from 1975 onwards. The tenor of the IHP was to be similar to that of the IHD, but with several additional aims, one being assisting member states in the organisation and development of their national hydrological activities. The IHP was to be executed through successive phases of 6 years duration. At this same Conference, WMO's Operational Hydrology Programme (OHP) was examined. Similar to the IHP in some respects, the OHP was fashioned to support the hydrological services of member states of WMO. Design of hydrological networks, flood forecasting, water resources assessment and other practical aspects of hydrology are the main thrusts of a programme overseen by the Commission for Hydrology (CHy) at its four yearly meetings. IH provided the UK lead for both the IHP and the OHP, a role continued by CEH.

In 1972, following the appointment of John Rodda to the post of Editor of the International Association of Hydrological Sciences (IAHS), Wallingford became the location of the publishing activities of the Association, now known as the IAHS Press (http://www.iahs.info). The Association's famous 'Red Books' and the *Hydrological Sciences Journal* and its other publications carry the name *Wallingford* to the global community of hydrologists.

In 1972, the UK acceded to the European Community, leading to the Institute's participation in the research programme promoted by the Commission's Directorate General XI, more recently DG Research. This brought IH into contact with a variety of hydrological bodies, governmental and quasi-governmental, within the Community, as well as with a number of universities.

1.9 Governmental Turbulence

In November 1971, a government green paper (Command 4814) was published containing the Rothschild and Dainton Reports on the restructuring of government research and development. The Rothschild Report proposed that part of the funding of government research in the research councils should be channelled through executive departments and that the programmes of research institutes should meet their needs on the customer/contractor basis. NERC (NERC, 1972) and the other councils objected strongly to the proposals; their misgivings were shared by much of the scientific community. However, the 'Rothschild Principle' was enacted (Command 5016) across the research councils, with the Department of the Environment (DoE) being the main customer for the transferred portion of NERC's research. Up to 35% of the science budget was to be transferred from the research councils to the relevant department or ministry over a three-year period. In the case of NERC, discussions with the different institutes, including IH, defined the research that was commissioned and the costs. For the Institute, this amounted to about £150 000 initially, but commissioned research also came for the Ministry of Overseas Development, so that by 1975–1976 some 56% of the IH income was derived from sources outside NERC.

The Dainton Report recommended that IH should be physically transferred, together with its staff, to DoE and fused with HRS. Fortunately for the integrity of IH, the proposed transfer was rejected by the government (Command 5016). In 1982, HRS was privatised as Hydraulics Research Station Wallingford Ltd (later renamed HR Wallingford Ltd.).

Turbulence has also characterised the history of the UK water industry since the end of WWII. Its structure has been radically altered a number of times, but with many of the provisions of the successive Acts of Parliament for England and Wales, Scotland and Northern Ireland still applying. Amalgamations of water undertakings in England and Wales reduced their number from about 1100 in 1945 to 286 by 1966. Twenty nine river authorities were established in England and Wales by the 1963 Water Resources Act (HMSO, 1963) along with the Water Resources Board (WRB). The Act promoted hydrometric schemes which led to a large increase in the number of observations of river flow and ground water. The Act's aim to conserve and ensure the proper use of water resources caused the Board to propose a strategy for water resources development in England and Wales to 2000AD and to embark on research to support its remit. IH was involved in certain aspects of this research. But then the 1973 Water Act (HMSO, 1973) dissolved the WRB and completely changed the complexion of the water industry by establishing 10 Regional Water Authorities for England and Wales, together with a National Water Council. This change demonstrated the government's preference for a regionally based policy for water, rather than continuing with a national water policy. The new water authorities were multifunctional with responsibilities ranging from the conservation of water resources, to recreation and to the control of pollution. The latter caused much concern as the authorities were required to regulate themselves – they were, in other words, both poacher and gamekeeper, something that had not happened previously. To replace the Water Resources Board, the Central Water Planning Unit and the Water Data Unit were established, the latter handling the collection, archiving and analysis of the UK's river flow and groundwater data, amongst other things. The water industries in Scotland and Northern Ireland have evolved over the same period in somewhat different ways.

1.10 An Expanding Role

During the 1970s, IH confirmed its place as leader in its field in the UK and as a global centre of excellence, based on the high quality and reach of its research (Anon, 1973). The global dimension prospered through projects conducted with European partners under the aegis of the EU, through international collaborative programmes and by development projects undertaken for technical assistance agencies, such as the World Bank, as well as with British consulting engineers. Cooperation with the British Antarctic Survey in South Georgia provided a new dimension. The Hydrogeology Unit

of the then Institute of Geological Sciences (now BGS) moved into the Maclean Building in 1977, opening the potential for fruitful collaboration.

In 1971, an investigation started of trace elements at Plynlimon, the first water quality study performed by IH (Anon, 1993). The results of the annual water balance experiment for the River Ray indicated that the storage and release of soil heat in the catchment accounted for the equivalent of about 100 mm of water. Inclusion of this term closed the water balance – a term usually ignored in such studies (Edwards & Rodda, 1970). Six newly developed AWS went into service at Plynlimon, Coalburn and South Georgia. IH staff numbers increased to about 100 by 1972, including three based abroad and three at Plynlimon. They were supported by a budget of £463,000. The 1971–1972 annual report showed that they published 17 scientific papers covering the 6 areas of the Institute's programme (IH, various years).

A UNIVAC 1100 series computer and various pieces of peripheral equipment were installed in 1973 to meet rising demands for handling large amounts of data and applications of increasingly complex and realistic mathematical models – lumped and distributed catchment models were being developed. Results emerging from Plynlimon showed an evaporation 'loss' of 17% of the annual rainfall for the grassland catchment and 29% forest catchment. To explain the cause of this difference, studies of hydrological *processes* were started at Plynlimon; detailed micrometeorological studies in a forest near Thetford project confirmed the importance of interception, leaf wetness and the biological controls on evaporation. Studies of soil water flow in the saturated and unsaturated zone were underway. In collaboration with the Met Office, the Water Research Centre and several other bodies, work commenced in 1974 to develop an operational flow forecasting system for the River Dee, using real-time data from radar and ground-based sensors. During the late 1950s, a series of East African catchment studies of land use change were initiated by the East Africa Forestry Research Organisation (EAFRO). In the 1970s, they were supported by IH funded by the Overseas Development Ministry (ODM). Instrument networks were upgraded and effort made to analyse the data. The very severe and extensive drought of 1975–1976 across the UK brought IH a project on low flows funded by the DoE increasing the total receipts from outside NERC to over £0.5 million.

Projects undertaken in Northern Oman, Botswana, Iran, India, Ecuador and Brazil contributed to this total, as did a number of consultancies carried out within the UK, such as a study to rationalise the national raingauge network.

Towards the end of the decade staff numbers had increased to over 130: scientific paper were being published in learned journals at a rate of more than 30 a year and 33 commissioned overseas projects had been completed over the previous 5 years. A substantial extension, the North Wing, was made to the Maclean Building in 1980. However, this apparent progress was sullied by the first of a series of cuts in science budget funding, whereas certain members of staff were encouraged to take early retirement. That the UNIVAC 1108 computer was moved from IH to NERC Computing Services at the Rutherford Laboratory did not assist the progress of mathematical modelling of hydrological systems.

1.11 Extending Hydrological Research into the Eighties

Studies of the hydrological consequences of land use change, previously centred on Plynlimon, were extended to Scotland, Scotland being the main area of expansion in the UK forestry industry. Monachyle Glen (heather, bracken and scrub) and Kirkton Glen (part forest) catchments in Perthshire, of similar size and geology were selected for investigation, in cooperation with a number of Scottish bodies. The instrumentation of these catchments completed in the autumn of 1981 was similar to those at Plynlimon, but with water quality monitoring being included from the outset. Meanwhile, hydrological modelling saw the development of more realistic physically based distributed models, namely the Institute of Hydrology Distributed Model (IHDM) and the Système Hydrologique Européen (SHE), both models representing significant advances. Water quality modelling was also progressing through studies of the River Thames and several other rivers, such as the Bedford Ouse, where a real-time forecasting system was developed.

Applications of remote sensing, assessment of evaporation from the Amazon rain forest and studies of the storage and movement of water in the Chalk in southern and eastern England were among the projects underway in the early 1980s. Groundwater studies included recharge estimates, pumping test analyses and modelling saline intrusion. Groundwater consultancies were undertaken in 12 countries, mainly in the Middle East. Recognition of the effects of atmospheric pollution impact on stream ecology through the type of land use and management led to the start of studies of hydrochemical balances at Plynlimon (Fig. 1.3). Land management, including field drainage, is also important to sediment yield, erosion and deposition and in turn to flood protection.

The 1975 Flood Studies Report (NERC, 1975) provided, for the first time, a set of techniques for estimating the flood of a given frequency on any river in the British Isles. The European Community then commissioned a similar study 'the European Flood Study' to provide the same facility for estimating floods across member states. An extension of this study was one on floods for some 33 countries 'the World Flood Study'. A GEC 4000 Series computer was installed at IH in 1983, the turnover rose to £2.5 million, but staff reductions were sought. With the closure of the Water Data Unit, its work of curating the data from the national river flow and groundwater networks was transferred to Wallingford. In 1985, a 6-man international team was established at IH to undertake the UNESCO FREND project (Flow Regimes from Experimental and Network Data), part of the programme for the third Phase of the IHP. The initial idea was to bring together data from the many studies of experimental basins established during the IHD with data from national networks in Western Europe, to develop better understanding of hydrological variability in time and space, through mutual exchange of data, knowledge and techniques at a regional level.

In the mid-1980s, studies of acid rain were initiated at Llyn Brianne and in the Cairngorms, while collaboration with several Brazilian institutes began in the Amazon on the water use of tropical forests. Michael Heseltine, who was at the time the local Member of Parliament and a government minister, inaugurated the IBM 4381/3 mainframe computer in 1986, but early personal computers, more commonly called 'microcomputers' in the UK were deployed at IH from an early date. These included the PET Commodore (1977 USA), Acorn (1978 UK) RM 380Z (1978 UK) and Sinclair ZX 80 (1980 UK) machines. Subsequently, IBM launched their 'Personal Computer' and thereby imposed a standardised architecture which has dominated ever since, Apple aside. Digitising the UK river

Fig. 1.3 The Workshops at Wallingford were essential in the development of novel hydrological instruments. (Source: Reproduced with permission of CEH.)

network at a 1:50,000 scale commenced prior to the development of digital terrain models.

A new instrument called the HYDRA was developed to measure evaporation directly by making high-frequency measurements of the flux of water vapour moving away from the land surface. IH was selected as the principal investigator in the FIFE Project for modelling and observing land–surface–atmosphere interactions on regional and global scales. A new series of year books and reports: Hydrological Data UK was launched, including the first occasional report. Its subject was the 1984 drought.

In 1988, the National Hydrological Monitoring Programme was instigated by IH in cooperation with BGS based on data on river flows, groundwater levels and reservoir levels collected from a variety of sources across the UK. The Hydrological Summary for the United Kingdom, published online monthly, also contains rainfall information from the Met Office and a hydrological outlook for 3 months ahead. It provides an immediate appreciation of the hydrological condition of the nation, valuable to decision makers at all levels, but particularly at times when floods or droughts are occurring. Few other nations produce such a publication.

Towards the end of the decade, work started with the Met Office to improve land surface inputs into global climate models. A separate study of the relationship between climate and hydrology was conducted, both endeavour enlarging the profile of IH research into the area of climate change. This enlargement was enhanced by a funded study of the effects of forest clearance in Brazil to obtain data for the calibration of global circulation models.

Box 1.2   The British Hydrological Society

In response to expressions of interest from professional research and engineering hydrologists, in the 1970s and early 1980s, the British Hydrological Society (BHS) was formed in 1983 – '*to promote interest and scholarship in both the scientific and the applied aspects of hydrology and to foster the involvement of its members in international activities directed to the promotion of such scholarship*' (BHS Statutes). The Institute of Hydrology (IH) and the Institution of Civil Engineers (ICE) together provided the coordination and steering necessary to establish the new Society. They became statutory stakeholders in BHS. Of the 15 BHS Presidents between 1983 and 2013, three were from IH/CEH. BHS is an Associate Society of ICE. The benefits are that membership and other administrative services are provided by ICE. The prestigious conference and meeting facilities at ICE, in Westminster, London are often used by BHS.

BHS has, in 2014, about 1000 members from universities, government departments and agencies and from consultants and practitioners. The National Committee is elected by the members. It includes representatives of the ICE Water Expert Panel, CEH, the UK Committee for IAHS and the Chartered Institution of Water and Environmental Management (CIWEM). There are four regional Sections (Midlands, South East, South West and Welsh). In addition, the north of England is covered by the Pennines Hydrological Group, and Scotland by the Scottish Hydrological Group. They organise regional events. BHS holds biennial National Symposia, with every sixth year being marked by an International Symposium.

Information about BHS, its activities and publication is given on its website (http://www.hydrology .org.uk/). The BHS quarterly Newsletter, *Circulation* provides reports of recent meetings and notices of future meetings. BHS offers an annual undergraduate dissertation prize, organises an annual event for early-career hydrologists (the Peter Wolf series) and administers applications for travel grants from its members seeking financial assistance to attend international conferences.

With the Nordic Association for Hydrology (NHF), Denmark, Estonia, Finland, Iceland, Latvia, Lithuania, Norway and Sweden, in January 2008, *Nordic Hydrology: An International Journal* was relaunched by BHS in cooperation with IWA Publishing as *Hydrology Research: An International Journal* (http://www.iwaponline.com/nh/)

Subsequently, the Italian Hydrological Society (IHS) and the German Hydrological Society (DHG) have adopted *Hydrology Research* as their research journal. *Hydrology Research* has six issues per year. Within this publishing schedule, there are Special Issues from the biennial BHS Symposia and NHF Conferences.

Following the appointment of Jim McCulloch to a post in NERC Headquarters in Swindon, in 1988 Brian Wilkinson took up the position of Director of IH supported by 160 staff and a budget of £5 million, £3.6 million coming from commissioned research. During the two years 1988–1989, over 150 scientific papers and reports were published.

The First Scientific Assembly of the IAHS took place in Exeter University in July 1982 with a strong IH input. This event led to the formation of the British Hydrological Society a year later (Box 1.2).

The 10 regional water authorities in England and Wales were privatised in 1989 (HMSO, 1989 and 1991), and their regulatory role, including pollution control and water resources management, vested in the newly formed National Rivers Authority (NRA). An economic regulator, OFWAT, was created and the Drinking Water Inspectorate established to monitor water safety and quality. The 16 small water supply companies in England and Wales remained in private ownership, whereas the water and sewerage services in Scotland and Northern Ireland continued as public services.

1.12   Into the Nineties

During this decade, many of the research initiatives launched earlier came to fruition. In a number of projects digital data, software advances and transformed communications radically changed the pace and scale of IH research. Computing turned gradually from the mainframe to desktop machines. Several Science Budget Community Research Programmes were commenced as NERC celebrated its twenty-fifth anniversary in 1990 (Sheail, 1992).

The first of these programmes was Terrestrial Initiative in Global Environmental Research (TIGER). With guidance from the NERC Terrestrial and Freshwater Sciences Directorate, the TIGER office at IH was responsible for a spend of £20 million over the seven years of activities from 1990. TIGER comprised four thematic areas of work – carbon cycling through soil and vegetation, biogenic trace greenhouse gas emissions and sinks, energy and water budgets for climate modelling and ecosystem impacts. These were supplemented by several cross-cutting areas concerning: monitoring environmental change, applications of remote sensing, palaeoclimatology via soil sampling (also managed from IH) and researching technical means of establishing sensitivities to carbon dioxide increase, change in drought and ambient temperature regimes, increase in ultraviolet radiation and functional aspects of biodiversity change. The TIGER office led by Max Beran plus a team of convenors supported the work of the top-level steering committee, the working groups for each of the themes and those for each of the cross-cutting topics from announcement of opportunity through project selection to the finalisation or extension of contracts. As well as this basic administration of research proposals, the office also looked after budget maintenance, organisation of meetings and conferences, setting up TIGER's own flagship field sites at Wytham and Moor House (parts of the Environment Change Network), the provision of publicity material, press releases, conference organisation, the TIGER quarterly newsletter and coordination with sister global change programmes with international links.

Studies were spread over more than 60 laboratories and university departments, the programme funding 300 scientists. IH's own expertise was required in many areas such as field monitoring and flux measurement, but most prominently in the energy and water budgets area (Oliver *et al.*, 1999). New insights were gained into how patchy terrain, such as Tiger Bush, could be represented in modelling transports through the soil–vegetation–atmosphere interfaces recognising strong non-linearities. Field measurement sites were set up in Niger and in Amazonia. Other IH activities in this TIGER area were for developing methods for incorporating lateral water movement at the land surface in a way appropriate to the requirements of global climate models.

Another collaborative programme named Land Ocean Interaction Study (LOIS) was carried out from 1992 to 2001, involving three hundred and sixty scientists from eleven research institutes and twenty seven universities. This unique multidisciplinary large-scale research study aimed to make major advances in understanding the processes and fluxes at the margins of continental shelf seas, from catchments and rivers to the coastal seas, and across the continental shelf edge into the deep ocean. The policy issues driving the investment in the LOIS programme were in part the need to further knowledge of the coastal zone in which over 60% of world cities lie. It was also developed at a time when the issues around coastal water quality were contributing to an impression, current in the media, of Britain as the 'Dirty man of Europe'.

The LOIS programme was at the forefront of the scientific efforts to link atmospheric, terrestrial, river and marine environmental systems. This work ran in parallel with similar initiatives at the global (e.g. the International Geosphere and Biosphere Programme), regional (e.g. The European equivalent to LOIS, known as ELOISE) and national levels. The research also received strong support from the water industry and environmental regulators.

The largest component of the LOIS programme considered the exchange of natural and man-made materials between the land and the sea in temperate regions. Chemists, physicists, biologists, mathematicians and engineers collaborated. The study focussed on catchments and estuaries of the UK's east coast, for example the Humber. This river drains a catchment area which is home to more than 20% of the UK population and is the site of a very significant proportion of the UK energy, industrial and agricultural production. Discharges from the estuary have an important impact on the water quality of the North Sea. The adjacent coast line is in rapid retreat and supplies large amounts of sediment. The Tweed estuary, also on the UK east coast, is a much smaller estuary with very little industrial activity within its catchment, providing an important contrasting environment.

The benefits of this large programme exploited the multidisciplinary nature of LOIS and a number of important innovations can be exampled. These include studies of tidal reaches, with their salinity variations, tidal flows and large fluctuations in river inputs. New insights were gained into the role of these regions as buffers between the fluvial and marine environments. In the intertidal region, a wide range of techniques and disciplines were deployed. They produced new process models for the erosion and deposition of fine sediments, incorporating the vital influence of biological activity at the surface. The cross-calibration of chemical measurements from river and marine samples made possible comparable measurements of nutrients from catchments draining into the adjacent North Sea. IH with other CEH institutes also developed new integrated systems for continuous and flow-related monitoring of river water quality and sediment transport, triggered by both level and turbidity thresholds.

The major research findings have been published widely in special volumes of science journals and in a book (Neal *et al.*, 1997, 1998; Leeks and Walling, 1999; Neal *et al.* 2000; Huntley, and Neal *et al.*, 2003).

In 1991, the report on the 'First Two Decades' of research at Plynlimon was published (Kirby *et al.*, 1991). This brought together the results of the catchment investigations to answer the question: '*Do trees use more water than grass?*' The comparison of the water balance between the forested Severn and the Wye in sheep pasture showed that, on an annual basis, runoff from the latter averaged nearly 200 mm more. Studies of the physical processes explained this difference in terms of the differences in the height and aerodynamic roughness of the vegetation, whereas work on water quality highlighted other contrasts between streams draining the different land covers. Several of the mathematical models developed by IH were employed to analyse the Plynlimon data. The results depicted the implications for land use strategy across the UK. Looking to the future, the report saw the Plynlimon catchments continuing as a valuable outdoor laboratory for long-term monitoring, coupled with new studies, such a geochemical cycling, adding a further dimension to the science. Hydrological studies similar to Plynlimon have been underway in different parts of the world, many stimulated by the IHP and other collaborative programmes. But few have portrayed the hydrology of the studied basins in such detail and with such precision as at Plynlimon. Plynlimon data remains the third most requested data set on the CEH online Information Gateway (https://gateway.ceh.ac.uk/home).

New instruments and techniques were developed and existing ones deployed more widely to the extent that IH became a world leader in sampling, sensors, software and satellite usage for hydrology (Fig. 1.4). One example is the capacitance probe (Dean, 1994) built for soil moisture measurement and, with tensiometers, incorporated in an automatic soil water station. Flow and quality sensors were combined in a system with an intelligent logger, whereas the widely used AWS pioneered at Wallingford was upgraded to measure 12 atmospheric variables at a rate of once every 10 seconds. The data recorded by these systems were transmitted by land-based telemetry or through METEOSAT. To support groundwater investigations, drilling techniques were improved and employed for contract work in the UK and abroad and in IH experiments.

Starting in 1994, the 1975 Flood Studies Report was updated in a 4-year project which produced the Flood Estimation Handbook (IH, 1999). Applications of research included a river flow forecasting system for timely warnings of floods. In 1993, IH was one of the 6 founding members of EUAQUA, a European network of freshwater research organisations, set up strengthen cooperation and transfer of knowledge. Later EURAQUA expanded to 22 members. Collaboration in a number of international programmes, such as the International Geosphere Biosphere Programme (IGBP) and the World Climate Research Programme (WCRP), took IH further into studies of global change. The Hydrosphere Atmosphere Potential Evaporation Experiment (HAPEX) in the Sahel and the Anglo-Brazilian Amazonian Climate Observation Study (ABRACOS) in the Amazon rainforest were two examples of work of this type. Another collaborative programme was the TIGER, a 5-year NERC Community Programme which was run from IH involving 50 different research groups and universities.

Completion of a digital terrain model for England and Wales, a river centre line network and several other developments permitted a flood risk map to be published in 1996 (Morris and Flavin, 1996) under an MAFF commission. The map shows the areas at risk from the 100-year flood, including urban areas; it is the first national map defining flood risk to a consistent standard and return period. A study (HyRAD) was underway in the Brue catchment in Somerset employing imagery from 3 weather radars and the records from over 50 recording raingauges to improve methods of

Fig. 1.4 One of the Chemistry Laboratories for rapid the in-house analyses of water samples. (Source: Reproduced with permission of CEH.)

network design, whereas images from radars and Meteosat were being employed in a rainfall–runoff model to upgrade flood forecasting. Seeking a method for measuring soil moisture over an area, rather than at a point, led to the testing of the European Space Agency's ERS 1 satellite's synthetic aperture radar.

Results from the study of the Coalburn catchment (Robinson *et al.*, 1998), into the long-term effects of establishing a forest, showed that ploughing prior to planting increased annual total flow and enhanced storm peaks. In contrast, the subsequent growth of trees reduced total flow, base flow and peak flows, while water chemistry (Fig. 1.5) was affected in a number of ways.

In 1994, IH was grouped with three other NERC institutes into the Centre for Ecology & Hydrology (CEH) to 'promote closer collaboration and to reinforce university links'. Brian Wilkinson was appointed CEH Director and Tony Debney became Director of IH. The previous year, a new wing was added to the Maclean Building (Fig. 1.5). By 1995, staff numbers had grown to some 200 (Fig. 1.6), with a budget of nearly £8 million, about £2 million coming from ODA, around £2.5 million from NERC and lesser amounts from other government departments and the EU. They contributed 126 papers to the scientific literature and produced 140 commissioned reports for clients. However, the IH income declined in the last years of the decade to less than £7 million. In 1997, Jim Wallace became Director of IH in place of Tony Debney who retired, then Brian Wilkinson retired in 1998 and was replaced by Mike Roberts as Director of CEH. In 1999, Lord Sainsbury, Minister for Science and Innovation, opened a new West Wing

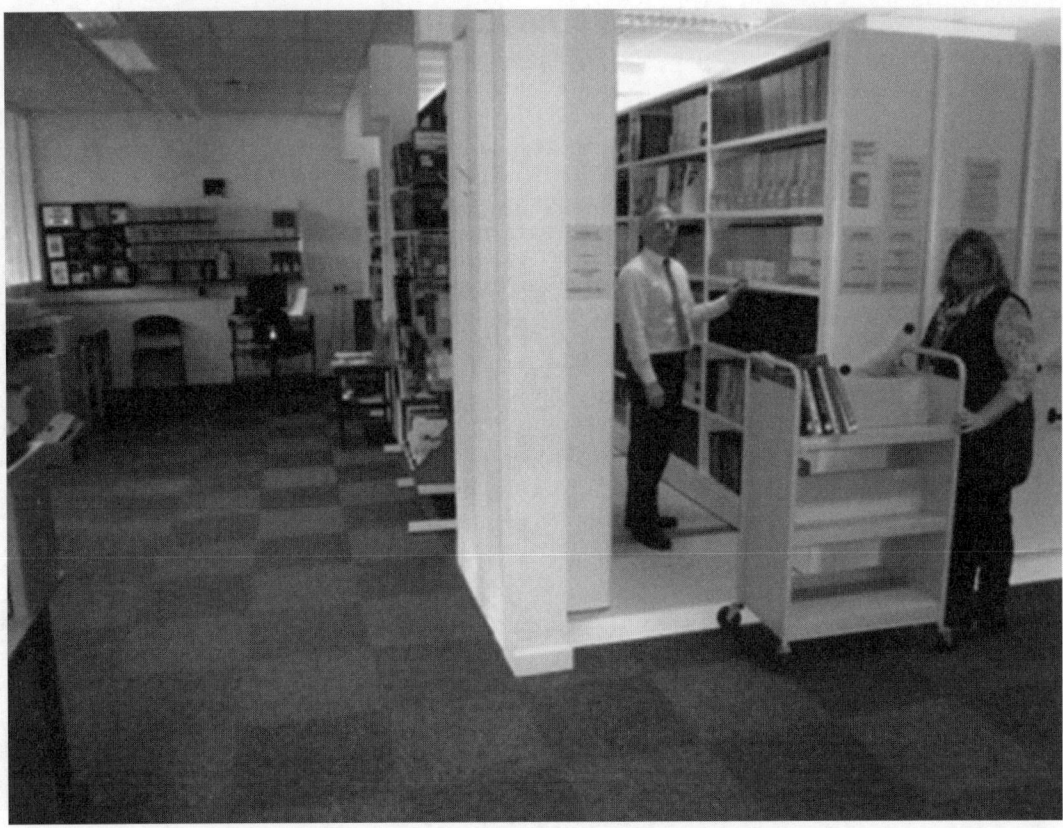

Fig. 1.5 The library. (Source: Reproduced with permission of CEH.)

and the remodelled front of the Maclean Building in the mid-1990s there was uncertainty about the future of IH due to the review of Government research establishments carried out under the 'Prior Options' initiative. There were threats of splitting the Institute and privatising parts of it, but nothing came of these moves.

In parallel to changes in NERC, on April 1, 1996 the EA came into existence, taking over the role of the NRA, Her Majesty's Inspectorate of Pollution (HMIP) and a number of other bodies mostly concerned with waste disposal (HMSO, 1995). Within a budget of over £1 billion, more than half is spent on water, including flood defences and the task of issuing flood warnings. Scotland and Northern Ireland established bodies equivalent to the EA. In 1999, 'An agenda for land surface hydrology research and a call for the Second International Hydrological Decade' was made by Enteyhabi

et al. (1999) which possibly led to the IAHS PUB Programme (Prediction in Ungauged Basins) a few years later.

1.13 Moving into the New Millennium

Further *consolidation* within NERC took place in 2000 AD when the Institute lost its separate identity and was designated part of CEH Wallingford, one of then 9 CEH sites. Chosen as the principal site for expansion within CEH and the location of the Headquarters, during the early years of the new millennium, the hydrologists at the Wallingford Laboratory were joined by ecologists with wide-ranging expertise from several of the other former institutes as they closed. As a result, Wallingford became a scientifically richer environment able to address a wider range

Fig. 1.6   Staff picture Institute of Hydrology in the 1990s. (Source: Reproduced with permission of CEH.)

of environmental problems than in the past. In terms of expertise, it could be argued that the strengthened linkage of hydrology and ecology complemented the long-standing links between hydrology and civil engineering. Further enlargement of the Maclean Building from the middle of the decade provided new accommodation for scientists and administrators from sites which closed as CEH was restructured to 4 laboratories, following adoption of a business plan in March 2006 (CEH, various years). Jim Wallace left CEH in March 2003 and Alan Jenkins became Water Science Director for the whole of CEH in the following September. The European Union's Water Framework Directive (WFD) came into force in 2000 requiring member states to establish targets for water quality in rivers and lakes (EU, 2000). It was to be implemented in stages over a period of 20 years, influencing the trend in certain aspects of CEH research. In 2001, CEH became a Partner in PEER (Partnership in

European Environmental Research) with 7 other large environmental research centres. PEER aims to follow a joint strategy in the environmental sciences to enhance research in all fields of the environment.

At the start of the decade, there was a growing concern in government and the water industry about steroid oestrogens disrupting fish hormonal systems and the likely effects on humans. A model was developed to predict the impact of these substances, assessing the human excretion of oestrogens, together with their transformations in the sewer and waste water treatment plant (WWTP). The model was used by the EA and water companies to provide concentrations throughout catchments, based on the distribution of the population and the number and location of WWTPs. A study of water fluxes in residential areas using a unique data set showed that if roof runoff were collected, depending on the rainfall of the locality, it would be sufficient to meet the grey water needs of the average household. This would reduce the demand for water and lessen domestic water charges. The 2003 Water Act (HMSO, 2003) seeks to improve water conservation, requiring water companies to publish water resources management plans and drought plans.

To complement the attention given to upland catchments, in 2000 a six-year NERC Lowland Catchment Research Programme (LOCAR) was commenced with a budget of over £10 million. A team of 75 scientists, including a number from IH, was established from 14 institutions to work on 12 projects on 3 chalk and sandstone river basins to study and model their hydrology, geology and ecology. The programme was based on the establishment of several data collection systems and a large number of new boreholes to give a valuable picture of how each basin functions. In 2003, IAHS launched PUB (Prediction in Ungauged Basins) aimed at improving understanding of hydrological processes. Members of IH took part in this programme, collaborating with a number of UK universities (O'Connell *et al.*, 2007). PUB continued to 2012 and has been succeeded by *Panta Rhei* ('everything flows') which started in October 2013.

The receipt, processing and display of weather radar products from HyRAD was developed as the interface to improve the EA's flood forecasting system, extending the lead time for flood warnings and improving decision making. Knowledge of soil moisture is important to many facets of hydrology, for example in flood forecasting, so that CEH work to model soil moisture to smaller grid sizes would result in better flood forecasts. Soil moisture was also studied for the Amazon Basin where previous work on rainforests suggested they were not sensitive to soil moisture deficits, but modelling fluxes for the Amazon revealed they react to even modest amounts of depletion. This gives rise to the conclusion that increasing levels of atmospheric CO_2 will alter rainfall patterns over the basin and to the possibility that Amazonia will change from being a sink for CO_2 to a source. There is concern that global warming has serious consequences for communities who depend on glacier-fed rivers for water. At the present rate of glacier recession in the Himalayas, there could be future widespread water shortages in northern India and Pakistan. A hydro-glaciological model predicts how the region will be affected by global warming. Although global warming is expected to bring significant changes to the world's climate, there is uncertainty about the extent of these changes, particularly at a local level and for extreme events. Using hourly rainfall series, a model was used to compare UK floods for the period 1961 to 1990 with those simulated for 2071–2100 in a number of catchments The results indicated that the current 10-year flood would occur once every 4 years by the end of the century.

Since 1985 when the FREND project began, IH has played a pivotal role – in 2006 a global perspective was published reviewing its geographical expansion and its contribution to the IHP (Servat and Demuth, 2006). When FREND started in Wallingford, its initial focus was the data-rich north-west Europe. By 2006, there were 8 regional projects within Flow Regimes from international Experimental and Network Data (FRIEND), including ones in the Hindu Kush Himalayas and the Nile. A Water Poverty Index was developed for DFID to offer a better understanding of the relationship between the extent of water availability, the ease of abstraction and the level of community welfare. Its key components are the physical availability of surface and groundwater taking account of its variability and quality, how easy it is to access this resource. In 2004, BHS celebrated its twenty-first birthday in a one-day symposium in the Institution of Civil Engineers in Westminster (Fig. 1.7).

In 2007, the Water and Global Change (Watch) Programme started as an integrated endeavour funded under the European Union's sixth Framework Programme. It brought together 25 internationally orientated hydrological, climatological and water resources communities from across the EU, headed by CEH. They analysed, quantified and predicted the components of the current and future global water cycle. They evaluated uncertainties and clarified the overall vulnerability of global water resources for key societal and economic sectors to better inform stakeholders and policy makers (Harding and Warnaars, 2011). See http://www.waterandclimatechange.eu/.

Research continued during the second half of the decade to improve the assessment of flood risk; research that became increasingly important with the growth of housing on flood plains. A prototype method based on a 1-km gridded hydrological and routing model resulted from research by the Joint Centre for Hydro-Meteorological Research setup at Wallingford by the Met Office and CEH

Fig. 1.7 BHS twenty-first birthday celebrations in 2004. (Source: Reproduced with permission of BHS.)

Box 1.3 Advancing hydrology

A special issue of Hydrology and Earth System Sciences (HESS) was published in 2007 (Neal and Clarke, 2007) dedicated to McCulloch (2007). It recognised his contribution to the science as an experimental physicist, as the first Director of the Institute of Hydrology and then subsequently as editor of the internationally acclaimed *Journal of Hydrology* and later as editor of Hydrology and Earth System Sciences. The special issue contains nearly 50 papers on contemporary topics grouped under catchment area research, process studies and modelling.

---

in December 2000. A newly developed cosmic ray soil moisture probe provides spatially integrated assessments of soil water content representative of conditions across a target area of 350 m in radius to a depth of 0.5 m. Deployment of a countrywide network of these probes and their operational use should ensure much improved flood forecasts and the better assessment of the nation's water resources.

Towards the end of the decade, NERC (2009) published a brief report 'Economic Benefits of Environmental Science'. It was based on extensive research by PricewaterhouseCoopers using an approach which developed a value chain for each of 10 case studies, identified a range of qualitative strategic benefits and wherever possible a quantitative benefits assessment. Detailed arguments were provided for assumptions and caveats. One case study was of the Flood Estimation Handbook which cost NERC £100,000 and produced benefits to the UK economy estimated to be between £7 and £34 million a year. By comparison, the summer

2007 floods, which hit a number of areas in England, were estimated by the EA to have cost £3.2 billion.

1.14 Looking Ahead

That the world's water problems will decline in the future is an unlikely scenario. A burgeoning world population, changes in climate and other forcing factors more or less guarantee a greater number in both the developing and the developed regions of the globe. Indeed, the 2014 World Economic Forum in its Global Risks Report (WEF, 2014) rated water crises third after fiscal crises and high unemployment as the most significant risks to humanity, with extreme weather events including floods as sixth. On this basis, the hydrological research conducted at Wallingford must grow in importance and became even more vital to humankind. Water issues will gather increasing complexity and sophistication, with greater interaction between them and they are very likely to be subject to growing social, political and economic pressures. Never before has the need been stronger for independent, multidisciplinary scientific research into water-related issues.

A settled structure for the UK research community with certain funding and little political interference is probably too much to expect. But if such turbulence can be minimised, that will help to maintain capabilities and their continued development to combat the uncertain future for water. Gathering fees for commissioned research and receiving core funding would continue the NERC financial pattern of the last 50 years, although this model is distant from the Haldane Principle of research independence that applied in much of the first part of the twentieth century. But obviously there are a number of factors which will determine the future well-being of hydrological research at Wallingford. The community of hydrologists must maintain those qualities that have sustained the past advances, and those that characterise the present, namely enthusiasm, dedication and commitment.

1.15 References

Anon (1973) *A view from the watershed, Institute of Hydrology*, Report No.20, 54.

Anon (1993) *The Concise History of the Institute of Hydrology*, Unpublished document to commemorate the 25th Anniversary of the Institute of Hydrology. 1 April 1993, 20.

CEH (Various years) Record of research. Centre for Ecology and Hydrology.

Command 4814 (1971) *A framework for government research and development*, Green Paper, (The Rothschild Report), HMSO London, 43.

Command 5016 (1972) *Framework for Government Research and Development*, White Paper, HMSO London, 15.

Dean, T.J., (1994) *The IH capacitance probe for measurement of soil moisture*. Report No. 125, Institute of Hydrology, 39.

Edwards, K.A. & Rodda, J.C. (1970) A preliminary study of the water balance of a small clay catchment. *Journal of Hydrology. New Zealand*, **9**, 202–218.

Enteyhabi, D., Asrar, G.R., Betts, A.K. *et al.* (1999) An agenda for land surface Hydrology research and a call for a second International Hydrological Decade (19). *Bulletin of the American Meteorological Society*, **80** (**10**), 2043–2055.

EU (2000) Directive 2000/60/EC of the European Parliament and the Council of 23 October establishing a framework for community action in the field of water policy. *Official Journal of the European Communities*, **L327**, 1–32.

Harding R. & Warnaars T. (2011) *Water and global change: the watch project outreach report*, CEH Wallingford, 40.

HMSO (1963) *Water Resources Act*, Chapter 38, HMSO London, 184.

HMSO (1965) *Science and Technology ACT 1965, Chapter 4*. HMSO, London.

HMSO (1973) *Water Act*, Chapter 37, HMSO London, 120.

HMSO (1989) *Water Act*, Chapter 15, HMSO London, 419.

HMSO (1991) *Water Resources Act*, Chapter 57, HMSO, 259.

HMSO (1995) *Environment Act*. Chapter 25, HMSO London.

HMSO (2003) *Water Act*, Chapter 37, HMSO London.

HMSO (2010) Flood and Water Management Act 2010. *Chapter*, **29**, 81.

HRU (Various years) *Annual Report, Hydrological Research Unit*.

IAHS/UNESCO/WMO (1970) *Symposium on Hydrometry*, IAHS Pub. Nos. 98 and 99, 892.

IAHS/UNESCO/WMO (1971) *Mathematical Models in Hydrology*, IAHS Pub. Nos. 100,101,102, 1357.

IASH/UNESCO/WMO (1972a) *Symposium on the Results of Research on Representative and Experimental Basins*, IAHS Pub. Nos. 96 and 97, 478.

IASH/UNESCO/WMO (1972b) Symposium on World Water Balance. Proc. Reading Symposium, IASH Pub. Nos,92, 93 and 94, 706.

IH (Various years) *Annual Report, Institute of Hydrology.*

IH (1999) *Flood Estimation Handbook,* Institute of Hydrology, **5** volumes.

Kirby C., Newson M.D. & Gilman K., (1991) *Plynlimon research: the first two decades.* Institute of Hydrology Report No. 109, 188.

Law, F. (1956) The effect of afforestation upon the yield of water catchment areas. *Journal of the British Waterworks Association,* **38**, 489–494.

Law, F. (1957) The effect of afforestation upon the yield of water catchment areas. *Journal of the Institution of Water Engineers,* **11**, 269–277.

Law, F. (1958) Measurement of rainfall, interception and evaporation losses from a plantation of Sitka spruce trees. *International Association of Scientific Hydrology, Proceedings General Assembly of Toronto,* **2**, 397–411.

Law, F.M. (2000) Role of the British Hydrological Society. In: Acerman, M. (ed), *The Hydrology of the UK: A Study of Change.* Routledge, London.

Leeks, G.J.L. & Walling, D.E. (1999) River basin sediment dynamics and interactions within the UK land-ocean interaction study. *Special Volume of Hydrological Processes,* **13**, 931–1179.

McCulloch, J.S.G. (2007) All our yesterdays: a hydrological retrospective. *Special Issue, Hydrology and Earth System Sciences,* **11**, 3–11.

Morris, D.G. & Flavin, R.W. (1996) *Flood risk map for England and Wales,* Report No. 130, Institute of Hydrology, 87.

Nash, J.E. (1958) Determining runoff from rainfall. *Proceeding of the Institution of Civil Engineers,* **10**, 163–184.

Nash, J.E. (1960) A unit hydrograph study, with particular reference to British catchments. *Proceeding of the Institution of Civil Engineers,* **17**, 249–282.

Neal, C., House, W.A., Leeks, G.J.L. & Marker, A.H. (1997) UK fluxes to the North Sea, Land Ocean Interaction Study (LOIS): river basins research, the first two years 1993–1995. *Special Issue Science of the Total Environment,* **194/195**, 1–4.

Neal, C., House, W.A., Whitton, B.A. & Leeks, G.J.L. (1998) Conclusions to special issue: Water quality and biology of United Kingdom rivers entering the North Sea: The Land Ocean Interaction Study (LOIS) and associated work. *Science of the Total Environment,* **210/211**, 585–594.

Neal, C., Leeks, G.J.L., House, A., Whitton, B. & Williams, R.J. (Eds.) (2000). *Rivers Research in East Coast British Rivers.* 3rd LOIS Special Volume of *Science of the Total Environment,* 704.

Neal, C., Leeks, G.J.L., Millward, G.E., Harris, J.R.W., Huthnance, J. & Rees, J.G. (2003) Land Ocean interaction: processes, functioning and environmental management: a UK perspective. *Special Issue, Science of the Total Environment,* **314–316**, 918.

Neale, C. & Clarke, R.T. (2007) A view for the watershed revisited. *Hydrology and Earth System Sciences,* **11**, 663.

NERC (1972) *Observations of the Natural Environment Research Council on "A Framework for Government Research and Development," Submission to the Secretary of State for Education and Science,* 17.

NERC (1975) *Flood Studies Report. Natural Environment Research Council,* Swindon, 5 Volumes.

NERC (2009) *Economic Benefits of Environmental Science.* Vol. **4**. Natural Environment Research Council, Swindon.

O'Connell, P.E., Quinn, P.F., Bathurst, J.C., Parkin, G., Kilsby C., Beven, K.J., Burt, T.P., Kirkby, M.J., Pickering, A., Robinson, M., Soulsby, C., Werrity, A., & Wilcock, D. (2007) *Catchment Hydrology and Sustainable Management, (CASM): an integrating methodological framework for prediction.* In: Proc. PUB Kick off Symposium Predictions in Ungauged Basins, IAHS Pub No. 309, 53–62.

Oliver, H.R., Gash, J.H.C., & Gurney, R.J. (Eds.) (1999) The tiger programme. *Special Issue, Hydrology and Earth Systems Sciences* **3**, 1–149.

Robinson, M., Moore, R.E., Nisbet, T.R. & Blackie, J.R. (1998) *From moorland to forest: the Coalburn catchment experiment,* Report No. 133, Institute of Hydrology, 64.

Robinson, M., Rodda, J.C. & Sutcliffe, J.V. (2013) Long-term environmental monitoring in the UK: origins and achievements of the Plynlimon catchment study. *Transactions of the Institute of British Geographers,* **38**, 451–463.

Rodda, J.C. (2006) On the British contribution to international hydrology – an historical perspective. *Hydrological Sciences Journal,* **51**, 1177–1193.

Royal Society (1971) *United Kingdom Hydrology Bibliography, 1960–1964.* Vol. **67**. The Royal Society, London.

Servat, A.E. & Demuth, S., (2006) *FRIEND – a global perspective. German IHP/HWRP National Committee/UNESCO.* 202.

Sheail, J. (1992) *Natural Environment Research Council: A History.* Vol. **105**. NERC Publishing Services, Swindon.

WEF (2014) *Global Risks 2014.* Vol. **58**. Ninth Edition, World Economic Forum.

2 Basin Studies and Instrumentation

IAN STRANGEWAYS[1], MARK ROBINSON[2], JIM HUDSON[3], JOHN C. RODDA[2], MALCOLM NEWSON[4], AND DAVID J. COOPER[3]

[1]Terradata, Wallingford, UK
[2]Centre for Ecology and Hydrology, Wallingford, Oxfordshire, UK
[3]Ex-Centre for Ecology and Hydrology, Wallingford, UK
[4]Tyne Rivers Trust, Corbridge, Northumberland, UK

Progress in Modern Hydrology: Past, Present and Future, First Edition. Edited by John C. Rodda and Mark Robinson.
© 2015 John Wiley & Sons, Ltd. Published 2015 by John Wiley & Sons, Ltd.

"When you can measure what you are speaking about, and express it in numbers, you know something about it;
when you cannot express it in numbers, your knowledge is of a meagre and unsatisfactory kind."
    Lord Kelvin, Lecture to the Institution of Civil Engineers, London 1883

2.1   Introduction

Measurements are central to all sciences and should be as free from errors as is practically possible. This is especially true in hydrology. From its beginnings, measurements were made and records kept, some of the earliest examples being the levels of the Nile floods observed on a number of Nilometers some 5000 years ago. The aim is said to have been prediction of the likely productivity of the harvest, but there are also ideas that the records were used as a basis for taxation. The Romans were very skilful water engineers, making measurements of levels and cross sections, but not rates of flow. These came much later, in the eighteenth century. Early measurements of rainfall probably began over 2000 years ago in Asia, but it was in the seventeenth century when the first gauges were installed in Europe. Measurements of evaporation and ground water levels date from the eighteenth century. However from the days of ancient Greece and Rome for the next 1500 years or so, rainfall was considered insufficient to cause rivers to flow: extraordinary theories were advanced about the origins of springs being fed by subterranean reservoirs. Conventional wisdom was that the hydrological cycle was internal to the earth. It was not until the seventeenth century, with Perrault's study of the headwaters of the Seine, and Mariotte's work on the basin above Paris, that scientific hydrology started (UNESCO/WMO/IAHS, 1974). Their observations showed that the basin's rainfall was more than sufficient to cause the Seine flow. Knowledge of the correct concept of the hydrological cycle was a key step towards understanding the motion of water (Strangeways, 2007). River basins have been and are a key tool in the development of the science of hydrology, particularly in defining the water balance through measurements of rainfall, river flow, evaporation and the storage in soil moisture and ground water.

The establishment of instrumented catchments has been vital to the development of knowledge of the hydrological impacts of land use on water resources. Small well defined, preferably water-tight, basins have been employed for more than 100 years to investigate relationships between land and water. Pioneer studies started in 1902 in Switzerland (Stähli *et al.*, 2011) with a 'static' comparison of paired basins in contrasting land covers. The first study where the land use was changed commenced in 1910 at Wagon Wheel Gap in the United States (Van Haveren, 1988), with the removal of vegetation from one of a pair of forested catchments. Since that time, many research basins have been studied across the world to quantify the hydrological changes resulting from alterations in land use (e.g. Penman, 1963; Rodda, 1976; McCulloch and Robinson, 1993; Herrmann and Schumann, 2010). A number of new studies were prompted by the UNESCO International Hydrological Decade (IHD) (1965–1974) Programme on Representative and Experimental Basins which aimed to understand the various aspects of a basin's hydrological behaviour, including the consequences of changing land use on: the volume of runoff, the storm hydrograph, low flows, the quantity and type of sediment and water quality. Concern to demonstrate the value of research basins caused the FREND (later FRIEND) Project into Flow Regimes from (International) Experimental and Network Data to be launched in 1985, as a contribution to the International Hydrological Programme (IHP). It aimed to widen the applicability of their results and to assist in solving local and regional problems. Now with so much attention given to long-term monitoring of the environment, the records and results from small research basins are invaluable in this world of global change.

Part 1 of this chapter describes the study of the Plynlimon catchments in some detail. It continues with discussion of work on catchments in East Africa and on some of the other small basins investigated by HRU/IH/CEH. Additional details of the science and its aims and objectives are described in later chapters of Progress in Modern Hydrology. Part 2 looks at the sensors, the means of logging their outputs and the techniques developed to measure the in-basin variables. Some of these have been deployed in other studies to gather data for problem solving across the range of water orientated projects. Without these advances in instrumentation hydrological research would have been sorely impeded and progress severely limited.

2.2   Part 1. Basin Studies

The development of hydrology as a scientific discipline in the latter half of the twentieth century, and the role of the Institute of Hydrology (IH), were very much aimed at making detailed measurements of key aspects of the hydrological cycle to better understand hydrological processes. Central to much of this work was the use of experimental catchments. In fact the original Wallingford laboratory, the Hydrological Research Unit (HRU) was set up in 1962 primarily to conduct hydrological studies of catchments with different land covers and particularly to answer a water balance question – what is the impact of land use on river flow, and specifically – 'do trees use more water than grass?' (Kirby *et al.*, 1991; Robinson *et al.*, 2013). In the mid-twentieth century there was acute concern about the nation's water security. Demand was growing rapidly and water shortages were predicted before the end of the century. There was concern that changes in land use, particularly the rapid expansion of commercial forestry that was then taking place, would adversely affect water yield (Law, 1956). Although there was some evidence from basin studies in North America, it was far from clear whether their results had relevance to UK catchments, where soils, tree species and climate differed considerably.

This required detailed measurement of the components of the hydrological cycle, including rainfall, soil moisture storage, evaporation losses as well as streamflow.

2.2.1   *Starting the Plynlimon study*

The search across England, Wales and Scotland for suitable catchments eventually led to the identification of the adjoining headwaters of the Severn and of the Wye draining the slopes of Plynlimon (Pumlumon) in Central Wales as a possible site for a comparative study (Fig. 2.1). The Upper Wye, 10.5 km² in area and mostly acid grassland, contrasted with the 8.7 km² Upper Severn, which had been afforested with commercial conifers (*Sitka spruce* and *Norway spruce*). This began about 25 years earlier and extended over 70–80% of the catchment. Both basins were equipped with flow gauges built in the 1950s as part of an investigation into hydroelectric power generation potential in mid-Wales. Although the average annual precipitation of about 2500 mm was much wetter than sites previously considered, the catchments fulfilled the other requirements, and the shallowness of the Plynlimon soils meant that a marked drought might result in a restriction of transpiration. The catchments had a similar altitude range of 320–780 m AOD and similar south easterly aspects. Their geology was Lower Palaeozoic hard mudstones, shales and slates, while the soils consist of blanket peat on the hilltops, podzols on the valley slopes and peats and gleys in the valley bottoms. The British Geological Survey (BGS) concluded that the catchments were watertight. Within each basin there were three sub-catchments, which could be gauged separately as a check on the internal consistency of the measurements. In addition, both the Wye and Severn River Authorities were keen to co-operate in the study

Meanwhile, UK environmental research was undergoing a major change and HRU was transferred to the newly formed Natural Environment Research Council (NERC) on April 1, 1965. NERC's Hydrology Committee confirmed the importance of Plynlimon as the research project 'for which the Unit was originally set up' and confirmed their belief that this long-term study would provide 'an outdoor laboratory of immense value to all sciences'. Agreements with the landowners were reached and the Plynlimon catchments became available as an experimental site on January 1, 1967. However there were some scientists who doubted that the right choice of site had been made, believing they would be too wet to show a water stress contrast between forest and grass but as events were to prove, the high and frequent rainfall

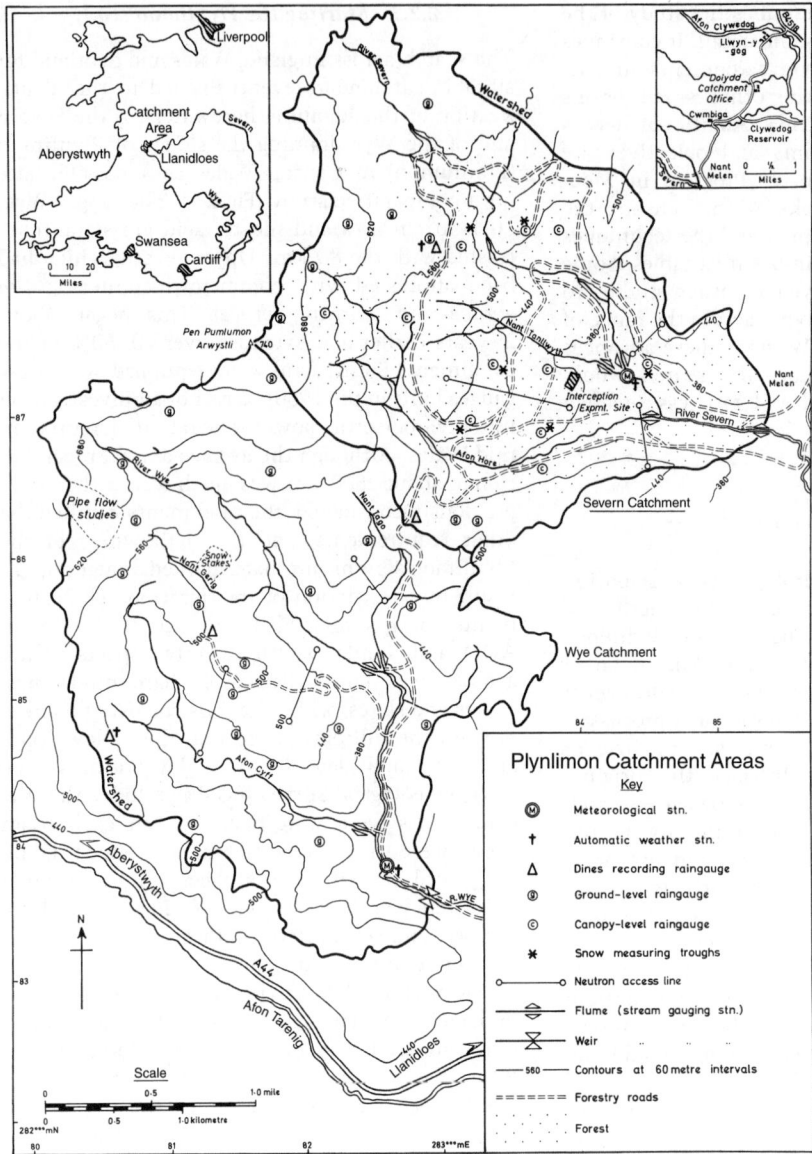

Fig. 2.1 Plynlimon catchments, comparing the mostly forested Severn and the grassland Wye. (Source: Reproduced with permission of CEH.)

would enable the extent of interception losses in forest evaporation to be clearly revealed.

2.2.2 Establishing the Plynlimon field experiment

The delays in establishing the Plynlimon catchments as a field site provided an opportunity to develop expertise in 'cutting edge' instruments and new technologies. This included pioneering work on neutron probes for soil water content measurement, automatic weather stations capable of operating at remote locations, and new raingauge designs for exposed sites. An aerial photogrammetric survey of both catchments was commissioned and detailed maps (1:5000) produced as a basis for the study

(Brandt *et al.*, 2004). It took several years to carry out the work to convert the catchments into an outdoor laboratory. The preparatory work included building an office and laboratory and recruiting staff to man it permanently to collect records and maintain instruments. A series of roads and bridges were constructed, new gauging structures were installed and exiting ones modified, offroad vehicles were tested and a range of novel instruments deployed. The initial focus was on measuring quantity: process studies and those on water quality came later, as did many variants of research on hydrological themes.

2.2.2.1 Precipitation

Accurate measurements of precipitation were considered essential to the study. Existing raingauge records indicated there was a great deal of spatial variation across Central Wales, while an analysis of an intensive raingauge network in the nearby River Rheidol showed evidence of strong topographic controls (Rodda, 1968). Accordingly, using the concept of stratified random sampling, a raingauge network was designed based on topography: namely altitude, aspect and ground slope to define raingauge domains and to locate gauges within each domain. Building on earlier research by the Meteorological Office into the accuracy of standard gauges at exposed sites, a comparative study was made of the catch of raingauges at the standard rim height of 1 ft (30.5 cm) and ground-level gauges protected from splash by metal grids. This revealed a systematic undercatch by the standard raingauges, an undercatch increasing with wind speed (Rodda, 1967). For the forested areas, experiments were conducted on different aerodynamic shapes of canopy gauges, and following wind tunnel experiments a simple funnel type gauge was found to have better agreement with ground-level gauges than any other design. Installation of the full raingauge network of ground-level and above-forest canopy-level gauges was completed by April 1971 comprising 39 gauges; 18 in the Severn and 21 in the Wye. Subsequent statistical analyses showed that in months free of snow, the difference in catch between ground level and canopy gauges was no greater than the spatial variations in the same altitude class (Clarke, 1976). Different techniques to provide areal average catchment rainfall values were compared and yielded similar results, (Clarke *et al.*, 1973). The catch of snow was a problem, for which the only practical solution was daily manual depth measurements and melting to compute water equivalents.

2.2.2.2 Streamflow

Whilst providing valuable background data, neither of the existing flow gauges at the main outfalls of the Severn and Wye was sufficiently accurate for research needs. The Severn weir was replaced by a critical depth trapezoidal flume and the compound weir on the Wye at Cefn Brwyn was heavily modified. Each of the main catchments contains three sub-catchments. To measure their flows a new form of gauging structure was designed for these small streams which are characterised by steep bed gradients, large sediment loads at high flows and great variability of flows (Harrison and Owen, 1967). Additional work was needed to ensure that the sub-catchment steep stream flumes performed properly. Detailed multi-point current meter measurements and dilution gaugings were performed on all the flumes to confirm, and where necessary modify, the theoretical stage-discharge ratings.

2.2.2.3 Evaporation

Given the importance of evaporation to the study, the measurement of weather variables was crucial for using the Penman method to assess this factor. A daily-read manually operated weather station was established in 1968 within an unplanted area in the lower part of the Severn catchment. The considerable range in climatic environments across the catchments meant there was also a need for stations at remote higher altitude locations. This was achieved with the development of an Automatic Weather Station (AWS) capable of operation in cold humid environments, measuring and logging solar and net radiation, air temperature and wet bulb depression, rainfall, wind speed and direction (Strangeways, 1972). Several of these stations were deployed across the catchments at contrasting sites.

2.2.2.4 Soil moisture

The development of an improved neutron moisture meter (Bell, 1969) provided portable equipment for routine soil profile measurements. Soil moisture measurements were made at 10 cm depth intervals down an access tube. The design of access tube networks hinged on the nature and distribution of the different soil types in the catchments. Detailed mapping of the soils in 1967–1968 made use of the exposures along the newly cut access roads and the forestry drains (Bell, 1968). The high relief but similar soil parent materials (glacial deposits derived from a uniform sequence of slatey mudstones) meant that differences in elevation, aspect and slope creating different microclimates for soil genesis had resulted in a strong relation between soil type and topography. Broad soil moisture 'domains' of hydrologically distinct soils could be defined from topographic maps – low angle hilltops, slopes and valley bottoms, which provided a basis for soil moisture network design. Soil water storage was estimated using three transects of access tubes in each catchment running from the watershed to the valley floor. These enabled seasonal patterns of storage changes to be estimated for catchment water budgets, but indicated that at a *local scale* the flow processes were extremely variable, with no consistent trends down hillslopes (Hudson, 1988).

2.2.3   The basins' water balance

Preliminary catchment water balance calculations for 1969 and 1970 showed lower flows (higher evaporation losses) from the forested Severn (Rodda, 1971). However caution was needed in reading too much into such a short-period of record and because there were uncertainties in some of the measurements. Subsequently, Clarke (1976) statistically analysed each element of the water balance for the period 1970–1975. This was crucial given that catchment evaporation is estimated as the difference between two much larger numbers. Even small errors in precipitation and/or streamflow could result in large differences in estimated catchment evaporation. He concluded that catchment precipitation (the largest water balance component) was estimated to within 5% of the true areal mean, and that the streamflow error was much smaller, yielding uncertainties in water loss that were far smaller than the difference between the two land covers (18% and 38% of annual precipitation for grass and forest respectively).

2.2.4   Finding the mechanism – why *forests use more water than grass*

Once it was shown that a coniferous forest reduced flows under British conditions, it was important

to explain why this happened by investigating the mechanisms responsible, principally so that the results could be extended to other areas. Developments in process studies (Rutter, 1975) indicated that to determine if the Plynlimon findings were applicable elsewhere, it was necessary to do far more than many contemporary catchment studies, that is, to simply draw up a water balance from precipitation and streamflow measurements, supplemented by potential evaporation estimates from meteorological measurements. A thorough understanding of the component processes of the hydrological cycle was essential for the findings to be applied elsewhere (McCulloch, 1975). This required additional measurements including forest interception. Early recognition of the need for process studies to *explain* the reason for *observed* differences has been one of the core strengths of Plynlimon which distinguished it from many contemporary watershed studies. This change in emphasis from the initial black box catchment approach commenced with small runoff pathway studies in the Wye, including those on soil pipeflow (Gilman and Newson, 1980). This move was encouraged by further developments in process studies, and Howard Penman, in his role on the NERC Hydrology Committee, was an enthusiastic supporter of process studies. Newson (1979) distinguished distinct phases of instrumentation (1968–1973) and process study intensification (1973 onwards) putting IH scientists versed in other disciplines at Plynlimon team to run process studies nested within the catchment water balance studies.

2.2.5   Process studies

From the early 1970s work commenced on detailed process studies. A forest lysimeter was constructed in the Severn catchment. It was similar to Law's installation at Stocks Reservoir, which had been designed to determine whether interception was responsible for the observed difference in evaporation between forest and grass. But to overcome criticisms of Law's work this was sited well away from the forest edge and augmented with intensive soil moisture measurements to enable transpiration to be accurately determined. Sheet gauges were installed to better assess net rainfall under the tree canopy and a 'tree cutting' experiment was conducted to examine the factors controlling stomatal resistance to transpiration. Combining these measurements with the Rutter canopy water balance model, Calder (1977) showed the importance of interception by which forests may use more water than grassland, producing evaporation losses consistent with the water balance results from the main Plynlimon experiment.

Additional intensive process studies were carried out on interception and transpiration to understand how forest evaporation depends on meteorological and physiological factors. It was crucial to determine the partition of energy into evaporation and heating to assess the importance of advection (downward transfer of sensible heat from the air) in comparison with direct solar energy on evaporation, particularly the intercepted rain on vegetation canopies. This required the application of very accurate micrometeorological measurements of temperature gradients, wind profiles and radiation to study the turbulent structure of windflow and the transfer properties at the boundary layer above the forest. The feasibility of instrumenting a flux tower within the forest in the Severn catchment was considered, enabling direct comparison with the catchment water balance data. But this plan was abandoned due to the very uneven airflow over the mountains. Instead a site was chosen close to the centre of the extensive, level and uniform Thetford Forest in eastern England. Although the climate was different to Plynlimon, the basic physics of wet canopy interception would be the same, and the conditions would be ideal for the high accuracy measurements required using pioneering instrumentation available. Detailed temperature and humidity measurements were made the forest on 30 m towers (Stewart and Thom, 1973), Jack Rutter worked closely with this project, measuring transpiration and interception losses (Chapter 5)

Work continued at Plynlimon to improve the accuracy and extend the measurements of the various components of the hydrological cycle in order to cover wider aspects of physical hydrology. As the collection of the basic data became more routine and instrument and logger reliability increased, effort was freed for increasingly diverse projects, with water quality studies being given particular attention (Chapter 8). And as originally envisaged, use was made of the infrastructure at Plynlimon to provide an 'Open Air Laboratory' for the many university research groups and individuals who came to work in the catchments.

These research projects included studies of geomorphology, soils and stream hydrochemistry, atmospheric deposition and acidification, fish populations and stream biota, and the fallout from the Chernobyl nuclear accident (Kirby *et al.*, 1991). Studies of erosion and stream sediment yields highlighted the high loads caused by poor forest drainage techniques, with the drains forming important sediment sources, only declining slowly over time (Fig. 2.2f). In the early years of Plynlimon *water chemistry* was used solely to provide natural 'tracers' to define water pathways to aid quantity studies. But regular water sampling began in the mid-1980s to study geochemical cycling and investigate the interaction of soil weathering and geology on stream chemistry. Hydrochemical studies expanded with the growing awareness and concern about acid deposition and the effect of forest management practices on stream chemistry (Neal *et al.*, 1997). New instruments enabling short-period or 'near-continuous' field measurements of streamflow chemistry provided fresh insights into the flowpaths and storages of subsurface water reaching the stream network (Kirchner *et al.*, 2004). Later, data from the hydrochemical measurements (Fig. 2.2g) carried out over nearly 30 years were made available on the CEH Information Gateway (https://gateway.ceh.ac.uk).

2.2.6 *The main findings of the Plynlimon research*

The Plynlimon results show that mature coniferous forest in the uplands of the UK will tend to increase the proportion of rainfall lost to the atmosphere through evaporation by comparison with short vegetation. Newson (1979) summarised the main findings of the first decade at Plynlimon, including preliminary water balances. He showed forest losses of up to double those from the grassland, with interception studies giving results of the same order. More comprehensive results for the first two decades were subsequently presented in Kirby *et al.* (1991). Calder and Newson (1979) published the outcome of several studies; suggesting that interception losses were equivalent to 30–35% of annual precipitation in areas of the UK receiving more than 1000 mm of precipitation. Plot studies measuring interception losses from plantations in the United Kingdom uplands confirmed that differences in evaporative loss from forest and grass

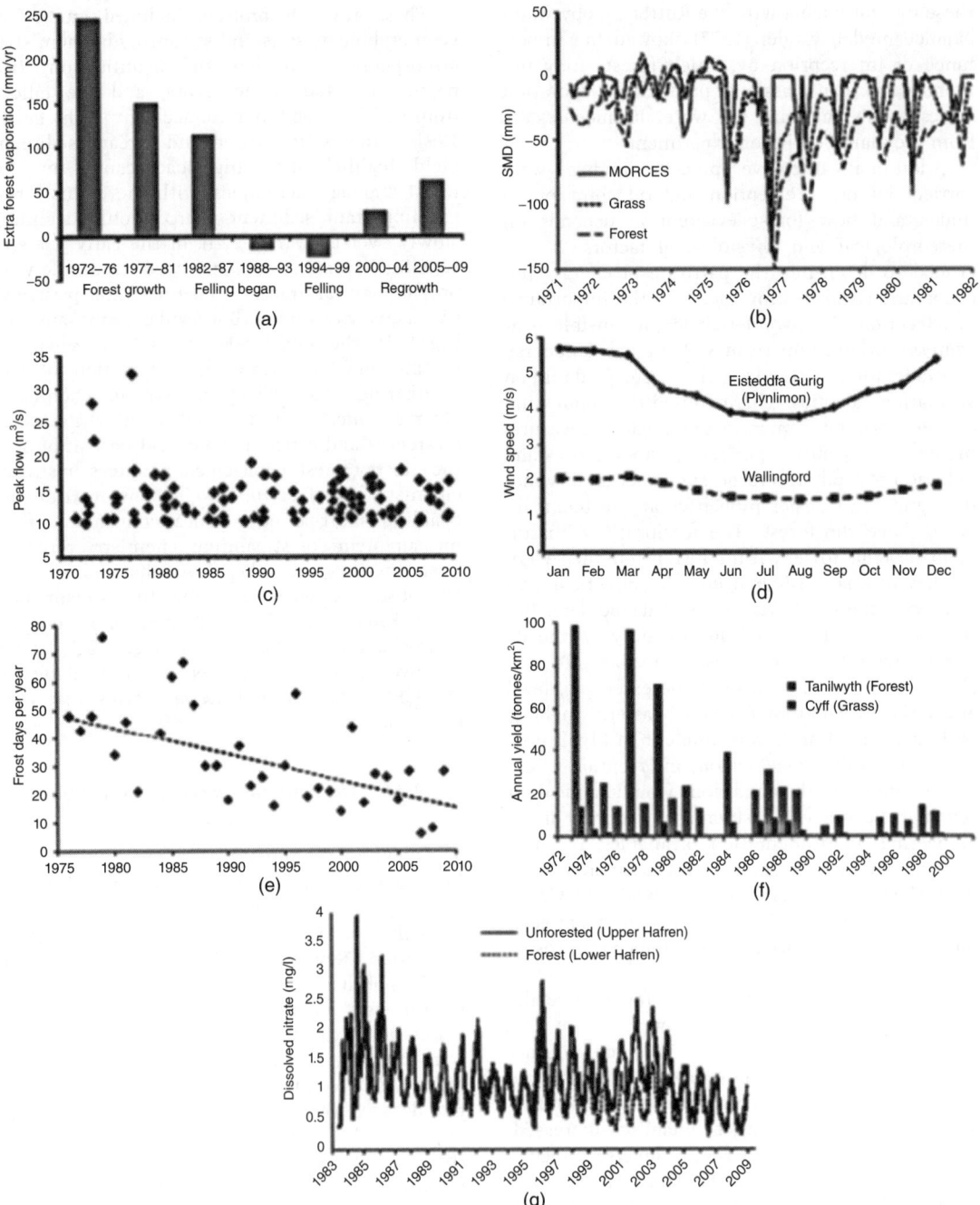

Fig. 2.2   Plynlimon overview of results, showing (a) Changing water use of forest relative to grassland, (b) Soil water under forest and grass, (c) Peak flows from forest, (d) Greater winds speeds in the uplands, (e) Reducing frost days over time, (f) Sediment loss from forest and grassland, (g) Changes in stream nitrate loss with tree felling. (a–f). (Robinson *et al.* 2013. Reproduced with permission of John Wiley & Sons.)

catchments were principally due to the high rates of *interception* loss from the wet forest canopy, rather than significant differences in *transpiration* (Calder, 1990). Studies of forest transpiration found it to be a rather conservative hydrological process with relatively small differences between different tree species (Roberts, 1982). Evidence from Plynlimon indicated the potentially dominant role that interception can play in high rainfall upland areas. The extended period of observations now available reveals clear differences in the hydrology at different stages of the rotation of the managed forest (Marc and Robinson, 2007). The changes through time are of sufficient magnitude that following recent extensive felling and with large areas of immature new forest, the forested Severn catchment evaporation losses are currently little different to the grassland Wye (Fig. 2.2a). In contrast, there is no evidence of changes in flood risk though the forest cycle, supporting evidence that forestry has only a limited impact of large floods (Fig. 2.2c). Soil water measurements have confirmed the drier conditions beneath the forest (Fig. 2.2b). The British uplands have relatively few weather observations and the combination of AWS and on-site field results have helped provide new information (Newson, 1979). These include the systematic raingauge undercatch varies with seasonal patterns in wind speeds (Fig. 2.2d), while long-term records of air temperature show clear evidence of a long-term warming trend (Fig. 2.2e).

The earlier belief that the evaporation from wet vegetation would be similar to when the vegetation was transpiring was based on an overestimation of the aerodynamic resistance of vegetation to vapour transfer relative to its stomatal resistance (Stewart, 2004). The extra energy had to come from cooling of the air, although the mechanism responsible remained controversial until the development of instruments capable of very accurate measurement of the small gradients of specific humidity and temperature. Sophisticated and carefully designed micrometeorological studies at Thetford Forest and elsewhere detected an appreciable downward flux of heat from the air, providing confirmation that loss rates due to interception could be much higher than the net radiation (Stewart and Thom, 1973). They also found that the latent heat fluxes due to transpiration, even when the trees were well supplied with water, could be significantly less than the available net radiation, demonstrating

that physiological controls were regulating transpiration rates. The finding that forest evaporation at Thetford could actually be *less* than the Penman short grass potential evaporation at first appeared to contradict both the work of Law and the lysimeter results at Plynlimon concerning the importance of interception, and it was only with further work that an appreciation arose of the critical importance of rainfall frequency and duration for interception losses (Gash and Morton, 1978). The trees at Thetford had only a quarter of the annual rainfall and far fewer rain days than at Plynlimon, which in contrast experiences frequent long-duration low intensity frontal rainfall, keeping the forest canopy wet for extended periods. When the regional patterns of increased losses were presented on a map of Britain (Calder and Newson, 1979) this had a major effect on water industry policy.

2.2.7 Plynlimon's legacy

The early Plynlimon records (even before the data were refined) showed lower flows from the forest, but there were uncertainties about the accuracy of the catchment measurements, doubts about the mechanism responsible and fundamental questions about the representativeness of the results – could they be applied to other areas with different, soils, geology and climate? The controversy was only solved through process studies at Plynlimon and elsewhere, principally Thetford, whereby the physical processes enabling high rates of evaporation from wet forest canopies were understood and quantitative predictive models produced (Gash and Shuttleworth, 2007) to extrapolate to other areas. The scientific basis for understanding the higher losses from forests came from the detailed process monitoring, rather than the catchment water balances, but crucially for many engineers and stakeholders the fact that the whole catchments provided confirmation contributed to the ultimate proof they required to justify changing their practices. And catchments remain the principal scale and unit of policy for land and water management to this day. The Plynlimon study provided a considerable stimulus to the development of the science of hydrology in Britain and directly influenced UK government forestry and environmental policy.

The Plynlimon basins are among the key European and world catchment studies, not just

Fig. 2.3 East African catchments. (Source: Edwards & Blackie 1981. Reproduced with permission of John Wiley & Sons.)

because of the 'length' of records (>40 years), but crucially because of the 'breadth' of measurements. The research carried out in these basins had been reported in well over 500 publications within its first two decades (Kirby *et al.*, 1991) and in many more since. The continuing importance of these data world-wide is evidenced by the range of use by researchers and unique data sets (e.g. Kirchner *et al.*, 2004). Much of the Plynlimon record has

been freely available to external researchers for decades.

2.2.8   Other catchment studies

The results from other long-term UK catchments studies conducted by IH/CEH, including Grendon Underwood in southern England, Balquhidder in Scotland (Whitehead and Calder, 1993) and

Coalburn in northern England (Robinson, 1998) have been extensively reported in the scientific literature, together with work at other research basins (Chapters 7 and 8) and operational basins (Chapters 3, 6, 7, 8, and 10). In this chapter the main findings of two groups of experimental studies are discussed, namely: water balance changes over a plantation forest cycle in the United Kingdom, and hydrological changes due to land use conversion from native vegetation to commercial farming in East Africa.

2.2.9 *Plantation forest cycle in UK*

Data were available from the process studies of forest evaporation in the Severn, with transpiration estimates coming from soil moisture balances on the forest natural lysimeter (Calder, 1976). The data were used to quantify these processes and to calibrate the various models of forest evaporation developed elsewhere, such as at Thetford Forest (Gash and Morton, 1978; Rutter *et al.*, 1971). Calder and Newson (1979) were able subsequently to link the process and catchment information to produce a semi-empirical model of forest evaporation that could be used to predict the effects of mature forest on water resources for other areas of the UK with broadly similar conditions to Plynlimon.

To augment and broaden the Plynlimon study, the Balquhidder study was started in the early 1980s, the aim being to improve water resource assessment in the Scottish Highlands, where the primary land use change was not from grazed pasture to forestry, but from heather-dominated grass moorland to forest. Here a paired catchment study compared a heather grass mix in the Monachyle 'control' catchment with the 39% forest covered Kirkton basin. A weighing lysimeter study (Wright and Harding, 1993), had suggested grass transpiration as low as 75% of potential evaporation but there was also a significant interception component from the heather to consider (Hall, 1987). The low grassland evaporation was a surprising finding, as Blackie and Simpson (1993) had reported higher than expected evaporation at the higher altitudes in the Kirkton and was probably due to the short growing season at these altitudes and soil moisture stress from thin soils.

The Balquhidder results showed a 200 mm/year extra water use from the heather grass mix in the Monachyle compared to the partly forest covered Kirkton, which at first glance seemed to contradict the generally accepted consensus, that trees used more water than moorland in most upland environments. Two land use modelling exercises (Blackie, 1993; Hall and Harding, 1993) and interception studies by Johnson (1990) indicated that the trees were freely intercepting rainfall at rates similar to those seen at Plynlimon. However, the overall evaporation from the Kirkton catchment was depressed by the very low transpiration rates from the grassland that dominated the higher altitudes. The high interception rates postulated from the modelling for the heather and other dwarf shrub vegetation community in the Upper Monachyle were therefore indicative that a change to forestry would not have such a major water resource impact in Scottish conditions as in upland grass-dominated areas south of the border.

With over 40 years data available from British forest catchment studies the hydrological picture became more complicated than first envisaged (Marc and Robinson, 2007). While the Wye showed no discernible trend in evaporation, there was however, a drop in the evaporation from the Severn catchment of about 200 mm per year. The low values from the forest area at the end of the study period were clearly affected by the large areas felled from 1985 onwards and the low evaporation from newly cleared areas was sustained almost entirely by interception from brash, the ground flora taking some years to recover or a second forest plantation cycle to be established. Clearly a managed forest comprising areas of different aged planting will behave differently to a single uniform block of even-aged trees.

The effects of the initial afforestation of virgin moorland on the hydrological character of upland headwater catchments has been studied at Coalburn in northern England (Robinson, 1998) and to a lesser extent at Llanbrynmair in mid-Wales (Hudson *et al.*, 1997) and demonstrated the complex processes at work during the ploughing and forest establishment phase, with dewatering contributing to streamflow and giving artificially low values of losses that can falsely be ascribed to low evaporation. Disruption of the vegetation can, however, cause low evaporation rates for a few years, but the Llanbrynmair records indicated an increase in catchment evaporation after just 5 years relative to the moorland control, that then rose rapidly for the next 8 years to well above moorland evaporation values. This was not just due to the growth of the forest but also to the rapid growth of dwarf shrubland vegetation on the unplanted and ungrazed parts of the catchment. The evidence from the longer-running Coalburn study was similar, with dewatering an even more obvious contributor to streamflow due to the deep drainage system, though the increase in evaporation took much longer due to the initially poor growth of the young forest. The replanting of felled areas for a second rotation is being covered by the re-afforestation of the Kirkton catchment at Balquhidder, which although no longer run by CEH is being monitored by the Scottish Environmental Protection Agency.

2.2.10   Land use change in East Africa

In the 1950s there was pressure on the Colonial governments in Kenya and Tanganyika (now Tanzania) to release more land from the high altitude forest reserves for cultivation by the rapidly increasing populations, for conversion to fast growing pine plantation by the Kenya Forestry Department and for further expansion of lucrative tea estates. The Water Development Departments opposed these proposals, believing that the indigenous cover provided the best possible protection for these stream source areas in terms of the quantity and quality of dry season water available to meet the rapidly increasing demand downstream.

In order to produce some hard facts in place of entrenched opinions Charles Pereira of the East African Agriculture and Forestry Research Organisation (EAAFRO) proposed, in 1956, a series of paired catchment studies in representative areas to quantify the effects of the proposed land use changes. This received the approval of the Governments and relevant Departments in each country and he and his team headed by Jim McCulloch proceeded to identify suitable sites. Those chosen were a pair of adjacent catchments in the South West Mau Forest near Kericho in Western Kenya, a pair in the montane bamboo forest on the Aberdare Mountains to the north of Nairobi at Kimakia and a pair in Southern Tanganyika near Mbeya (Fig. 2.3).

In the Kericho study the land uses compared were indigenous rain forest and tea estate, at Kimakia it was indigenous bamboo forest versus fast growing *Pinus Patula* and at Mbeya it was indigenous rain forest versus peasant cultivation.

The approach adopted in these studies was to instrument each catchment to quantify as accurately as possible the water use of its vegetation and compare this with potential evaporation computed from meteorological data obtained from sites within the catchment using the Penman equation (Penman, 1948). This was a radical departure from the traditional comparison of rainfall/streamflow relations, as used in earlier such studies in US and elsewhere, and was partly inspired through the close ties between EAAFRO and Rothamsted Agriculture Research Station where Penman was based.

The chosen catchments were assessed geologically to ensure that they were 'watertight', that is, that water left only as streamflow or evaporation. Water use could then be quantified from the water balance equation,

Rainfall was measured using stratified random sampling networks of gauges mounted at canopy level with one or more recording gauges to provide a means of time distribution. Streamflow measurements came from water level recorders on accurately rated weirs and soil moisture from gravimetric sampling over the rooting depth at carefully chosen sites. Initial volumetric sampling at these sites provided a means of converting the sampling results to equivalent depths of water in the profiles. The change in annual groundwater storage between dry seasons was calculated by integrating under the baseflow recession curve. A detailed description of the catchments, the methods adopted and the initial results are provided by Pereira (1962).

In 1964 Jim McCulloch left EAAFRO to take up the post as the first Director of what became the UK Institute of Hydrology. With the agreement of, and generous funding from, the Overseas Development Ministry he initiated a programme of cooperation with EAAFRO which provided useful overseas experience for many IH staff and made possible the upgrading of catchment instrumentation. Digital logging systems replaced chart recorders, automatic weather stations provided continuous meteorological data for Penman computation and neutron probe soil moisture measurement replaced tedious and destructive sampling. Digitisation of the long runs of catchment data transformed the methods of quality control, storage and analysis and made them more accessible to interested parties.

Work on the Mbeya catchments ceased in 1969 and on the detailed studies at Kericho and Kimakia in 1974, although rainfall and streamflow measurements continued on these catchments. After a period of intensive analysis a summary of the data was published (Edwards and Blackie, 1981) and a final report on all aspects of the studies was published as a Special Issue of the East African Agriculture and Forestry Journal (Blackie *et al.*, 1979).

After 18 years of dedicated work by the field observers and the EAAFRO staff, the cooperation of Forestry and Water Development Departments in the two countries and the UK Government funding for the staff from the IH, hard facts could replace entrenched opinions. There was direct evidence on the effects of these land use changes on the quantity, quality and timing of flows in the rivers emerging from these upland, high rainfall areas of East Africa (Blackie and Robinson, 2007).

The Mbeya catchments differed significantly from those in Kenya in terms of seasonal rainfall distribution, with over 90% falling between November and April. They were also much steeper sided and the soils were derived from volcanic ash rather than from weathered lava. Not unexpectedly the water use of the cultivated catchment where the land lay fallow from June to November was significantly lower than that of the forested one, averaging 64% of Penman as opposed to 92% for the forest where the moisture storage within the extensive rooting range was sufficient to maintain transpiration rates throughout the long dry seasons. The surprise came in the remarkably low erosion rates from the steep sides of the cultivated catchment. Despite appearing to be at high risk of degradation these light, highly porous soils stabilised remarkably quickly after the onset of the rains. In these circumstances peasant cultivation would seem to increase the streamflow available downstream without materially affecting its quality, although it was probable that this would deteriorate in the long term if more robust soil conservation measures were not implemented.

At Kimakia the rainfall patterns in both catchments were very similar with mean annual totals (1958–1973) of 2325 and 2198 mm for the control and experimental catchments respectively. During the initial stages of planting and establishment water use by the plantation pines was significantly lower than that by the indigenous bamboo but the difference diminished as the pines grew. From the time of effective canopy closure the water use by the two covers was virtually identical with a mean value of 76% of Penman. A marginal increase in the very low level of sediment yield was noted during the clearing and planting phase but thereafter the differences were negligible. Thus the Forest Department's contention that this zone could be used to produce much needed softwoods without damaging the water resource, appeared to be justified.

The results of the Kericho study were rather more complex. Although the long-term mean water use of the control catchment with indigenous forest was 92% of Penman, similar to that of the forested catchment at Mbeya some 1500 km away, the inter-yearly range was much greater. Interpretation of the behaviour of the experimental catchment was complicated by the fact that only 54% was converted to tea estate and a substantial subcatchment of 26% was almost entirely under bamboo forest. The water use of the complete experimental catchment was 84% of Penman, somewhat lower than in the control catchment, whilst that of the bamboo sub-catchment was very similar to the complete catchment both in magnitude and trend. Detailed process studies and conceptual modelling investigated the reasons for the difference in overall water use between the catchments and its variability in time. They concluded that the differences in the transpiration rates and interception characteristics of the vegetation types, arose directly from the great contrast in aerodynamic roughness between the uneven forest canopy and the smooth 'plucking surface' of the tea estates. From a water resources viewpoint the results showed that, with well-designed soil conservation measures, the land use change from montane forest to tea estate did not result in any reduction in streamflow volume or significant increase in sediment loads although there could be some changes in the time distribution of flow.

These studies provided valuable guidance on the effects of the land use changes for Government planning decisions as well as extending scientific knowledge and were considered to have justified the considerable investment involved. They were also notable as being the first examples *anywhere in the world* where cutting edge research techniques were applied to land use problems in developing countries. This was also reflected in the conscious decision to publish their results in an African science journal rather than in a western journal (e.g. Blackie *et al.*, 1979). This well-meaning policy, however, unfortunately meant that their results took a number of years before they were widely appreciated by the international scientific community.

2.3 Part 2. Instruments

Until the 1960s the instruments available for measuring meteorological and hydrological (hydrometeorological) variables had changed little since the nineteenth century. They needed observers to read them, or they recorded their measurements as pen traces on paper strip-charts, all of which was labour-intensive. A typical conventional 'Met Site' is illustrated in Figure 2.4 (Strangeways, 1995). When IH began its catchment research during this period these were the only available instruments. They were unsuited to unattended operation at remote upland sites that could not be visited daily, and an essential part of the Institute's early work was to develop new instruments that would allow measurements of all the basic hydrometeorological variables to be made automatically and frequently. Without these developments, catchment research at the Institute could not have progressed as rapidly and as successfully as it did.

The instruments described in part 2 of this chapter are those used in basic catchment research, are also an integral part of many of the other studies described in this book, and indeed in many environmental studies around the globe. Especially important are precipitation, streamflow, soil moisture content and the meteorological variables needed to compute potential evaporation.

Development of these general-purpose instruments was accompanied by the development of other specialised instruments for very specific purposes. At various times during the history of Plynlimon and other experimental catchments, process studies were introduced to identify the reasons behind the measurements. These included a lysimeter study of interception losses, tree cutting experiments on transpiration as well as detailed localised observations of evaporative flux, and the

movement of river bedload. These are described in the relevant chapters.

In addition to high tech instrumentation there was also a need for more basic solutions for particular purposes. IH published a widely used report for other researchers detailing practical methods for field studies including building ground-level raingauge grids, forest canopy-level raingauges, interception sheets and pipeflow meters (Anon, 1977).

This new generation of automatic instruments was made possible by the emergence in the 1960s of solid-state microelectronics, in particular of chips that handled logical processes digitally. To put this in context, in the era just prior to this, in the 1950s, only 'thermionic valves' were available and these required a great deal of power. Nevertheless, it was this 'vacuum tube' technology that had enabled the first digital computers to be developed at Bletchley Park in the United Kingdom in the 1940s to crack the German Enigma and Lorenz codes in the Second World War. The arrival of solid-state electronics was the essential key to moving digital electronics out into the field and away from mains power, their crucial advantages being that they operated off low voltages, consumed only milliwatts of power and were very small. They were largely

Fig. 2.4   A typical traditional Met Site. *From left to right:* Campbell-Stokes Sunshine recorder, Stevenson screen containing thermometers for recording maximum and minimum air temperature and 'wet and dry bulb' thermometers for measuring relative humidity, with a five inch manual raingauge. In right photo are the wind direction and windspeed sensors. (Source: Reproduced with permission of Ian Strangeways.)

unaffected by damp conditions and extremes of temperature.

2.3.1   Data logging

Central to modern automatic meteorological and hydrological instrumentation is the data logger, its key component being memory for storing the measurements. In the 1960s, and for several decades thereafter, the only practical recording medium was magnetic tape, the first data loggers using reel-to-reel ¼ in. tape (Fig. 2.5). Loggers using 'Compact Cassettes' were soon developed, jointly by IH and Microdata Ltd, which served all the logging needs of IH for the next 25 years (Fig. 2.5). But the problems with tape are that it has

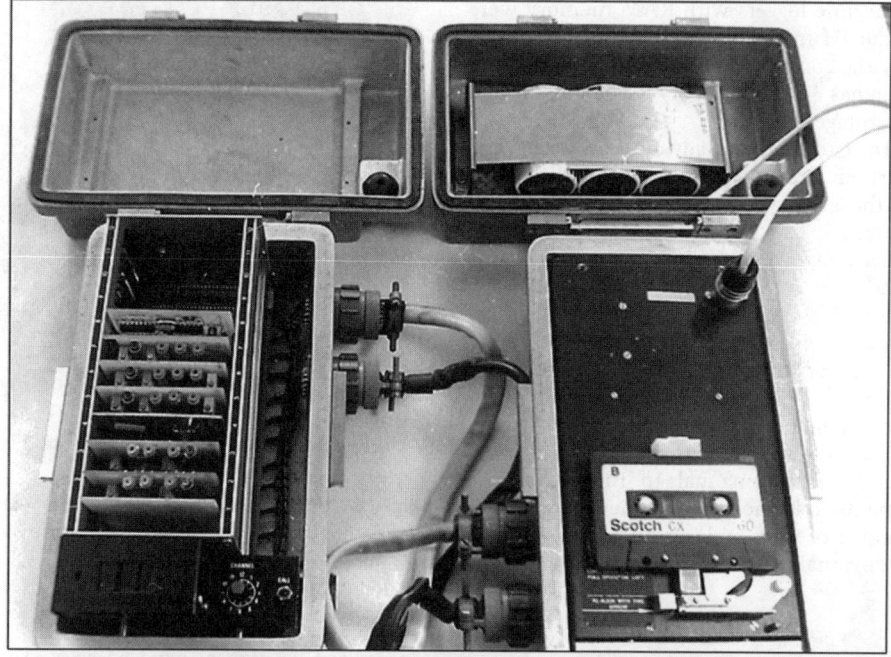

Fig. 2.5   Two different data loggers. *Top:* Microdata logger used in the 1970s with data stored on compact cassettes. *Bottom:* A modern solid-state logger. (Source: Reproduced with permission of Ian Strangeways.)

limited storage capacity (typically under 0.1 MB) on a C90 Compact Cassette) and is subject to failure, especially at the low field temperatures because of its mechanical nature. In the mid-1980s solid-state memory started to become available: now the storage capacity of modern memories is truly impressive. Such memory is also extremely cheap and can operate over a wide range of temperatures – from the tropics on Earth to the cold of outer space – and its power consumption is negligible. While loggers with RAM memory were developed at IH in the 1980s, modern loggers are typified by the commercial Campbell Scientific CR 1000 which has 4 MB of memory for processing and final data storage.

In contrast to the rapid advances in logger and microelectronic technology, sensors have changed little over the last 50 years and most still use exactly the same principles today. Long before solid-state loggers became available, hydrometeorological sensors had been available that produced electrical signals, alongside the older traditional mechanical and manual devices. Their measurements were recorded on paper strip charts or by electromechanical counter. When loggers became available it was convenient in many cases simply to adopt these existing electrical sensors with little modification. The transition from manual to fully-automatic measurements was, however, achieved gradually over a number of years and in some cases the conventional, manual instruments remained adequate.

2.3.2 Precipitation

There are literally thousands of different designs of raingauge in use around the world, manual and mechanical-recording. Figure 2.6 illustrates the UK's standard 5 in. diameter storage gauge. Manual gauges continue to be employed in large numbers in national networks across the world today. The tipping bucket raingauge is the most common type of automatic gauge now used (see later below). Raingauges that weigh the collected water date back into the nineteenth century, but with the availability of accurate electronic scales, weighing raingauges are becoming popular again although they tend to be more expensive.

2.3.2.1 Wind errors

The main difficulty with measuring rain, using any type of gauge, is that the gauge acts as an obstruction to the wind which speeds up as it

Fig. 2.6 The UK standard five-inch manual rain-gauge. (Source: Reproduced with permission of Ian Strangeways.)

passes over the gauge, causing catch to be lost; losses can be high at exposed sites. To overcome the problem, gauges are best exposed with their rim set at ground level (Fig. 2.7). Where this is not practicable, surrounding the gauge with a wind shield (Fig. 2.8) or using a gauge with an aerodynamic profile (Fig. 2.9) are alternatives. The ground-level pit gauge was evaluated in detail at IH and was used throughout all of its catchments. To maintain the aerodynamic advantages of pit gauges, the gauge orifice and grid had to be parallel to the slope of the ground in mountainous areas. This resulted in the area presented to vertical rainfall being smaller than the nominal orifice size and a simple cosine correction was applied to the catch. Rainfall measured without some wind protection can be in considerable error, around 5% annually in lowland parts of the UK and 25% and more in the uplands and highlands. The problem is much greater when measuring snowfall. The WMO

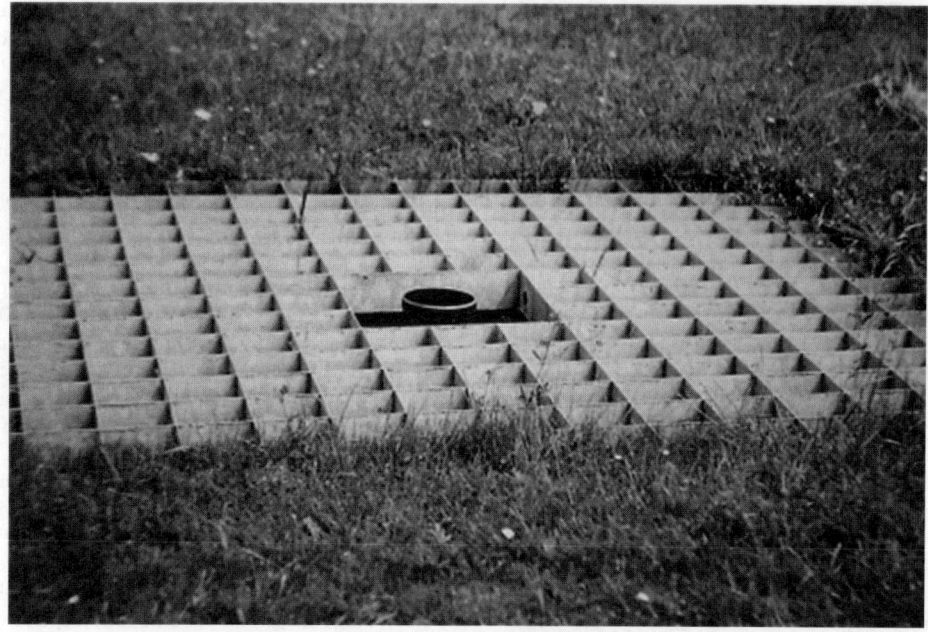

Fig. 2.7 Ground-level raingauge (pit gauge). (Source: Reproduced with permission of Ian Strangeways.)

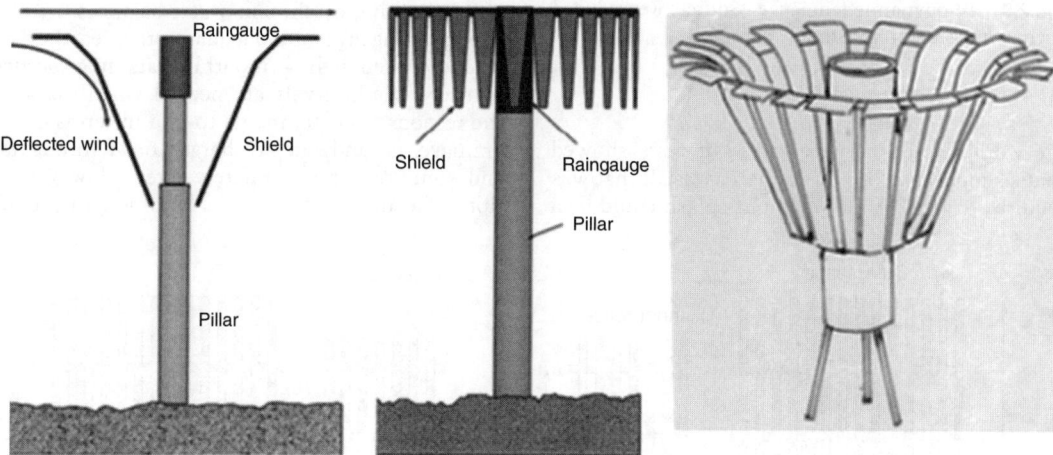

Fig. 2.8 Wind effects on catch can be reduced by surrounding a raingauge with a wind shield: *From left to right* – Nipher, Alter and Tretyakov wind shields. (Source: Reproduced with permission of Ian Strangeways.)

solution to snowfall measurement is the large double fence snow shield (Fig. 2.10).

For forested areas a canopy-level raingauge had to be developed that would have the correct aerodynamic characteristics so as not to undercatch, while providing the required flexibility in construction to allow the funnel to be raised to the optimum position in the forest canopy as the trees grew. Initially it was felt that an aerodynamic shield, of the type used on standard gauges around

Fig. 2.9 Aerodynamic raingauge. (Source: Reproduced with permission of Ian Strangeways.)

problem in heavy rain. Provided the funnel is not raised above the general canopy, the canopy acts as a shield, akin to the turf wall or the anti-splash grid, as used on the ground. The main problem with the canopy gauges was that it was difficult to prove they would catch the same amount as an equivalent ground-level gauge in the same location in the absence of the trees. Newson and Clarke (1976) devised a statistical approach that showed that the two gauge types caught the same amount in all rainfall conditions, though not during snow.

More recently Robinson *et al.* (2004) set up an experiment in the Severn catchment to compare the actual catches of canopy-level and ground-level gauges installed close to a grass/forest edge, at the same time checking the impact of gauge-orifice-height relative to the forest canopy. This new study, although extending over only a short time-scale, confirmed that the two types of gauges gave similar catches. Moreover the tests showed that the canopy gauge catch was relatively insensitive to the height of the orifice relative to the average height of the surrounding tree canopy.

2.3.3 Storage and mechanical gauges

For estimating catchment water balances, manual storage raingauges are adequate provided they are read often enough – monthly data may suffice. Shorter time-intervals are needed when assessing the response of catchments to rain in terms of soil drainage, groundwater recharge, forest throughfall and stemflow and ultimately streamflow generation. The first attempts at Plynlimon to provide

the world, would be necessary, but tests showed that a simple cone of 5 in. diameter was all that was required (Fig. 2.11), although out-splash could be a

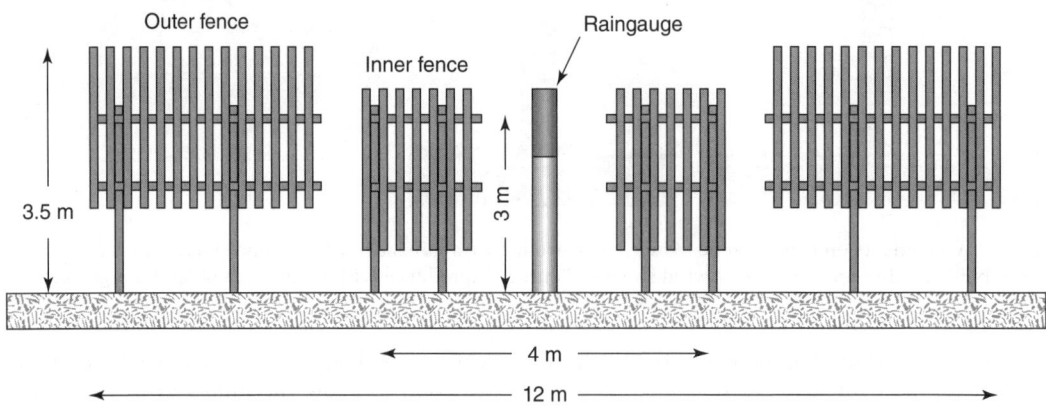

Fig. 2.10 Wind effects are even more severe for snowfall, and the The World Meteorological Organisation (WMO) recommends a double-fence snow shield. (Source: Reproduced with permission of Ian Strangeways.)

Fig. 2.11 Five inch funnel gauge over forest canopy. (Source: Reproduced with permission of CEH.)

daily rainfall was a standard raingauge feeding a clockwork mechanism that moved a distribution arm once per day over a series of collecting bottle, providing 32 individual days' catches. But the technology was not sufficiently reliable in the harsh conditions, and this technique was abandoned. The Met. Office-approved Dynes tilting-syphon hyetograph was used in its place at a limited number of sites to distribute the catches of the monthly gauges to shorter period values. This led to a considerable workload for scaling, calibrating and digitising the data, which provided the impetus for the development of electronic logging systems.

2.3.4 Tipping bucket gauges

This type of gauge works on the principle of a two-sided 'bucket' which tips from side to side, as first one side and then the other becomes filled with rainwater collected in the funnel (Fig. 2.12). As it tips, a switch generates a pulse, the total number over a pre-set interval being logged. Or the time and date of each tip can be recorded to give an estimate of intensity. The 'Rimco' tipping bucket gauge replaced the Dynes at Plynlimon, logging on a single channel Compact Cassette data logger that just recorded tips; its appearance was the same as the multi-channel logger in Figure 2.5. Tipping bucket gauges alone are not suitable for all purposes due to errors in bucket calibration or to evaporative loss from the bucket between tips. Calibration is carried out by some manufacturers by passing a known amount of water slowly and at constant rate through the gauge. This 'dynamic' calibration allows for the small loss of water during the finite time it takes the bucket to tip from one side to the other. This loss varies with the rainfall intensity, but correction factors can be produced to allow for the losses at different intensities and it is not difficult to achieve a fair calibration accuracy with modest equipment. Gauges can malfunction, particularly in winter conditions causing a complete

Fig. 2.12 Tipping bucket raingauge (showing the interior). (Source: Reproduced with permission of Ian Strangeways.)

loss of data. To allow for this it is advisable that storage gauges are also used alongside recording gauges, both as a check on calibration and as a backup in case of failure of the recording gauge.

2.3.5 Estimating areal rainfall from a raingauge network

Having chosen the best precipitation instrumentation, the next stage was to estimate the areal precipitation into the catchments. Gauges had to be positioned within a catchment in a way that allows for the perceived spatial controls on precipitation. The 'domain theory' of raingauge distribution was developed at Plynlimon in which the catchments are divided into zones with similar altitude, slope and aspect. A single gauge was then installed randomly in all domains with an area greater than 2% of the whole catchment. The same technique was used at Balquhidder. At Plynlimon, several methods – isohyets, Thiessen polygons and domains – all gave similar results, suggesting strongly that the raingauge locations were well chosen and that the network was also denser than strictly necessary. An uncertainty-analysis suggested that around 10 gauges would be required in each main catchment to give an areal mean consistently within 2% of the true value (Clarke *et al.* 1973). This 'excess' of gauges was useful to infill data when gauge catches were missing in any particular month. The method adopted assumed

long-term 'stationarity' in catch to calculate long-term ratios between gauges in each network. Provided not too many gauge readings were missing in any one month, the ratios could then be used to estimate the missing catches. In 1999, after a reduction in funding for upland research, the size of the Plynlimon storage raingauge networks was reduced from 48 to 21 (Marc and Robinson, 2007).

The isohyetal method can be useful visually to indicate how the controlling factors may vary, although it can be labour-intensive and subjective. Both Thiessen and domain weighting methods, on the other hand, could be achieved using fixed representative areas for each gauge, providing a simple and objective calculation of areal rainfall.

2.3.6 Automatic weather stations

AWS measure all the basic meteorological variables in one package using a multi-channel data logger. They are very versatile tools with many applications and are widely used by hydrologists to estimate potential evaporation (atmospheric 'demand'), for comparison with water balance estimates of actual water evaporation. The early development of reliable AWS was an important achievement at IH, setting the general path for the development of AWSs right up to the present. Figure 2.13 illustrates one of the first IH stations, along with the production design that followed on from it. IH designed and made some of the first modern AWSs to be developed anywhere in the world, using the new technology which was just becoming available in the mid-1960s. Indeed the AWS seen in Figure 2.13, designed in 1965, was most probably the first of its kind (Strangeways and McCulloch, 1965; McCulloch and Strangeways, 1966, Strangeways, 1972). This same design was still in use at Plynlimon in 2015, although the earlier cassette loggers have been replaced by a modern Campbell Scientific CR 1000 data logger. The on-going use of this AWS demonstrates its reliability and the soundness of its manufacture. It also demonstrates how the sensors have remained unchanged while the data loggers have developed rapidly. Details of AWS design, their sensors and the importance of ensuring correct exposure is described by Strangeways (2003, 2007). Here the sensors are briefly considered.

Fig. 2.13 Automatic Weather Stations: *Left*: Early prototype developed at IH in the late 1960s. One of the first modern AWSs. *Right*: Didcot Instruments production model – still in use. (Source: Reproduced with permission of Ian Strangeways.)

2.3.6.1 Temperature

Arguably air temperature is, meteorologically, the single most important variable of all those measured today, considering the concerns about climate change. It is generally measured in the first couple of metres above the land surface. The electrical sensor used most widely to measure environmental temperatures is the platinum resistance thermometer (PRT). The electrical resistance of all metals changes with temperature, but platinum wire is preferred for its relative linearity and stability. For measuring air temperature, thermometers of any type, mercury-in-glass or PRT, have to be shielded from radiation, both direct solar, reflected solar and terrestrial infrared. Historically for the last 160 or so years shielding has been achieved by exposing mercury-in-glass thermometers in wooden screens of slatted construction, very often of Stevenson's double-louvered design. But when mercury-in-glass thermometers were replaced by PRTs, which were small and did not need to be read

by an observer, a miniature compact screen was designed at IH (Fig. 2.13) and this now has many derivatives.

For the most accurate results, however, an 'aspirated' screen is necessary in which a fan draws air over the sensors. All naturally ventilated screens, which rely on the wind to move the air through them, can give large errors under low wind and high radiation conditions which cause the temperature of the screen and its contents to increase by up to several degrees above actual air temperature; nevertheless they suffice for catchment research work which does not require as high an accuracy as the detection of climate change demands. However, it is preferable, whenever possible, to use aspirated screens, the main deterrent being their need for power for the fan, although this can be provided with solar panels.

Work done at Thetford required the measurement of very small temperature (and humidity) profiles up though the depth of the forest from ground to canopy top. This was micrometeorology, needing temperature measurements to fractions of a degree at different heights. This was highly specialised equipment for a very specific project and is described in Chapter 5.

2.3.6.2 Humidity

Atmospheric humidity has been measured traditionally by the 'wet-and-dry-bulb' method in which a thermometer has a wick around its bulb dipping into distilled water. Evaporation from the wick cools the thermometer, the amount of cooling giving a measure of the relative humidity (RH). Aspirated screens produce much more reliable and consistent results.

Although electrical RH sensors have been available since the 1980s, they were not to the standards of today. They included capacitive and conductive sensors, but it was not certain how stable their calibration was over the long term and they were also liable to sudden calibration shifts if kept at high humidity for several hours. Thin-film capacitive RH sensors were developed initially for short-term use (a few hours only) in radiosondes, but soon found a use at surface-based stations. They have the advantage of not requiring a water supply and can operate below freezing point. The latest design has built in miniature heaters to drive off any water vapour molecules periodically to maintain calibration. This heating was necessary for radiosondes to ensure that after passage through cloud the calibration was quickly re-established. What the long-term (five years +) stability of these latest sensors is not known, but it might be prudent to replace them periodically. While used at some research sites, this type of sensor has not been used on AWSs at Plynlimon at any time, the well-proven wet/dry method being maintained throughout the period from the 1960s to the present so that continuity is maintained. It is important in operating any data collection network to keep accurate records of the changes, of recalibrations or of the moving of instruments to a new location (known as *Metadata*).

2.3.6.3 Wind

The well-proven cup anemometer remains the most common windspeed sensor (Fig. 2.13, top arm on left). Many designs have evolved but all work on the same principle that the cups (generally three) cause a shaft to rotate as the wind blows past them, the speed of rotation being more or less linearly proportional to the windspeed. Some are large and might have high starting torques, while others are miniaturised and start rotating at low windspeeds. A magnetic or optical switch produces pulses that are counted, in the case of a modern system by digital counter chips. While sonic anemometers are now becoming more commonplace (see Chapter 5), the cup anemometer is still the first choice even today.

Wind direction is sensed by a vane activating a potentiometer or shaft encoder (Fig. 2.13, top arm on right). Knowing the direction can be useful for interpreting spatial variations in rainfall in hilly areas (e.g. 'rainshadows') and also in pollution deposition studies to identify likely sources.

2.3.6.4 Solar radiation and sunshine duration

Since the nineteenth century, sunshine duration has been measured with the Campbell-Stokes sunshine recorder. These instruments continue to be used in the twenty-first century, but are now augmented by electrical sunshine sensors that aim to simulate the characteristics of the traditional device, allowing automatic logging, and thereby maintaining continuity of this long-standing measurement.

But during the first half of the twentieth century, electrical sensors of solar radiation energy were developed for recording on paper strip charts. These 'pyranometers' measure the total solar energy received on a horizontal surface, the sensor being a black disc of a few square centimetres in area exposed to the sun under glass domes. The disc's temperature rises in the presence of solar radiation in comparison to the body temperature screened from the sun the difference being measured with a thermopile (Fig. 2.13, at top of pole). It measures both direct solar radiation and diffuse solar radiation from the sky and by reflection from clouds. The output of the thermopile is a signal in the low millivolt range which is amplified and logged every few seconds, the average over a period giving the total incoming radiation during that period – be it a minute, an hour or a day. It is also possible to derive 'pseudo-sunshine' duration from these sensors. More recently, silicon photo-diodes have become available that can be used to measure solar radiation, although they do not sense the full spectral band of the sun equally. They have the advantage of cheapness and of being very small.

2.3.6.5 Net radiation

Net radiation is the difference between the total incoming solar radiation together with the total incoming infrared radiation from the atmosphere above and the total outgoing solar radiation (the reflected solar radiation or 'albedo') and the total outgoing infrared radiation from the surface. This difference represents the energy available to evaporate any available water (the latent heat flux) and to warm the air in contact with the surface (the sensible heat flux). Net radiation is sensed by 'net pyrradiometers' which are similar to solar pyranometers in that they sense the radiation levels using a black disc, but in this case one disc looks upwards another downwards, the difference in temperature between the two discs (measured with a thermopile) being a direct measure of the net energy available at that point (Fig. 2.13, lower arm). However, since glass is opaque to infrared radiation, the domes of a net radiometer are made of polyethylene which passes the full spectral range from the shortest solar to the longest infrared radiation involved in the energy exchange (0.3–70 μm). But this plastic material ages and is vulnerable to mechanical damage.

2.3.7 *Potential evaporation (Penman method)*

While temperature, windspeed, humidity, solar and net radiation may seem remote from hydrology, they are not, for evaporation can be calculated using the relationships between these variables as derived by Penman (1948) in his well-known equations and their many subsequent derivatives. Indeed this was one of the driving forces behind the development of AWSs at IH. Evaporation had traditionally been measured by evaporation pans or, more accurately, by lysimeters (Fig. 2.14), but all have limitations of uncertain representativeness.

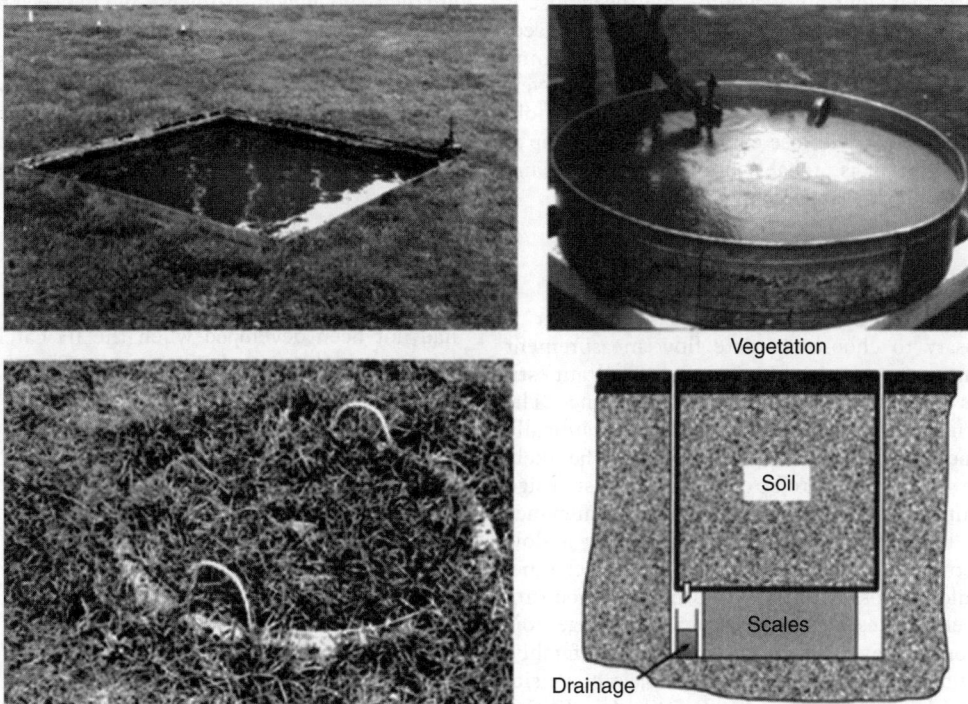

Fig. 2.14 *Top left*: UK Evaporation pan sunk in the ground. *Top tight*: 'Class A' evaporation pan, normally exposed above the ground. *Bottom left*: A lysimeter. *Bottom right*: Drawing of a lysimeter components. (Source: Reproduced with permission of Ian Strangeways.)

Automatic weather stations were installed in the Plynlimon catchment in the late 1960s, to provide estimates of evaporation as quantified by the Penman (1948) formula for short grassland and the Penman-Monteith variation on the formula to allow for the aerodynamic effect over forest cover. AWSs in the forested Severn catchment were mostly set over grass, either in clearings or near the edge of the forest, so did not represent the true forest canopy conditions. However, a few were installed on towers at canopy height for specific studies. AWSs of the IH design were also used in many projects both in the United Kingdom and overseas, ranging from agricultural investigations in the Sahara desert to the South Georgia Islands in Antarctica.

For continuity reasons, early AWS data were (rightly) not initially accepted by the Meteorological community because of their unknown accuracy. Also, even if it were accepted that they were more accurate than the old manual readings, there was concern that continuity would have been lost if there was a sudden step-change to AWSs. Indeed, the change over to AWSs that is now on-going around the globe, unless accompanied by an overlap period of at least a year – comparing the old with the new – will lose continuity. An overlap is essential; yet it is probably not occurring in many cases. Again this is where metadata are vital.

2.3.8　River level and discharge

When setting up a catchment experiment it is necessary to choose accurate flow measurement techniques to minimise errors in any resultant estimates of evaporation and in the water balance. The chosen method must be well suited hydraulically to cope with the range of flows thought to be likely at the site, as well as to be of suitably robust design to minimise the operational and maintenance effort to run and maintain them. Ideally a flow 'structure' will be used. This is a regular channel of uniform cross section with a hydraulic control that eliminates hysteresis from any stage-flow relationship developed for the structure. Preferably, its calibration will have been the subject of considerable laboratory research (and codified in British Standards or ISO) and verified locally in the field.

How river flow is measured depends very much on the size and nature of the river, which extend from the smallest upland stream to the largest rivers of the world. River depth is closely related to flow rates and is usually measured in a sheltered stilling well to the side of the smallest flume or largest river, the depth traditionally being recorded on a float-driven paper chart recorder. Logging, as opposed to strip-chart-recording, has the advantage that the record is directly in computer-compatible form, while a pen-trace on paper has to be converted subsequently into a digital record. At IH strip-chart recorders were augmented in the 1970s by a new electrical sensor in which the float rotated a series of potentiometers (Fig. 2.15) their outputs being logged at pre-set intervals. More recently 'shaft-encoder' sensors, driven by a float, which produce digital signals directly, have replaced the earlier analogue potentiometer sensors because of better resolution and stability. Shaft encoders are now used at all Plynlimon stations, with a Campbell Scientific logger.

Today, pressure sensors can replace floats and these can often eliminate the need for stilling wells, simplifying the overall construction. These are relatively new instruments and may be prone to drift. Stilling wells with float sensors are generally more accurate over the long term. A more complex method of measuring water level by pressure is to bubble nitrogen gas from a cylinder through a tube fixed in the river and to measure the pressure required to maintain a constant stream of bubbles. There are also a number of alternative more complex methods of measuring river flows such as by the electromagnetic and ultrasonic methods but these were not used at IH. Electromagnetic methods are now available for small streams but they had not been developed when the IH catchment instrumentation was being set up. They require power, however, which would not be available at many remote sites, and they would probably not have suited the steep streams. Ultrasonic methods of measuring river discharge using ultrasonic pulses or continuous waves sent diagonally across the river from bank to bank are now available. The difference in Doppler shift of the sound, or in the transit-time of a pulse, between the upstream and downstream paths, gives a measure of the average velocity of the stream. This technique is best suited to large rivers. These techniques are described in specialist texts (e.g. Strangeways, 2003).

2.3.9　Weirs and flumes

Most experimental catchments are either located on headwater streams where there are few instruments

Fig. 2.15
Potentiometer river
level sensor developed
at IH to replace
chart-recording
instruments. (Source:
Reproduced with
permission of Ian
Strangeways.)

and structures for flow measurement already in place, or where the existing instruments must be extensively modified to cope with the strict accuracy criteria required for experimental flow measurement. At Plynlimon for example, neither an existing compound Crump weir on the River Wye nor a rectangular sharp crested weir on the Severn were thought suitable in their original form. The Severn weir capacity was too low for the likely peak flows and it had the propensity to fill with sediment after floods. The Wye weir, although of sufficient capacity for floods, did not meet the required degree of accuracy. The Wye weir was fitted with divide piers and a lowered centre crest to increase precision at low flows, essentially splitting the structure into three separate parallel Crump weirs (Fig. 2.16). The use of compound structures such as this, had been approved to British Standards (BS3680). Considerable work was put into verifying the modifications made to the structures. The results highlighted a number of generic problems concerned with operating flow measurement stations on upland streams. It should never be taken for granted that the laboratory calibration will fit without independent verification, and the crests/inverts of any structure need a regular survey to check that the zero for the stage measurement

equipment remains correct. A general programme of data quality control was applied, with particular attention paid to the maintenance and clearing of tapping pipes, stilling wells, stilling pools, approach channels and sediment traps upstream of each structure.

2.3.10 *Flow gauging in steep streams*

Steep channel slopes in the Plynlimon sub-catchments precluded the use of conventional flumes as it would have been impossible to reduce approach Froude numbers to a level (<0.5) that ensured a stable water surface for head measurement and calibration coefficients within BS limits. The Hydraulics Research Station (now HR Ltd) which designed the flow measurement networks developed a special modification of the critical depth flume (Harrison and Owen, 1967). This new 'Steep Stream' flume (Fig. 2.17) had vertical side walls, and was configured to provide a level water surface from the natural channel to the approach section of the flume, thus maintaining the channel velocity through the flume by effectively turning the natural channel on its side and so keeping a constant cross-sectional area from channel to flume. This required a ramp to

Fig. 2.16   Compound Crump weir in the Wye. (Source: Reproduced with permission of CEH.)

lower the channel bed but causes flow to accelerate down the ramp, giving supercritical flow in the measurement section. An arrangement of kinetic-energy-dissipating-baffles on the floor and walls of the ramp forces a hydraulic jump to form at the base of the ramp, increasing the depth and maintaining subcritical flow. The steep stream flume has proved an ideal solution, with good stage–discharge sensitivity and maintenance of a constant velocity to move sediment through the flume without deposition. Much has been learned over the years about the design, construction and operation of steep stream flumes. Having been designed specifically for use at Plynlimon, modified over their lifetime and run successfully on the sub-catchments, they have given highly accurate flow records for all but the lowest flows, where volumetric and current metering data can be used for infill, and for the highest flood flows, where hydrograph reconstruction has occasionally been necessary. The technique has been used elsewhere, in steep upland areas.

2.3.11   Independent checks and rated sections

It is beneficial occasionally to check the ratings using independent means, with the warning that when the two do not agree it should not always be assumed that it is the frontline technique that is at fault; it could just as easily be miss-application of the independent check. Current metering has proved to be excellent in the regularised concrete channels upstream of structures with stable water surfaces. Most of the uncertainty in calculation of flow from the current meter readings of velocity is down to imprecision in the measurement of depth and channel-survey errors, rather than inherent inaccuracies in the sensors themselves. Not all upland streams are suitable for applying the current metering technique, due to their shallow depths, uneven channels and highly turbulent flow. Once the agreement between conventional propeller current meters and structures in the Plynlimon catchments had been established, the process was usefully reversed, with the structures used as

Fig. 2.17 Steep stream flume at Plynlimon. (Source: Reproduced with permission of CEH.)

test-beds to check the efficacy of new techniques such as the electromagnetic current meters (Smart, 1977; Strangeways, 2003).

Fortunately, the very turbulent conditions that limit the current metering technique are conducive to using 'dilution gauging' in its various forms (Strangeways, 2003). Attempts were made at Plynlimon to develop new chemical, flourimetric and colourimetric tracing techniques, mainly because background salt (NaCl) concentrations were high and variable in the maritime climate of the western uplands of Britain. Combined with current metering for moderate flows and the latest Acoustic Doppler Current Profiling (ADCP) technology for very high flows, these techniques would allow rating through the whole of the stage range to 'bank-full' discharge, and even beyond in compound natural channels.

2.3.12   Soil water measurement

Before the advent of the neutron probe in the 1960s, the routine measurement of soil water in the field was very difficult. Many soil samples had to be taken and their water content determined by oven drying at 105°C. Hydrologists need water content on a volumetric basis and this requires additionally the *in situ* volume of each sample to be determined – a complicated and difficult procedure. Furthermore, although topsoil is easily accessed, the deeper profile requires the digging of pits, which is impracticable as a routine procedure. Water content is frequently found to vary from one point to another, even at the same depth, to such an extent that this swamps the change from one sampling occasion to the next. This made the taking of sufficient samples to monitor changes of water content extremely laborious and time-consuming. It is also very destructive of the monitoring sites. The neutron probe changed all that, enabling repeated, non-destructive measurements to be made of the entire profile at the same point, thereby overcoming much of the spatial variability problem.

From the 1960s onwards HRU/IH was in the forefront of these studies in the field that became known as Soil Hydrology. This started quite modestly, as an evaluation of the neutron probe as a means for improving the determination of soil water storage in experimental catchments.

2.3.13   The neutron probe

The principle of measuring the water content of porous media by the neutron method was invented in the United States during the 1940s, initially as a laboratory technique. Instruments for field use started to appear in the 1950s and by the early 1960s several commercial instruments had become available. Among these were the American Troxler and Campbell Nuclear Pacific, the French Lepaute, the British Nuclear Enterprises, the EAL and the Danbridge (Fig. 2.18) probes. Most users of these instruments, especially in America, were farmers with irrigated crops, so accuracy and stability were less important than in the context of research studies. Although the technique was promising, the first commercially available designs of neutron probe were not tools that could be used easily in hydrological research, being electronically unstable and too heavy and bulky for routine field

Fig. 2.18 Neutron probe 1 the Danbridge equipment showing scaler and transport shield. (Source: Reproduced with permission of CEH.)

Fig. 2.19 Neutron probe. (Source: Reproduced with permission of CEH.)

applications. There was a need for a better design that overcame these problems. After evaluation of available systems, IH designed its own probe (in association with the Atomic Energy Research Establishment). This probe was licensed to a commercial company, Pitman Instruments, and enjoyed world-wide success as 'The Wallingford Probe'. The Wallingford probe became standard equipment for soil hydrology studies both in the United Kingdom and abroad (Fig. 2.19). It was adopted by many hydrological agencies world-wide and although manufacture has now ceased, many probes are still in use. An improved design, the 'IH Probe' was later produced by Didcot Instruments, who became the licensed manufacturer, and enjoyed a reputation for exceptional reliability and stability.

Neutron probes of all types are used to determine the profile of volumetric water content by taking readings non-destructively at a series of depths within an aluminium access tube installed permanently in the ground at each site. Although these early instruments varied in design and appearance, they shared fundamental common features: a probe that is lowered down an 'access tube' houses a sealed source of 'fast' neutrons (typically, 1.1 or 1.85 GBq Americium/Beryllium) with a 'slow' neutron detector (boron trifluoride or ^{3}He) and a scaler or rate-meter that counts and displays the pulses generated by the detector. The reading at each depth takes typically between 10 and 60 seconds, so that, with time taken to reposition the probe at each depth, an entire profile can take typically between 5 and 20 minutes, depending on the maximum depth measured and the counting time at each depth.

Fast neutrons are emitted continuously into the surrounding soil where they are scattered by collisions with the nuclei of the atoms of the soil, being transformed into 'slow' or 'thermal' neutrons, which exist briefly before being absorbed by various nuclear reactions. Thus a 'cloud' of scattered slow neutrons is formed in a spheroidal zone within the soil surrounding the source position. Most of the elements making up the soil matrix exert only a small effect on the number of slow neutrons passing into the detector, the only element that has a major scattering effect is hydrogen, which is almost entirely present in water (H_2O). It therefore follows that the density of the slow neutron cloud around the source and counted by the detector is directly related to volumetric soil water content. The higher the count rate, the wetter the soil, and vice versa). The resulting data are normalised by reference to the count rate in a pure water standard (the 'Standard Count').

The development of the neutron probe gave a great boost to field process studies in soil hydrology. Other uses for the probe soon appeared, particularly research into soil water conditions under conifer forest versus grass, and subsequently for a range of forest species and crops. The neutron probe could also be used in conjunction with tensiometers of various types, which opened up the means to measure soil water processes directly in the field. Experimental methods were developed for studying water flux and storage in the unsaturated zone. Water fluxes could be partitioned between upward and downward components, so enabling direct measurement of evaporation (e.g. crop water use) and recharge to groundwater. The methods could also be used to measure the unsaturated hydraulic conductivity characteristics *in situ*. The techniques were applied at a number of sites to determine water use of agricultural crops, grassland and trees, plus matrix and macropore flow to groundwater such as the Chalk aquifers, where water resources are of crucial importance.

In total about 250 Wallingford and IH probes were made, many of which were exported world-wide. Many are still in use today, although ever-increasing radiological safety constraints have made the use of neutron probes a much less attractive solution to soil water measurement. There has therefore been increasing interest in developing non-radioactive alternatives.

2.3.14 The capacitance probe

All versions of the neutron method have some limitations, including the fact that being radioactive

they cannot be left unattended in the field and thus cannot be automated. Also being radioactive (albeit at a low level), they are subject to many regulations for their storage, transport and use. An alternative method was therefore sought to overcome these limitations and restraints. The method chosen used the measurement of soil relative permittivity (dielectric constant). The relative permittivity of water, around 80, is much greater than that of most other compounds, so if this could be measured *in situ*, conveniently and cheaply this could provide the alternative method.

Several properties of electrical circuits are sensitive to the permittivity of material between electrodes. These depend either on the capacitance of the electrodes or on the properties of a transmission-line formed by those electrodes. The term 'dielectric' implies a perfect insulator and damp soil is usually anything but. However, the effects of electrical conduction diminish at high frequencies relative to those of the real part of the permittivity. The probe was therefore designed to operate at around 100 MHz to avoid most of the interference caused by electrical conduction. The 'Depth Probe' was designed to operate in an access tube, similarly to a neutron probe but, unlike that of the neutron probe, this had to be made of plastic to allow the permittivity to be sensed through the tube. The probe was also made of plastic and was typically 2 m long. Most of the length was empty, acting as a handle. Within the lowest end was the electronic circuitry, batteries, a magnetic on/off switch and the two annular electrodes mounted coaxially one above the other spaced typically 49 mm apart. Data were transmitted up to the top of the handle of the probe by a fibre optic cable to de-couple the sensor electrically from the display assembly at the top end of the handle. Full details of the depth probe, including the physical theory of the method can be found in Dean *et al.* (1987), and Bell *et al.* (1987).

The probe was found to be sensitive to a relatively small volume of soil. This meant that small gaps around the access tube were likely to have a disproportionately large effect on the readings, but also that the depth resolution of the instrument was about 40 mm, much less than the neutron probe. The small volume of soil monitored perhaps affected the representativeness of the measurement. The good depth resolution also required a much more precise location of the probe in the soil. This was achieved by a small assembly placed on top of the access tube. The sensitivity to gaps around the access tube was accommodated by an enhanced installation procedure, involving a baseplate anchored firmly to the ground and a rigid guide for the access tube (Bell *et al.*, 1987). The procedure also allowed calibration samples to be recovered from inside the access tube using an adapted auger and precise depth location stops.

The measurement precision of the neutron probe is limited by random counting error, implying a significant count time at each measurement depth, whereas the capacitance probe gives effectively an instantaneous reading with very little scatter. This allows data to be collected very much more quickly, even at 20 mm depth intervals. The biggest drawback of the new instrument proved to be the delicate fibre optic link between the sensor and reader units.

The probe design was licensed to Didcot Instrument Company, who sold a modest number of units before going out of business, which effectively sealed the end of the instrument as a commercial product. Adaptations of the design were, however, marketed by Troxler Inc. of the US for a number of years and by Sentek of Australia. The latter are still available in several forms and widely used in both the research community and for irrigation scheduling.

2.3.14.1   The surface probe

The technical success of the depth probe led to development of individual sensors which could be buried directly in the soil or set at a specific depth in an access tube. These could be attached to a logger to provide time series data or potentially used to control irrigation or to give continuous records of changing catchment soil water content to aid flood forecasting (Robinson and Stam, 1993).

A second version was designed for manual determination of soil water in the surface layer of the soil. In this version the electrodes were two sharp-ended stainless steel rods 6 mm in diameter, 50 mm apart and 100 mm long. In use, the electrodes were pushed into the ground and a reading taken from the integral display. In this way reconnaissance transects of surface soil water could be obtained rapidly.

2.3.15   *Cosmic ray neutron technique*

A new instrument that uses naturally occurring cosmic rays to generate fast neutrons instead of having

to use an artificial radioactive source has recently become available. It allows continuous and unattended monitoring of soil water content of the upper few decimetres of soil over an area of some 330 m around the instrument location.

Cosmic rays consist of high energy charged particles from space, and as they enter the Earth's atmosphere, collisions with air molecules give rise to showers of secondary particles. As a result of these processes, so-called 'fast' neutrons are generated. These fast neutrons are slowed down or moderated in the upper soil, predominantly by the hydrogen present in water. The number of fast neutrons detected by the probe can thus be related to soil and other stores of water, such as that contained within plants, snowpacks, or in depression storage. A higher neutron count implies drier conditions. The data need to be adjusted for factors such as variations in the incoming neutron flux due to changes in solar activity, atmospheric humidity and the elevation of the site. As a consequence, the COSMOS measurements have lower count rates, and hence greater measurement uncertainty, for sites at low altitudes and near the equator. Accuracy is also reduced during conditions of high pressure, high humidity or high soil water content. The latter is particularly regrettable for flood studies and to compensate, larger detectors or a longer averaging time can help to increase the signal to noise ratio. The sampling volume changes with water content and atmospheric conditions. The horizontal extent is affected mainly by atmospheric humidity (Desilets and Zreda, 2013) but only weakly dependent on soil water, whilst the depth sampled is much smaller in wet than dry conditions, varying between about 0.10 and 0.70 m for conditions of saturated and dry soil, respectively (Zreda *et al.*, 2008). Once a corrected count rate is obtained, this is related empirically to the soil water content by field calibration (Zreda *et al.*, 2012).

The COSMOS probe represents the combined effects of all water sources and so in certain environments there may be other significant sources of near surface water which also contribute to the depletion of neutrons. These include when there is snow on the ground (in which case the instrument can potentially usefully measure the water equivalent of the snow) and at sites where there is an appreciable amount of water within the biomass (Bogena *et al.*, 2013).

Currently there are about 70 operational COSMOS sites around the world: mainly in the United

States, Europe, southern Africa and Australia. Monitoring networks using cosmic ray neutron detectors have been established in the United States (www.cosmos.hwr.arizona.edu), Australia (www .ermt.csiro.au/html/cosmoz.html), the United Kingdom (www.ceh.ac.uk/cosmos/) and Germany.

2.3.15.1 Soil water potentials

Soil water content measurements are extremely useful, but hydrologists often need information on soil water movement, which is controlled by potential energy (primarily gravity and soil water suction). It was realised that soil water fluxes, as a function of depth and time, could be measured by using the neutron probe in combination with water potential measurements. Measurements of soil water potential are usually made by tensiometers. These consist of a porous barrier, usually of ceramic, embedded in the soil and bonded to a water-filled reservoir. The porous barrier is permeable to water, but because it has very small pores, does not allow air to pass through it in its saturated state. Water in the reservoir comes into equilibrium with the water in the soil, so that the soil water potential can be determined by measuring the pressure of water (usually below atmospheric) in the reservoir. Tensiometers are limited, however, in the range of suction they can work over from about 0 to 0.8 bar; higher vacuum pressures result in air bubbles starting to form through cavitation. The water pressure can be measured by manually read mercury manometers (once common but now rare due to health concerns) or by pressure transducers, which allow automatic recording by a field data logger.

2.3.15.2 Mercury manometer tensiometers

A tensiometer design was refined over several years to cope with problems of frost, ultraviolet radiation, thermal disturbance to readings, servicing and manufacturing challenges (Fig. 2.20). The design could accommodate up to 15 tensiometers connected to mercury manometers with a common mercury reservoir, facilitating intercomparison of readings from different depths. A porous pot filled with water is connected by a plastic tube also filled with water to a mercury manometer. When the pot is installed in the soil lower than the manometer on the surface, a negative pressure is indicated, being made up from the difference in height between pot and manometer reservoir (due

Fig. 2.20 Manually read tensiometers. (Source: Reproduced with permission of CEH.)

to gravity) and the tension caused by the suction of the soil (Strangeways, 2003).

2.3.15.3 Pressure transducer tensiometers

An electronic pressure sensor can be substituted in place of the manually read manometer. This allows the transducer to be close to the cup, removing the gravity component of the negative pressure, thereby extending the range of soil tensions that can be measured (Strangeways, 2003) and almost eliminating thermal disturbance to the readings. Pressure transducers also have fast response times and are very suitable for automatic recording by data loggers. The principal limitation is in refilling of the system with water periodically. IH was an early user of pressure transducers for tensiometer applications and developed a novel solution to this problem, which avoided the need to remove the instrument from the ground and replace it – with all the attendant time demands and problems of re-establishing good contact.

2.3.15.4   Deep borehole tensiometers

Cooper (1980) and Wellings (1984) described a technique that allowed multiple tensiometers to be installed inside an open, unlined borehole. The early version was not capable of being serviced in situ, but a modified design was used by Rutter *et al.* (2012), which could be refilled without removal from the ground. The tensiometer is mounted on a sectional aluminium mast via a small screw jack. Installation of the system is by lowering the mast into the borehole, section by section with the jacks retracted. When all tensiometers are at the desired location, each tensiometer is screwed out on its jack until the ceramic plate is pressed firmly against the borehole wall. The operating rod must be turned sufficiently to ensure good contact with the formation, but excessive force risks breaking the ceramic plate. A cap is placed on the borehole and locked into position. By this means, measurements of water potential have been made at a number of sites, mainly in chalk formations, to a depth of more than 60 m (Fig. 2.21). Wellings (1984) modified the design by adding a gypsum resistance block embedded in Plaster of Paris onto the tensiometer mounting, so allowing measurement beyond the range of tensiometers.

2.3.16   Telemetry

Telemetry has several uses. If measurements are logged on-site, the data have to be collected periodically, requiring a visit to each logging station; in some cases this is inconvenient or impracticable. Telemetry reduces the need to visit sites as often, although the need for periodic checks and maintenance remains. If the data are needed in real-time, for example in flood warning systems or water resources management, telemetry is essential.

In the 1980s the telemetry of data via satellites was pioneered at IH on an experimental basis. Ten 'Data Collection Platforms' (DCPs) were purchased and installed at a range of UK sites from which DCPs transmitted river level, AWS and precipitation measurements via Meteosat (Fig. 2.22). A satellite receiving station was also installed at

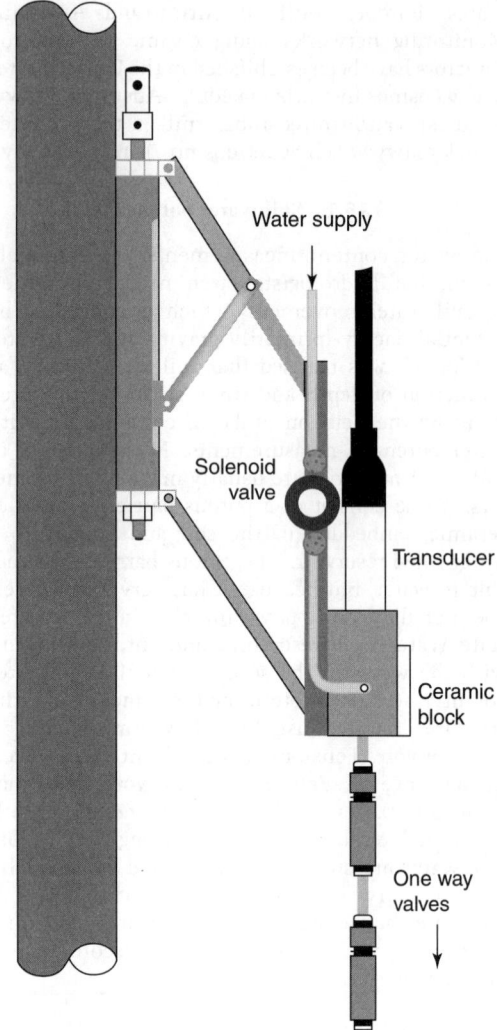

Fig. 2.21 Borehole tensiometer. (Source: Reproduced with permission of CEH.)

Wallingford. Tests were continued for several years and the high reliability of the method was well demonstrated. An IH experimental cold regions AWS was installed in Antarctica at the BAS base at Faraday (Fig. 2.23), its data being telemetered back to Wallingford hourly using a DCP, demonstrating the ability of satellite telemetry to be used over great distances. All of these systems were solar powered. Satellite telemetry is possible through geostationary and polar-orbiting weather satellites

 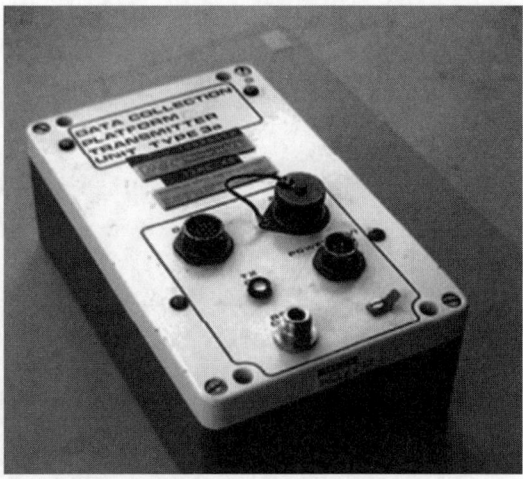

Fig. 2.22 Data collection platform satellite telemetry station (left). Data collection platform which processed the measurements (right). (Source: Reproduced with permission of Ian Strangeways.)

Fig. 2.23 Cold regions AWS at Faraday. (Source: Reproduced with permission of Ian Strangeways.)

as well as through commercial systems such as Inmarsat, Orbcomm, and Iridium. Less commonly used is the 'meteor-burst' technique (Strangeways, 1990, 2003). Telemetry is also possible through telephone, UHF and VHF line-of-sight radio links, as well as over very short distances by low-power UHF links which do not require a licence (in some countries). The Institute developed a comprehensive telemetry network of river level and precipitation stations transmitting their data through a mix of UHF and telephone telemetry links for Anglian Water.

IH also undertook a telemetry development project for UNESCO in the Pantanal region of north-east Brazil (Fig. 2.24). Its purpose was to monitor the Upper Paraguay River for navigation purposes and for flood warning. This was done just prior to satellite telemetry becoming available using, instead, HF (short wave) radio links to span the large distance from field stations to base (up to 600 km). While this network used commercially available equipment, the system-design by IH staff was new and unique and operated satisfactorily for many years (Strangeways and Lisoni, 1973).

Currently CEH operates about 30 telemetry sites in the United Kingdom, including 10 COSMOS sites and 10 for flux measurement.

Fig. 2.24 IH undertook the development of a telemetry network for the Pantanal (Brazil). (Source: Reproduced with permission of Ian Strangeways.)

Acknowledgements

Chapter 2 has drawn on the work of many others in addition to the chapter authors. John Smart did much to raise the standards of data collection, and Simon Grant, Sean Crane and Jim Hudson worked hard to maintain them, together with Phil Hill and Alan Hughes. Thanks are due to these and to others for their efforts in supporting catchment experiments.

2.4   References

Anon (1977) *Selected measurement techniques in use at Plynlimon experimental catchments*. Institute of Hydrology Report No. 43.

Bell, J.P. (1968) The soil hydrology of the Plynlimon catchments. Institute of Hydrology report 8.

Bell, J.P. (1969) A new design principle for the neutron soil moisture gauges: the 'Wallingford' neutron probe. *Soil Sci.*, **108**, 160–164.

Bell, J.P., Dean, T.J. & Hodnett, M.G. (1987) Soil moisture measurement by an improved capacitance technique, Part II. Field techniques, evaluation and calibration. *J. Hydrol.*, **93**, 79–90.

Blackie, J.R. (1993) The water balance of the Balquhidder catchments. *Journal of Hydrology*, **145**, 239–257.

Blackie, J.R. & Robinson, M. (2007) Development of catchment research with particular attention to Plynlimon and its forerunner, the East African catchments. *Hydrology and Earth System Sciences*, **11**, 26–43.

Blackie, J.R. & Simpson, T.K.M. (1993) Climatic variability within the Balquhidder catchments and its effect on Penman potential evaporation. *Journal of Hydrology*, **145**, 371–387.

Blackie, J.R., Edwards, K.A. & Clarke, R.T. (1979) Hydrological research in East Africa. *East African Agriculture and Forestry Journal, Special Issue*, **43**, 313pp.

Bogena, H.R., Huisman, J.A., Baatz, R., Hendricks Franssen, H.-J. & Vereecken, H. (2013) Accuracy of the cosmic-ray soil water content probe in humid forest ecosystems: the worst case scenario. *Water Resources Research*, **49**, 5778–5791.

Brandt, C., Robinson, M. & Finch, J.W. (2004) Anatomy of a catchment: the relationship of physical attributes of the Plynlimon catchments to variations in hydrology and water status. *Hydrology and Earth System Sciences*, **8**, 345–354.

Calder, I.R. (1976) The measurement of water loss from a forested area using a 'natural' lysimeter. *Journal of Hydrology*, **30**, 311–325.

Calder, I.R. (1977) A model of transpiration and interception loss from a spruce forest in Plynlimon, central Wales. *Journal of Hydrology*, **33**, 247–265.

Calder, I.R. (1990) Evaporation in the Uplands. Wiley.

Calder, I.R. & Newson, M.D. (1979) Land use and upland water resources in Britain - a strategic look. *Water Resources Bulletin*, **16**, 1628–1639.

Clarke, R.T. (1976) *Water balance of the headwater catchments of the Wye and Severn*. Institute of Hydrology Report 33, Wallingford.

Clarke, R.T., Leese, M.N., & Newson, A.J. (1973) *Analysis of data from Plynlimon raingauge networks*, April 1971-March 1973. Institute of Hydrology Report 27, Wallingford.

Cooper, J.D. (1980) *Measurement of moisture fluxes in unsaturated soil in Thetford Forest*. Institute of Hydrology Report 66, Wallingford, 97pp.

Dean, T.J., Bell, J.P. & Baty, A.J.B. (1987) Soil moisture measurement by an improved capacitance technique, Part I. Sensor design and performance. *Journal of Hydrology*, **93**, 67–78.

Desilets, D. & Zreda, M. (2013) Footprint diameter for a cosmic-ray soil moisture probe: theory and Monte carlo simulations. *Water Resources Research*, **49**, 3566–3575. doi:10.1002/wrcr.20187

Edwards, K.A. & Blackie, J.R. (1981) Results from the East African catchment experiments, 1958–1974. In: Lal, R. & Russell, E.W. (eds), Tropical Agricultural Hydrology. John Wiley and Sons, pp. 163–188.

Gash, J.H.C. & Morton, A.J. (1978) An application of the Rutter model to the estimation of the interception loss from Thetford Forest. *Journal of Hydrology*, **38**, 49–58.

Gash, J.H.C. & Shuttleworth, W.J. (eds) (2007) Evaporation. In: Benchmark Papers in Hydrology. Vol. **2**. IAHS Press, Wallingford, pp. 521.

Gilman, K. & Newson, M.D. (1980) Soil pipes and pipeflow; a hydrological study in upland Wales. Br. Geomorphological Research Group. In: *Research Monograph 1*. Geo Books, Norwich, pp. 110pp.

Hall, R.L. (1987) Processes of evaporation from vegetation of the uplands of Scotland. *Earth and Environmental Science Transactions of the Royal Society of Edinburgh*, **78**, 327–334.

Hall, R.L. & Harding, R.J. (1993) The water use of the Balquhidder catchments: a processes approach. *Journal of Hydrology*, **145**, 285–314.

Harrison, A.J.M. & Owen, M.W. (1967) A new type of structure for flow measurement in steep streams. (incl appendix). *ICE Proceedings*, **36** (**2**), 273–296.

Herrmann, A. & Schumann, S. (2010) *History and present status of small hydrological research basins*. IAHS-AISH Publication **336**, 316pp.

Hudson, J.A. (1988) The contribution of soil moisture storage to the water balances of upland forested and grassland catchments. *Hydrological Sciences Journal*, **33**, 289–309.

Hudson, J.A., Crane, S.B. & Robinson, M. (1997) The impact of the growth of new plantation forestry on evaporation and streamflow in the Llanbrynmair catchments, mid-Wales. *Hydrology and Earth System Sciences*, **1** (**3**), 463–476.

Johnson, R.C. (1990) The interception, throughfall and stemflow in a forest in Highland Scotland and the comparison with other upland forests in the U.K. *Journal of Hydrology*, **118**, 281–287.

Kirby, C., Newson, M.D. & Gilman, K., (Eds.) (1991) *Plynlimon research: the first two decades*. Rept. No. 109, Institute of Hydrology, Wallingford, Oxon, UK.

Kirchner, J.W., Feng, X., Neal, C. & Robson, A.J. (2004) The fine structure of water-quality dynamics: the (high-frequency) wave of the future. *Hydrological Processes*, **18**, 1353–1359.

Law, F. (1956) The effect of afforestation upon the yield of water catchment areas. *Journal of the British Waterworks Association*, **38**, 484–494.

Marc, V. & Robinson, M. (2007) The long-term water balance (1972–2004) of upland forestry and grassland at Plynlimon, mid-Wales. *Hydrology and Earth System Sciences*, **11**, 44–66.

McCulloch, J.S.G. (1975) Hydrology– the science of water. *NERC News Journal*, **12**, 14–15.

McCulloch, J.S.G. & Robinson, M. (1993) History of forest hydrology. *Journal of Hydrology*, **150**, 189–216.

McCulloch, J.S.G. & Strangeways, I.C. (1966) *Automatic weather stations for hydrology*. Proc. WMO Tech. Conf. – Automatic Weather Stations, Geneva. 262–264.

Neal, C., Wilkinson, J., Neal, M. *et al.* (1997) The hydrochemistry of the headwaters of the River Severn, Plynlimon. *Hydrology and Earth System Sciences*, **1**, 583–617.

Newson, M.D. (1979) The results of ten years' experimental study on Plynlimon, mid-Wales, and their importance for the water industry. *Journal of the Institution of Water Engineers and Scientists*, **33**, 321–333.

Newson, A.J. & Clarke, R.T. (1976) Comparison of the catch of ground-level and canopy-level raingauges in the upper Severn experimental catchment. *Meteorological Magazine*, **105**, 2–7.

Penman, H.L. (1948) Natural evaporation over open water, bare soil and grass. *Proceedings of the Royal Society (A)*, **193**, 25pp.

Penman, H.L. (1963) Commonwealth Agricultural Bureaux, Farnham Royal, Technical Communication. *Vegetation and Hydrology*, **53**, 124pp.

Pereira, H.C. (1962) Hydrological effects of changes in land use in some East African catchments. *East African Agriculture and Forestry Journal, Special Issue*, **27**, 131pp.

Roberts, J.M. (1982) Forest transpiration: a conservative hydrological process? *Journal of Hydrology*, **66**, 133–141.

Robinson, M. (1998) Thirty years of forest hydrology changes at Coalburn: water balance and extreme flows. *Hydrology and Earth System Sciences*, **2**, 233–238.

Robinson, M. & Stam, M.H. (1993) A study of soil moisture controls on streamflow behaviour: results for the Ock basin, United Kingdom. *Acta Geológica Hispánica*, **28**, 75–84.

Robinson, M., Grant, S. & Hudson, J.A. (2004) Measuring rainfall to a forest canopy: an assessment of the performance of canopy level raingauges. *Hydrology and Earth System Sciences*, **8**, 327–333.

Robinson, M., Rodda, J.C. & Sutcliffe, J.V. (2013) Long-term environmental monitoring in the UK: origins and achievements of the Plynlimon catchment study. *Trans Institute of British Geographers*, **38**, 451–463.

Rodda, J.C. (1967) The systematic error in in rainfall measurement. *Journal of the Institute of Water Engineers*, **21**, 173–177.

Rodda, J.C. (1968) *An approach to raingauge network design*. Unpublished Report. Hydrological Research Unit, Wallingford.

Rodda, J.C. (1971) *Progress at Plynlimon – problems of investigating the effect of land use on the hydrological cycle*. Paper to British Association for the Advancement of Science, Section K (Forestry). Swansea.

Rodda, J.C. (1976) Basin studies Chapter 10 In: Rodda, J.C. (ed), Facets of hydrology. Wiley, Chichester, pp. 257–297.

Rutter, A.J. (1975) The hydrological cycle in vegetation. In: Monteith, J.L. (ed), Vegetation and Atmosphere. Vol. **1**. Academic press, London.

Rutter, A.J., Kershaw, K.A., Robins, P.C. & Morton, A.J. (1971) A predictive model of rainfall interception in forests I. Derivation of the model from observations in a plantation of Corsican Pine. *Agricultural Meteorology*, **9**, 367–384.

Rutter, H.K., Cooper, J.D., Pope, D. & Smith, M. (2012) New understanding of deep unsaturated zone controls on recharge in the Chalk: a case study near Patcham, SE England. *Quarterly Journal of Engineering Geology and Hydrogeology*, **45**, 487–495.

Smart, J.D.G. (1977) *The design, operation and calibration of the permanent flow measurement structures in the Plynlimon experimental catchments*. Institute of Hydrology Report 42. Wallingford.

Stähli, M., Badoux, A., Ludwig, A., Steiner, K., Zappa, M. & Hegg, C. (2011) One Century of hydrological monitoring in two small catchments with different forest coverage. *Environmental Monitoring and Assessment*, **174**, 91–106.

Stewart, J.B. (2004) Review of forest evaporation studies, primarily in the UK. In: Mencuccini, M., Grace, J. & McNaughton, K.G. (eds), Forests at the Land-Atmosphere Interface. CAB Internastional, Wallingford, pp. 159–73.

Stewart, J.B. & Thom, A.S. (1973) Energy budgets in pine forest. *Quarterly Journal of the Royal Meteorological Society*, **99**, 154–170.

Strangeways, I.C. (1972) Automatic weather stations for network operation. *Weather*, **27**, 403–408.

Strangeways, I.C. (1990) *The telemetry of hydrological data by satellite*. Institute of Hydrology Report No **112**.

Strangeways, I.C. (1995) Back to Basics: The met. enclosure: Part 1 – Its background. *Weather*, **50** (6), 182–188.

Strangeways, I.C. (2003) Measuring the Natural Environment, 2nd edn. Cambridge University Press ISBN 0 521 82205 X hardback. 0 521 52952 2 paperback.

Strangeways, I.C. (2007) Precipitation – Theory, Measurement and Distribution. Cambridge University Press ISBN 13 978-0-521-85117-6 hardback.

Strangeways, I.C. & Lisoni, L. (1973) Long-distance telemetry of data for flood forecasting. UNESCO. *Nature and Resources*, **IX**, 18–21.

Strangeways, I.C. & McCulloch, J.S.G. (1965) A low-priced automatic hydrometeorological station. *Bulletin IAHS*, **10**, 57–62.

UNESCO/WMO/IAHS (1974) *Three centuries of scientific hydrology*. Key papers from the celebration of the Tercentenary of Scientific Hydrology, UNESCO, Paris

Van Haveren, B.P. (1988) Notes: a reevaluation of the Wagon Wheel Gap forest watershed experiment. *Forest Science*, **34**, 208–214.

Wellings (1984) Recharge of the Upper Chalk aquifer at a site in Hampshire, England: 1. water balance and unsaturated flow. *Journal of Hydrology*, **69**, 259–273.

Whitehead, P.G. & Calder, I.R. (eds) (1993) The Balquhidder experimental catchments. *Journal of Hydrology*, **145**, 215–480.

Wright, I.R. & Harding, R.J. (1993) Evaporation from natural mountain grassland. *Journal of Hydrology*, **145**, 267–283.

Zreda, M., Desilets, D., Ferré, T.P.A. & Scott, R.L. (2008) Measuring soil moisture content non-invasively at intermediate spatial scale using cosmic-ray neutrons. *Geophysical Research Letters*, **35** (21), L2140 doi: 10.1029/2008gl035655.

3 Risks and Extremes

LISA STEWART[1], MAX BERAN[2], FRANK FARQUHARSON[3],
DUNCAN FAULKNER[4], DAVID JONES[5],
THOMAS KJELDSEN[6], MALCOLM NEWSON[7],
ENDA O'CONNELL[8], AND JOHN SUTCLIFFE[2]

[1]Centre for Ecology and Hydrology, Wallingford, Oxfordshire, UK
[2]Ex-Institute of Hydrology, Wallingford, UK
[3]Water Resource Associates, Henley on Thames, UK
[4]JBA Consulting, Skipton, UK
[5]Ex-Centre for Ecology and Hydrology, Wallingford, UK
[6]Department of Architecture and Civil Engineering, University of Bath, Bath, UK
[7]Tyne Rivers Trust, Corbridge, Northumberland, UK
[8]Dept. Civil Engineering & Geoscience, University of Newcastle upon Tyne, Newcastle upon Tyne, UK

Progress in Modern Hydrology: Past, Present and Future, First Edition. Edited by John C. Rodda and Mark Robinson.
© 2015 John Wiley & Sons, Ltd. Published 2015 by John Wiley & Sons, Ltd.

3.1 Overview

This chapter describes highlights of the research activity on risks and extremes at Wallingford. This research area has been a primary focus of CEH and its predecessor organisations for more than 40 years, targeted particularly at flood risk modelling, but also encompassing low flow studies, digital mapping, climate change impacts and uncertainty estimation. The publication of the Flood Studies Report (FSR – NERC, 1975) was a major milestone in hydrology, providing an unprecedented set of generalised methods for hydrological frequency estimation in the UK, as well as detailing all aspects of the analyses and underlying data. The FSR was followed by other major studies, notably the Low Flow Studies report (Institute of Hydrology, 1980), the Flood Estimation Handbook (FEH – Institute of Hydrology, 1999) and Low Flows 2000 (Young et al., 2003). The methods developed during this time have become the national standard for flood risk estimation in the UK.

3.2 The UK Flood Studies Report (FSR)

3.2.1 Inception of the FSR

Until 1965, the only guidance on flood estimation issued by the Institution of Civil Engineers (ICE) was the 1933 Interim Report (ICE, 1933) which was intended to provide guidance on the design of spillways for reservoirs in upland areas with moderate catchments of less than 100 km^2. The report was based on a simple graph of maximum recorded floods related to upstream area, which included an enveloping 'normal maximum curve'. It was suggested that larger 'catastrophic floods' could be caused by severe rainfall, with peak discharges at least twice the normal maximum.

In 1959, a meeting was held at the Institution to consider the report *Floods in the British Isles* by the Sub-Committee on Rainfall and Runoff (ICE, 1960). After discussion of the Interim Report and subsequent research, the subcommittee recommended

that the 1933 report should be reprinted with the additional data included. It was also agreed that the recommendations of the subcommittee should be published with the discussion of the meeting, and that there should be preparation of another report dealing with aspects of floods other than those dealing with reservoir practice. The ICE Council accepted this proposal and a further recommendation that papers should be invited describing current methods of flood estimation, and a symposium on River Flood Hydrology was held in 1965 (ICE, 1965). The Institution set up a Revision Committee under the chairmanship of Sir Angus Paton in 1965, which formed a subcommittee in 1966 to outline the studies of British floods which would use existing hydrological and meteorological data and knowledge for engineering design purposes. This led to the publication by ICE in February 1967 of the report *Flood Studies for the United Kingdom* (ICE, 1967).

This report recommended that a new investigation of all aspects of flood hydrology should be undertaken. Meteorological records should be studied to understand the causes of floods, extend flood records, assist in flood frequency analyses and provide estimates of probable maximum precipitation. All available flood records should be assembled and reviewed, and frequency analyses of flood peaks and volumes should be carried out. Regional analyses and correlation with catchment characteristics were also recommended to improve single station frequency distributions and to estimate flood frequencies at ungauged sites. Unit hydrographs, soil infiltration properties and snowmelt should be studied to derive models for use with meteorological studies. Flood routing techniques should be reviewed and tested. The proposed programme of studies should be undertaken as a three-year project, with the Meteorological Office requested to undertake the meteorological studies, while the hydrological studies should be undertaken by an independent full-time team of a leader, four hydrologists, a statistician and supporting staff. The project team should have its own offices. After the importance of the project had been discussed, the Natural Environment Research Council undertook to fund the research based on the existing Hydrological Research Unit/Institute of Hydrology. The Hydraulics Research Station agreed to carry out the flood routing studies. The announcement of the project coincided with the incidence of serious flooding in South East England in July 1968. A Flood Studies Steering Committee was set up under Mr. Marshall Nixon to supervise the project.

The investigation was planned as a three-year programme following a period from October 1969 to March 1970 for recruitment of staff. A suitable office was located in Wallingford, and a team of fifteen was supplemented by temporary staff when required, for example for extraction of flood peaks from microfilm. The collection and checking of flood flows took a large proportion of the time available.

Some 1200 gauging stations were visited with local staff from the gauging authorities to confirm rating curves and prepare station reports; the collection and microfilming of river level charts resulted in an archive of about half a million charts, from which peak flow records could be extracted. Early discussion led to the view that the study should be extended to the British Isles like the previous ICE report, in which nearly a quarter of the flood records had been provided by the Irish authorities. This was agreed by the Steering Committee, which was joined by the Office of Public Works, Dublin, which carried out an identical review of flood records and basin characteristics in Ireland.

The programme was based on the ICE committee report and was approved by the Flood Studies Steering Committee, which met regularly during the period of study. The Meteorological Office, which also recruited staff to undertake their share of the programme, carried out statistical analysis of daily and recording gauges. Maps were produced of point rainfall of 2-day duration and 5-year return period, and falls of other return periods were related to these. After it emerged during the investigation that antecedent soil moisture conditions were a more useful indicator of runoff percentage than rainfall intensity, they also extended the current estimation of soil moisture deficit backward in time to compare with historical events. The Hydraulic Research Station undertook a survey of flood routing methods for British rivers at the request of the Steering Committee. The Water Resources Board collaborated with the quality classification

of flow records and the collection and microfilming of gauge level charts.

3.2.2   *Publication of the FSR*

With the agreement of the Steering Committee, the FSR (NERC, 1975) was published in 1975 in five volumes, containing the hydrological recommendations, the meteorological and flood routing studies and the hydrological data and maps. Because the assembly and extraction of the basic flood records had taken so high a proportion of the time spent, the data volume ensured that the available data for the UK and Ireland were published for future use. It was also accepted that a full account of the analysis should be given to support the recommendations.

The two main routes recommended for estimation of design floods were through statistical analysis of peak flows and through unit hydrograph synthesis of the flood corresponding to a design storm. The choice between these routes depended largely on circumstances: was the complete hydrograph required in addition to the peak flow, and was the estimate of the maximum flood required rather than the flood of a given frequency. If either was needed, then the unit hydrograph approach was appropriate; otherwise a choice of methods was available. The detailed approach would depend on whether flow records were available near the design site and whether these were long or short (Fig. 3.1).

The FSR was discussed at a meeting (ICE, 1975) held at the Institution of Civil Engineers on 7/8 May 1975, where each component of the report was introduced by the individual author. Many attended the meeting and a large number of comments were made, but in general the report was accepted as a basis for engineering design, and incorporated into the ICE report on Reservoir Flood Standards.

3.2.3   *Details of FSR methods and data*

3.2.3.1   Flood frequency analysis of peak discharge

The flood frequency approach to flood estimation leads to a graphical or algebraic relationship between a flood peak, Q, and its return period, T; the Q:T relationship. Despite being well established as a tool of hydrological research, this statistics-based approach to flood prediction and design was not well embedded in the engineering practice of the time. Hence Chapter 1 of the FSR devotes considerable space to explaining the underlying principles

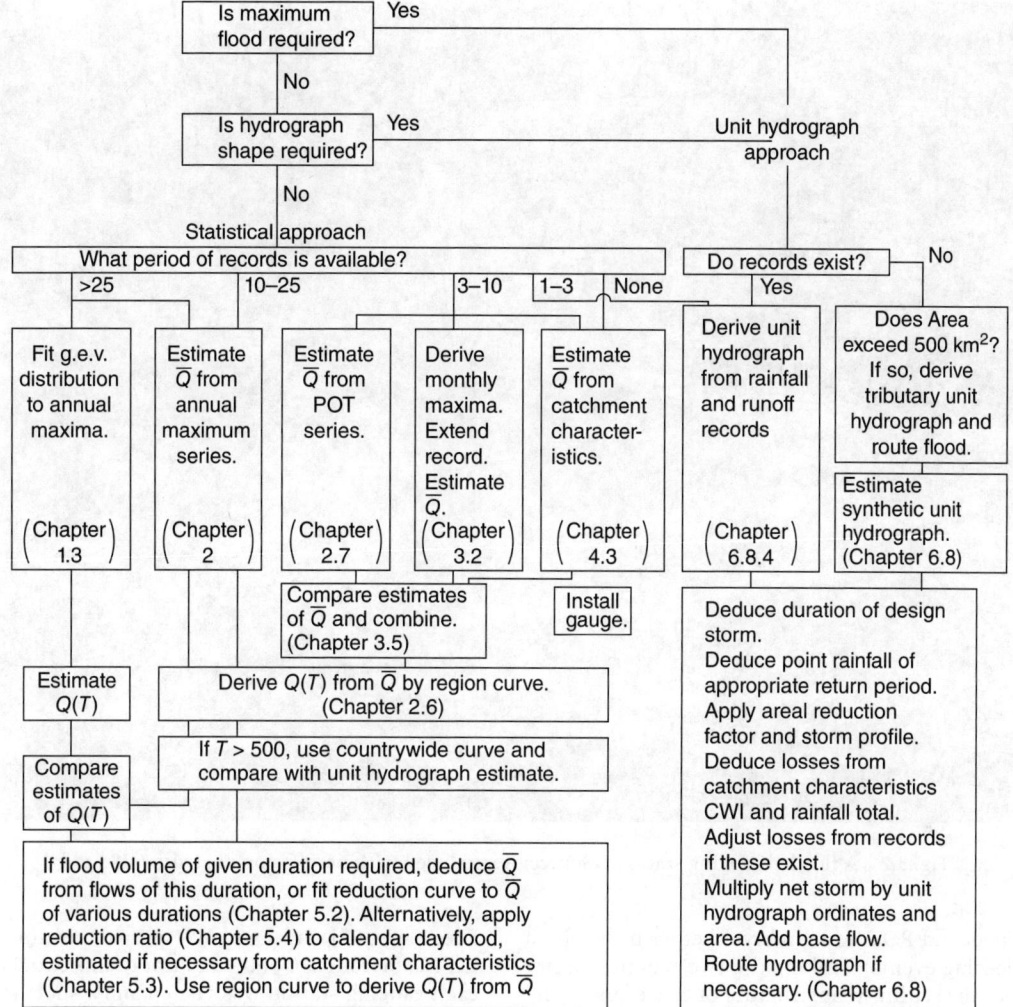

Fig. 3.1 Estimation of the design flood (from FSR Vol. I.). (Source: Reproduced with permission of CEH.)

which are carried forward to Chapter 2, where they are applied to a database of annual maxima (AM) and peaks over a threshold (POT). The conclusion is a structured set of procedures for estimating the T-year return period flood peak (along with its likely accuracy in terms of error of estimate) according to location, data availability and value of T.

Explaining basic concepts Concepts to be explained included the geometries of distributions such as: lognormal, Pearson Type III and Gumbel and their appropriateness for describing

extreme events; what happens when a finite sample is drawn from a population, requiring appreciation of the distinction between a parameter of a population and its sampling analogue, and how this provides the error bars on flood estimates; and how and why methods of parameter estimation such as moments, maximum likelihood and graphical procedures via plotting position differ.

Return period, despite its ubiquity in flood estimation, also needed its subtleties to be explained – especially how it differs between annual maximum data (the waiting time between years containing

Fig. 3.2   A flooded gauging station. (Source: Reproduced with permission of Hydro-GIS Ltd.)

a flood) and POT data (interval between threshold exceeding events). Linking the two requires clarity about the within-year structure of exceedances and provides the all-important equation relating return period to risk and design life.

This barely scratches the surface of the material presented in the FSR and it was a novelty to see concepts hitherto appreciated only as empirical formulae given rigorous treatment enabling users to judge positive and negative features of alternative models. An example of this is the concept of a plotting position, treated in most prior hydrology texts almost as an art form, but elucidated in the FSR in terms of order statistics and fitting criteria, then tabulated for ease of application.

Flood peak data   Members of the team reviewed more than 800 gauging stations across the British

Isles (Fig. 3.2) and extracted annual maxima and POT data from 530 of them. Prior to this extraction, each gauging station was visited and a short report written, concluding with a categorisation of its performance at high flow measurement from A to D, according to the degree of by-passing and modularity of the gauging structure. Rating curves (stage–discharge relationships) had frequently to be extended and both this and the categorisation were discussed with the gauging authority while the full set of water level charts were sent for microfilming. Extracting POT data from multi-peak events required an independence rule to be applied. This depended on the timing between candidate peaks and the depth of the recession between peaks. Assistants prepared an initial extraction of water levels from the microfilm which was then checked by one of the team of hydrologists. A system of coding was used and the data, by then on punched cards, were held on a data base for conversion to discharge units.

Deriving the Q:T relationship Core questions were which statistical distribution to use to describe flood peaks, how to estimate the parameters of the distribution from the available data, and how to express the uncertainty in the resultant flood estimates. Regarding the choice of distribution and parameter estimator, station data were first screened using tests for serial independence and trend, and then compared for quality of fit to the contending distributions – Gumbel, Gamma, lognormal, GEV, Pearson III and logPearson III. Tests were based on the linearity of the data when plotted on the relevant graph paper for the distribution, as well as more formal Chi-squared and Kolmogorov–Smirnov statistical tests. Despite exhaustive testing, no clear victor emerged. Shortage of long records reduces the power of such tests – only 37 stations contributed to the main recommendations concerning statistical distribution and parameter estimation. It was very evident though that no two-parameter distribution had the flexibility to capture the variety of flood frequency relationships, so a three-parameter distribution would be essential.

Of the contending distributions, the generalised extreme value (GEV) distribution was selected for most subsequent analyses. This has theoretical advantages in being designed for data which are themselves maxima from a generating mechanism, and also link tidily to logical models for fitting POT peaks and times. The GEV distribution includes the two-parameter Gumbel distribution as a special case, but has a third parameter giving the flexibility to cope with heavier tailed cases (increasing gaps between first, second and third ranking events) or lighter tail cases, where the flood regime is such as to approach an upper bound on the magnitude. The same distribution family had been selected by the counterpart study of extreme rainfalls at the UK Met Office – indeed the leader of their team, A. F. Jenkinson, had pioneered the distribution through unifying the three extreme value distributions into a single algebraic structure.

The statistical concept of the standard error of estimate, translated in this context to error bars surrounding a flood estimate, was also one introduced into British practice by the FSR. Here again the report does double service by explaining the underlying theory and also providing a convenient

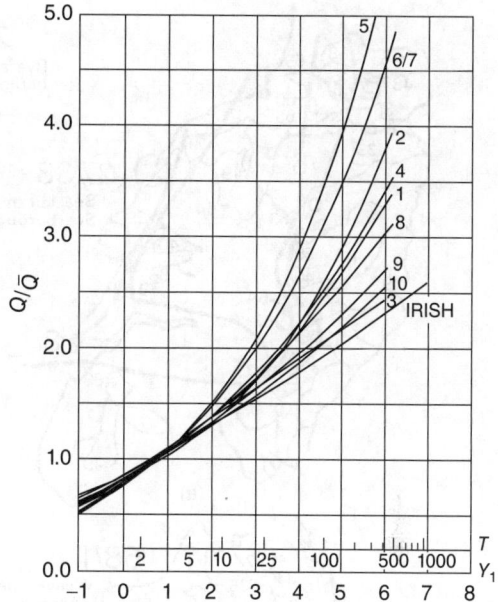

Fig. 3.3 Flood Studies Report regional growth curves. (Source: Reproduced with permission of CEH.)

formulation that could be applied irrespective of distribution and parameter estimator.

Because most individual station records were too short for reliable estimates of scale and shape parameters of the three-parameter GEV distribution, it was decided to adopt a regional approach. This involved standardising each station record by dividing by its own sample average and then combining data from several stations in a semi-graphical manner. The regional approach was adopted as it had been previously observed that there was a tendency for geographic clustering. As examples of the outcome, the 100-year flood in East Anglia is about 3.6 times the mean annual flood (MAF), while for the north of England the scaling factor is 2.1 (Fig. 3.3).

Other statistical approaches In addition to the basic method outlined above, a number of other statistical applications were trialled and reported on in the FSR. A statistical approach to volume floods was considered, which explored the relationship between the T-year return period instantaneous peak and the same return period D-hour average maximum discharge. This gave rise to reduction factors but proved difficult to regionalise.

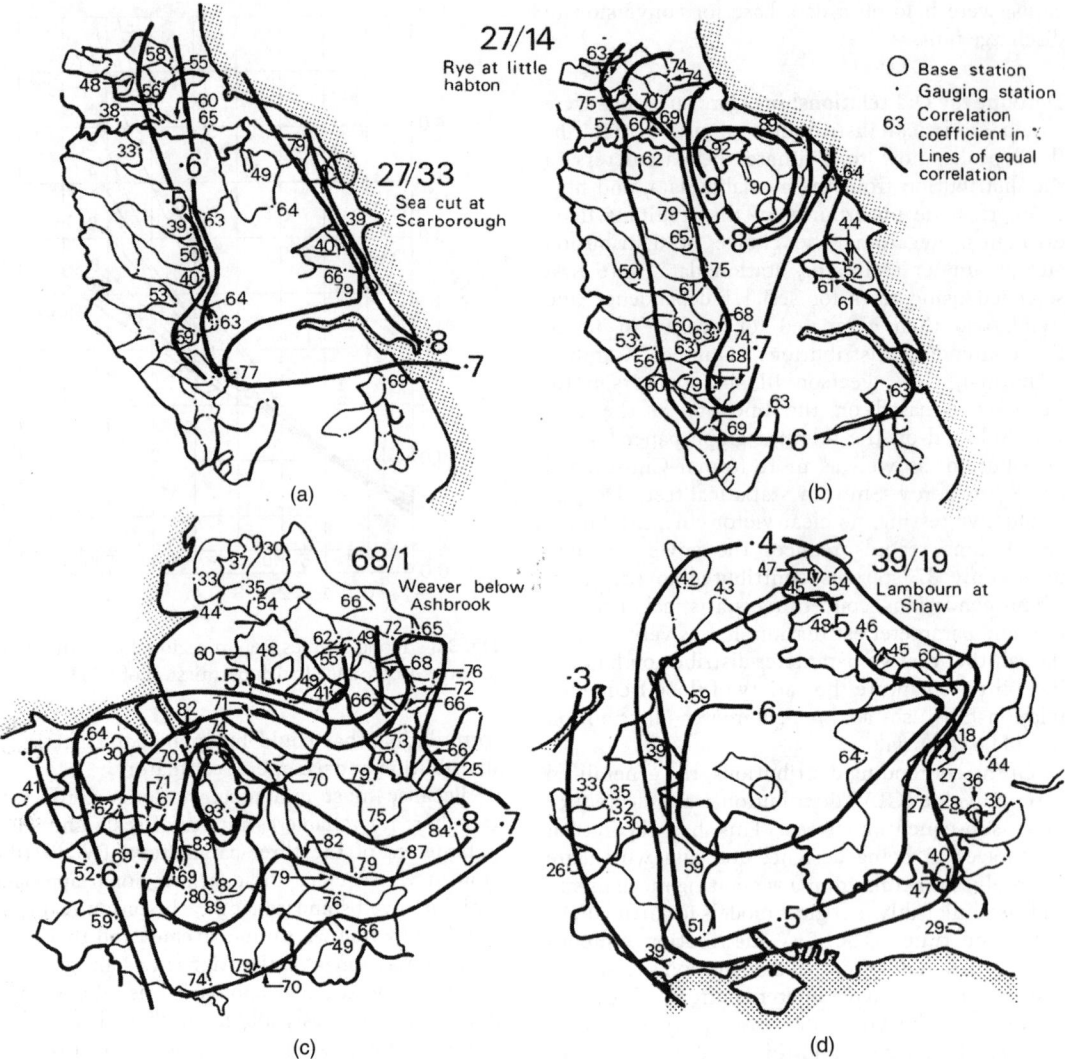

Fig. 3.4 Examples of correlation maps (from FSR I.). (Source: Reproduced with permission of CEH.)

In addition to the AM and POT approaches, a continuous time series model was also studied based on the shot-noise process – randomly occurring exponentially distributed increases followed by an exponential decay. More complex parameter estimation procedures were also explored such as Bayes estimators, where the flood observations were considered as adjustments to an underlying 'prior' distribution. Extensive use was made of data extension using inter-station correlation (Fig. 3.4).

This contributed to improved estimates of the MAF rather than the Q:T relationship.

The most significant statistical add-on though involved methods for incorporating historical data (flood peaks pre-dating the instrumental record such as flood marks or press records; see Fig. 3.5). These data points were collected at the same time as the basic data and were included in the database. An objective approach to merging these sources with the pooled regional data was a key element in the

27/01/1809

15/11/1894
25/12/1821
14/12/1768
17/03/1947
17/11/1852
11/12/1929

15/11/1875

Fig. 3.5 Historical flood marks at Shillingford Wharfe. (Source: Reproduced with permission of Hydro-GIS Ltd.)

construction of the 'national' $Q{:}T$ relationship for return periods beyond 500 years.

3.2.3.2 Estimating the flood peak at an ungauged location

The flow regime and the river basin have co-evolved over geological time so it is not surprising that there should be exploitable relationships between the $Q{:}T$ relationship and characteristics of the catchment area. Such links were researched by the Flood Studies team from which emerged a two-stage method for estimating a design flood for an ungauged river. The first step was to estimate the MAF from catchment characteristics, the second was to scale up to the T-year flood using the appropriate regionally pooled dimensionless flood frequency relationship just described.

Classification of catchment characteristics The rise of quantitative geomorphology in the form of morphometry, driven by military applications in the USA (e.g. Melton, 1957) offered a means of using catchment dimensions and features as independent variables to predict catchment outputs: water, solutes and sediments. The United States Geological Survey carried out predictive exercises

of this type in the 1960s using such variables as catchment area, catchment slope and drainage density (Benson, 1962).

From the outset of the Flood Studies exercise it became clear that to gather a full set of potentially successful independent variables for 1100 catchments would be challenging to manual derivation: measurements from maps had been very low tech until that point. It was obvious that some trials of potential variables should be carried out before a huge amount of cartographic effort was potentially wasted. There were also several issues of map scale and availability and of the inevitable cross correlations between catchment dimensions. Newson (1975, 1976, 1978) describes these preliminary investigations before moving on to retro-analyse the established British Isles database and re-map the variables.

Progress in computing had, by the 1970s, initiated the rapid rise of 'automated cartography'; NERC established an Experimental Cartography Unit, conveniently close to Wallingford (at Oxford). 'Digitizing' catchment variables proved to be extremely useful in the case of channel slopes; this variable had proved of key value in the Benson (1962) US flood study. For the remaining 'independent' (highly inter-correlated!) variables the manual approach remained the only option, for example in deriving an index of drainage density. Rather than measure the length of every stream segment of all orders for 1100 catchments a reliable surrogate, stream frequency, was derived by counting stream junctions. The OS 1:25,000 map was deemed, after investigation to be the most accurate depiction of channels (Newson, 1975).

Many authors seeking to predict flood flows from unguaged catchments had suggested that the big missing element was the characteristics of catchment soils; the role of soils in runoff had been vividly revealed by the 1960s and 70s research on throughflow, saturation and piping (see Chapter 2).

The official soil survey organisations across the British Isles rose to a challenge issued from many joint meetings of IH and SSEW staff at Rothamstead to design, test and publish a 'Winter Rain Acceptance Potential' (WRAP) map (Farquharson *et al.*, 1978).

Many options exist for indexing a catchment's climate, however for flood study purposes it is most natural to use ones representing precipitation input, for example the standard period annual average rainfall (SAAR). In addition, the FSR explored indices more closely linked to flood formation, one being the five-year return period two day rainfall, a mapped statistic which was used as a base for other durations and return periods in the meteorological studies, and another combining short term rainfall with a soil moisture deficit into a single number index of effective rainfall (RSMD). The mapping of RSMD required separate monthly rainfall and soil moisture data at a network of stations which were statistically combined to extract the 5-year return period difference.

Regional regression equations As stated above, the first step in estimating the $Q{:}T$ relationship for an ungauged river location is to calculate the MAF from catchment characteristics. Chapter 4 of the FSR provides the theory and the practice leading to the recommended regression equations for performing this calculation. Regression analysis was already well established in the hydrological literature, indeed several previous flood studies had used it to obtain simple equations based on catchment area including in the UK, but the FSR was at a much more ambitious scale and included some novel variations.

The simplest case is a two-variable regression of MAF on catchment area (A) which one may picture as a best-fit line through a scatter plot of the two variables on logarithmic graph paper. In the case of the FSR dataset the equation of this line is

$$MAF = 0.68 \ A^{0.77}$$

but inspection of the departures of the points above and below the best-fit line reveals the need to include other characteristics. For example, where there are two catchments of similar area, one with steep terrain, impermeable geology and high rainfall, it would experience larger flood discharges than a similar size permeable lowland catchment in south-east England. This raised many questions to be addressed:

1 Which among the several estimators of MAF to use?
2 Some catchment characteristics could be indexed in several ways – e.g. climate and slope – which of several contending measures to use?
3 Whether to use the entire 533 station dataset for which MAF values were available or to concentrate on the longer and more accurately gauged stations?
4 How to deal with characteristics such as lake and urban areas that are present on a minority of

catchments in the dataset and geographically not well distributed?
5 The form of the regression – multiplicative or additive – and alternative robust estimators of regression coefficients?
6 Could the same regression approach be used to estimate quantiles, such as the 5-year flood, or higher moments, such as coefficient of variation, of the flood frequency distribution?
7 Would the same regression equation suit the entire study area or would accuracy be improved by partitioning it into regions?

It was decided to work in the log domain, and a preliminary principal component analysis reinforced the north-west south-east gradient that characterises the climate and morphometry of the British Isles through geology and adjacency to mainland Europe. This also accounts for quite high levels of intercorrelation between different characteristics which adds to the difficulty of the statistical analysis. A database handling computer programme enabled subsets of the data to be readily extracted and also enabled mapped graphical output, for example of residuals from a regression line, to be produced on the line printer. It was found that record length and station category did not strongly affect coefficient values, the reason probably being that lack-of-fit in the model cloaked any differences due to error of measurement arising from short records or inadequate rating curves. So for many studies – in particular the selection of dependent and independent variables – the entire dataset was used.

Several estimators were available for the dependent variable – the MAF – based on annual maxima and peak-over-threshold approaches, a graphical $Q{:}T$ fit and where the record had been extended from regional flood and rainfall data. The version adopted at any given station, named BESMAF, used whichever was considered the most accurate, a frequent issue being the strength of the influence in a short record of the (then) recent 1968 flood event.

Choosing among catchment characteristics required individual treatment – e.g. lake and urban fractions, and inspection of combinations of different slope and climate variables (Fig. 3.6). The set carried forward to the final analysis was catchment area, stream frequency (measuring the density of the stream network in terms of stream junction number), main channel gradient, a numerical index of soil type, climate indexed by the RSMD rainfall

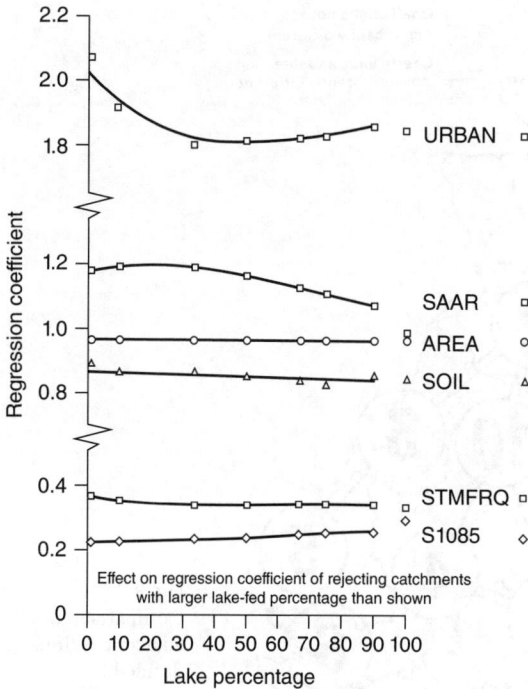

Fig. 3.6 Effect of eliminating lake catchments on regression coefficients (FSR I.). (Source: Reproduced with permission of CEH.)

soil moisture deficit combination, and lake and urban area where present.

The baseline for model building was a regression equation with coefficients fitted to the entire British Isles dataset. When the residuals from this equation were plotted on a map there were strong suggestions of geographic clustering. Analysis of Variance (ANOVA) was used to test formally whether the geographical patterns were real. Obviously the closest fit, or more specifically the lowest error sum of squares, is when each region has its own individual regression equation. However this is at the expense of a large loss of degrees of freedom. So it was possible that combining regions into larger groupings would gain more in saved degrees of freedom than lost in increased error sum of squares, and so would provide the more reliable predictions. This is the basis of the ANOVA which distinguished between entirely separate regressions, regressions which share the same coefficients but differ in terms of their multipliers (parallel regression), and regressions

which share both coefficients and multiplier (Fig. 3.7).

The outcome was a set of regression equations which divided the study area into six regions, five of which shared coefficients but differed in multiplier, and an entirely separate equation for the Essex, Lee and Thames hydrometric areas with their stronger continentality and predominance of urbanised catchments on a permeable chalk geology. For the rest of the area the chosen regression equation was

$$BESMAF = c\ AREA^{0.94}\ STMFRQ^{0.27}\ SOIL^{1.23}$$
$$RSMD^{1.03}(1 + LAKE)^{-0.86}\ S1085^{0.16}$$

where the multiplier, c, varied from 0.014 in East Anglia to 0.029 in the south-west. Scaling characteristics – area, rainfall and soil type – have coefficients close to unity; morphometric variables which determine the speed of response rather than its magnitude have lower sensitivities. The effect of lakes is to approximately halve the predicted peak for a point close to the lake outfall.

Attempts to extend the methodology beyond the MAF to higher moments or the distribution tail did not produce useful predictions, so step 2 of the statistical procedure remained the scaling of the MAF from the regression equations to any required return period using the regional dimensionless $Q{:}T$ relationship described earlier. Again this description only scratches the surface of the material presented, omitting matters like estimation error and finer points of catchment characteristic selection and map derivation.

3.2.3.3 Flood hydrograph prediction

The statistical approach described above provides an estimate of the T-year return period peak discharge, however many practical design cases require the volume of runoff and its variation through time – how the hydrograph builds up to and decays away from the peak. Examples are flood plain management, sizing off-stream storage, and evaluating reservoir safety. The FSR responded to the requirement with a two-step procedure involving first the estimation of effective rainfall – the fraction of the precipitation that reappears in the river during the flood event, then transforming this via a unit hydrograph to the total flood response hydrograph. Commonly the output flood hydrograph would become the input to a hydraulic

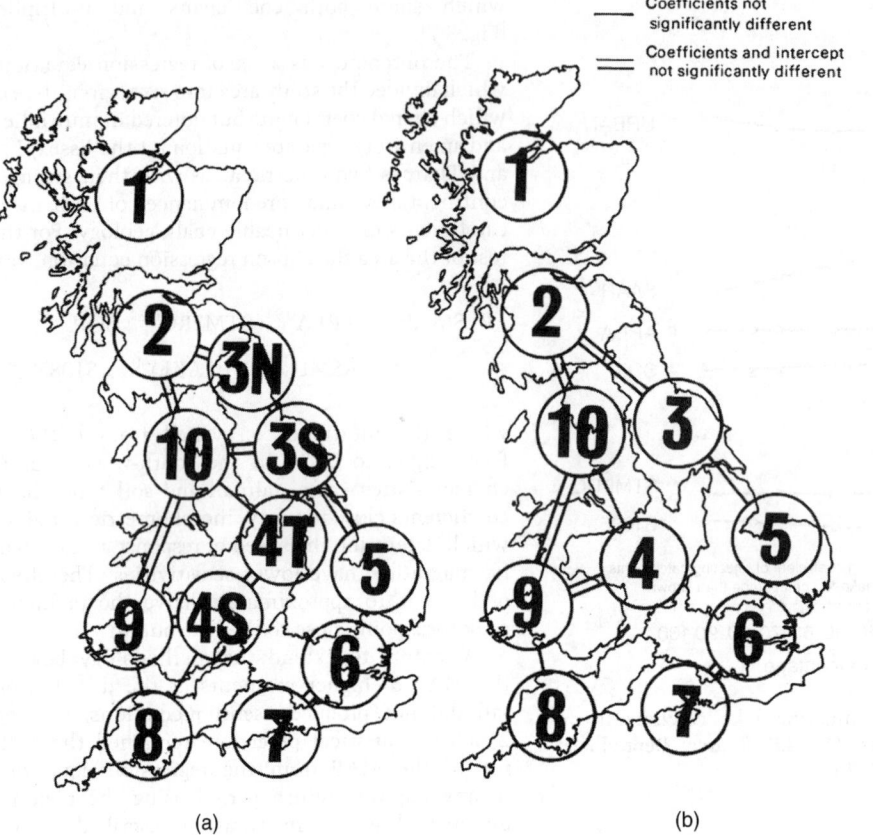

—— Coefficients not
significantly different

══ Coefficients and intercept
not significantly different

(a)

(b)

Fig. 3.7
Comparison of
adjacent regions: (a)
divided basic
regions. (b) basic
regions (from
FSR I.). (Source:
Reproduced with
permission of
CEH.)

model for calculating the depth of inundation (FSR volume III).

The rainfall–runoff approach to flood estimation had a century-long pedigree in water engineering practice through the 'rational formula'

$$Q = ciA$$

where the output flood discharge is deemed proportional to the input precipitation rate, i, scaled by the catchment area, and by an empirical coefficient c. The great advance over this by the FSR flood hydrograph method was that it made explicit the processes that control the value of c – the time distribution of rainfall, the antecedent condition of the catchment, and the way incident rainfall is routed down to the point of interest – to allow their impacts to be separately quantified.

Some key research steps are summarised in the following paragraphs, many being new at the

time to UK application. A major data collection and processing stage preceded analyses comprising derivation of unit hydrograph and runoff percentage predictions at gauged sites, their generalisation for application at ungauged locations, and a design package that ensured the return period of the peak discharge matched what would be estimated from the flood frequency approach.

Flood event data Gauged catchments were selected for the study on the basis of the quality of their rating curve and the availability of a nearby recording and daily-read raingauges. The study was limited to catchments below 500 km² area in order to ensure uniformity of input and response. The criteria were met for about 160 catchments providing from 2 to 25 events per catchment all screened for size of peak flow and separation from confusing interactions with preceding and following events. The event database that was created contained the water level hydrograph from the start of the rising limb through to the tail of the recession digitised at 0.2 or 0.5 hour intervals and converted to discharge, hourly rainfall data from autographic and daily-read raingauges from the vicinity, and soil moisture deficit updated to the start of the event, a measure of how close the soil is to field capacity. The latter two variables were combined into an antecedent catchment wetness index (CWI), that proved to be the primary control over effective precipitation. The rainfall data were combined to produce a single storm hourly hyetograph for each event. River flow and rainfall data were closely scrutinised to ensure simultaneity and avoid extraneous factors such as when snowmelt also contributed to the runoff, also to separate multi-peak events. In total 1631 flood events were available for analysis, some retained for validation.

Event analysis Insights from this prodigious quantity of event data led to many pragmatic decisions demanded of any rainfall–runoff study (Fig. 3.8):

• a practical measure of lagtime based on time difference between the centroid of the rainfall and the peak of the flood hydrograph
• which formed the basis of an objective procedure for separating quick response runoff from baseflow
• development of an index of CWI based on the soil moisture deficit and antecedent rainfall
• which fixed a loss rate function to partition the hyetograph between effective rainfall and infiltration
• matrix deconvolution of hydrograph and hyetograph with smoothing to derive the 1-hour 10 mm unit hydrograph

The final tally was 1500 events from 138 catchments with data available for unit hydrograph and event runoff analysis. The unit hydrographs for each catchment were averaged and characterised by a simplified set of geometric parameters – peak discharge Q_p, time to peak T_p, and width at half-peak W. It was found that these three were interdependent and a single 'waisted' triangle would provide sufficient reconstruction of the original hydrograph by fixing $Q_p T_p = 220$ and $W/T_p = 1.2$. A supplementary finding was a close correlation between T_p and the lagtime, $T_p = 0.9$ LAG suggesting a viable prospect of estimating the unit hydrograph shape from, say, a single season's field observations.

Partitioning raw rainfall into effective and infiltration fractions can be based on water balance,

infiltration and the contributing area principles and broadly translate to a total, a percentage or a loss rate. The procedure adopted by the FSR – the percentage approach – was finalised after inter-catchment analysis involving both storm and catchment characteristics as summarised in the following section.

Regionalisation for use at ungauged sites Just as with the flood frequency approach, the ability to apply it at an ungauged location was central. The approach was similar – in this case using the flood event and catchment characteristic databases to calibrate regression equations for the three components of the flood hydrograph procedure:

• Synthesising the unit hydrograph
• Computing effective rainfall
• Adding the baseflow component

As mentioned, the simplified unit hydrograph could be reconstructed from just the time to peak T_p. A considerable international literature concerning the speed of catchment response would lead one to anticipate that the size and steepness of the catchment would dominate plus perhaps a 'nod' to the drainage network density. The FSR data indicated otherwise; catchment slope was found to have the largest explanatory power with useful additional variance explained by a 4-variable equation involving also the urban fraction, rainfall climate through RSMD, and the length of the main channel. This last size-related variable carried much less than the expected weight due to intercorrelation with catchment slope in the study catchments. No significant event varying sources were identified, for example no consistent tendency for larger storms to be associated with more rapid response.

Catchment characteristics and event variables were used to examine two issues relating to runoff volume: (a) the form of the partition between effective rainfall and infiltration, and (b) an equation to predict effective rainfall (i.e. runoff volume). Both approaches were new to UK hydrology. Comparing different algebraic formulations, the first was found to be best expressed as a percentage runoff – quickflow runoff/total storm rainfall. The form of the regression equation for predicting percentage runoff gave rise to the new concept of 'standard percentage runoff' (SPR), dependent only on soil type and urban fraction and defined for fixed CWI set to 125 mm (corresponding to the catchment at soil moisture capacity) and a 10 mm

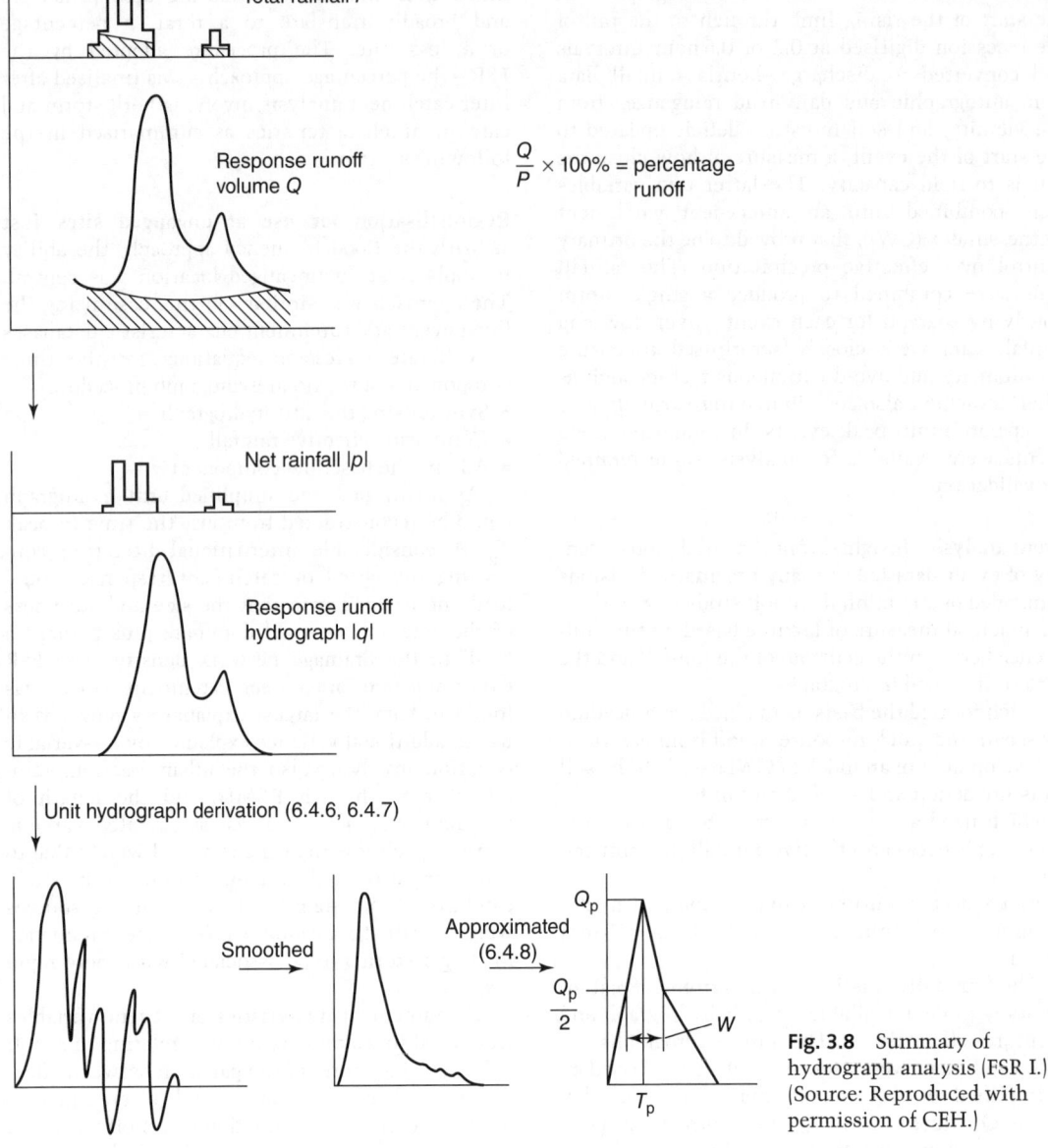

Fig. 3.8 Summary of hydrograph analysis (FSR I.). (Source: Reproduced with permission of CEH.)

rainfall. Typical values of SPR were from below 15% for a highly permeable catchment on chalk derived soils to near 50% for a clay catchment. The final percentage runoff for an individual flood event, PR, is adjusted from the SPR according to actual catchment wetness and storm rainfall depth, for example, 50 mm of antecedent rain and a 50 mm storm rainfall would uplift PR to SPR + 15 per cent.

These are the bare bones of the procedure; other details relating to definition of the hyetograph and the time unit to express it with implications to the unit hydrograph, shortcut for peak discharge, and substitution of local data can be read in Section 6.8.2 of the FSR. Also a regression equation was provided for baseflow addition, a relatively minor component in most instances. There is an uncertainty in the method of at least 15% much of it deriving from the percentage runoff estimation. This is indicative of the complexity of the catchment processes leading from rainfall through to response in the river, and the importance of obtaining local data to firm up estimates based on regression with catchment characteristics.

Assignment of return period Historically the outcome of a rainfall–runoff approach to flood estimation amounted to 'a flood following a T-year rainfall'. While storm characteristics and catchment conditions would be specified so as to ensure a severe event within the spectrum of possibilities, there could be no guarantee that the peak discharge would share the same return period as the input rainfall (from the depth-duration-frequency diagram) or indeed have any pre-specifiable return period. A novel aspect of the FSR was that it specified storm and catchment wetness conditions with the intention of outputting a flood peak of the desired return period.

This was achieved though a two-step approach. The first step was to build up the entire flood frequency curve by embedding the loss/unit hydrograph within a sampling framework that generated rainfall depths, durations, temporal profiles and catchment wetness from across their frequency distributions. It was established that such a scheme simulated a peak flood distribution close to that observed. The second step was to identify a path through this sampling scheme that led to the desired return period with reasonable consistency. The variables that dominated were, unsurprisingly, those that determined the runoff volume in the model – depth of rain and catchment wetness; rainfall duration and profile being of secondary importance.

The conclusion was a recommended set of model inputs – a rainstorm of neutral profile and duration that maximised the peak discharge, a CWI close to its median, and a storm depth which in combination with the selected duration gave a stable return period relationship with that desired. This latter aspect gave rise to some consternation as some users took it to imply, for example, that 80-year return period rainfalls always cause 50-year return period floods. This is of course a misinterpretation as a rainfall labelled 80-year return period could produce flood peaks spanning the very common to the extremely rare depending on the other causative factors.

3.2.3.4 Application to reservoir safety

Very high standards of safety have long been required for reservoir design and dam maintenance which, for flood hydrology, meant a spillway capacity capable of passing an extremely rare flood event without prejudice to the integrity of the dam. This had been achieved prior to the FSR by 'envelope' or 'maximum of experience' methods applied to the reservoir inflow. A major advantage of the rainfall–runoff approach was the ability to particularise the individual processes so that the impact of each assumption could be objectively assessed. The fundamental philosophy was however retained in that the ultimate driver of the flood – the storm rainfall – still compounded of storm-causing factors set to their maximum of experience. This Estimated Maximum Precipitation (now termed probable maximum precipitation, PMP) is described in FSR volume II, Meteorological Studies.

The FSR procedure for calculating the 'Estimated Maximum Flood' (or probable maximum flood, PMF) took each step as described above for general design flood hydrograph estimation but substituted more extreme albeit still plausible conditions, the more significant being:

• Unit hydrograph: a more peaked version than for normal use, leading to a revised storm duration, D
• Snowmelt: allowance made for snowmelt
• Rainstorm: D-hour Estimated Maximum Precipitation set within a $5D$ EMP used also to increase antecedent CWI
• Effective precipitation: CWI allowed to increase through the event with corresponding increase in the value of PR at each time step.

These basic methods have continued to be recommended with minimal modification in several editions of the ICE publication *Floods and Reservoir Safety* (ICE, 1996) and will form the basis of the PMF estimation method in the forthcoming 4th edition.

3.3 Post FSR Developments

Research into flood hydrology entered something of a golden era in the years following the publication of the FSR thanks principally to the support of the Ministry of Agriculture, the government department then responsible for land drainage and flood protection. Many of the new directions made possible by this support were summarised during a review meeting organised by the Institution of Civil Engineers, 'FSR – Five Years On'.

3.3.1 *European and world flood studies*

Following the publication of the FSR, the combination of the assembly of annual maximum floods (AMFs) from reasonably homogeneous areas and the derivation of regional flood frequency curves, first by averaging of groups of bands of *y* values and later by application of the General Extreme Value (GEV) distribution, was used in applied studies in a number of overseas investigations. It was felt that the techniques used could be applied to other countries, but that these should be applied to local data if they were to be useful. As the methods of estimating flood frequencies by relating the results of analysis of single stations to wider areas had been applied in overseas investigations, the decision was made, and supported by ODA, to expand the study on a more systematic basis. Sets of AMF series were sought from different countries or acquired from available hydrological yearbooks, while additional areas were included by ongoing consultancies in other countries.

In a discussion of The FSR-Five Years On, 1980, studies of NE Botswana, Central Iran, Mahweli basin, Sri Lanka, Malawi and Blue Nile basin Sudan were compared in a discussion of the use of the FSR overseas (Sutcliffe, 1981) In the discussion on how the FSR might be made more applicable overseas, it was pointed out that there was a choice between a gradual building up of experience and examples from overseas countries, or a concerted attempt to analyse records, which depended partly on the financial support provided. Thanks partly to funding from ODA, both approaches were attempted, resulting in a series of papers: floods in arid areas (Farquharson *et al.*, 1992); in West Africa, 1993; in many regions 1987; worldwide (Meigh *et al.*, 1997; in 17 European countries (Sutcliffe and Farquharson, 1996).

The methods of analysis initially followed the techniques followed in the FSR, but as new methods of statistical analysis were developed at Wallingford or elsewhere, these new techniques were adopted during the study. The studies were based on two components, with the MAF at a single station being treated as the basic flood indicator and this was correlated with the basin characteristics found useful in the FSR. In practice, the only basin characteristics which could be measured in many overseas countries were basin area and mean annual precipitation over the basin. Although correlations were sought in most of the regional studies, it was found that a short period of record at the site was more precise than a regional equation. In most cases a more useful result was the regional curve which described the relation between the MAF and the flood of a given return period, which could be deduced by combining the flood frequency curves at the single station over an area or a group of stations in a homogeneous region. The basic frequency distribution used was the GEV distribution and the Gringorten plotting position. At first the frequency curves were combined by arithmetic means over ranges of return period, but later the method of probability weighted moments (PWMs) was used to estimate the parameters of the GEV distribution.

As a result of the initial 'stamp-collecting' approach of selecting study areas, certain wider areas than single countries were covered by regional investigations and patterns of regional curves were observed. Two early investigations in Botswana and Guinea revealed contrasting regional curves, with the Botswana curve being steep and curved upwards, while the curve from Guinea was flat and curved slightly downwards. Flood records from other semi-arid regions, including Saudi Arabia and Yemen, South Africa and Botswana and the arid parts of Queensland, gave rise to very similar curves, while those from a wide area of West Africa from Senegal to Cameroun with a large number of stations resulted in very flat curves, distinguished only slightly by rainfall averages. Interesting exceptions from the pattern emerging were from Kenya and Sri Lanka, where the defining difference appeared to be the variety of dates when the maximum events occurred. Another study based on the available flood records across Europe showed a fairly consistent pattern of increasing variability from west to east, or from maritime to continental climates, with the lowest variability in Ireland and Norway and the highest in Romania and Bulgaria.

3.3.2 *Low Flow Studies*

The statistical methods developed during the FSR analysis were subsequently adapted and applied to the estimation of low flows in both gauged and ungauged catchments (Institute of Hydrology, 1980). The main applications of the Low Flow Studies report were abstraction and effluent licensing, the setting of environmental flows, irrigation and the estimation of drought frequency. Further details are given in Chapter 6.

3.3.3 Further methodological developments

3.3.3.1 The birth and evolution of L-moments

The origins of PWMs and L-moments can be traced to a paper published in 1975 entitled 'Regional skew in search of a parent' (Matalas *et al.*, 1975). In that paper, the authors showed that the relationship between the mean and standard deviation of regional estimates of skewness for annual maximum streamflow data from the western United States could not be explained by corresponding relationships for the then conventional frequency distributions. A plot of these relationships exhibited the phenomenon of 'separation', in that the observed data exhibited much higher standard deviations of regional skews than conventional distributions. This finding intrigued Professor Harold A Thomas, the brains behind the Harvard Water Programme, who derived a five-parameter distribution that could mimic separation. In 1976, he communicated his finding to J. R. Wallis through a letter written from Cape Cod, where he had been holidaying alongside Wakeby Pond, which is how the Wakeby distribution got its name (Houghton, 1978). This distribution was expressed in inverse form, and did not lend itself to parameter estimation using conventional moments or maximum likelihood methods, so a new fitting method was needed. Around this time, J. R. Wallis met J. A. Greenwood, an American statistician, at an Austrian statistical symposium, and together with J. M. Landwehr and N. C. Matalas, they derived PWMs, and demonstrated how they could be used to fit the Wakeby distribution and others expressed in inverse form (Greenwood *et al.*, 1979). PWMs were immediately attractive in that the fitting method, based on order statistics, was easy to implement and had extremely good small sample properties (Landwehr *et al.*, 1979a,b), and avoided the numerical difficulties associated with maximum likelihood.

During 1983/84, J. R. Wallis spent a sabbatical year at the Institute of Hydrology (IH) in Wallingford, where Enda O'Connell introduced him to J. R. M. Hosking. They worked together (and with E. F. Wood who was also on a sabbatical visit to IH) on an appraisal of the regional flood frequency analysis procedure used in the UK FSR (Hosking *et al.* 1985a), and derived PWM fitting relationships for the Generalised Extreme Value (GEV) distribution (Hosking *et al.* 1985b).GEV/PWM quantile estimates were shown to be more efficient than maximum likelihood estimates in small samples. This was also shown to be the case for the generalised Pareto distribution (Hosking and Wallis, 1987). Using Monte Carlo simulation, Hosking *et al.* (1985a) demonstrated that a computerised version of the graphical fitting procedure used in the FSR gave quantile estimates that were more biased and variable than estimates provided by regional GEV/PWM and Wakeby/PWM procedures. They recommended that, since more flood data were then available, the FSR regional growth curves should be updated, and suggested that the GEV/PWM procedure could provide more accurate and consistent estimates than the FSR procedure.

Following this initial collaboration, J. R. Wallis invited J. R. M. Hosking to spend sabbatical periods at IBM Research in Yorktown Heights. This led to J. R. M. Hosking joining IBM Research on a permanent basis, forming a scientific partnership that was to transform the field of regional frequency analysis. Subsequent papers focussed on the use of historic (Hosking and Wallis, 1986a, 1986b) and paleoflood data in flood frequency analysis, and on the effect of intersite dependence on regional flood frequency analysis (Hosking and Wallis, 1988). A major theoretical milestone was the introduction of L-moments and their properties (Hosking, 1990) which are direct analogues of conventional moments, and which provide similar, but much more robust, measures of the location, scale and shape of probability distributions. L-moments, which are a linear combination of PWMs, offer the simplicity of conventional moments but can surpass the efficiency of maximum likelihood estimates in small samples for distributions typically used in flood frequency analysis (Hosking *et al.*, 1985b). The main advantage of L-moments over conventional moments is that L-moments, being linear functions of the data, are less affected by outliers, and enable more robust inferences to be made from small samples about an underlying probability distribution (Hosking, 1990). They formed the basis for a comprehensive theory for the description, identification and estimation of probability distributions.

L-moments were to transform research on regional frequency analysis, with the title of one subsequent paper asserting that '*L-moment diagrams should replace product moment diagrams*' for identifying a probability distribution, on the basis that product moments were subject to substantial bias and variance (Vogel and

Fennessey, 1993). Hosking and Wallis (1997) produced a complete and well integrated treatise on regional frequency analysis in a book that was to become the key reference on the subject in the coming years, both in research and practice. The book described, clearly and succinctly, the basic L-moment methodology and its practical application to regional frequency analysis, covering the identification of homogeneous regions, the choice and estimation of a regional frequency distribution, and applications to annual rainfall and AMF data from the US. It also provided a critical assessment of the impacts of several factors on the performance of the overall approach using extensive Monte Carlo simulation experiments, as well as practical advice for users.

A notable milestone in the uptake of regional L-moment procedures in practice was their adoption in the preparation of a US national drought atlas underwritten by the U.S. Army Corps of Engineers (Guttman, 1993; Guttman *et al.*, 1993). Regions that had homogeneous statistical distributions of annual precipitation were defined through the iterative use of clustering and an L-moment-based homogeneity test; this was one of a set of statistics based on L-moments developed by Hosking and Wallis (1993; 1997) for use in regional frequency analysis. At IH/CEH, research was underway to produce the FEH (Institute of Hydrology, 1999), the successor to the FSR, and a regional flood frequency estimation procedure was adopted based on the Logistic distribution and its estimation using pooled L-moment ratios for hydrologically similar groups of catchments, marking another milestone in the uptake of L-moment methodology in practice. Since then, there have been continuing research contributions from CEH on L-moment procedures and their application (e.g. Kjeldsen and Jones, 2004; 2006).

Starting in the early 2000s, the Hydrometeorological Design Studies Center of NOAA's National Weather Service embarked on a major upgrade of the mapped US Intensity-Duration-Frequency relationships that had been developed in the 1960s and 1970s (Bonnin, 2002); this upgrade is still ongoing (e.g. Perica *et al.*, 2013). Regional frequency curves are estimated using regionally determined L-moments, based on the most appropriate probability distribution. Svensson and Jones (2010) carried out a review of national approaches to rainfall frequency estimation in nine countries (Canada; Sweden; France; Germany; United States; South Africa; New Zealand; Australia) which demonstrated that PWMs/L-moments are the preferred approach to distribution selection, the delineation of homogeneous regions, and the estimation of regional frequency curves in these countries.

In conclusion, the published literature and web-searches currently show that, in both research and practice, L-moment methodology currently occupies a dominant position in both rainfall and flood frequency analysis across the world, and must rank as one of the most significant developments in hydrological research in the last half-century. And the highly successful Hosking–Wallis partnership that provided this enduring legacy was formed at IH some thirty years ago!

3.3.3.2 The two-component extreme value distribution

During the 1980s, work continued at IH to investigate the application of a range of statistical distributions to the analysis of UK flood frequency. One of the most notable studies took the two-component extreme value (TCEV) as described by Rossi *et al.* (1984) and modified by Fiorentino *et al.* (1985) and applied it to UK flood data (Arnell and Beran, 1988; Beran *et al.*, 1986). The TCEV distribution is characterised as the maximum of two independently distributed extreme value type 1 (EV1) variables: a 'basic' series and an 'outlier' series.

Beran *et al.* (1986) developed an estimation procedure based on PWMs for the TCEV and demonstrated that the distribution had an intuitive appeal and offered a practical approach to regional flood frequency estimation.

3.3.3.3 Studies of complex events and joint probability

The analysis of extreme events in practical hydrological applications may not be straightforward because of difficulties arising from the definition of the events of interest, that is those events whose frequency of occurrence is to be calculated. Events of practical interest are not always defined in terms of simple extremes of a single series of values. One example of a more complicated analysis of a single series was examined by Jones (2003), where a flood or drought event might be declared to be extreme if the running cumulative total for one or more different accumulation-durations of rainfall or river flow exceeds the total having a given return period for the particular duration; such extreme events would be found to occur rather more frequently than suggested by the nominal return period used in the definition. An example of multiple simultaneous series of data was provided by Reed (1995), who found that a 1 in 100 year flood or drought (judged on a single-site basis) occurs somewhere in the UK nearly every year. A different problem involving multiple series concerns defining an extreme event in terms of the largest value across all available rainfall series in a region. Dales and Reed (1989) examined the problem of using single-site rainfalls to derive the properties of spatial maxima, while FEH volume 2 and Stewart *et al.* (2013) make use of the reverse relationship to allow spatial maxima to contribute to the estimation of single-site rainfall frequency curves. Problems that involve more complex combinations of separate series are typically discussed under the heading of 'joint probability'. Methods for treating such problems have been summarised by Reed and Jones (1999). Examples of joint probability problems in hydrology include river levels downstream of the confluence of two or more tributaries (Reed and Dwyer, 1996); levels on tidal rivers and estuaries deriving from either extreme river flows or sea levels, or both (Beran *et al.*, 1988); and levels on reservoirs arising from inflows and wind effects (Reed and Anderson, 1992). The immediate target of these studies is typically the properties of the derived variable in particular cases, but others such as those by Svensson and Jones (2002; 2004) and Keef *et al.* (2009) examine measures of statistical dependence between the extremes in different series and look at how variations in the strength of dependence can be explained.

3.3.3.4   The region-of-influence approach

Another important development was the attempt to define homogeneous regions based on hydrological similarity rather than geographical coherence (Acreman and Sinclair, 1986) which resulted in the development of the region-of-influence (ROI) approach proposed by Burn (1990) where homogeneous groups of catchments are formed based on similarity in catchment descriptor space.

3.3.3.5   Digital catchment data

Finally, progress in computing power and digital mapping technology enabled the development of an integrated hydrological digital terrain model (IHDTM) from which new digital catchment descriptors could be derived, effectively replacing the catchment characteristics used in the FSR and derived manually from 1:50,000 paper maps.

3.4   The Flood Estimation Handbook (FEH)

A government-funded project was initiated at the Institute of Hydrology in 1994 under the leadership of Duncan Reed with the aim of bringing together these developments to form a new set of flood estimation tools. The results were published by the Institute of Hydrology in 1999 in the FEH and associated software (primarily the FEH CD-ROM, which provided a user-friendly interface for identifying catchments and extracting relevant catchment descriptors, and the WINFAP-FEH package for statistical frequency analysis). The FEH was adopted by most of the UK water industry and regulators and, just like the FSR, also had a significant impact on academic and practical flood estimation far beyond the UK.

3.4.1   Background

By the mid 1990s, the principal methods of rainfall and flood frequency used in the UK had been based on the FSR for 20 years, and it was recognised that there were significantly more rainfall and flood data available that should be incorporated into estimation methods. It was also considered that, in a period when the use of computers had become routine in engineering design, there might be better methods to access spatial data than paper maps.

In addition, updating of the FSR by FSSRs in the intervening period had become progressively more intricate, and the effective dissemination of revised guidance less assured. This included the practical difficulties of succinctly explaining updates to updates, and of ensuring that supplementary reports reached all those holding copies of the earlier material. The difficulties were exacerbated by the successful penetration of FSR methods into other guides, including hydrology textbooks, engineering guides, teaching notes and the internal documents of agencies and consultants. A further reason for developing the FEH was to pull together a coordinated programme of research, generalised and validated for UK-wide application, with the clear goal of improving readily available methods of flood estimation.

Following external review (Ackers, 1992) and subsequent discussions, the Ministry of Agriculture Fisheries and Food (MAFF) adopted a new strategy for flood estimation research. Central to this strategy was the consolidation of research in rainfall and flood frequency estimation, to develop new procedures for application in river flood defence. The FEH research programme was conceived to meet this objective, and an advisory group was formed in 1994. The five-year programme gained financial support from MAFF, the National Rivers Authority (NRA, which subsequently became the Environment Agency (EA)), the Department of Agriculture Northern Ireland (DANI) and a consortium led by the Scottish Office. Related work was also funded by NERC. The budget for the programme totalled about £1.7 m. In parallel with this work of consolidation, a long-term programme of MAFF and NERC-funded research was also begun to advance the feasibility of flood frequency estimation based on continuous simulation modelling (e.g. Calver and Lamb, 1996).

3.4.2 Objective

The overall aim of the FEH was to improve the accuracy and reliability of flood estimates in the UK primarily for the design and appraisal of flood defence works, but also for other catchment management activities. It achieved this by creating a comprehensive and easy-to-use handbook based around a set of methods making maximum use of nationally available data. With few exceptions, the methods offered choice to the hydrologist in pursuit of that elusive best estimate, yet did not detract from the important roles of professional judgement and expertise when decisions had to be made relating to public safety and/or large investments.

The initial plan for the FEH research (Reed, 1994) was to concentrate new developments on statistical methods of rainfall and flood frequency estimation to make best use of the extended datasets available since the FSR had been released. During the research it was found that digital terrain models, and digital thematic data, had matured sufficiently to allow their full use in the FEH. Several changes flowed in consequence, notably development of the FEH CD-ROM to provide descriptive data for every catchment of $0.5\,\text{km}^2$ or greater in the UK. It also allowed the comprehensive technical restatement

of the FSR rainfall–runoff method in volume 4 to provide for the use of digital catchment data.

3.4.3 Structure of the FEH

Although the FEH mirrored the presentation of the FSR in the form of a five-volume publication, the handbook attempted to be more direct than the FSR, and aimed to provide details of the recommended methods, rather than simply a comprehensive report of flood studies. Thus the FEH did not dwell on the alternative methods and options that might have formed the basis of new guidance. Instead it presented a unified set of procedures for rainfall and flood frequency estimation that had general application. Though long, the FEH sought to provide a relatively easy read, presenting the procedures first and the supporting theory and results second. The handbook drew attention to the inherent uncertainty in estimating event magnitudes that are rarely observed, and the user was encouraged to query data quality, to refine generalised estimates of flood frequency by reference to local data, to appraise historical information and to think clearly about the uses to which flood estimates are put.

Following the general philosophy of its predecessor, the FSR, the FEH presented two main approaches to flood frequency estimation: the statistical analysis of peak flows (volume 3) and the FSR/FEH rainfall–runoff method (volume 4). Table 3.1 summarises the structure and main elements of the FEH.

3.4.3.1 The FEH philosophy

The FEH, particularly the procedures for statistical flood frequency (volume 3), contained many new features and an attendant new vocabulary. Perhaps the single most influential change was the way in which flood peak data were 'pooled' for growth curve derivation. The 'pooling-group' was introduced as a set of catchments judged to be hydrologically similar to the subject catchment, that is the catchment to the site for which flood estimates are required. The number of flood peak data pooled was taken in proportion to the target return period. The procedure provided a dynamic framework for flood frequency estimation, with flood peak datasets being published in full and a recommendation being made to incorporate additional data when available. A key feature of the statistical method is the concept of expert

Table 3.1 The Flood Estimation Handbook, an indication of structure

Overview (volume 1)
• Guidance is provided on the choice of procedures and uses to which the estimates are put.

Rainfall frequency estimation (volume 2)
• A generalised procedure is provided for rainfall depth-duration-frequency estimation.
• The method is prescriptive and, in general, the incorporation of local data is not recommended.

Statistical Procedures for Flood Frequency Estimation (volume 3)
• The T-year flood is estimated as the product of the index flood, QMD and a growth factor, x_T in a single site as well as a pooled analysis.
• On urbanised catchments flood frequency estimates are derived as if the catchment were rural and then adjusted for urbanisation.
• A special procedure is presented for growth curve estimation on permeable catchments.
• Data transfers are encouraged with special emphasis given to transfers from donor catchment directly upstream or downstream of the subject site.
• Methods are provided to construct a design hydrograph to be consistent with preferred estimate of the T-year peak flow.

Restatment and Application of the FSR rainfall–runoff method (volume 4)
• All information about the FSR rainfall–runoff method and its application is presented in a single volume.
• Data transfers are encouraged with special emphasis given to transfers from donor catchment directly upstream or downstream of the subject site.

Catchment descriptors (volume 5)
• The FEH CD-ROM provides catchment descriptors for mainland England, Wales, Scotland and Northern Ireland and Anglesey and the Isle of Wight.
• The catchment descriptors are based on digital data and use the IH digital terrain model (IHDTM) to define catchment boundaries.
• Descriptors for catchments draining to sites in Northern Ireland are adjusted to reflect the different mapping conventions and formats used there.

application, and encouragement is given to the analyst at all stages to think about the problem in hand and explore all the attendant assumptions. The statistical procedures were (and still are) supported by the WINFAP-FEH software package, which provides diagnostic information to allow the experienced analyst to refine the membership of the pooling-group.

Key aspects of the FEH were:
• The introduction of digital catchment descriptors, giving the potential to automate calculations, and to summarise and present flood estimates in new ways.
• The publication of flood peak datasets in full, with the aim of revitalising interest in gauged flood records.
• New dynamic statistical procedures for flood frequency that utilise all data available at the time the estimate is made.
• A new model of rainfall frequency estimation.

• A comprehensive restatement of the FSR/FEH rainfall–runoff method including new catchment descriptor equations.

3.4.3.2 Return period

The FEH retained the term 'return period' – generally defined as the average interval between years containing exceedances of a given peak flow – as the primary measure of rarity/risk. An alternative would have been to express rarity in terms of annual exceedance probability (AEP or ARI), defined as the probability of exceedance of a given flow within a period of one year (and thus the reciprocal of return period) as in Australian practice (Pilgrim, 1987). There are conflicting views as to whether or not such an approach makes it easier to explain risk and probability to decision-makers and members of the public but, in a text aimed at professionals used to the FSR procedures, the advantages of the change were judged to be minimal.

3.4.3.3 Maxims for flood frequency estimation

FEH volume 1 provides the following set of maxims for practitioners (Box 3.1)

3.4.4 *FEH catchment descriptors*

One of the most innovative aspects of the FEH was the development and application of digital catchment descriptors to quantify the physical, hydrological and climatological properties of catchments for use in estimating flood frequency at ungauged sites. The map-based procedures required by the FSR were time-consuming and required skill to define catchment boundaries accurately by hand, and were also subjective and often inconsistent. The FEH catchment descriptors are based on the Institute of Hydrology Digital Terrain Model (IHDTM) which evolved over many years starting with Morris and Heerdegen (1988). The IHDTM is derived principally from contour data, spot heights and blue-line (i.e. river and lake) data and coastline information taken from 1:50,000 OS maps (Morris and Flavin, 1990) and was specifically designed for hydrological applications. Based on the steepest route to neighbouring grid nodes, the IHDTM includes a $50\,m \times 50\,m$ grid of drainage path directions from which catchment boundaries can be generated automatically. This drainage path grid allows the computation of a range of physical and network descriptors and, when overlain on other thematic maps, descriptors such as soil type, land cover and average annual rainfall. A particular improvement over the FSR was the use of the Hydrology of Soil Types (HOSTs) classification (Boorman *et al.*, 1995) to define the dominant soil moisture characteristics of catchments, SPRHOST and BFIHOST (see Fig. 3.9).

Volume 5 of the FEH (Bayliss, 1999) describes the derivation of the catchment descriptors and provides values for a large number of gauged catchments. The FEH CD-ROM (now in its third incarnation which includes additional descriptors) provides catchment descriptors for any catchment of at least $0.5\,km^2$ in the UK (Fig. 3.10).

3.4.5 *FEH statistical method*

Volume 3 of the FEH (Robson and Reed, 1999) presents details of the FEH statistical method which is based on the analysis of annual maximum series of instantaneous peak flow. The concept of hydrological similarity is used to allow data from different sites to be pooled together to enhance confidence in the resulting frequency estimates. The two key elements of the FEH statistical method are the estimate of the index flood (the median annual flood, QMED) at the subject site, together with the definition of a growth curve giving the ratio (growth factor) between QMED and the flood of the required return period (Q_T). Perhaps the greatest advance over the FSR method is the replacement of fixed regional growth curves with those found by fitting a statistical distribution on a site-by-site basis to data pooled from multiple hydrologically similar catchments. These are catchments primarily with similar area, annual rainfall and baseflow index values, but also with other similar properties such as seasonality. L-moments are used to weight the parameters of the pooled growth curve according to the similarity of individual sites in the pooling-group. Explicit consideration is given to urbanisation

---

Box 3.1 Six maxims for flood frequency estimation (from FEH volume 1)

1 Flood frequency is best estimated from gauged data.
2 While flood data at the subject site are of greatest value, data transfers from a nearby site, or a similar catchment, are also very useful.
3 Estimation of key variables – such as the index flood (QMED, volume 3) or the unit hydrograph time to peak $(T_p,$ volume 4) – from catchment descriptors alone, should be a method of last resort; some kind of data transfer will usually be feasible and preferable.
4 The most appropriate choice of method is a matter of experience, and may be influenced by the requirement of the study and the nature of the catchment; most importantly, it will be influenced by the available data.
5 In some cases, a hybrid method – combining estimates by statistical and rainfall–runoff approaches – will be appropriate.
6 There is always more information; an estimate based only on readily available data may be shown to be suspect by a more enquiring analyst.

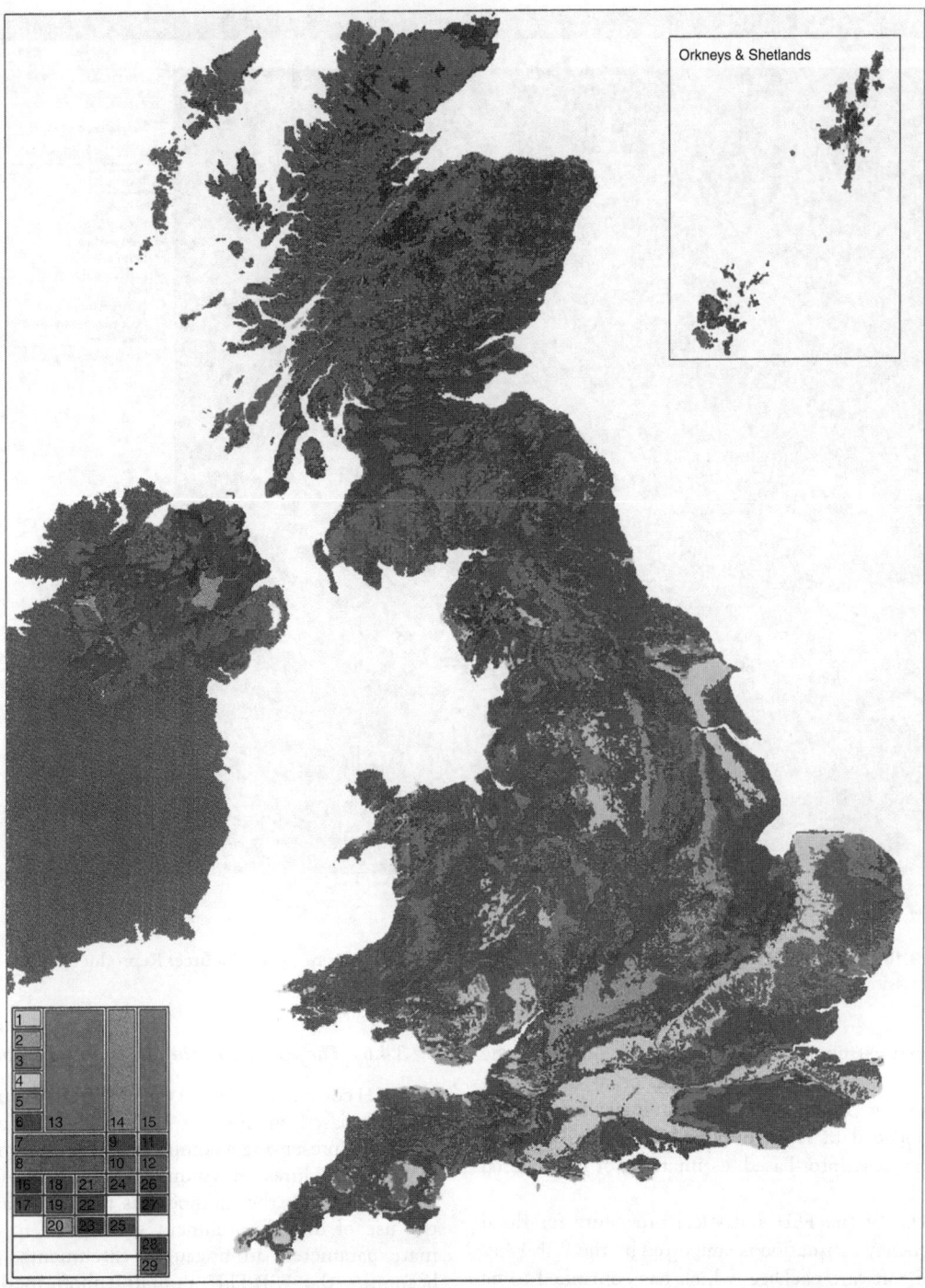

Fig. 3.9 HOST class for each kilometre square. (Source: Reproduced with permission of CEH.) *(See insert for colour representation of this figure.)*

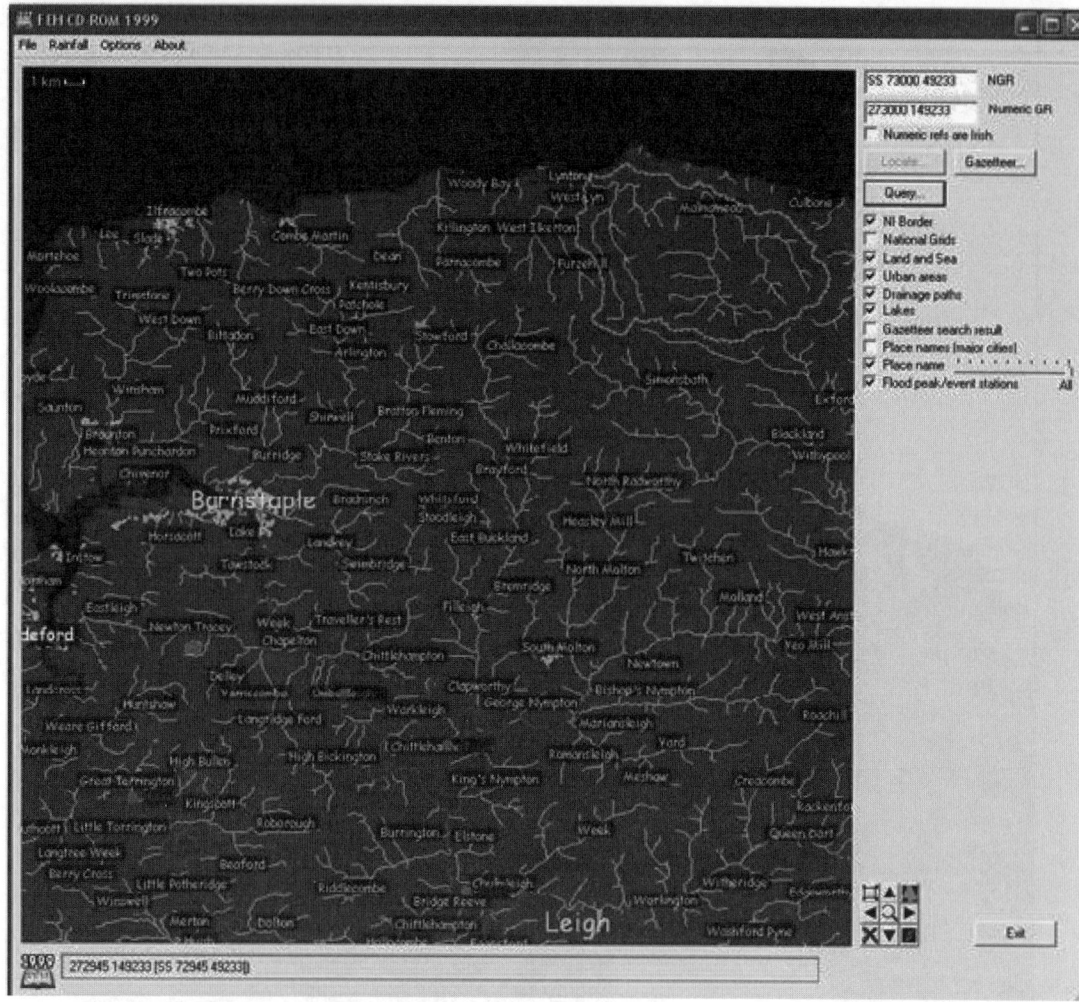

Fig. 3.10 FEH CD-ROM screen shot showing the IHDTM derived river network. (Source: Reproduced with permission of CEH.) *(See insert for colour representation of this figure.)*

in estimating the index flood and adjusting the pooled growth curve. The use of at site, local donor or more distant, hydrologically similar, analogue data is encouraged, with purely catchment descriptor-based estimates not considered viable.

Use of the FEH statistical procedure for flood frequency estimation is supported by the WINFAP-FEH software package, which has continued to be updated as modifications and improvements have been made to the set of basic methods (Fig. 3.11).

3.4.6 *The FSR/FEH rainfall–runoff method*

The FEH consolidated use of the FSR rainfall–runoff method, based on the unit hydrograph and losses model, by presenting a comprehensive restatement of the procedures in volume 4 (Houghton-Carr, 1999). Although the method was revised to allow the use of digital catchment descriptors to estimate parameters in ungauged catchments (thus becoming the FSR/FEH rainfall–runoff method), and the use of new design inputs based on the FEH rainfall depth-duration-frequency model was

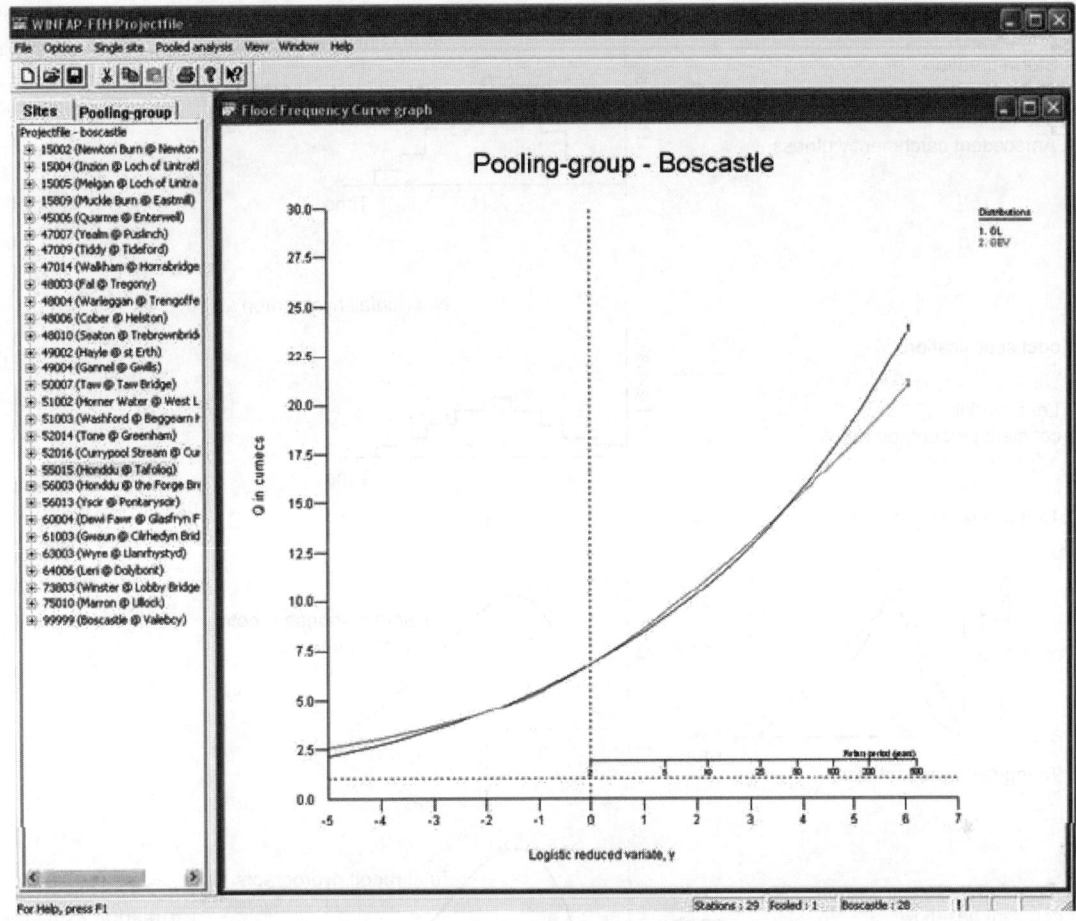

Fig. 3.11 Screenshot from the WINFAP-FEH software. (Source: Reproduced with permission of CEH.)

recommended, the overall design event package was not recalibrated, and its key features remained unchanged (see Fig. 3.12).

FEH volume 4 brought all information about the FSR/FEH rainfall–runoff method together, including relevant aspects of the basic FSR methodology, supplementary research and recommendations, and specialist guidance on particular uses. Details are given of how to estimate the key model parameters: unit hydrograph time to peak, percentage runoff and baseflow. Once the model parameters have been derived for a catchment, the method may be used to estimate the total flow from any rainfall event in the form of a hyetograph, defined by a duration, depth and temporal profile. The rainfall may be a statistically-derived design event to produce a

flood of a specified return period (the *T*-year flood), or a PMP to produce a PMF. Alternatively, the rainfall may be an observed event with the aim of simulating a notable flood.

More recently, the FSR/FEH model has been replaced by the Revitalised Flood Hydrograph (ReFH) design model (Kjeldsen and Jones, 2007; Kjeldsen and Prudhomme, 2005), further details of which are discussed in Section 3.5.2.

At the time of publication of the FEH, the FSR/FEH rainfall–runoff method was supported by the use of the Micro-FSR software package. As this method has been largely superseded by the ReFH method, Micro-FSR has been withdrawn. However, the method remains the principal means of deriving a design hydrograph in situations where storage is

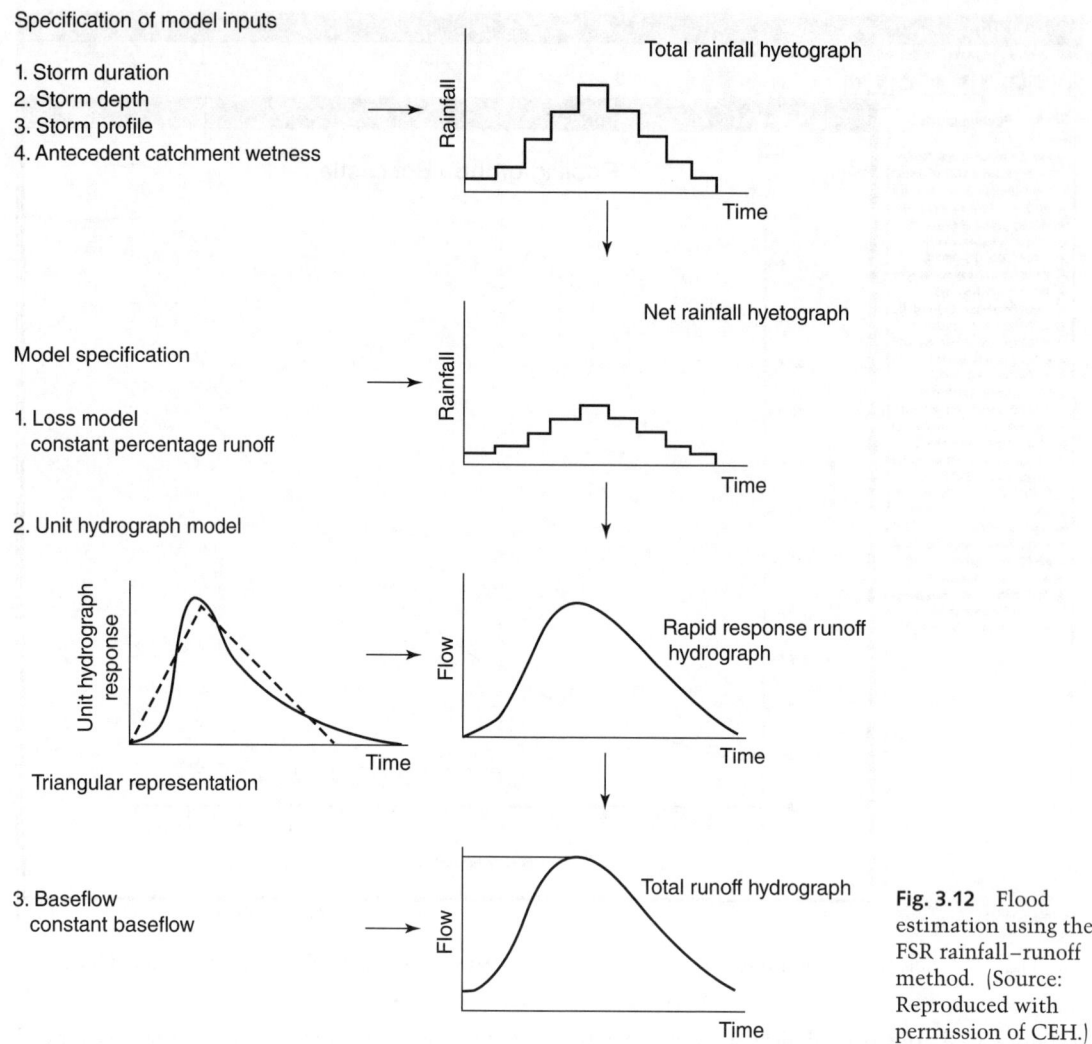

Specification of model inputs

1. Storm duration
2. Storm depth
3. Storm profile
4. Antecedent catchment wetness

Total rainfall hyetograph

Model specification

1. Loss model
   constant percentage runoff

Net rainfall hyetograph

2. Unit hydrograph model

Triangular representation

Rapid response runoff hydrograph

3. Baseflow
   constant baseflow

Total runoff hydrograph

Fig. 3.12 Flood estimation using the FSR rainfall–runoff method. (Source: Reproduced with permission of CEH.)

involved including reservoirs, flood storage ponds and pumping station design, as well as some tide lock situations, and appropriate computer code has been incorporated into widely used river modelling software systems.

3.4.7 *The FEH model of rainfall depth-duration-frequency*

Volume 2 of the FEH (Faulkner, 1999) presented a new method for rainfall frequency estimation. The new procedures were intended to replace the use of FSR volume II, which had formed the basis

of UK rainfall frequency estimation for more than 20 years and had a worldwide influence. The FEH rainfall model was based on an additional 25 years of data and addressed some of the weaknesses of its predecessor, for example the concern that the FSR's limited regionalisation led to serious underestimation of the frequency of 2-day rainfalls in Somerset and the surrounding area (Bootman and Willis, 1981). The FEH rainfall frequency analysis was based on annual maximum rainfalls abstracted from raingauge records throughout the UK and aggregated over durations ranging from 1 hour to 8 days. The analysis was undertaken in two stages, involving the mapping of an index variable and the derivation of rainfall growth curves. In this way, the analysis was consistent with the overall philosophy of the FEH as it introduced a new method of pooling data flexibly, rather than relying on the definition of fixed regions.

FEH vol. 2 adopted RMED, the median annual maximum rainfall of a specified duration, as the index variable. Values of RMED were interpolated between raingauge sites using topographic information to provide 1-km grids covering the UK using the technique of georegression, an extension of the kriging method of spatial interpolation (Faulkner and Prudhomme, 1998). The georegression used variables including distance from the coast, continentality (indexed by the distance in kilometre from Lille in northern France), elevation and other topographical variables. Figure 3.13 shows a map of RMED for the 1-day duration.

Rainfall growth curves were estimated for a number of key rainfall durations using the FORGEX (Focused Rainfall Growth Extension) method (Faulkner, 1999) which evolved from earlier research to develop the FORGE technique (Reed and Stewart, 1989) to address some of the FSR's shortcomings in South West England. FORGEX pools data from a hierarchy of expanding circular regions centred on the point of interest (Fig. 3.14). Data from smaller networks are used to estimate the growth curves for short return periods and data from the larger networks are used for the longer return periods. An upper limit of 200 km was used for the largest network radius. Growth curves were constructed using a complex empirical method which combined a regional frequency analysis with the use of network maximum points, the plotting positions of which were shifted to allow for the spatial dependence of rainfall extremes.

A depth-duration-frequency model was fitted to the rainfall frequency curves which ensured consistency between design rainfall depths of different durations and return periods. The model was implemented on the FEH CD-ROM to provide the design inputs to the FSR/FEH rainfall–runoff method and, subsequently, the ReFH method (see 3.5.2).

3.4.8 *The FEH procedures in practice*

The FEH was released during a period when there was an upsurge in interest and demand for flood estimation, particularly within the EA. The year before the Handbook was published, severe floods had affected central and eastern England and parts of Wales in Easter 1998. The EA commissioned an independent review of its performance during the floods, and the resulting 'Bye Report' (Bye and Horner, 1998) recommended preparation of national guidelines on flood probability estimation for use by the EA and its consultants, to be based on the FEH. Within months of the release of the FEH, the EA issued a statement recommending use of the new methods for all calculations at a new stage of work from August 2000. FSR methods, for most applications, were to be phased out by 2001. Other flood management authorities across the UK followed suit, and so FEH methods were rapidly adopted across the industry. The widespread floods of October to November 2000 served to push flooding further up the political agenda and created a renewed impetus for understanding and predicting the expected effects of floods.

There was some resistance to use of the new methods, for example for rainfall design in the urban drainage industry or for design flood estimation for reservoir safety. In part these reflected the limits of the FEH research, in particular the rainfall frequency analysis which was intended principally for rainfall durations no shorter than 1 hour and return periods up to around 2000 years. Some potential FEH users were initially put off by the need to buy new software and learn new methods, but most of these pockets of resistance disappeared when it was realised that flood management authorities would no longer accept designs or flood risk assessments produced using superseded methods.

These guidelines recommended in the Bye Report were commissioned by the EA and the Rivers Agency and issued in 2000. Their adoption was limited at first, but subsequent re-writes and updates, along with templates for recording calculations, were welcomed both within the EA and its consultants. The value of the Flood Estimation Guidelines increased as user experience accumulated and the document was updated to cover revised FEH methods as discussed in the following sections.

Training courses in FEH methods were developed by both CEH and consultants, and hundreds of users attended within the first couple of years. As of 2014, JBA Consulting, for example, had trained over 2000 course delegates in use or awareness of FEH methods.

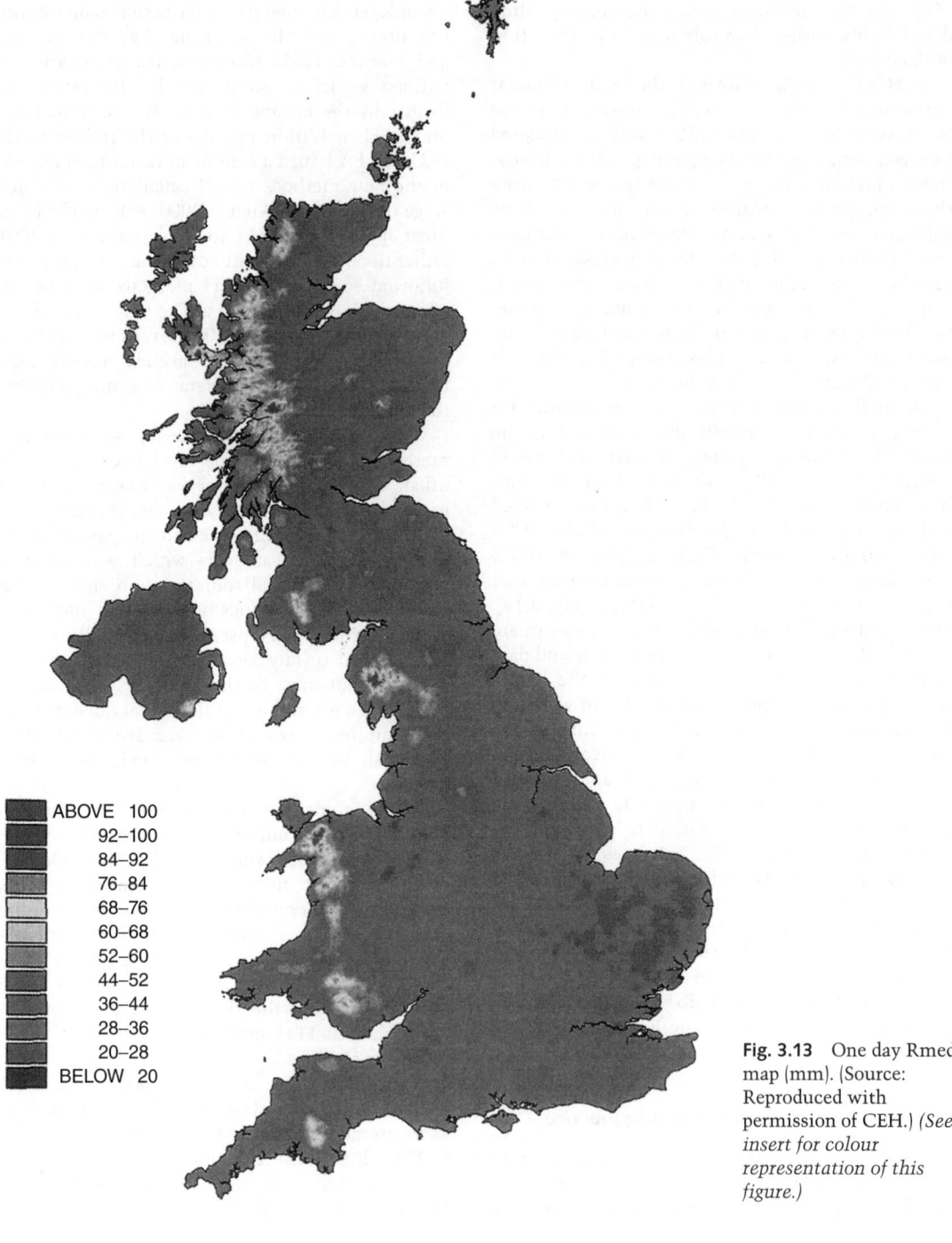

ABOVE 100
92–100
84–92
76–84
68–76
60–68
52–60
44–52
36–44
28–36
20–28
BELOW 20

Fig. 3.13 One day Rmed map (mm). (Source: Reproduced with permission of CEH.) *(See insert for colour representation of this figure.)*

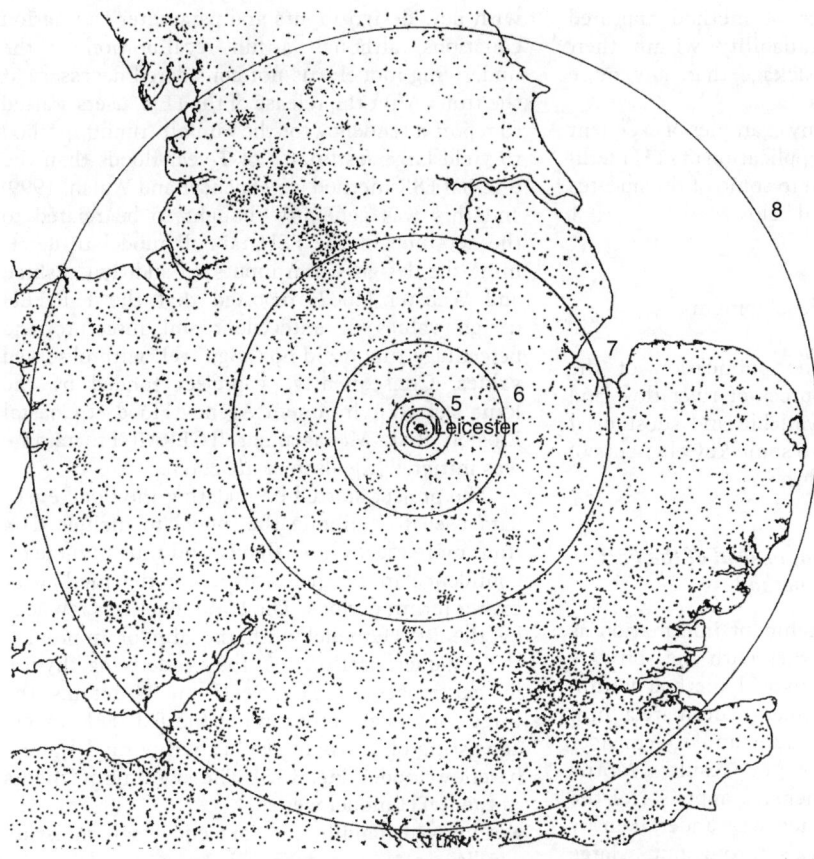

Fig. 3.14 Networks of daily gauges focused on Leicester (from FEH Vol. 2.). (Source: Reproduced with permission of CEH.)

Another aspect of the auspicious timing of the launch of the FEH was a rapid increase in demand for flood mapping across the UK. In England and Wales this was driven by Section 105 of the Water Resources Act 1991 which required definition of flood risk areas. 1999 saw the development of the first national flood map of England and Wales, the Indicative Floodplain Map. Numerous detailed local flood mapping studies were commissioned, and from 2000 onwards these ensured widespread use of the FEH. The availability of digital catchment descriptors and the FEH software led to application of flood estimation methods on a scale far greater than had been seen with the FSR. Eventually the FEH methods were being applied at a national scale to produce flood maps for regulators and insurers (see Section 3.5.1 for an example of automated application).

Such prolific application of methods that were originally intended to be applied in the context of detailed local hydrological investigations inevitably led to concerns over the quality of the flood estimation. The dangers of a 'point and click' mentality were recognised during development of the FEH procedures and software, with the realisation that convenience could prove to be a double-edged sword. Despite the wide uptake of training and the development of guidelines that emphasised the need to think carefully about choice of method, examples of poor application of FEH methods could be found, with some users appearing not to realise that the FEH is a book to be read and heeded rather than a software application with buttons to be clicked. Aspects of the methods that are not implemented in software packages (such as consideration of longer term flood history) tended to be neglected. The FEH rainfall–runoff method, and the newer ReFH method (see Section 3.5.2) found their way into commercial river modelling software suites such as ISIS and MIKE 11, and for some users the choice of method appeared to be based more on availability within their chosen hydraulic model package than any more hydrological considerations.

However, there were many examples of excellent and sometimes innovative application of FEH methods. Demand from users led to some of the updates of FEH approaches discussed below.

3.5 Post-FEH developments

In the years following the publication of the FEH, a number of new sources of river flow and catchment data became available and substantial improvements were made to some key elements of the flood estimation procedures.

3.5.1 *Automation and appraisal of the FEH statistical method*

The adoption of digital catchment information as the basis of the FEH, together with the software implementation of the statistical method within WINFAP-FEH, made it feasible to automate a form of the pooling-group procedure and to consider the spatial consistency of the results. Morris (2003) describes a comprehensive appraisal of the FEH statistical method which was undertaken by applying it at 50-m intervals throughout a large part of the national river network for several key return periods. The results showed that the FEH statistical method performed well in most places, providing spatially consistent flood peak estimates. Problems were found to occur at a minority of confluences, for example where catchments of very different permeability converge. The analysis led to the development of a number of modifications to the procedures, primarily to avoid or reduce the spatial inconsistencies that would have been generated by adhering to FEH rules and guidance, and a number of recommendations for updating the FEH statistical method were made.

3.5.2 *Revitalisation of the FSR/FEH rainfall–runoff method*

The first element of the FEH to be significantly revised was the rainfall–runoff method of design flood estimation. Volume 4 of the FEH had consolidated the use of the FSR's flood event package

with new design inputs and parameter estimation equations, although a full recalibration of the underlying model was not considered necessary at the time. After the release of the FEH, users started to report a tendency for the rainfall–runoff method to yield larger estimates of T-year floods than the original FSR method (e.g. Spencer and Walsh, 1999) and this was generally thought to be related to the adoption of the FEH rainfall model to determine the design storm inputs. In addition, Ashfaq and Webster (2002) reported that the FSR/FEH design values of antecedent soil moisture and percentage runoff did not align well with observed values. Consequently, a project funded by the Joint Defra/Environment Agency Flood & Coastal Erosion Risk Management (FCERM) Programme was initiated (Kjeldsen *et al.*, 2005).

The project started by updating the flood event data used to calibrate the model to include data from more recent large floods, and thus to increase confidence in the performance of the method at higher return periods. The FSR model was replaced by the ReFH model with the aim of improving the characterisation of the underlying hydrological processes. The ReFH model retains the three-component structure of the FSR/FEH model consisting of a loss model, a routing model and a baseflow model (Fig. 3.15). The key improvements in the ReFH model were:

- improved handling of antecedent soil moisture conditions
- a loss model based on the uniform PDM of Moore (1985)
- a more flexible unit hydrograph shape
- a new baseflow model which provides a more objective means of separating total runoff into baseflow and direct runoff.

The design storm inputs were revised to include a seasonal correction factor and the design package was calibrated so that the resultant hydrograph peaks corresponded to those derived from the FEH statistical method. A spreadsheet package to facilitate the application of the ReFH design method was released, and a comprehensive software package for ReFH flood modelling was added to the FEH tools. The ReFH software allows the estimation of local model parameters based on observed flood events as shown in Figure 3.16.

Current research is updating and recalibrating the ReFH method to incorporate new research on urban flood flows and to extend the range of return periods to which the model can be applied.

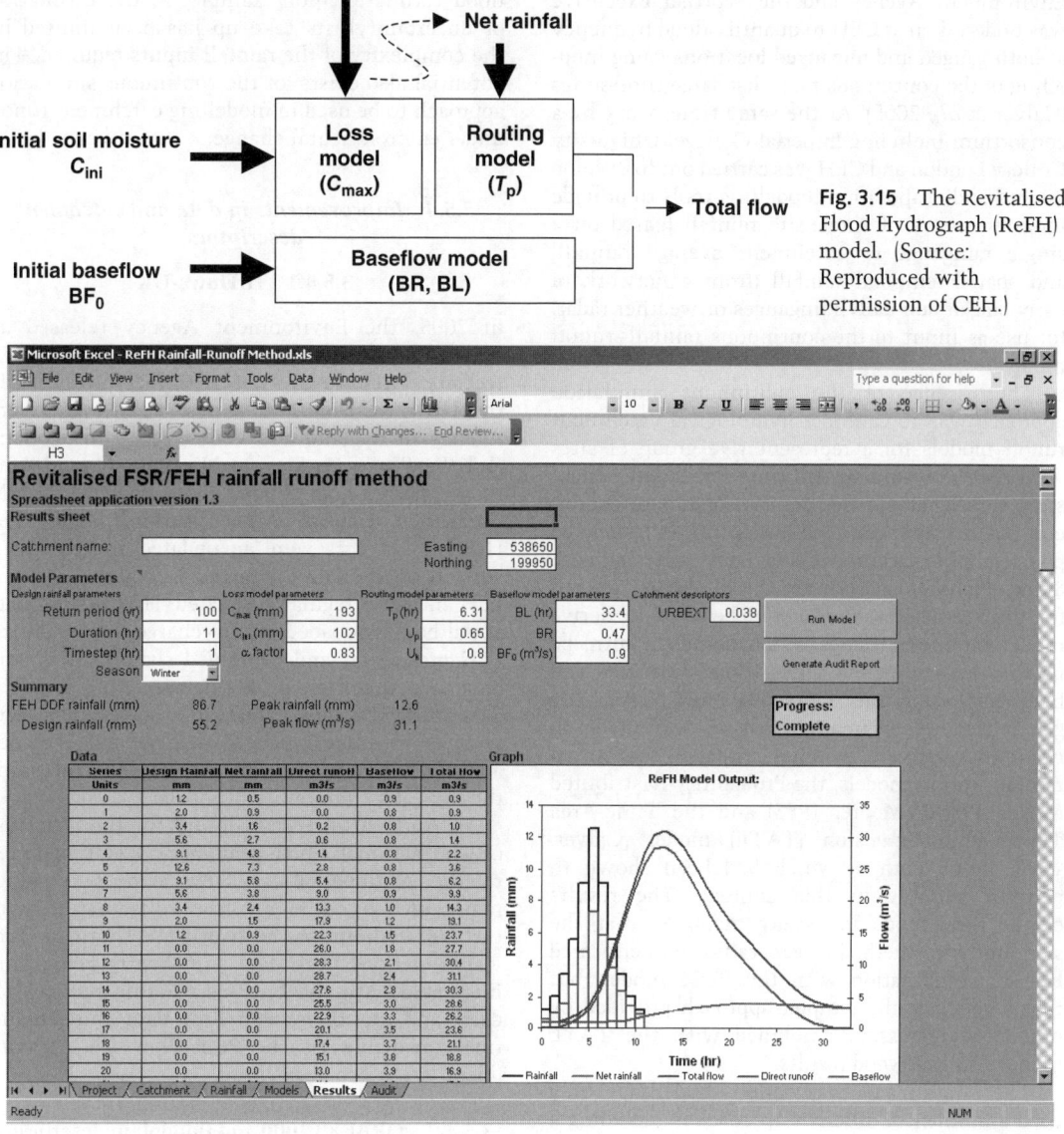

Fig. 3.15 The Revitalised Flood Hydrograph (ReFH) model. (Source: Reproduced with permission of CEH.)

Fig. 3.16 Screen shot of the spreadsheet based ReFH software. (Source: Reproduced with permission of CEH.)

3.5.3 *Continuous simulation modelling for flood frequency estimation*

Volume 1 of the FEH (Reed, 1999) defends the decision not to develop the rainfall–runoff design method further, citing controversy about whether the design event concept is convenient and valuable or constricting and unrealistic. At the time, an alternative approach to deriving design hydrographs based on whole catchment continuous simulation modelling was seen as a more viable future prospect which offered the added potential of assessing the likely influence of climate and land use change on future flows. Building on earlier work, a major study funded by Defra, the Environment Agency and the Scottish Executive was undertaken at CEH to quantify flood frequency at both gauged and ungauged locations using modelling of the continuous river discharge time series (Calver *et al.*, 2005). At the same time, work by a consortium including Imperial College, University College London and CEH was carried out to develop rainfall and evaporation modelling tools to provide long sequences of single-site rainfall (based on a single raingauge or catchment average rainfall) and spatial-temporal rainfall (from a network of daily and/or sub-daily raingauges or weather radar) for use as input to the continuous rainfall–runoff models.

The essence of the continuous simulation approach was to calibrate hydrological catchment runoff models for a representative group of sites with river flow and rainfall time series data. These sites were characterised by sets of model parameter values and catchment descriptors (such as physiographic, geometric and land cover indices). Since catchment descriptors could also be derived for ungauged sites, they could be used to derive model parameter values which could, in turn, be used to generate flow series, flood statistics and hydrographs and thus the results could be spatially generalised. A pragmatic decision was made to use two relatively parameter sparse conceptual rainfall–runoff models, the Probability-Distributed Model (PDM) (Moore, 1985) and the Time-Area Topographic Extension (TATE) model (Calver, 1993, 1996), both of which had been shown to perform suitably in this context. The results of the flow modelling study indicated that the site-similarity method of generalisation performed best in combination with the PDM model, but that a conceptually simpler approach provided by univariate regression combined with the TATE model also gave good results.

The rainfall and evaporation modelling study used a number of different approaches including a Poisson process model for single-site rainfall simulation, a daily spatial model based on Generalised Linear Models, a simple spatial-temporal disaggregation technique and a complex spatial-temporal model based on radar data. The simpler single-site and disaggregation models were shown to perform well on most catchments, but critical evaluation of model performance for each individual application was recommended. Therefore, although in general the continuous simulation approach was

shown to offer advantages over other methods of flood estimation (for example, in the estimation of uncertainty), its take up has been limited by the complexity of the rainfall inputs required. The potential also exists for the continuous simulation approach to be used in modelling catchment runoff under environmental change.

3.5.4 Improvements in data and catchment descriptors

3.5.4.1 HiFlows-UK

In 2005, the Environment Agency released an updated set of flood peak data via the HiFlows-UK website. HiFlows-UK was a partnership between the UK measuring authorities – the Environment Agency in England and Wales, SEPA in Scotland and the Rivers Agency in Northern Ireland – and the initiative was funded by a grant from HM Treasury's Capital Modernisation Fund. The HiFlows-UK data were an updated and enhanced version of the data contained in volume 3 of the FEH and were regularly updated. HiFlows-UK data could be downloaded free of charge for use in the FEH procedures and contained significantly more data than the original FEH dataset. Moreover, the data incorporated the results of subsequent data reviews to improve data quality, and provided additional information on each station, including the indicative suitability of the data.

In April 2014, responsibility for the provision of UK national flood peak data was transferred to the National River Flow Archive (NRFA), maintained by CEH in close collaboration with the UK hydrometric measuring authorities. Provision of flood peak data is currently being fully integrated with the NRFA's existing services for daily and monthly mean flow data and remains freely available via the Peak Flow Data Service (http://www.ceh.ac.uk/nrfa).

3.5.4.2 URBEXT2000 and floodplain descriptors

Urbanisation often has a considerable effect on the downstream flow regime and the catchment descriptor URBEXT provided a basis for taking account of this effect within the FEH methods. The land cover data used in the derivation of URBEXT were based on satellite imagery dating back to around 1990. The release of the CEH Land Cover Map 2000 provided an opportunity to update the indexing of catchment urbanisation and a new catchment descriptor, $URBEXT_{2000}$, was developed (Bayliss *et al.*, 2006). The research project also led to an upgrade to the FEH CD-ROM (version 2.0) which provided the $URBEXT_{2000}$ descriptor, and proposed a number of revised urban adjustment procedures.

One of the key FEH catchment descriptors is flood attenuation from lakes and reservoirs (FARLs). The importance of attenuation from flood-plains was also recognised in the FEH methods, but no accompanying descriptor was developed. The SC050050 project (see next section) developed a new catchment descriptor of flood plain extent (FPEXT) for all UK catchments, defined as the fraction of the total catchment area covered by the 100-year flood extend as defined by the first generation flood risk maps of England and Wales developed by Morris and Flavin (1996) in the IH130 report, and subsequently extended to Scotland and Northern Ireland. Accompanying descriptors of floodplain location (FPLOC) and mean flood depth (FPBAR) were also developed, but only FPEXT found its way into the final operational procedures.

3.5.5 *Improving the FEH statistical method*

The launch of HiFlows-UK combined with method-ological developments at CEH resulted in the Environment Agency funding a R&D project in 2005 aiming to use these methods to improve the FEH statistical method. Also, aspects of risk and uncertainty were becoming more important in discussions of flood management (e.g. Reed, 2002), opening up new research avenues and leading to further methodological developments.

The Environment Agency commissioned the R&D project SC050050 'Improving the FEH sta-tistical procedures for flood frequency estimation' in 2005 and the final report was published in 2008 (Environment Agency, 2008). The methods developed as part of the project have largely been adopted by the Environment Agency as best prac-tice for flood frequency estimation, and have also been implemented in the latest version of the WINFAP-FEH 3 software package (WHS, 2009). A number of new developments were introduced in SC050050, with the most important of these discussed below.

3.5.5.1 Estimation of the index flood

Through a series of methodological developments and an extensive exploratory analysis, a new model

was developed enabling the prediction of the index flood (QMED) in ungauged catchments using catch-ment descriptors only. The new QMED model is of the form

$$QMED = 8.3062 \ AREA^{0.8510} \ 0.1536^{(1000/SAAR)}$$

$$FARL^{-3.4451} \ 0.0460^{BFIHOST^2}. \qquad (3.1)$$

This model has a factorial standard error (*fse*) of 1.431 which is an improvement over the *fse* value of 1.549 reported for the equivalent QMED model from the original FEH publication.

What is less obvious from the model in Eq. (3.1) is the close link between the structure of the underlying statistical model and the benefit of data transfer from nearby gauged donor catchments. The FEH strongly encouraged the use of data transfer from nearby and hydrologically similar gauged catchments when estimating flood frequency in an ungauged catchment. The use of the donor transfer technique is valid and can be viewed as compen-sating for the inability of the catchment-scale lumped catchment descriptors used in Eq. (3.1) to capture more local factors controlling flood response. But research by Kjeldsen and Jones (2007; 2010) showed that the donor transfer scheme suggested by the FEH could potentially result in estimates of the index flood with inflated levels of uncertainty when compared to estimates obtained using the FEH regression model only. They went on to propose a revised transfer scheme where the influence of a donor site is weighted according to the geographical distance between the centroids of the catchments draining to the subject site and the donor site, respectively. The revised donor transfer method has become an integral part of the improved FEH methodology.

3.5.5.2 Pooled analysis

As already discussed, for estimating the growth curve at ungauged catchments and for sites where only a short record is available, the FEH had developed a procedure combining the index flood method with the ROI method to create site-specific homogeneous regions – the pooling-group. In prac-tice, a pooling-group was created for the target site by selecting gauged catchments from the database of 1000 gauged UK catchments considered to be hydrologically similar to the target site. Hydrologi-cal similarity was defined by a similarity distance measure based on a comparison of catchment area, standard annual average rainfall and soil type as determined by the BFIHOST dataset (Boorman *et al.*, 1995). Catchments were added to the pooling-group until the sum of AMAX events from all pooling-group members exceeded five times the target return period (the 5T rule), that is a minimum of 500 AMAX events for a 100-year design flood. The pooling procedure was retained in SC050050, but the measure of hydrological similarity was updated; the soil factor was replaced by measures of attenuation from lakes and reservoirs (FARL) and FPEXT.

Another important development introduced in SC050050 was a new method for assigning weights to the L-CV and L-SKEW estimates from the each of the pooling-group members. Where the original weights introduced in the FEH accounted for record length and rank of catchment within the pooling-group, the new weights more explicitly relate to the record length and the actual value of the distance similarity measure used for judging hydrological similarity. Consequently a user is no longer required to consider and revise the ranking of pooling-group members. The new pooling-group method was found to perform better than the original FEH method when considering ungauged sites. However, the biggest innovation was the introduction of a separate set of weights when considering a gauged catchment, where the at-site estimate of L-CV was given more weight than in the previous FEH weighting scheme. This results in enhanced single-site growth curves that are more akin to the single-site growth curves, but derived using all the pooled data and thus associated with lower level of uncertainty.

3.5.6 The new FEH13 rainfall depth-duration-frequency model

Research funded by Defra and CEH has led to the development of a new rainfall DDF model, currently known as FEH13, which provides rainfall frequency estimates for durations ranging from 1 hour to 8 days and return periods from 1 in 2 to over 1 in 10,000 years. The FEH13 model was developed through a complex statistical analysis of an extensive dataset of annual and seasonal maximum rainfall depths from raingauges throughout the UK. The basic approach taken mirrored that of the FEH rainfall analysis, but with key revisions: (i) the standardisation of the rainfall maxima is now more complex, making the rainfalls at the different sites

more similar prior to data pooling; (ii) the model of spatial dependence has been revised; and (iii) changes were made to the pooling methodology to overcome anomalous behaviour observed across a wide range of test cases. Full details of the statistical analysis are given by Stewart *et al.* (2013), and subsequent work has refined the structure of the model and generalised and smoothed the results across the UK on a 1-km grid.

From comparisons with the FEH rainfall DDF model across the full range of durations and return periods, several notable features emerge. Firstly, the estimates from the new model are higher over most of Scotland at the shortest durations (<6 h). Secondly, the estimates from the FEH13 model tend to be lower than the FEH at higher return periods (>200 years) and this is thought to be due mainly to the improved model of spatial dependence. At extremely high return periods, estimated rainfalls from the FEH13 model are often considerably lower than the FEH model because the extrapolation of the new model is an approximate straight line on the Gumbel scale whereas the FEH model curves upwards (an exponential extrapolation). Finally, whilst FEH 1 in 10,000-year rainfall estimates commonly exceeded FSR PMP, this is less often the case with estimates from the FEH13 model.

As in the FEH analysis, the development of FEH13 did not include the development of new PMP estimates for the UK. Areal reduction factors and design rainfall profiles also remained outside the scope of the work.

A new FEH web service is currently under development and this will deliver the new FEH13 model results, as well as the results of the FSR and FEH models. The web service will also provide access to the FEH catchment descriptors and will replace the existing FEH CD-ROM.

3.6 Future challenges and opportunities

Applied research into flood estimation continues to be an important topic. There are several aspects of current methods that are in need of critical evaluation and innovative new approaches.

3.6.1 Uncertainty estimation

There remains a need for better information about the true level of uncertainty associated with estimates obtained using generalised methods of flood and rainfall frequency estimation. In the past, the main aim of research in this area was to develop new datasets and methods to enable the most accurate prediction of the flood frequency relationship in cases where little or no hydrometric data were available. Less attention has been paid to the quantification of the uncertainty associated with frequency estimates, even though it can be substantial. Notably none of the methods presented in the FSR, the FEH, or the SC050050 report provides an assessment of the uncertainty of the *T*-year estimate derived from a regional flood estimation procedure in an ungauged catchment.

3.6.2 *Non-stationarity*

Quantifying the effects of both land use and climate change on the flood frequency relationship is an important, but also very difficult, task. The effects of land use change on flooding characteristics in UK catchments have been studied both from the perspective of rural land management and increasing urbanisation. In a comprehensive review of the link between rural land management and flooding, McIntyre *et al.* (2013) concluded that the current ability to quantify the impact of rural land use change on the water cycle is limited and cannot provide consistently reliable evidence to support planning and policy decisions. In contrast it has long been recognised that the expansion of urban areas will result in a reduction of catchment lag-times and increased flood volumes, both acting to increase the downstream flood risk (Hollis, 1975). Urban effects are most pronounced for smaller events and less important during large events (Robson and Reed, 1999). The result is flood frequency curves that are shifted upwards more for lower return periods than for higher return periods, that is less steep flood frequency curves, which again is synonymous with a reduction in the variability of the flood records. In a comparison of flood frequency in 200 urban UK catchments with expected flood behaviour in similar rural catchments, Kjeldsen (2010) found that while a general tendency for increased flood magnitude and reduced variability, contrasting behaviour could be detected, that is decreased flood magnitude and increased variability. These results indicate that more research is needed to provide better tools for enabling hydrologists to predict land use effects (rural and urban) on flood frequencies. The potential

effect of climate change on flood risk is a topic that has occupied researchers and policy-makers for a considerable time. The problem has been approached from two different directions. A large number of studies have used downscaled predictions of future climate scenarios in combination with rainfall–runoff models (Prudhomme *et al.*, 2010; Wilby *et al.*, 2008). These projections have generally found that future flood risk is likely to increase, and Defra (2006) advised that a precautionary 20% should be added to estimates of design flood made which are based on use of historical flood data, such as the FEH methodologies.

More recent studies based on UKCP09 scenarios have suggested that the 20% factor might insufficient (Prudhomme *et al.*, 2010). The other approach is based on trend analysis of historical records of flood flow series such as (Hannaford and Marsh, 2008; Robson and Reed, 1999). These studies generally find no or little evidence of trend in the existing flow series. In a recent study, Prosdocimi *et al.* (2014) concluded that for most catchments, the trends observed in available AMAX series from UK catchments were not sufficiently significant to support the 20% addition by 2085. Clearly, there is more work required to bring together the results obtained from the different methods, and to turn the disparate scientific evidence into actual practical guidance for flood estimation.

3.7 Looking to the future: flood protection investment in a highly variable climate

As the hydrological cycle is perceived to be intensifying due to global warming, governments are under pressure to invest proactively in flood protection infrastructure, while those at risk from flooding are becoming more aware of the need to compete for scarce funding. This has created a complex societal problem of decision making under uncertainty. Here, one possible approach to this problem is outlined.

The traditional approach to flood protection investment involves fitting an appropriate probability density function to AMFs (at site or regional), combining this with a damage function, and determining the level of investment by optimizing a Cost-Benefit function. A key assumption underlying this approach is that the AMFs are identically and independently distributed (IID) which in turn implies stationarity for the underlying stochastic process. Today, the problem of estimating the *T*-year flood and its uncertainty is fraught with difficulty due to the uncertainty over a changing climate. It has been suggested that the assumption of stationarity inherent in the IID assumption underlying classical flood frequency analysis should be discarded in favour of non-stationarity (Milly *et al.*, 2008). However, while the existence of global warming cannot be disputed, the evidence for an anthropogenic climate change signal (and non-stationarity) in precipitation extremes and flood records has proved difficult to find, based on the findings of the IPCC 2012 special report on extreme events and other extensive analyses of global flood records. For example (IPCC, 2012):

> There is *limited* to *medium evidence* available to assess climate-driven observed changes in the magnitude and frequency of floods at regional scales because the available instrumental records of floods at gauge stations are limited in space and time, and because of confounding effects of changes in land use and engineering. Furthermore, there is *low agreement* in this evidence, and thus overall *low confidence* at the global scale regarding even the sign of these changes.

GCMs are not sufficiently reliable in reproducing the properties of precipitation extremes in control climates to provide a reliable basis for predicting how these properties will change in the future. Also, they may not reproduce the natural long-term variability in the climate (Brown and Wilby, 2012; Johnson *et al.*, 2011).

O'Connell and O'Donnell (2014) have suggested that the implications of a more variable climate for investment in flood protection should first be explored within the limits of the stationarity assumption. There is increasing evidence that the IID assumption for AMFs is not valid due to natural long-term variability in the climate, which manifests itself as 'flood rich' and 'flood poor' periods (e.g. Ntegeka and Willems, 2008; Villarini *et al.*, 2013). O'Connell and O'Donnell (2014) argue that, in the absence of clear and unequivocal evidence of non-stationarity that can be incorporated into modelling non-stationary hydrological variables, invoking non-stationarity presents somewhat intractable challenges. Rather, it would seem prudent to explore the limits of stationarity in the first

instance, particularly in representing the long-term natural climatic variability that pre-existed global warming. By increasing the memory in a stationary stochastic model, the resulting increase in long-term variability can be indicative of the increased variability to be expected under global warming, and under which adaptation investment decisions will have to be made.

In an exploratory Monte Carlo study of adaptation investment decision making in a highly variable climate, O'Donnell and O'Connell (2014) have generated an ensemble of AMF series using a stationary stochastic model with increasing levels of long-term persistence. A classical Cost-Benefit (CB) approach has been used to estimate the appropriate level of protection for a hypothetical site with a specified damage function. This indicates that, as persistence increases, there is a higher probability of having very high levels of damage in a design-life period than under the IID case, reflecting temporal clustering of highly damaging floods which is of great concern to the insurance industry. Investment approaches to account for this increased risk have been explored by O'Donnell and O'Connell (2014), including the Conditional Value at Risk (CVaR) risk assessment approach which is often used to reduce the probability that a financial investment portfolio will incur large losses. The public's view of flood risk can significantly differ from that of flood risk management professionals, and hence cost-benefit analysis may not adequately address the society's priorities.

The above investment approach reflects the traditional top-down engineering approach to flood protection investment. As the impacts of floods on society have grown, this narrow top-down flood protection paradigm has evolved into a broader, people-centred flood risk management approach that considers the economic, social and environmental dimensions of sustainability, and a portfolio of both structural and non-structural measures for addressing flood risk (Fig. 3.17).

The non-structural measures typically focus on the need for more structured approaches to land use management/development in floodplain areas, better institutional functioning, better flood warning and emergency service operation, and the development of flood resilience (McEwen and Jones, 2012; McEwen *et al.*, 2012). Moreover, there a shift in responsibility for flood risk management downwards and outwards which means that those affected by flooding have an increasing role to play

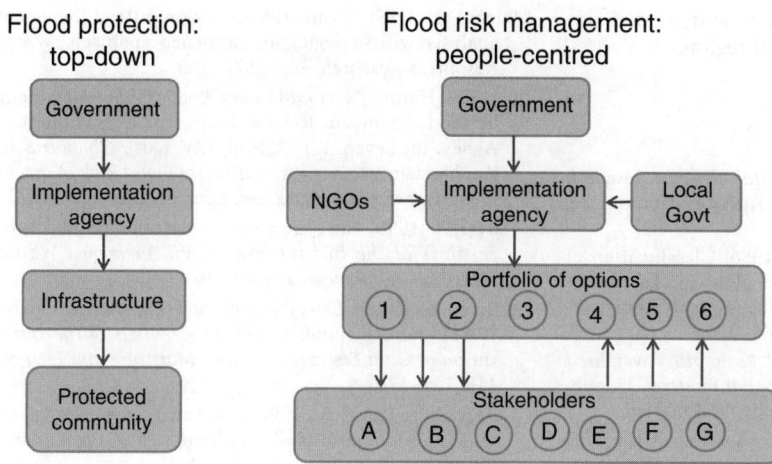

Fig. 3.17 Top down flood protection scheme. (Source: Reproduced with permission of P E O'Connell.)

in flood risk management, and which presages increasing, and highly desirable, cooperation between the state, operating agencies, public bodies and citizens. The social science literature on the complex socio-economic dimensions of flooding has therefore grown, and encompasses institutional analysis, the social impacts of flooding and how to address them, the evolution of flood protection investment policies, and of reactive institutional responses to major flood events. While there is evidence of increasing engagement between engineers and social scientists in developing interdisciplinary approaches to flood risk management, it is still the case that there is something of a 'paradigm lock' between the quantitative modelling approaches of flood hydrology, and the more qualitative approaches that characterise the social sciences. To address this, O'Connell and O'Donnell (2014) have proposed that flood protection investment should be explored within a Coupled Human and Natural System modelling framework that can represent the interaction between the natural hydrological system that generates flood events and flood damage, and the human system whereby funds for investment in flood protection are secured. The natural system is characterised by a stationary stochastic model of AMFs within an increasingly variable climate, and this is then coupled to a human system in the form of a virtual city that is impacted by flood events. As the traditional top-down flood protection approach has evolved into a more participatory people-centred flood risk management paradigm, the human system governing flood protection investment has become more

complex. The structure of this human system, and the various actors that influence decision making on flood investment within it are identified for a typical UK region (O'Connell and O'Donnell, 2014). An Agent-based Modelling approach to modelling this human system is proposed, and the functioning of social networks/flood action groups is analysed. This reveals how opinion dynamics and election campaigns respond to, and might be influenced by, a flood event in a virtual city, and how dominant stakeholder groups might secure flood protection investment funds at the expense of other groups, thus raising a social justice issue. It is argued that an Agent-based Modelling approach can help make stakeholders aware of the consequences of their actions for other stakeholders, thus helping to promote equity and social justice.

Further development of this framework will need to explore how an ABM representation of decision making can play out in terms of investments in the virtual region considered above, and what the value for money and social justice consequences can be. In the first instance, selected flood rich and flood poor realisations from a stochastic model of AMFs exhibiting long-term persistence will be used to gain insight into how the coupling between the Human and Natural Systems affects investments as the level of persistence/interannual variability increases. Investment decisions will be realised as outcomes of the interactions between multiple stakeholders, and assessed in terms of flood damage, economic efficiency, equity and social justice. These scenarios will be purely exploratory, and validating them will be a major future challenge, involving a real region.

3.8   References

Ackers, P. (1992) *Flood and coastal defence research and development*. Report of the Advisory Committee, MAFF, London.

Acreman, M.C. & Sinclair, C.D. (1986) Classification of drainage basins according to their physical characteristics – an application for flood frequency analysis in Scotland. *Journal of Hydrology*, **84**, 365–380.

Arnell, N.W. & Beran, M.A. (1988) *Probability-weighted moments estimators for TCEV parameters*. Unpublished report, Institute of Hydrology, Wallingford.

Ashfaq, A. & Webster, P. (2002) Evaluation of the FEH rainfall-runoff method for catchments in the UK. *Water and Environment Journal*, **JO16**, 223–228.

Bayliss, A.C. (1999) *Catchment descriptors. Volume 5 of the Flood Estimation Handbook*. Institute of Hydrology, Wallingford.

Bayliss, A.C., Black, K.B., Fava-Verde, A. & Kjeldsen, T.R. (2006) URBEXT2000 – A new FEH catchment descriptor. Joint Defra/EA Flood and Coastal Erosion Risk Management R&D Programme, R&D Technical Report FD1919/TR, 49p.

Benson M.A. (1962) Factors influencing the occurrence of floods in a humid region of diverse terrain. *USGS Water Supply Paper*, 1580B, Washington DC

Beran, M.A., Hosking, J.R.M. & Arnell, N.W. (1986) Comment on 'Two-component extreme value distribution for flood frequency analysis'. *Water Resources Research*, **22**, 263–266.

Beran, M.A., Jones, D.A., Harpin, R. & Smith, A.P.L. (1988) Stage frequency estimation for the tidal Thames'. *Proc. 5th IAHR Int. Symp. on Stochastic Hydraulics*, Birmingham.

Bonnin, G.M. (2002) *Updating NOAA Rainfall Frequency Atlases, Preprint Volume, 2002 Conference on Water Resources Planning and Management*, May19-22, Roanoke, VA (Published by the Environmental and Water Resources Institute of the American Society of Civil Engineers) (http://www.nws.noaa.gov/oh/hrl/presentations/wrpm02/pdfs/webpdf.pdf).

Boorman, D.B., Hollis, J.M. & Lilly, A. (1995) *Hydrology of Soil Types: A hydrologically-based classification of the soils of the United Kingdom*. IH Report No. 126, Institute of Hydrology, Wallingford.

Bootman, A.P. & Willis, A. (1981) Discussion of Paper 5 (Folland *et al.*). In: Flood Studies Report – Five Years On. Institution of Civil Engineers, London, pp. 62–63.

Brown, C. & Wilby, R.L. (2012) An alternate approach to assessing climate risks. *Eos, Transactions of the American Geophysical Union*, **93** (41), 401. doi:10.1029/2012EO410001.

Burn, D.H. (1990) Evaluation of regional flood frequency analysis with a region of influence approach. *Water Resources Research*, **26**, 2257–2265.

Bye, P. & Horner, M. (1998) Easter Floods Final Assessment by the Independent Review Team, vol 1, Environment Agncy. In: Beven, K.J. & Hall, J.W. (eds), *Chapter 8 in Applied Uncertainty Estimation for Flood Risk Management*. Imperial College Press, London, UK, pp. 500p.

Calver, A. (1993) The time-area formulation revisited. *Proceedings of the Institution of Civil Engineers: Water, Maritime and Energy*, **101**, 31–36.

Calver, A. (1996) Development and experience of the 'TATE' rainfall-runoff model. *Proceedings of the Institution of Civil Engineers: Water, Maritime and Energy*, **118**, 168–176.

Calver, A. & Lamb, R. (1996) Flood frequency estimation using continuous rainfall-runoff modelling. *Physics and Chemistry of the Earth*, **20**, 479–483.

Calver, A., Crooks, S., Jones, D., Kay, A., Kjeldsen, T. & Reynard, N. (2005) National river catchment flood frequency method using continuous simulation. Joint Defra/Environment Agency Flood and Coastal Erosion Risk Management R&D Programme. R&D Technical Report FD2106/TR. 135 pp plus appendices.

Dales, M.Y. & Reed, D.W. (1989) *Regional flood and storm hazard assessment*. Institute of Hydrology, Wallingford, pp. 159. (IH Report No.102).

Defra (2006) *Flood and coastal defence appraisal guidance (FCDPAG3), Economic appraisal supplementary note to operating authorities – climate change impacts*. Department for Environment, Food and Rural Affairs, London, 9 pp.

Environment Agency (2008) *Improving the FEH statistical procedures for flood frequency estimation*. Final research report R&D Project SC050050. Environment Agency, Bristol.

Farquharson, F.A.K., Mackney, D., Newson, M.D. & Thomasson, A.J. (1978) *Estimation of runoff potential of river catchments from soil surveys*. Soil Survey of England & Wales, Spec. Survey 11, pp. 27p.

Farquharson, F.A.K., Meigh, J.R. & Sutcliffe, J.V. (1992) Regional flood frequency analysis in arid and semi-arid areas. *Journal of Hydrology*, **138**, 487–501.

Faulkner, D.S. (1999) *Rainfall frequency estimation. Volume 2 of the Flood Estimation Handbook*. Institute of Hydrology, Wallingford.

Faulkner, D.S. & Prudhomme, C. (1998) Mapping an index of extreme rainfall across the UK. *Hydrology and Earth System Sciences*, **2**, 183–194.

Fiorentino, M., Versace, P. & Rossi, F. (1985) Regional flood frequency analysis using the two-component extreme value distribution. *Hydrological Sciences Journal*, **30**, 51–64.

Greenwood, J.A., Landwehr, J.M., Matalas, N.C. & Wallis, J.R. (1979) Probability weighted moments: definition and relation to parameters of several distributions express-able in inverse form. *Water Resources Research*, **15**, 1049–1054.

Guttman, N.B. (1993) The use of L-moments in the deter-mination of regional precipitation climates. *Journal of Climate*, **6**, 2309–2325.

Guttman, N.B., Hosking, J.R.M. & Wallis, J.R. (1993) Regional precipitation quantile values for the continen-tal United States computed from L-moments. *Journal of Climate*, **6**, 2326–2340.

Hannaford, J. & Marsh, T.J. (2008) High flow and flood trends in a network of undisturbed catchments in the UK. *International Journal of Climatology*, **28** (**10**), 1325–1338.

Hollis, G.E. (1975) The effect of urbanization on floods of different recurrence interval. *Water Resources Research*, **11** (**3**), 431–435.

Hosking, J.R.M. (1990) L-moments: analysis and estima-tion of distributions using linear combinations of order statistics. *Journal of the Royal Statistical Society, Series B*, **52**, 105–124.

Hosking, J.R.M. & Wallis, J.R. (1986a) The value of histor-ical data in flood frequency analysis. *Water Resources Research*, **22**, 1606–1612.

Hosking, J.R.M. & Wallis, J.R. (1986b) Paleoflood hydrol-ogy and flood frequency analysis. *Water Resources Research*, **22**, 543–550.

Hosking, J.R.M. & Wallis, J.R. (1987) Parameter and quan-tile estimation for the generalized Pareto distribution. *Technometrics*, **29** (**3**), 339–349.

Hosking, J.R.M. & Wallis, J.R. (1988) The effect of intersite dependence on regional flood frequency anal-ysis. *Water Resources Research*, **24** (**4**), 588–600. doi:10.1029/WR024i004p00588.

Hosking, J.R.M. & Wallis, J.R. (1997) *Regional frequency analysis: an approach based on L-moments*. Cambridge University Press, Cambridge, UK.

Hosking, J.R.M., Wallis, J.R. & Wood, E.F. (1985a) An appraisal of the regional flood frequency procedure in the UK Flood Studies Report. *Hydrological Sciences Journal*, **30** (**1**), 85–109.

Hosking, J.R.M., Wallis, J.R. & Wood, E.F. (1985b) Estima-tion of the generalized extreme-value distribution by the method of probability-weighted moments. *Technomet-rics*, **27**, 251–261.

Houghton, J.C. (1978) Birth of a parent: The Wakeby Dis-tribution for modeling flood flows. *Water Resources Research*, **14**, 1105–1109.

Houghton-Carr, H.A. (1999) *Restatement and application of the Flood Studies Report rainfall-runoff method. Vol-ume 4 of the Flood Estimation Handbook*. Institute of Hydrology, Wallingford.

Institute of Hydrology (1980) Low Flow Studies Report No. 1. Wallingford, UK.

Institute of Hydrology (1999) *Flood Estimation Handbook*. Vol. 5. Institute of Hydrology.

Institution of Civil Engineers (1933) *Interim Report of the Committee on Floods in Relation to Reservoir Practice*. Institution of Civil Engineers, London.

Institution of Civil Engineers (1960) *Floods in relation to Reservoir Practice*. Institution of Civil Engineers, London.

Institution of Civil Engineers (1965) River Flood Hydrol-ogy. ICE Symposium, 1965.

Institution of Civil Engineers (1967) *Flood Studies for the United Kingdom*. Report of the Committee on Floods in the United Kingdom,. Institution of Civil Engineers, London.

Institution of Civil Engineers (1975) *Flood Studies Confer-ence*. Institution of Civil Engineers, London.

Institution of Civil Engineers (1996) *Floods and Reservoir Safety*, 3rd edn. Thomas Telford, London.

IPCC (2012) Summary for Policymakers. In: Field, C.B., Barros, V., Stocker, T.F., Qin, D., Dokken, D.J., Ebi, K.L., Mastrandrea, M.D., Mach, K.J., Plattner, G.-K., Allen, S.K., Tignor, M. & Midgley, P.M. (eds), *Managing the Risks of Extreme Events and Disasters to Advance Cli-mate Change Adaptation. A Special Report of Working Groups I and II of the Intergovernmental Panel on Cli-mate Change*. Cambridge University Press, Cambridge, UK, and New York, NY, USA.

Johnson, F., Westra, S., Sharma, A. & Pitman, A.J. (2011) An Assessment of GCM Skill in Simulating Persistence across Multiple Time Scales. *Journal of Climate*, **24**, 3609–3623.

Jones, D.A. (2003) Use of several indices of event severity for floods and droughts. *Hydrology and Earth Systems Sciences*, **7**, 642–651.

Keef, C., Svensson, C. & Tawn, J.A. (2009) Spatial depen-dence in extreme river flows and precipitation for Great Britain. *Journal of Hydrology*, **378** (**3–4**), 240–252.

Kjeldsen, T.R. (2010) Modelling the impact of urbanisation on flood frequency relationships in the UK. *Hydrology Research*, **41** (**5**), 391–405.

Kjeldsen, T.R. & Jones, D.A. (2004) Sampling variance of flood quantiles from the generalised logistic distribution estimated using the method of L-moments. *Hydrology and Earth System Sciences*, **8**, 183–190.

Kjeldsen, T.R. & Jones, D.A. (2006) Prediction uncer-tainty in a median-basedindex flood method using L moments. *Water Resources Research*, **42**, W07414. doi:10.1029/2005WR004069

Kjeldsen, T.R. & Jones, D.A. (2007) Estimation of the index flood using data transfer in the UK. *Hydrological Sci-ences Journal*, **52**, 86–98.

Kjeldsen, T.R. & Jones, D.A. (2010) Predicting the index flood in ungauged UK catchments: on the link between data-transfer and spatial model error structure. *Journal of Hydrology*, **387**, 1–9. doi:10.1016/j.jhydrol.2010.03.024

Kjeldsen T.R. & Prudhomme, C.l. (2005) *Continuous flow modelling in Britain with hourly rainfall series generated by a random-parameter Bartlett-Lewis model ReFH report Geophysical Research Abstracts*, 7, 1 (Conference of the General Assembly of the European Geosciences Union, Vienna 2005, Session HS38)

Landwehr, J.M., Matalas, N.C. & Wallis, J.R. (1979a) Probability weighted moments compared with some traditional techniques in estimating Gumbel parameters and quantiles. *Water Resources Research*, 15, 1055–1064.

Landwehr, J.M., Matalas, N.C. & Wallis, J.R. (1979b) Estimation of parameters and quantiles of Wakeby distributions. *Water Resources Research*, 15, 1361–1379.

Matalas, N.C., Slack, J.R. & Wallis, J.R. (1975) Regional skew in search of a parent. *Water Resources Research*, 11, 815–826.

McEwen, L. & Jones, O. (2012) Building local/lay flood knowledges into community flood resilience planning after the July 2007 floods, Gloucestershire, UK. *Hydrology Research*, 43, 675–688.

McEwen, L., Krause, F., Hansen, J. G. & Jones, O., 2012. Flood histories, flood memories and informal flood knowledge in the development of community resilience to future flood risk, *BHS Eleventh National Symposium, Hydrology for a changing world*. BHS, Dundee.

McIntyre, N., Ballard, C., Bruen, M. *et al.* (2013) Modelling the hydrological impacts of rural land use change. *Hydrology Research*, 45, 737–754.

Meigh, J.R., Farquharson, F.A.K. & Sutcliffe, J.V. (1997) A worldwide comparison of regional flood estimation methods and climate. *Hydrological Sciences Journal*, 42, 225–244.

Melton, M.A. (1957) *An analysis of the relations among elements of climate, surface properties and geomorphology*. Office of Naval Research, Project NR389-042, Tech Rept 11.

Milly, P.C.D., Betancourt, J., Falkenmark, M. *et al.* (2008) Stationarity Is Dead: Whither Water Management? *Science*, 319, 573–574.

Moore, R.J. (1985) The probability-distributed principle and runoff production at point and basin scales. *Hydrological Sciences Journal*, 30 (2), 273–297.

Morris, D.G. (2003) *Automation and appraisal of the FEH statistical procedures for flood frequency estimation*. Final report. Centre for Ecology and Hydrology, pp. 207p.

Morris, D. G. & Flavin, R. W. (1990) *A digital terrain model for hydrology. Proc. 4th International Symposium on Spatial Data Handling*, Zurich, 1, 250–262.

Morris, D. G. & Flavin, R. W. (1996) *Flood risk map for England and Wales*. IH Report No. 130, Institute of Hydrology, Wallingford.

Morris, D.G. & Heerdegen, R.G. (1988) Automatically derived catchment boundaries and channel networks and their hydrological applications. *Geomorphology*, 1, 131–141.

Natural Environment Research Council (1975) *Flood Studies Report*. Vol. 5. NERC, London.

Newson, M.D. (1975) *Mapwork and flood studies: 1 Selection and derivation of indices*. IH Report 25, 52p.

Newson, M.D. (1976) *Mapwork and Flood Studies: 2 Analyses of indices and re-mapping*. IH Report 25, 33p.

Newson, M.D. (1978) Drainage basin characteristics, their selection, derivation and analysis for a flood study of the British Isles. *Earth Surface Processes*, 3, 277–293.

Ntegeka, V. & Willems, P. (2008) Trends and multidecadal oscillations in rainfall extremes, based on a more than 100-year time series of 10 min rainfall intensities at Uccle, Belgium. *Water Resources Research*, 44. doi:10.1029/2007WR006471

O'Connell, P.E. & O'Donnell, G. (2014) Towards modelling flood protection investment as a coupled human and natural system. *Hydrology and Earth System Sciences*, 18, 155–171.

O'Donnell, G. & O'Connell, E. (2014). *Decision-making on the level of flood protection under high climatic uncertainty. Proc. Dooge Nash International Symposium*, Dublin, pp. 285–294.

Perica, S., Martin, D., Pavlovic, S. *et al.* (2013) *NOAA Atlas 14 Volume 8 Version 2, Precipitation-Frequency Atlas of the United States, Midwestern States*. NOAA, National Weather Service, Silver Spring, MD.

Pilgrim, D.H. (ed) (1987) *Australian Rainfall & Runoff – A Guide to Flood Estimation*. Institution of Engineers, Australia, Barton, ACT.

Prosdocimi, I., Kjeldsen, T.R. & Svensson, C. (2014) Non-sttionarity in annual and seasonal series of peak flow and precipitation in the UK. *Natural Hazards and Earth System Sciences*, 14 (5), 1125–1144. doi:10.5194/nhess-14-1125-2014

Prudhomme, C., Wilby, L.R., Crooks, S.M., Kay, A.L. & Reynard, N.S. (2010) Scenario neutral approach to climate change impact studies: application to flood risk. *Journal of Hydrology*, 390, 198–209. doi:10.1016/j.jhydrol.2010.06.043

Reed, D.W. (1994) *Plans for the Flood Estimation Handbook. Proc. MAFF Conference of River and Coastal Engineers 1994*, Loughborough, 8.3.1-8.3.8. MAFF, London.

Reed, D.W. (1995) *Rainfall assessment of drought severity and centennial events. Proc CIWEM Centenary Conf.*, October 1995, Chartered Inst. Wat. Environ. Mang., London 16.1–16.7

Reed, D.W. (1999) *Overview. Volume 1 of the Flood Estimation Handbook*. Institute of Hydrology, Wallingford.

Reed, D.W. (2002) Reinforcing flood-risk estimation. *Philosophical Transactions of the Royal Society, A*, 360 (1796), 1373–1387. doi:10.1098/rsta.2002.1005

Reed, D.W. & Anderson, C.W. (1992) A statistical perspective on reservoir flood standards. BDS Conf. In: Parr, N.M., Charles, J.A. & Walker, S. (eds), *Water Resources and Reservoir Engineering*. Thos. Telford Ltd., London, pp. 229–239.

Reed, D.W. & Dwyer, I.J. (1996) *Flood estimation at conflu-ences: ideals and trials. '31st MAFF Conference of River and Coastal Engineers'*, Keele University, 3–5 July 1996, Flood and Coastal Defence Division, Ministry of Agriculture, Fisheries and Food, 3.2.1–3.2.10.

Reed, D.W. & Jones, D.A. (1999) *Joint Probability Problems. Appendix B of Flood Estimation Handbook.* Vol. **1**. Institute of Hydrology, Wallingford.

Reed, D.W. & Stewart, E.J. (1989) Focus on rainfall growth estimation. In: *Proc. Second National Hydrology Symposium*, Sheffield, British Hydrological Society, 3.57-3.65.

Robson, A.J. & Reed, D.W. (1999) *Statistical procedures for flood frequency estimation. Volume 3 of the Flood Estimation Handbook.* Institute of Hydrology, Wallingford.

Rossi, F., Fiorentino, M. & Versace, P. (1984) Two-component extreme value distribution For flood frequency analysis. *Water Resources Research*, **20** (7), 847–856. doi:10.1029/WR020i007p00847

Spencer, P. & Walsh, P. (1999) The Flood Estimation Handbook: Users' perspectives from North West England. In: *Proc. 34th MAFF Conf. River and Coastal Engineers*, Keele, UK.

Stewart, E.J., Jones, D.A., Svensson, C. et al. (2013) *Reservoir Safety - Long Return Period Rainfall. Final report.* Centre for Ecology and Hydrology, Wallingford.

Sutcliffe, J.V. (1981) Use of the Flood Studies Report overseas. In: *Flood Studies Report – Five Years On.* ICE. Thomas Telford, London.

Sutcliffe, J.V. & Farquharson, F.A.K. (1996) Flood frequency studies using regional methods. In: *Conference for Jacques Bernier.* UNESCO Press, Paris.

Svensson, C. & Jones, D.A. (2002) Dependence between extreme surge, river flow and precipitation in eastern Britain. *International Journal of Climatology*, **22** (10), 1149–1168.

Svensson, C. & Jones, D.A. (2004) Sensitivity to storm track of the dependence between extreme sea surges and river flows around Britain. In: Webb, B., Arnell, N., Onof, C., MacIntyre, N., Gurney, R. & Kirby, C. (eds), *Hydrology: Science and Practice for the 21st Century.* Vol. **1**. British Hydrological Society, pp. 239a–245a.

Svensson, C. & Jones, D.A. (2010) Review of rainfall frequency estimation methods. *Journal of Flood Risk Management*, **3**, 296–313. doi:10.1111/j.1753-318X.2010 .01079.x

Villarini, G., Smith, J.A., Vitolo, R. & Stephenson, D.B. (2013) On the temporal clustering of US floods and its relationship to climate teleconnection patterns. *International Journal of Climatology*, **33**, 629–640.

Vogel, R.M. & Fennessey, N.M. (1993) L-moment diagrams should replace product-moment diagrams. *Water Resources Research*, **29**, 1745–1752.

Wallingford HydroSolutions (WHS) (2009) WINFAP-FEH 3 software package.

Wilby, R.L., Beven, K.J. & Reynard, N.S. (2008) Climate change and fluvial flood risk in the UK: More of the same? *Hydrological Processes*, **22** (14), 2511–2523. doi:10.1002/hyp.6847

Young, A.R., Grew, R. & Holmes, M.G.R. (2003) Low Flows 2000: a national water resources assessment and decision support tool. *Water Science and Technology*, **48**, 119–126.

4 Terrestrial Hydrological Processes

DAVID J. COOPER[1], JOHN BELL[2], MARTIN HODNETT[1],
KEITH BEVEN[3], KEVIN GILMAN[2], ATUL HARIA[1],
CATE GARDNER[4], MARK ROBINSON[5], JON EVANS[5], AND
HELEN WARD[6]

[1]Ex-Centre for Ecology and Hydrology, Wallingford, UK
[2]Ex-Institute of Hydrology, Wallingford, UK
[3]Lancaster Environment Centre, University of Lancaster, Lancaster, UK
[4]IAHS Press, Wallingford, Oxfordshire, UK
[5]Centre for Ecology and Hydrology, Wallingford, Oxfordshire, UK
[6]Department of Meteorology, University of Reading, Reading, UK

Progress in Modern Hydrology: Past, Present and Future, First Edition. Edited by John C. Rodda and Mark Robinson.
© 2015 John Wiley & Sons, Ltd. Published 2015 by John Wiley & Sons, Ltd.

4.1 Introduction

Soil hydrology research at the Institute of Hydrology (IH) at Wallingford began with quite modest ambitions, limited to making direct measurements of soil water storage for catchment water balance calculations (see Chapter 2). Traditional methods to determine soil water content by gravimetric sampling were not suitable. They are destructive, time-consuming and suffer from poor time resolution. The neutron scattering technique which was becoming established in the 1960s offered a new approach. It involved using an artificial radioactive source to supply fast neutrons which are slowed down or moderated in the upper soil, mainly by the hydrogen present in water. The number of moderated neutrons detected by the probe can thus be related predominantly to soil water content (Chapter 2).

Commercially available designs of neutron probe were unsuitable for routine field applications and so IH designed its own probe ('The Wallingford Probe') which enjoyed worldwide success with a reputation for exceptional reliability and stability.

Field measurements of soil water content have many potential uses beyond merely monitoring changes in water storage of a soil profile. In addition to farmers and growers interested in soil water availability for their crops, they were quickly adopted by river engineers to aid making flood forecasts. Other uses for the neutron probe include measuring the water use of different vegetation types, particularly in comparison of soil water storage under conifer forest versus grass. This was subsequently extended to a range of forest species and crops. The probe could also be used in conjunction with tensiometers to measure soil water processes and hydraulic properties directly in the field. Experimental methods were developed to measure water flux in the unsaturated zone, so enabling direct measurement of evaporation (e.g. crop water use) and recharge to groundwater. The methods were also used to measure the unsaturated hydraulic conductivity and water retention characteristics *in situ*. These techniques were applied at a number of sites to determine water use of agricultural crops, grassland and trees, plus matrix and macropore flow to groundwater in situations where water resources are of crucial importance such as the English Chalk aquifers. By combining water flux measurements with measurement of solute concentration, the flow of agricultural and other chemicals could be measured, enabling the monitoring of pollutant transport to groundwater bodies.

4.2 Soil Water Under Different Land Covers

The neutron probe method has been applied widely to measure the effect of differences in evaporation losses from different vegetation on soil water balances. Examples include trees versus grass (Plynlimon in mid-Wales, Black Wood in southern England), tropical rainforest versus cleared land (Brazil), Eucalyptus plantations (India), sugarcane (Mauritius) and tea plantation (Kenya). The aluminium access tubes for the probe can be installed by manual methods to at least 10 m depth under favourable soil conditions and deep bedrock. Deeper installation is possible with mechanical assistance. Many tubes can be installed to represent soil water changes within a particular landscape element, given precautions to ensure that the soil surface and the vegetation around the tubes are not disturbed by trampling during both installation and observation (see Chapter 2 for more details on the methodology).

4.2.1 *Coniferous forest versus grass*

Detailed soil water studies were conducted at Plynlimon where a series of neutron probe access tubes were installed across the catchments to compare conditions under contrasting land covers of coniferous forest and grassland, and to assist with the 'closure' of the catchment water balances by providing measurements of seasonal changes in catchment water storage. This was important, because the primary purpose of the hydrological research at Plynlimon (Chapter 2) was to estimate a small quantity, that is, evaporation from the difference between two much larger quantities, rainfall and streamflow. The aluminium access tubes were sealed at the bottom, so measurements could be made in profiles within which the water table fluctuates. Once installed, repeated readings can be made in the tube at any time, but a reliable and careful human operator is required as data cannot be logged automatically. The access tubes were installed in a series of downslope lines in the expectation that soil water conditions would be linked to position on the hillslope. In fact, this proved to be incorrect and the neutron probe results, later confirmed by process studies, indicated that patterns of water flow were complex and not necessarily related to surface features (Hudson, 1988). Nevertheless, the work demonstrated clear systematic differences in water content and field capacity between soil and topographic groupings identified by Bell (2005): hilltop peat soils, slope podsols and valley bottom mires. At Plynlimon, blanket peat, often degraded by erosion, tends to occupy the interfluves, whereas gentle slopes with slowly permeable subsoils, especially glacial till, are covered by more minerotrophic mire dominated by purple moor grass. The steeper slopes have shallower soils, often unstable, with strongly layered peaty podsols.

The soil water data showed a clear difference between forest soils and soils under grassland. There was a tendency for forest soils to dry out more in summer periods than the soils under shorter vegetation. It is now known that this is due to the generally higher evaporation losses from forest than from grassland resulting from rapid evaporation of water intercepted by the forest canopy. This is a result of advection of energy from air passing over the forest in the windy climate of upland Britain (see Chapter 5). It is also compatible with micrometeorological studies on conifers in Thetford Forest in Eastern England, which revealed enhanced evaporation as a result of rapid evaporation of intercepted rainfall. Early work in the Plynlimon catchments showed that, for most years, change in catchment storage was

not a significant component of the annual water balance, although it could become important in extreme years (Hudson, 1988). Soil water measurements indicated that estimates of annual water use (catchment rainfall minus streamflow) for the grassland Wye catchment agreed well with Penman potential evaporation during average years, which was subsequently confirmed by eddy covariance measurements (Marc and Robinson, 2007).

4.2.2 Broadleaved woodland and grass

Although there have been consistent results indicating that upland coniferous forests in Britain result in higher evaporation losses than from short vegetation, there have been uncertainties whether this may also apply to deciduous woodlands in lowland areas. A study was conducted on broadleaf woodland at Black Wood and nearby grassland at Bridgets Farm in southern England overlying a chalk soil (Roberts and Rosier, 2005a). This study combined soil water content measurements to 9 m depth with measurements of evaporation by eddy covariance. They concluded that whilst there were seasonal differences in the amount of soil drying, on an annual basis differences in the overall water balance were in fact small. They pointed out that a number of earlier studies, which found higher water use by broadleaf woodland than grassland, were on soils that had less capacity to support transpiration to the grass during drought conditions than the chalk subsoil at Black Wood, and had also been conducted in small woodlands or at woodland edges, which might therefore have experienced enhanced evaporation losses (Roberts and Rosier, 2005a, 2005b).

4.2.3 Tropical rainforest and cleared land

Following the success of the field studies in the United Kingdom, attention turned to investigating the impact of land use changes in other countries, particularly to address concerns where native forests were replaced by farming and by commercial plantations.

One such study was the Anglo Brazilian Amazonian Climate Observation Study (ABRACOS). A part of this work involved the collection of soil water and climatic data from undisturbed rainforest and cleared (pasture) areas to provide more accurate data for the land surface sub-models of general circulation models (GCMs) used in climate prediction. Pairs of forest and pasture sites were selected in Eastern, Central and Western Amazonia and equipped with an automatic weather station to monitor climate, eddy covariance devices to measure water vapour, and neutron probe access tubes and tensiometers to measure soil water. The soil water sites included plateau, slope and valley floor locations.

The measurement sites were preselected on micrometeorological, rather than soil, criteria as they had to be fairly level, with adequate fetch over the same type of vegetation for valid atmospheric flux measurements. The main site in Central Amazonia was at Fazenda Dimona, about 75 km north of Manaus. The pasture site there had been created in 1980 by felling the trees, then burning the trunks and debris in the dry season and sowing grass. A second set of measurements were made in an adjacent forest site, with minimal differences in rainfall and soil type, so that water storage changes under the two vegetation types were directly comparable. Fifteen access tubes were installed along a transect in the direction of the prevailing fetch of the micrometeorological site to represent the plateau, slope and valley floor under both land covers. Nests of 10 mercury manometer tensiometers were also installed (0.2–2.4 m depths), one each on the pasture and forest plateau, and one on the forest valley floor (maximum depth 1.5 m) (Fig. 4.1).

Suitable areas with immediately adjacent pasture and forest unfortunately could not be found for sites in Eastern and Western Amazonia; the sites eventually selected for micrometeorological study were more than 100 km apart. Consequently, less ambitious soil water instrumentation was installed near Marabá in Para (Eastern Amazonia) and near Ji-Paraná in Rondônia (Western Amazonia). This comprised 6–8 access tubes and a nest of tensiometers sited close to the micrometeorology tower, on plateau or gently sloping sites (Figs. 4.2 and 4.3).

Previous work near Manaus (Nepstad *et al.*, 1994) had indicated that forest water uptake from >1 m depth was unlikely. Accordingly, 2 m was selected as the maximum reading depth for the neutron probe, but measurements quickly showed significant changes at this depth indicating that the forest was taking up water from greater depth (Hodnett *et al.*, 1995), and hence deeper access tubes to 3.6 m were installed at all sites.

Early results indicated that drainage from the oxisol soils at the Central Amazonian site was

very rapid, and that water availability was low; neither are properties normally associated with very clayey (>85%) soils. These soils have strong aggregation and interpedal pore space leading to rapid bypass flow after rainfall events (Renck and Lehmann, 2004). Hodnett *et al.* (1995) examined the depth distribution of storage changes, the spatial variability between access tubes, and the causes of this variability for the first 17 months of data from the Manaus (Central Amazonia) forest and pasture. The mean monthly rainfall in the dry season exceeds 80 mm, and monthly potential evaporation is ~110 mm. Large peaks in soil water storage were not observed even soon after heavy rainfall, indicating that drainage was very rapid. The variability of water content between the tubes in the pasture was greater in the wet season, mainly due to rainfall redistribution by localised surface runoff. This seasonal change in variability was not seen in the forest, where there was no surface redistribution.

Hodnett *et al.* (1996) analysed the seasonal water content changes for all three site pairs. The average range at Manaus was 132 mm (pasture) and 200 mm (forest). Soil water storage changes were much larger at the other paired sites because of a longer and more pronounced dry season, and also higher available water capacity at those sites. However, at all three sites, storage changes under pasture were less than under forest. Interestingly, at the Eastern Amazon (Marabá) forest site, the profile soil water storage did not return to a typical wet season value (field capacity) every year, showing that even in the humid tropics, soil water deficits can be carried over from year to year.

4.2.4 Water use of Eucalyptus in India

Eucalyptus accounts for about half of all plantation forestry in the tropics and subtropics. It provides a fast-growing source of timber, firewood and pulp and can help reduce the pressure on indigenous forests as wood sources. They have, however, been claimed to cause excessive transpiration losses, leading to 'mining' of soil water and groundwater. Studies at low rainfall sites in India (Calder *et al.*, 1997) found that on medium depth soils, the water use of eucalyptus was similar to that of indigenous forest and approximated the annual rainfall.

In contrast, on deep soil sites (>8 m) water use over three years of measurement was greater than the rainfall. Deep neutron probe access tubes (>7 m)

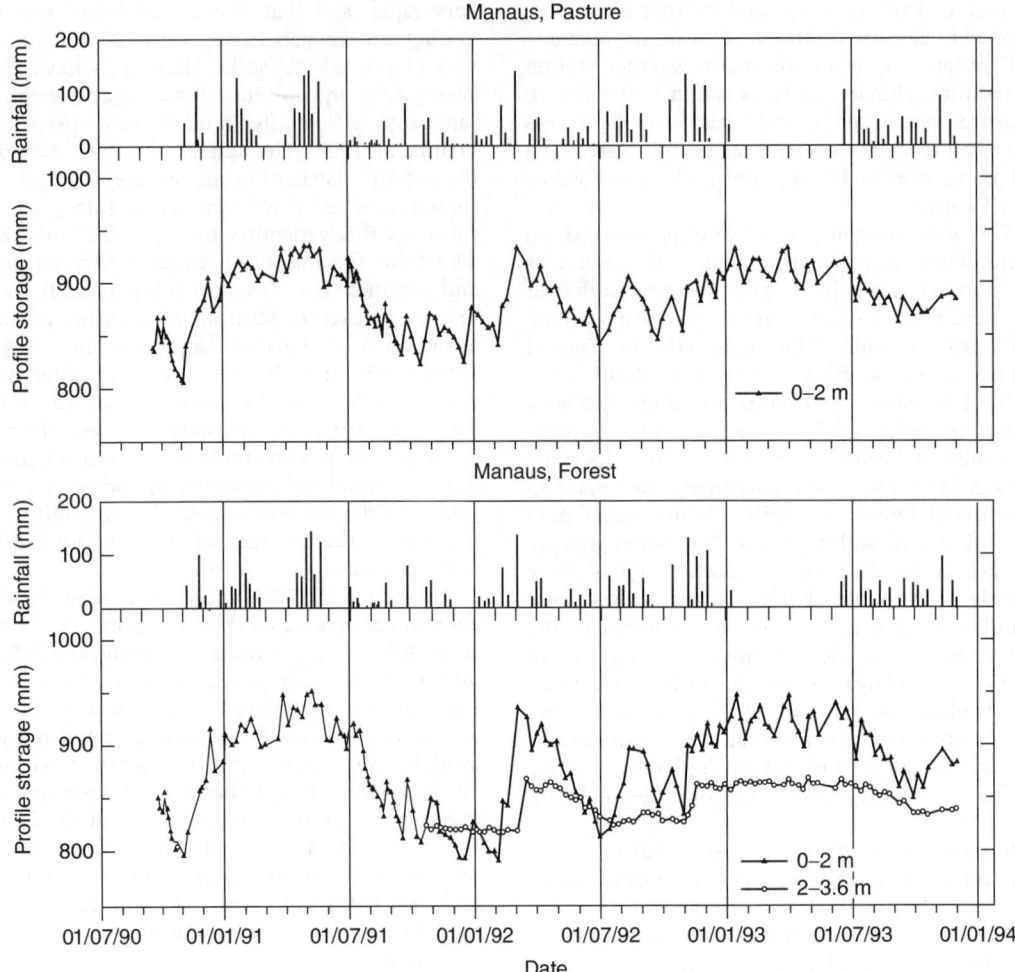

Fig. 4.1 Manaus: Pasture (top) profile storage in top 2 m. Forest (bottom) profile storage in top 2 m and 2–3.6 m depth. (Source: Hodnett *et al.*, 1996. Reproduced with permission of CEH.)

were installed to measure soil water contents to investigate this phenomenon. The measurements revealed that over time the eucalyptus was extracting water from progressively deeper layers of the soil. Together with observations from deep soil pits, the eucalyptus roots were estimated to be extending downwards by more than 2 m per year. This finding contradicted the local belief that eucalyptus roots only extend to about 2.5 m depth, and raised concerns about the validity of other reported studies of their water use; if the full rooting zone depth had not been monitored, then the

reported soil water depletion could have seriously underestimated the actual total water use.

4.2.5 Drip irrigation of sugar cane in Mauritius

Agriculture competes with industry and domestic requirements for finite water resources. Irrigation by overhead spray or furrow irrigation is wasteful of water and energy, and both tend to have deleterious side effects. For these reasons, drip irrigation is becoming increasingly important as the control of water applications can be very precise and has

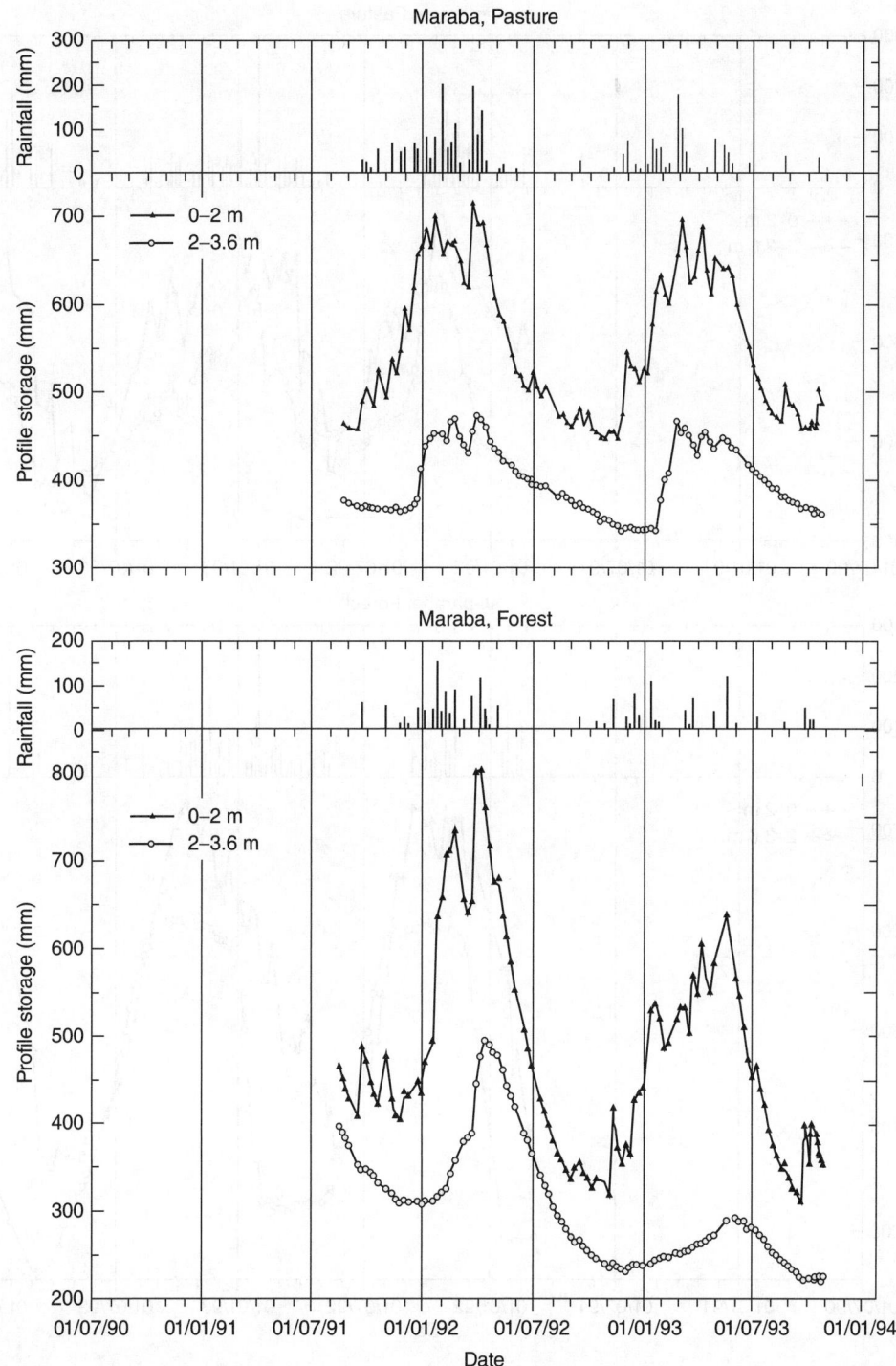

Fig. 4.2 Marabá: Pasture (top) profile storage in 0–2 and 2.0–3.6 m layers. Forest (bottom) profile storage in 0–2 and 2.0–3.6 m layers. (Source: Hodnett *et al.*, 1996. Reproduced with permission of CEH.)

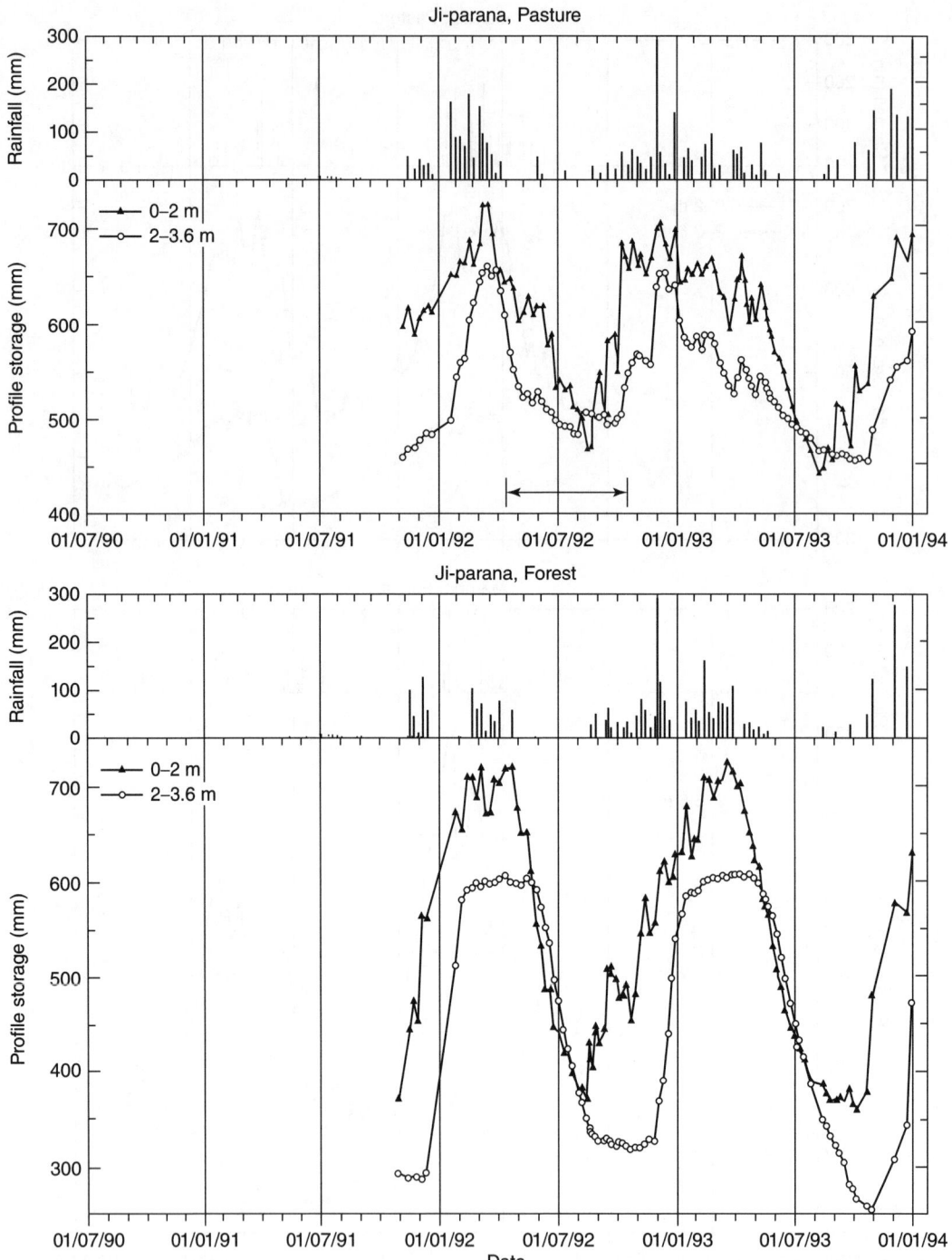

Fig. 4.3 Ji-Paraná: Pasture (top) profile storage in 0–2 and 2.0–3.6 m layers. Forest (bottom) profile storage in 0–2 and 2.0–3.6 m layers. (Source: Hodnett *et al.*, 1996. Reproduced with permission of CEH.)

Fig. 4.4 Photo of each array before crop emergence – sugar cane can be seen in the distance. (Source: Reproduced with permission of CEH.)

a low energy and labour requirement. The method also permits precise inputs of fertiliser directly into the root zone.

A major project was carried out in Mauritius between 1982 and 1988 to assess the viability of drip irrigating sugar cane to reduce water consumption and energy requirements. Drip irrigation is particularly suited to cane growing in Mauritius because the crop is replanted only every 10 years or so, compared with 2 or 3 years in most other cane growing areas. After harvest, the cane regrows from the old roots and each regrowth is known as a 'ratoon'. The long period between replanting favours the use of buried drip lines, which would otherwise be destroyed by ploughing for replanting.

The project was performed within a newly cleared area. The drip lines and associated controls, together with soil the physical instrumentation, were installed before planting of the cane in April 1983. Fourteen plots, each representing a different treatment, were set up within the 10-ha experimental area. Each was replicated randomly within the trial area. Data were collected continuously from the first crop and four subsequent ratoons.

The instrumentation was designed to study the distribution of soil water content and soil matric potential in the upper 1.5 m. of the soil in relation to drip line position, and cane row spacing in the different treatments over the crop cycle. In each treatment, a 6×6 array of manually read mercury manometer tensiometers was installed in a cross section of the soil profile at right angles to the crop row. Close to the tensiometers was a parallel array of 6 neutron probe access tubes to measure soil water content.

This instrumentation was installed prior to the planting of the new crop. Figure 4.4 gives some idea of each array before crop emergence, it not being possible to photograph much once the cane had grown up. Over 500 tensiometer and soil water content readings were taken daily throughout the growing season for five consecutive years. The methods used to measure soil water content and potential were not in themselves an advance. The advance lay in revealing the size and dynamics of the wetted zone along the drip line and the interaction of the sugar cane crop with the soil and the irrigation applications. These data were used to optimise water use while maintaining the best

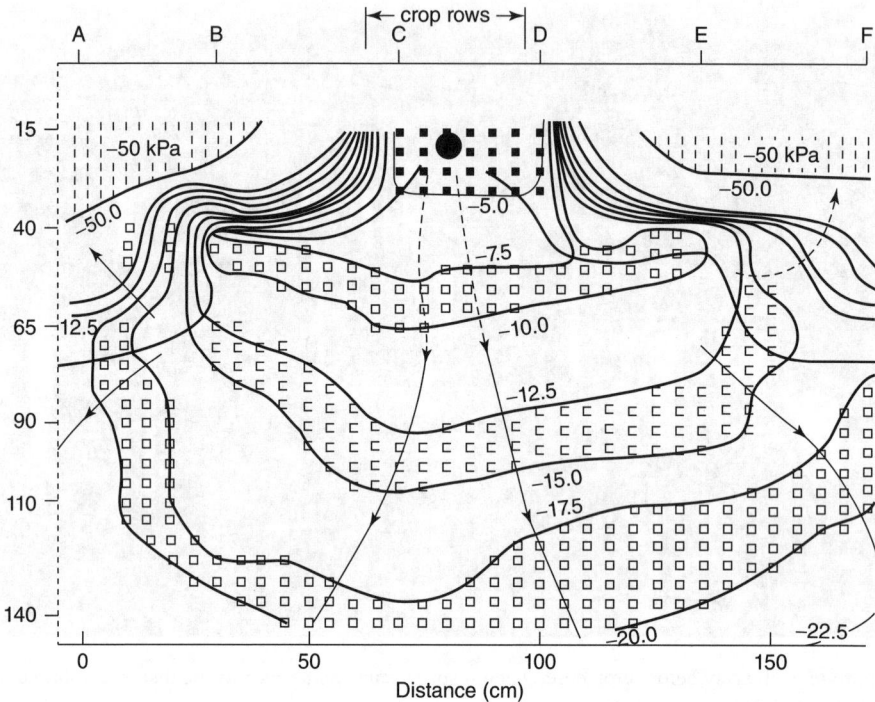

Fig. 4.5   2-D distribution of soil water potentials (kPa) beneath a drip irrigated row crop, based on tensiometer profiles at A–F. across the dripline (●). (Source: Reproduced with permission of CEH.)

growing conditions and yields. During this period, the concept of soil water status for characterising the 2-D distribution of soil water conditions beneath a drip irrigated row crop (Fig. 4.5) was developed (Bell *et al.*, 1989).

In commercial agriculture, such intense instrumentation would be impractical, but once the interactions between root uptake and water inputs from the drip line were understood it would be possible to control water applications by means of a relatively small number of 'Index' tensiometers located at key positions (Hodnett *et al.*, 1989a). A system comprising a few index pressure transducer tensiometers linked to an automated irrigation control system would aid optimal irrigation for that particular crop, soil and climate situation. In the case of regularly spaced crop rows with drip lines beneath the row, two or three tensiometers at a depth of 0.65 and 0.25 m from the drip line could be sufficient to regulate the water input without the need to measure rainfall, or estimate evaporation losses using crop factors.

The aim was to maintain a wet zone at a certain size based on a target potential (in this case, −8 kPa). The amount of water to apply was determined using simple rules, requiring only the previous day's index readings and the amount of input on the previous day. The index readings showed whether the wetted zone around the drip line was increasing or decreasing in size and the application rates were adjusted to keep the index potential as close to −8 kPa as possible.

By this means, losses by drainage to groundwater could be minimised and the duration of saturated conditions would be brief, avoiding the creation of anaerobic conditions that inhibit root function. Yields and water use efficiencies using this approach were very good. Findings are presented by Ah Koon *et al.*, (1989); Batchelor *et al.*, (1989); and Hodnett *et al.*, (1989b).

4.2.6   *Soil water databank*

Field measurements of soil water contents have potential uses in many contexts. Whilst it is possible to take measurements in a soil profile at a given point, obtaining measurements to give a sense of how soil water content changes in time and space in a selection of soils was, until more recent developments in remote sensing of soils, problematic. In the late 1970s, soil water content measurements were required to evaluate a model newly developed by the UK Met Office that provided weekly estimates of precipitation, evaporation and soil water deficit for a 40 ×40 km grid across Britain. An important study involved cooperation with most neutron probe users in Britain to create a database of soil water measurements in a wide range of soils and land use environments for comparison with soil water deficits predicted by the MORECS model (Meteorological Office Rainfall and Evaporation Calculation System, Thompson *et al.*, 1981).

The model was an ambitious development from the former Estimated Soil Moisture Deficit bulletin (ESMD bulletin of Grindley, 1967), and its aim was to provide information to water authorities, flood forecasters, river engineers, agricultural advisers, farmers, growers and others for the monitoring and prediction of river flows, evaporation losses, planning and understanding drainage, subsidence and drought issues for specific soils and crops within each square, and also a real land use composite for the square.

Evaluation of MORECS soil water deficit estimation required data measured over extended periods from a wide range of soils in different locations. The IH was aware of many organisations using a neutron probe for various purposes across the country. Accordingly, in 1979 a project was initiated to identify what data were available and, if the originator was amenable, to transfer a copy to a soil water databank at the IH for use in evaluating MORECS. Invitations were issued to about 60 organisations in Britain known to own a neutron probe, inviting them to contribute data to the databank. The response was very positive and a programme of visits was conducted to assess the sites from which data were available.

It was soon apparent that different users had different protocols regarding use of the probe, for example, whether or not there was a calibration, and how measurements near the surface were dealt with. To circumvent such problems, the calibrations that had been established by IH for clay, loam and sand soils in Britain were applied to all the neutron count data after standardising with water

counts for the probe used. Because the focus was on differences in soil water content – a soil moisture deficit (SMD) is the difference between the given water content of the soil profile and the field capacity water content of that profile – rather than absolute soil water content, this was acceptable. It was necessary to determine the field capacity water content of each soil. In MORECS, field capacity is conceived as the water content when the soil profile has drained freely for two days after saturation; the soil profile depth is not defined and does not need to be, as the deficit calculated is the net balance of fluxes into and from the soil. The field capacity water content of the field soils was defined as the lowest decile of profile water content measured in January–March each year, that is, it assumed that for 90% of the time during those winter months soils were wetter than field capacity. A standard profile depth of 1 m was used.

The data were transferred to Wallingford providing a total of 104 sites mostly having two or more individual access tubes. It included data for different crops/vegetation and soils from sites across Britain (Fig. 4.6; Gardner, 1981a). Necessarily, much of the data pre-dated MORECS. To undertake the evaluation, MORECS had to be run in retrospect for these earlier periods. In practice, MORECS was run for the meteorological station (i.e. a station used in the operational MORECS) and the raingauge located closest to a given soil water measurement site, that is, there was no spatial interpolation element in the estimates and only data from sites with a grass crop were used, yielding a set of 50 grass sites. The deficits estimated by MORECS were compared against the measured deficits using the root mean square of the difference (RSMD) between the measured and estimated deficits and the bias (Gardner, 1983). Figure 4.7 illustrates the results for four sites on different soils; it is clear that the trend of the estimates replicate the measured data and that the periods of no deficit correspond at each site. The mean RSMD and bias across the 50 sites were 29 and −16 mm, the negative bias indicating a tendency for overestimation of the deficit by MORECS.

These results gave confidence to the developers of MORECS and users of the deficit estimates that, whilst in absolute terms they might not be truly accurate, they did reflect what was happening on the ground. Development of MORECS continued subsequent to this project (Hough and Jones, 1997) and MORECS has operated continuously since, still providing real-time data on evaporation and

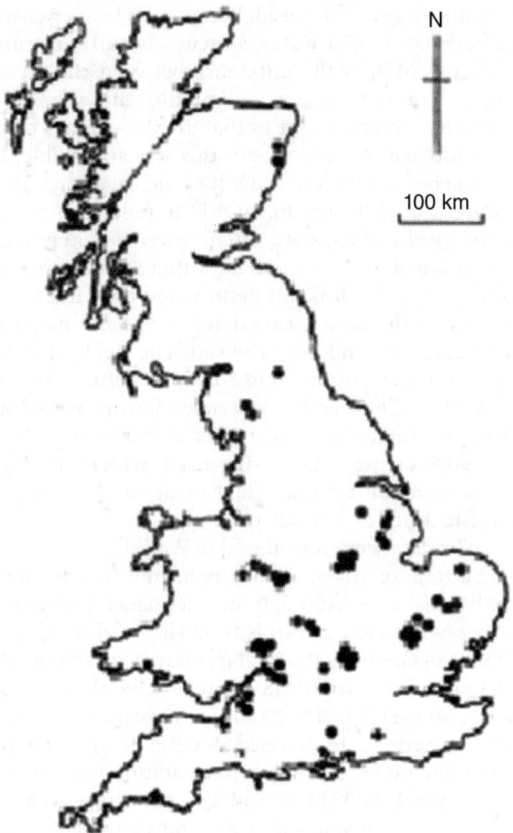

N

100 km

Fig. 4.6 Location of the soil moisture databank sites. (Source: Gardner 1981a. Reproduced with permission of CEH.)

soil water status, information important for a large number of users (http://www.metoffice.gov.uk).

The soil water databank, which includes more data than used for the MORECS assessment, was archived on computer tapes and in 2011, some 25 years later, it was resuscitated for a project, not dissimilar to the MORECS, that required soil water data to evaluate soil water estimates across the UK. From 2015 the data are freely available from the CEH website.

4.3 Unsaturated Zone Water Balances

A number of key studies of unsaturated zone water balances in the UK were undertaken, with particular attention to the Chalk, which is the country's most important aquifer. These investigations included a study of the effect of land cover on the water balance of an area where annual rainfall and evaporation were similar (Thetford), and a study using soil physics methods as well as a large lysimeter to assess water movement under controlled conditions (Fleam Dyke).

With the availability of soil water content measurements as a function of depth and time using the neutron probe, it was realised that, in combination with water potential measurements using tensiometers, the soil water fluxes could also be determined. The methodology chosen was the *Zero Flux Plane* (*ZFP*) method (see Fig. 4.8), which allows for independent measurement of evapotranspiration and drainage from a soil profile without requiring information on the hydraulic properties of the soil. The ZFP method is applicable only to periods in which net evapotranspiration is greater than rainfall, that is, late spring, summer and early autumn in temperate regions. For periods when no ZFP could be observed because drainage occurs throughout the profile, evapotranspiration was estimated from meteorological data to compute the net flux at the soil surface and drainage from the profile was calculated from a simple water balance using this and water storage changes. This is less restrictive than it appears, particularly in areas of relatively low rainfall, as evaporation at such times is usually quite low and the soil is sufficiently wet as not to restrict water uptake by plant roots, so that meteorologically based estimates of evaporation are expected to be reasonably accurate, and any errors will be fairly small.

4.3.1   *Thetford Forest*

The site chosen for initial work was the existing IH experimental site in Thetford Forest in Norfolk (Cooper, 1980). As well as benefiting from the existing infrastructure associated with the ongoing micrometeorological studies (see Chapters 2 and 5), the opportunity to make comparisons with independent measurements of evaporation was a major bonus.

The soil at Thetford is extremely heterogeneous, comprising a shallow layer of wind-blown sand over Chalk. Cryoturbation and later weathering produced a mixing of the sand and chalk and podsolisation with a very irregular, clayey B_t horizon.

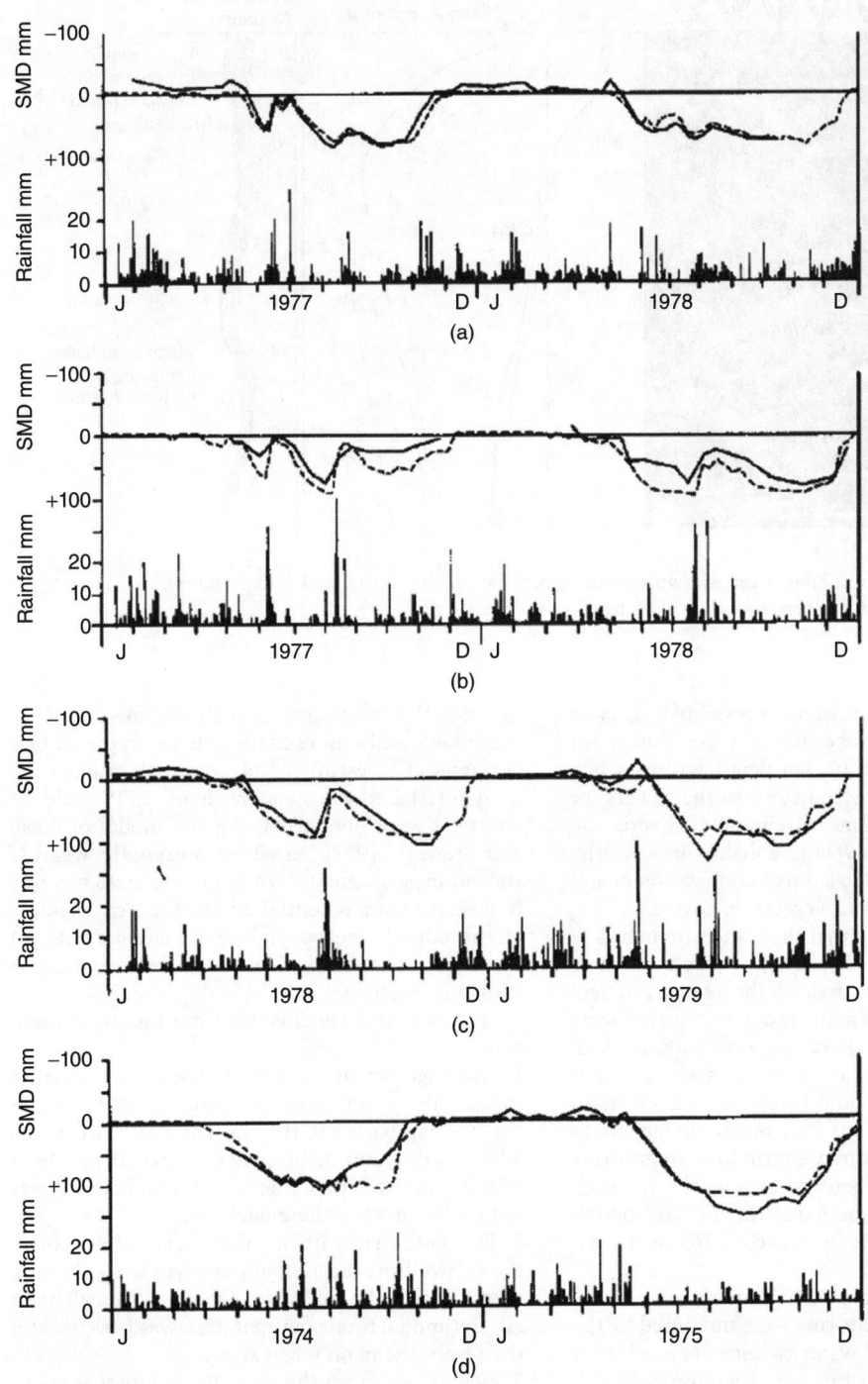

Fig. 4.7
Comparison of
MORECS SMD
estimates (- – -) and
measured SMDs (—)
for four sites on
different soils with
grass vegetation for:
(a) sandy soil in
Staffordshire, RSMD
13.4, bias −5; (b)
loam in
Gloucestershire,
RMS 38.5, bias
−32.5; (c) clay in
Gloucestershire,
RSMD 27.0, bias
−17.8 and (d) chalk
in Lincolnshire,
RSMD 18.4, −1.3.
Deficits estimated
by MORECS were
compared against
the measured
deficits for four sites
on different soils.
(Source: Gardner &
Field 1983.
Reproduced with
permission of
Elsevier.)

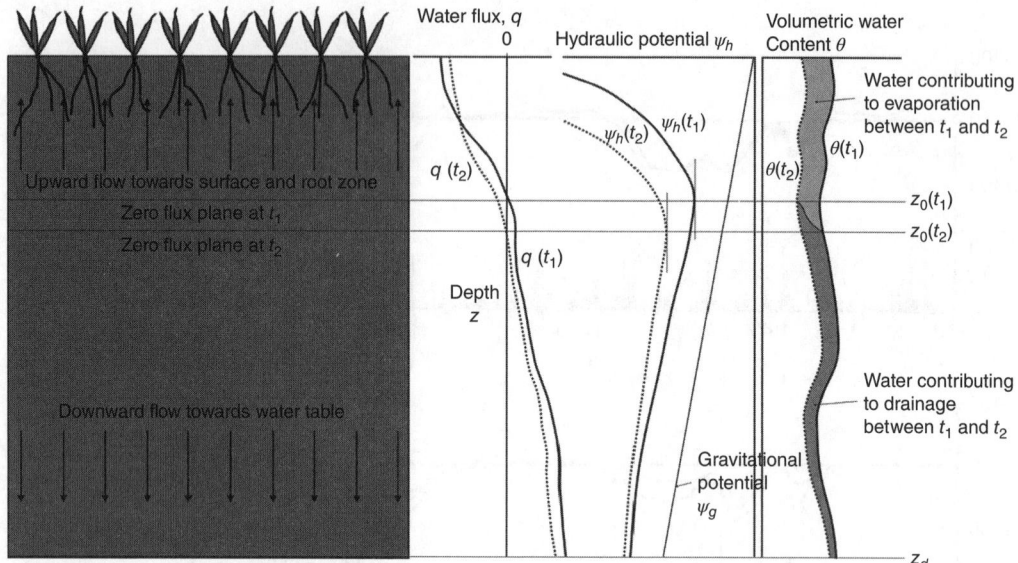

Fig. 4.8 Principle of the ZFP method, showing water flux, hydraulic potential and water content changes as a function of depth. (Source: Reproduced with permission of CEH.)

Four replicate micro-sites were established, each with four neutron probe access tubes and a set of tensiometers to about 3 m depth within a few metres of one another, and close to the towers for micrometeorological measurements (Chapter 5). A further similar site was established in a nearby grass clearing to provide a direct comparison of soil water under the different vegetation covers.

It was soon realised that these measurements of soil water fluxes could be used to assess recharge to groundwater in an area of the country where the magnitude of rainfall and evaporation were similar and land use effects may be quite critical. In such places even small errors in assessing either rain or evaporation would result in large errors in estimated recharge. This lent much sharper focus to the work and required continuous monitoring with a frequency of about twice a week; the work was conducted over a period of three years, including the severe drought from early 1975 to August 1976.

Drainage and evaporation from the forest site and the nearby grass clearing were measured by the combination ZFP and water balance approach over a period of three years from the beginning of 1974. Additionally, a second forest site (Feltwell) was

measured for one year. This had a shallower (~10 m) water table and a more chalk-rich soil in the surface horizons. The estimates of potential evaporation used for the forest sites when no ZFP could be observed were obtained using the model of Gash and Stewart (1977), based on automatic weather station measurements, while for the grass clearing it was Penman potential estimates. An example of the cumulative water balance components for the main forest site during 1976, a very dry year, is shown in Figure 4.9.

The principal conclusions from the experiment were:

1 Good agreement was found between evaporation measured by soil physics using the ZFP–water balance approach and the Gash and Stewart model when soil water deficits were less than about 60 mm. Beyond this, measured evaporation was reduced relative to the model.

2 The more chalky nature of the near-surface soil at the Feltwell site meant that there was less reduction of evaporation by soil water deficits, although overall, the annual total of evaporation was little greater than from the main forest site.

3 Evaporation from the main forest site was about 49% greater than from the grass clearing, and

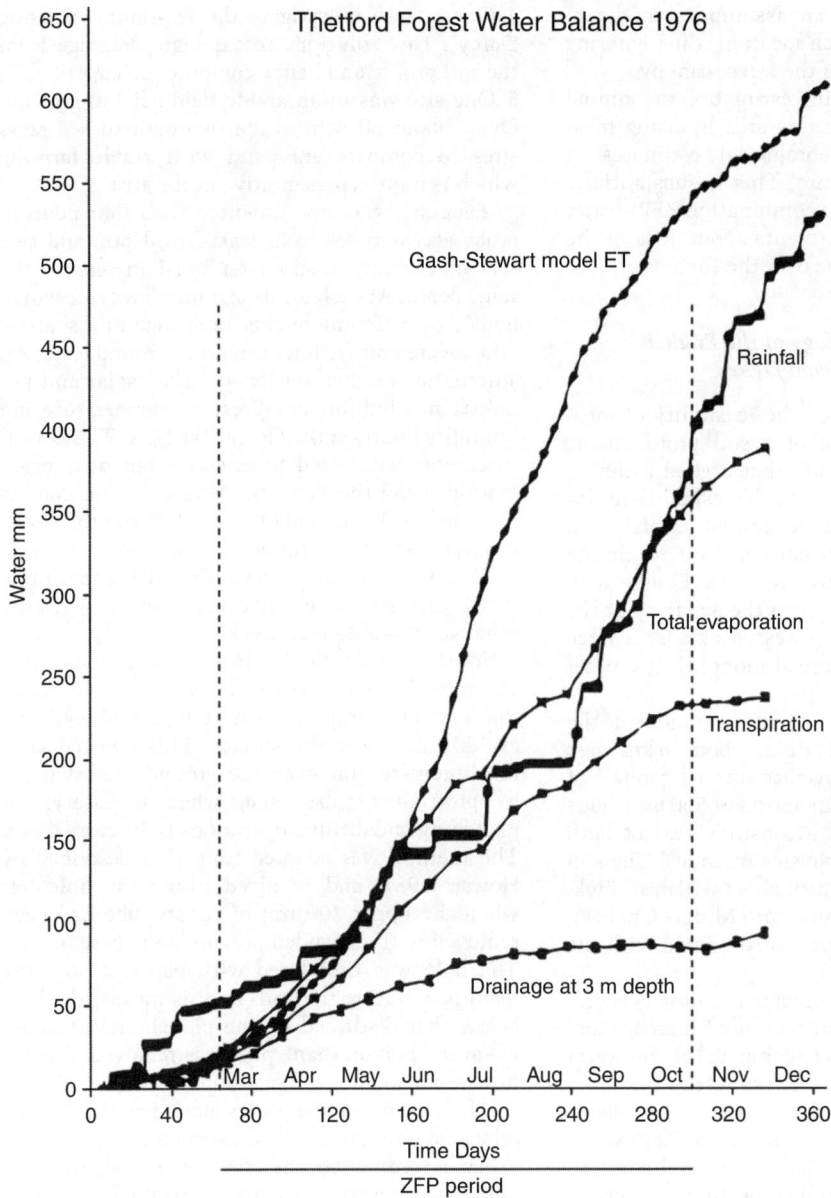

Fig. 4.9 Cumulative components of the water balance for a stand of Scots Pine in Thetford Forest during the very dry year, 1976 (Cooper, 1980). Total evaporation (transpiration plus evaporation of intercepted water) was predicted well by the Gash and Stewart (1977) model until about Day 155, after which increasing water stress limited transpiration. Note the large contribution of interception (the difference between total evaporation and transpiration) to the overall evaporation. (Source: Cooper 1980. Reproduced with permission of CEH.)

drainage from the soil at 3 m depth was 44% less as a consequence. The largest contribution to this difference was evaporation of intercepted precipitation, which comprised 41% of forest evaporation.

4 A substantial amount of drainage from both forest (43%) and grass (19%) sites occurred during periods when there was a soil water deficit. This finding conflicts with established concepts of field capacity.

5 Differences of annual totals of evaporation or drainage measured using replicate sets of instruments in both forest and clearing were about 20 mm, despite extreme heterogeneity of the soil.

This is due in part to an assumption of equal precipitation input to each location, thus ignoring the redistributive effect of the forest canopy.

6 The probable error in estimates of annual groundwater recharge for this area by using measured rainfall and meteorological estimates of evaporation is about 50 mm. This is substantially greater than that using the combination ZFP–water balance approach and represents about 30% of the average measured recharge over the three years.

4.3.2   Soil hydrology of the English Chalk: Fleam Dyke

The Thetford study proved the feasibility of measuring the water balance of a soil profile using soil physics techniques, and that a good estimate of groundwater recharge can be established for areas of the country where annual rainfall and evaporation are closely balanced. As a result the Department of the Environment (DOE) agreed to fund an investigation in which the accuracy of the method could be further assessed against a large (5 m × 5 m × 5 m) undisturbed monolith lysimeter (Kitching and Shearer, 1982).

The site chosen was in the grounds of the Fleam Dyke Pumping Station, about 6 km east of Cambridge, with an average annual rainfall of 600 mm and potential evaporation of 560 mm. This was expected to provide a sensitive test of both meteorological and soil physics methods. The soil was a calcareous brown earth about 300 mm thick, grading over another 200 mm into Middle Chalk.

A total of six sites were instrumented with the following five objectives:

1 One was within the lysimeter to allow evaporation to be calculated from measured drainage and water storage changes and to compare the soil water status within and outside the lysimeter.

2 Two were 30 m from the lysimeter and about 16 m apart to compare the combination ZFP–water balance method with the lysimeter results and to investigate the reproducibility of different sets of observations in close proximity to one another.

3 One was on undisturbed grassland (a golf course) 5 km from Fleam Dyke with very similar soil and climate conditions and another (the Stud Farm) was also on grassland a similar distance from Fleam Dyke, but with a much thicker and clay-rich soil (1.2–2.2 m thick) to investigate the spatial variation of water balances at similar sites in the neighbourhood.

4 One site to investigate the feasibility of using Darcy's Law solutions to calculate drainage from the soil profile and hence compute recharge.

5 One site was in an arable field adjoining Fleam Dyke, about 60 m from the two undisturbed grass sites to compare grassland with arable farming which is more representative of the area.

Each site was instrumented with four neutron probe access tubes to at least 3 m depth, and two sets of mercury manometer tensiometers to the same depth. At each location, rainfall was measured hourly by a tipping bucket raingauge in a shallow grid-covered pit with its rim set at ground level. An automatic weather station recorded solar and net radiation, wind run and direction, temperature and humidity hourly at the Fleam Dyke site. These measurements were used to calculate potential evaporation using the Penman–Monteith equation, as modified by Thom and Oliver (1977) and these values were used in periods when no ZFP was observed.

One neutron probe access tube and the tensiometers at each site were read every weekday, with the other access tubes read weekly.

In the arable field, farming operations were accommodated by access tubes and tensiometers that could be temporarily split at ground level and at 200 mm below the surface. This allowed farm machinery to run over the ground unobstructed by protruding tubes and, when necessary, for ploughing and drilling operations to be carried out. The method was adapted from that described by Howse (1981) and involved filling the hole left when the upper 200 mm of access tube had been removed with expanded polyurethane cast *in situ*. The hole was first lined with paper to stop the foam penetrating the soil. By this means, the hole below that disturbed by the plough stayed intact when the polyurethane plug was removed. Careful measurement from fixed points at the edge of the field allowed the instrument positions to be relocated with about 15 mm accuracy.

To accommodate the growing crop, lightweight metal stands were constructed, on which a portable platform could be placed to allow the operator to stand above the crop. The neutron probe access tubes and tensiometers were similarly extended by about 1 m.

The saturated hydraulic conductivity of the profile to 3 m depth was measured using the method of Poulovassilis *et al.* (1974), which allowed the conductivity to be measured over a range from about 1000 mm/d down to about 0.1 mm/d. This method uses steady irrigation at relatively high rates of water flux, down to about 1 mm/d. Following this, the profile was allowed to drain naturally to attain measurements at flux rates down to about 0.1 mm/d. The whole area of about 10 m square was covered by a temporary plastic shelter to exclude rainfall and suppress evaporation from the soil surface.

Monitoring over a period of three years with a variety of weather conditions showed that:

1 Annual totals of evaporation and drainage between the estimates from the combination ZFP–water balance method and the lysimeter were not distinguishable. There was a small difference in the timing of drainage caused by the fact that saturated conditions must be established at the base of the lysimeter before drainage could be initiated from it. This meant that drainage from the lysimeter ceased in early summer, while it continued at the same depth outside, but the difference was very small.

2 There was minimal difference between the different water balance components calculated from the measurements of the replicate instruments on each site, or between the two replicate sites at Fleam Dyke.

3 There was little difference between the water balance of the Golf Course site and the similar Fleam Dyke grass sites when small differences in annual rainfall (up to 45 mm/year) were taken into account.

4 Evaporation from the arable field followed a different course from that of the adjacent grass site, although there was little difference between the overall annual totals, except for when a crop of peas was grown. This had a longer growing season than the previous and subsequent barley crops. The arable crops evaporated water more slowly in the early part of the growing season when there was incomplete ground cover. Once the crop became better established and its height exceeded that of the adjacent mown grass, the more efficient aerodynamic transfer of water vapour ensured that evaporation from the cropped field became progressively greater. Cumulative evaporation from the arable crop comfortably exceeded that from the grass by mid-June before the crop started to senesce followed by harvest and a dramatic fall in evaporation rate.

5 The deeper soil site showed quite different behaviour from that of the sites with a shallow soil cover. Evaporation was more restricted in dry summer periods, leading to about 50% more annual drainage. Significant drainage from the soil profile continued much longer into late spring and early summer.

6 There was poor agreement between the cumulative drainage calculated from the ZFP–water balance method and Darcy's law solutions based on the separately measured hydraulic conductivity on the same site using the same instruments. Darcy's law estimates were consistently greater than the measured drainage, in one case exceeding the annual rainfall. It was concluded that since the hydraulic conductivity is extremely sensitive to water content or potential, using measured values of hydraulic conductivity with Darcy's law to estimate water flux would not be sufficiently reliable, even for the conditions of this experiment, which were close to ideal.

7 The hydraulic conductivity relation with soil water matric potential showed a progressive transition with depth. Close to the soil surface, there was a steady reduction of hydraulic conductivity with falling matric potential from a value of about 1000 mm/d close to saturation. At depths from 2.0 m and below, there was a very sharp reduction in conductivity from in excess of 1000 mm/d to around 1 mm/d at matric potentials of −5 kPa or above. Below this matric potential, the hydraulic conductivity appeared to remain constant to the limit of tensiometer measurements (about −70 kPa).

The constant value at lower potentials is consistent with reported values of the saturated hydraulic conductivity of samples of English Chalk (1–10 mm/day – Edmunds *et al.*, 1973). Its constant value over a wide range of water potentials is consistent with an air entry value of −10 to −100 kPa suggested by mercury intrusion porosimetry (Price *et al.*, 1976).

The Chalk is known to be broken into roughly rectangular blocks of size varying from a few cm to several tens of cm, generally becoming larger with depth and with fine fractures between them. It seems natural to associate the rapid reduction of hydraulic conductivity close to saturation with flow within this system of fine fractures. Analysis of the rate of change of conductivity with matric potential, assuming that they are planar voids, produced an estimate of the width of the cracks centred on 50 μm.

The more gradual reduction of hydraulic conductivity higher in the profile is more reminiscent of the characteristics of normal soil and is interpreted as being due to weathering processes which have partially broken the Chalk blocks down into individual particles leading to a wider range of pore sizes.

Similar experiments were carried out in other settings, described in less detail below. All sites were grass-covered and regularly cut, unless otherwise stated.

4.3.3 Bridgets Farm

A series of experiments similar to those conducted on the Middle Chalk of Cambridgeshire was carried out at an Upper Chalk site at Bridgets Farm in Hampshire (Wellings and Bell, 1980; Wellings, 1984a,b). The soil was a similarly thin brown earth rendzina and many features of the soil's hydrological behaviour were very similar. In particular, the saturated hydraulic conductivity versus matric potential curve showed the same rapid reduction close to saturation followed by a near constant value to low matric potential. The Chalk matrix exhibited a higher hydraulic conductivity of between about 2 and 6 mm/d than at the Cambridgeshire sites. As a consequence, and in spite of a somewhat higher annual rainfall than in Cambridgeshire, flow in the fractures rarely occurred and matric potential over the winter period rarely rose above −20 kPa, which is well within the range in which flow is expected to be wholly within the chalk matrix.

At Bridgets Farm, the water balance studies were accompanied by a solute balance study, with a focus on nitrate. This was in response to rising concerns about pollution of groundwater supplies by nitrates derived from agricultural practices.

The low matric potentials often experienced as a consequence of the small and quite uniform pore size distribution of the Chalk matrix precluded non-destructive methods of pore water sampling. The Chalk was therefore sampled by collecting cores using hand-driven core sampling tubes and the solution extracted by centrifugation with an immiscible fluid (Kinniburgh and Miles, 1983).

Wellings (1984b) also backed up the nitrate measurements with artificially applied deuterium. After 2 years, a clear seasonal pattern of deuterium concentration with depth could be seen, supporting the view that piston-like displacement of the tracer occurred, with no evidence of the forward tailing which would be expected if there were a significant amount of flow through the fractures (Wellings, 1984b).

Low potentials in the summer months meant that conventional tensiometers could not be used to measure water potentials in this period, nor to detect a ZFP, nor to measure the hydraulic gradient. However, individually calibrated gypsum resistance blocks were found to be adequate in this calcium-rich environment to measure all three.

As at Fleam Dyke, there was little difference between annual totals of evaporation from a barley crop and grass, although the seasonal pattern was different (Wellings, 1984a).

4.3.4 West Ilsley and Warren Farm

To investigate the applicability of the results from the Cambridgeshire and Bridgets Farm sites to other areas of the English Chalk outcrop, and especially to improve understanding of the transport of nitrates from artificial fertiliser towards the water table, a site was set up near West Ilsley in Berkshire on another thin soil over Upper Chalk in association with the then ICI Agricultural Chemicals Division. In addition to neutron probe access tubes and tensiometers, artificial tracers in the form of deuterium, chloride and N-15-enriched calcium nitrate fertiliser were applied to a grass sward. The unsaturated hydraulic conductivity of the profile was measured *in situ* by the instantaneous profile method (Watson, 1966).

The hydraulic conductivity measurements were similar to those found at the Cambridgeshire and Bridgets Farm sites, with a rapid reduction in conductivity from a high value near saturation to a much lower one of a few mm/d at −5 kPa, which was essentially constant to values beyond the tensiometer range (Gardner *et al.*, 1990; Cooper *et al.*, 1990).

A nearby clay-with-flints covered site at Warren Farm, 13 km distant, with a soil thickness of 1.0–1.8 m, showed similar soil hydrological behaviour to that at the Stud Farm Site in Cambridgeshire, with more restriction of evaporation by the soil than on the thin soil at West Ilsley, leading to greater annual drainage, more evenly spread over the year. Quantitative comparison is made more difficult by a *ca.* 70 mm difference in annual rainfall between the two sites (Cooper *et al.*, 1990).

4.3.5 Shallow unsaturated zone behaviour of Chalk in southern England

All the sites on thin soils (Fleam Dyke and Golf Course, Cambridgeshire; Bridgets Farm, and West

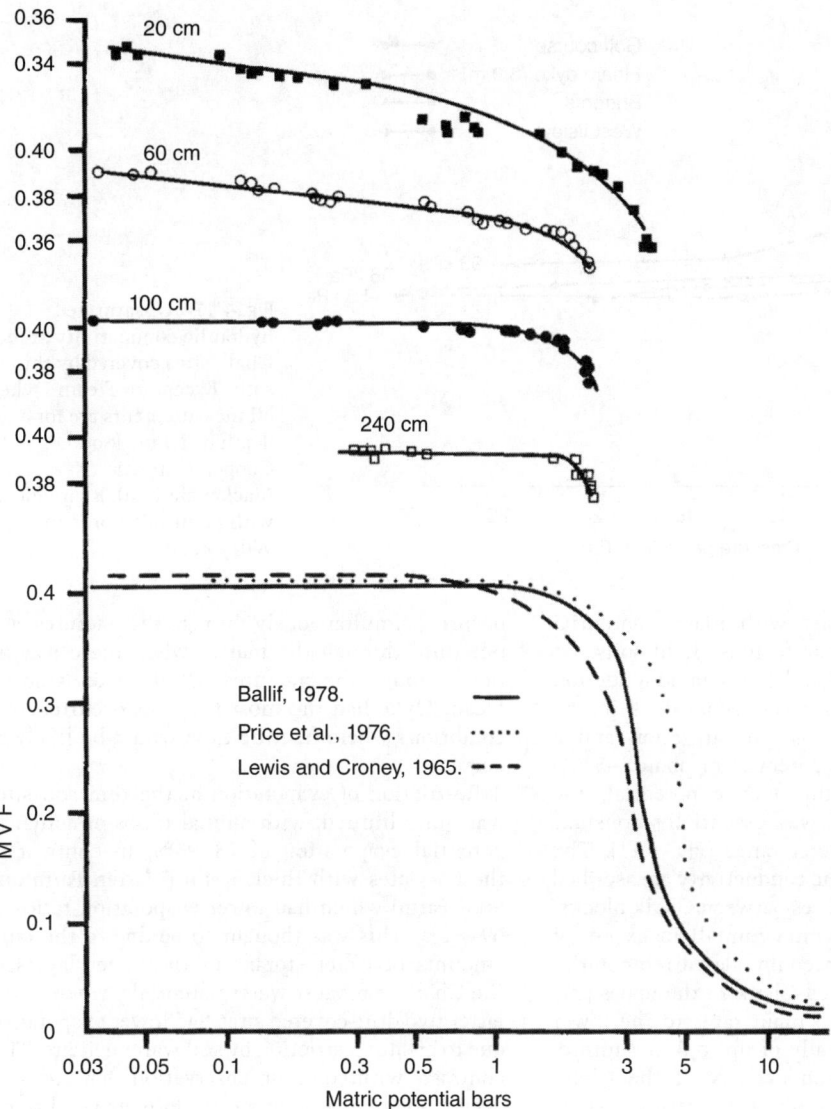

Fig. 4.10 In situ measurements of soil water characteristics for four depths at Bridget's Farm, Hampshire, showing a transition from typical soil behaviour at shallow depth to undisturbed Chalk at deeper depths. Published soil water characteristic curves for Chalk samples determined in the laboratory by other workers are shown for comparison in the lower graph. (Source: Wellings & Bell 1980. Reproduced with permission of Elsevier.)

Ilsley) had brown earth soils of thickness not more than 300 mm overlying weathered Chalk, which graded into apparently unweathered blocky Chalk at depths between 0.8 and 2 m.

Results from the Cambridgeshire, Hampshire and Berkshire Chalk sites show several consistent features:

1 On these sites, the hydraulic properties of the shallow subsurface horizons showed a gradual transition from those typical of most soils to a more extreme behaviour at depth. Close to the surface, a gradual reduction of both water content (Fig. 4.10) and hydraulic conductivity with falling matric potential from saturation was observed. In the relatively unweathered Chalk, little, if any, change in water content with matric potential could be detected due to its enormous air entry value (see Wellings, 1984a; Gardner *et al.*, 1990; Cooper *et al.*, 1990). Hydraulic conductivity variation mirrored this, with a fairly gradual reduction

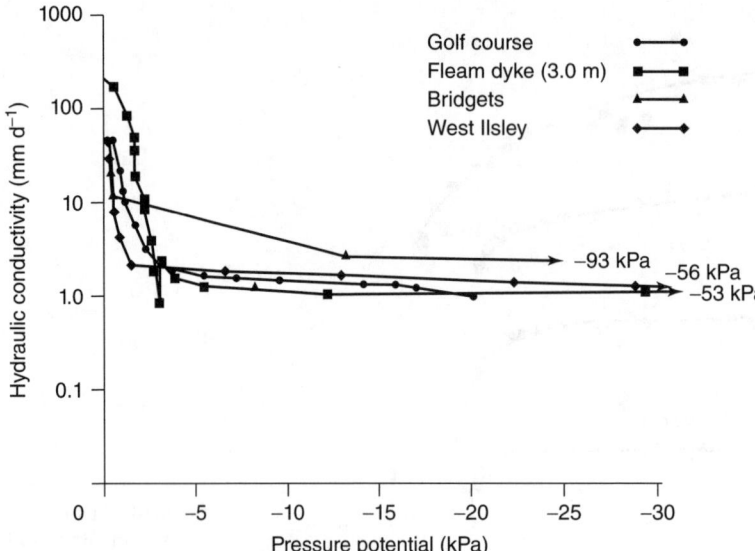

Fig. 4.11 Unsaturated hydraulic conductivity at four Chalk sites covered by thin soils. Except for Fleam Dyke, all measurements are for a depth of 2.1 m. (Source: Cooper, Gardner & Mackenzie 1990. Reproduced with permission of John Wiley & Sons.)

of hydraulic conductivity with matric potential at shallow depth, similar to most field soils. At depths between about 2 and 3 m, a rapid reduction in hydraulic conductivity between values in the region of 1000 mm/d close to saturation and a few mm/d at a matric potential of about −5 kPa was observed. Below this matric potential, the hydraulic conductivity was essentially constant to beyond the tensiometer range (Fig. 4.11). The higher values of hydraulic conductivity are ascribed to flow within the fractures between Chalk blocks.

2 For the most part, winter rainfall in excess of a few mm/d could be accommodated temporarily by the soil and weathered Chalk in the upper part of the profile, leading to input rates to the lower part of the profile typically of up to 3 or 4 mm/d. Where the hydraulic conductivity of the Chalk matrix was above this (Bridgets Farm), matric potential remained low throughout the year, rarely rising above −20 kPa, well below the *ca.* −5 kPa threshold at which flow in fractures would be initiated. At the other sites, the behaviour was more mixed, with periods of low rainfall leading to matric potential ≤5 kPa (matric flow) and those of high rainfall rising above −5 kPa at which part of the flow was in the matrix and part *via* the fractures (Cooper *et al.*, 1990). For the Fleam Dyke grass site, Jones and Cooper (1998) found that about one-third of drainage flux occurred *via* the fractures, about one-third through the matrix whilst flow

occurred simultaneously through the fractures, and one-third through the matrix when there was no flow through the fractures. Of the sites studied, Fleam Dyke had the most frequent occurrence of conditions where fracture flow would be likely to occur.

3 Restriction of evaporation at the thin soil sites was quite limited, with annual ratios of actual to potential evaporation of 78–93%, in contrast to the two sites with thicker soil (Warren Farm and Stud Farm) which had lower evaporation ratios of 69–81%. This was thought to be due to the large amounts of water storage in the upper layers of the Chalk at modest water potentials, whereas the latter two drift-covered sites had lower evaporation due to greater restriction by soil water deficits. This contrasts with common observation that crops on deeper soils are less prone to drought conditions. The West Ilsley site with a low ratio did not fit this pattern, owing to its very exposed location and high wind speeds resulting in a very high potential evaporation (~750 mm/yr) that was greater than the annual rainfall (Gardner *et al.*, 1990).

4 Substantially higher drainage (>120%) from the drift-covered sites was primarily a consequence of the lower evaporation (Fig. 4.12). Drainage in these permeable areas can be identified with aquifer recharge.

5 Drainage from the drift-covered sites was more evenly spread over the year than from the thin soil

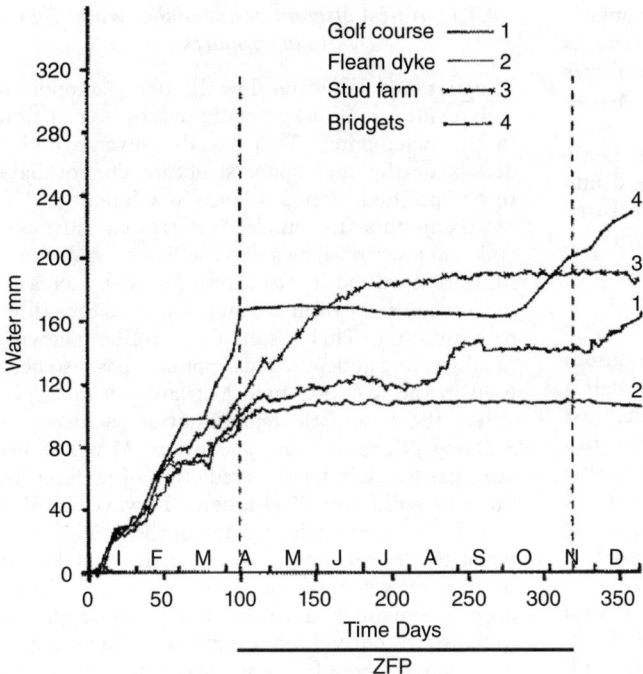

Fig. 4.12 Cumulative drainage at 3 m depth at four sites on Chalk during 1980 (Wellings and Cooper, 1983). The Stud Farm site in Cambridgeshire was beneath a deep soil, whereas the other three were under shallow soils. The Golf Course and Fleam Dyke sites were close to and had similar rainfall to the Stud Farm, whereas Bridget's Farm is in Hampshire and had higher rainfall. The period marked ZFP shows when a zero flux plane was observed at the Golf Course site. (Source: Wellings & Cooper 1983. Reproduced with permission of Elsevier.)

sites, where drainage almost ceased once a soil water deficit was established.

6 Good recovery of conservative solutes was found where tracer experiments were conducted on thin soil sites. At Bridget's Farm and West Ilsley, relatively little dispersion of peaks was observed (Wellings, 1984b; Barraclough *et al.*, 1994). At both sites, the data were consistent with a piston flow model of solute movement. At Bridget's Farm, the peak displacement was consistent with the measured accumulated drainage and the porosity of the saturated Chalk matrix (Wellings, 1984b). However, at West Ilsley, the solute peaks moved at approximately twice the expected speed based on an average drainage flux of 176 mm/year and porosity of 40% (0.8 m/year v 0.44 m/year), suggesting that only about half of the porosity was involved in solute transport (Barraclough *et al.*, 1994).

4.3.6 Other water balance studies

Additional water balance studies were also conducted, including:

• A tea estate in Kenya, (Cooper, 1979) where the ZFP method was used to measure crop water use over 93 days of mostly dry weather. Transpiration by the crop was an almost constant 3.36 mm/d, with no apparent response to water stress, and evaporation of intercepted water accounted for 2.5 mm in each storm of over 8 mm. Deep rooting of the crop necessitated the installation of neutron probe access tubes to 6 m and tensiometers installed from a 6-m deep pit. The ZFP was observed to reach a depth of 4.5 m. Unsaturated hydraulic conductivity of the soil was measured on a nearby plot by the method of Poulovassilis *et al.* (1974). As at Fleam Dyke, it was found to be a strong function of depth and attempts to replicate the drainage results by application of Darcy's Law gave poor agreement.

• A chalk Coombe site (chalk cobbles, in a matrix of chalk debris and loamy soil) (Hodnett and Bell, 1990), where the ZFP technique was used to estimate water flux and hence unsaturated hydraulic conductivity. This showed an initially rapid drop in conductivity with falling matric potential to a few mm per day, similar to the behaviour of undisturbed Chalk. However, rather than remain constant to lower potentials, it continued to drop steadily, accompanied by a very small change in water content. This was interpreted as film flow over the surface of the cobbles. Use of pressure transducer tensiometers connected to an automatic data logger showed rapid response to rainfall events at depths to 3 m, implying large drainage fluxes occurring over a short space of time as bypass flow.

• A loamy sand over sand (Cooper *et al.*, 1990), where both water balance of a grass sward and unsaturated hydraulic conductivity were measured *in situ*.

4.4   Role of Macropores

Much of the classical work in soil physics assumes that water flows in a regular fashion through a soil matrix which is homogeneous, or at least comprises a series of identifiable layers with uniform properties. There has been increasing evidence of preferential flow occurring in cracks or fauna-related channels in soil profiles, bypassing the soil matrix and resulting in water reaching depths much faster than expected.

The visit to Wallingford of Peter Germann on study leave from the Federal Institute of Technology (ETH) in Zürich provided a focus for work on macropores and water flow in soils. His thesis work in a forest soil in Switzerland had suggested that preferential flows might be important in the hydrological response of hillslopes. Work on modelling the data collected by Weyman (1973) at the East Twin catchment, Mendip, UK, had suggested that preferential flows might be one reason why a physically based hillslope model had failed to reproduce the field observations (see Beven, 2001). A review of research on macropores and water flows in soils was published in 1982 (Beven and Germann, 1982) and is now one of the most cited papers in hydrology (recently updated in Beven and Germann, 2013).

Laboratory experiments were conducted on large undisturbed soil cores and the first attempt made at producing a process-based model of infiltration into macroporous soils. This work was published as a series of papers in *Journal of Soil Science* (Germann and Beven, 1981a,b; Beven and Germann, 1981) after an interesting conflict with reviewers who clearly thought that the Darcy–Richards equation description was a sufficient description of the physics of flow in soils. On reflection, more progress might have been made in soil physics if the inappropriate equilibrium conditions of the original Richards (1931) experiment had been recognised earlier (Beven, 2014).

4.4.1   A first attempt at a model of water flow in macropores

The first model of water flow through macroporous soils made use of a kinematic description of flow in the macropores. This has the advantage that details of the macropore structure do not have to be specified; only a storage-flux function. The matrix in this first model was treated only as a sink, parameterised by a loss coefficient. A distance limitation was built into the model, with increased infiltration away from the macropores as they filled to saturation. This resulted in profiles showing rapid wetting at depth. This approach has also been used in the MACRO model of Jarvis *et al.* (1997) where the kinematic representation is linked to a Darcy–Richards matrix solution. MACRO has been used widely for the prediction of preferential flows of pollutants. The kinematic wave solution can also be used for drying and for the prediction of when the (faster) drying wave might overtake the macropore wetting front if the soil is sufficiently deep (Germann and Beven, 1985). The approach has also recently been extended to fingering as well as macropore flows by making use of laminar Stokes flow as films in arbitrary larger pore spaces (Hincapié and Germann, 2009a,b).

That first model of water flow in macropore systems worked quite well at the laboratory scale and for short times. However, there was clearly a need to allow for the potential heterogeneity of macropore distribution, and to be more explicit about interactions with the matrix, especially in situations where the macropores were not continuous but had a limited depth of penetration into the soil. In a first attempt to build in more complexity, a similar kinematic approach was taken, but the parameters of the flux relationships, and the loss to the matrix, were assumed to be random variables. This resulted in significant variability in infiltration rates that could be integrated over time to give an areal average infiltration estimate (Beven and Germann, 1985). In Beven and Clarke (1986), the distribution of cylindrical macropores of different diameters and depths in a soil was considered as a purely statistical distribution. These assumptions allowed an analytical solution (due primarily to Robin Clarke) for infiltration into the matrix, and at the surface between the macropores, to be developed by assuming that flow in the macropores was not limiting, but that the flux would start to create a head in the macropores as they started to fill. The result again was soil water profiles showing distributed infiltration throughout the profile affected by the macropores, something quite different to profiles produced by solutions of the Darcy–Richards equation.

While many tracing experiments (mostly destructive, and often after high application rates) have demonstrated the importance of preferential flows in the infiltration of water into the soil, the significance of preferential flows at the hillslope scale has proven more difficult to assess. Destructive sampling has also shown that macropores and pipes can extend for some distance downslope (e.g. Sidle *et al.*, 1995) and it has been suggested that preferential flows over a bedrock surface might be important once connectivity becomes established (e.g. Graham and McDonnell, 2010). Models of the impact of macropores at the hillslope scale include those that depend on the specification of specific preferential flow links between parts of the hillslope and others that treat preferential flows as local zones of high hydraulic conductivity in a Darcy–Richards solution (see Beven and Germann, 2013, for a review). A more recent attempt to represent preferential flow on hillslopes is embodied in the Multiple Interacting Pathways (MIPs) model of Davies *et al.* (2011, 2013) where water on a hillslope is represented as many particles moving with velocities chosen from random distributions. This can include both relatively slow velocities in the matrix and fast velocities in preferential flow pathways.

4.4.2 Pesticide movement through macropores in clay soils

A hydrological study at Rosemaund in Herefordshire was conducted from 1989 to 1992 to examine soil water pathways through the silty clay loam soil profile to determine how they influence pesticide movement (Williams *et al.*, 1996). Soil water content was measured by neutron probe and tensiometers were used to determine soil water potential and the direction of water movement. This revealed the crucial hydrological role of soil macropores, which form the dominant flow pathways in the very low conductivity soil matrix. These comprise both large shrink/swell cracks that vary seasonally in size and depth according to changes in soil water content, and smaller macropores created by biological activity such as worms and root holes that remain active throughout the year. The summer period is characterised by

progressive drying of the soil accompanied by the development of a complex network of shrinkage cracks, some of which may persist throughout the following winter. In general, autumn rewetting may move slowly downwards because of the poor conductivity of the soil, but if heavy rainfall occurs this may generate flow down the cracks, resulting in the soil profile wetting from the bottom upwards. This pattern may be altered by cultivation of the topsoil destroying the natural cracks structure.

4.4.3 Natural pipes, a runoff process on upland catchments

In the mid-1970s, the main Plynlimon catchment experiment was providing data on the major water balance components, but it became apparent that more specialised studies of the internal workings of the catchments would be needed to understand the dynamics of runoff and the storage of water on the catchment slopes and in the soils. Given that the natural processes of storage and flow of water in the soil were likely to have been disturbed by drainage and ploughing in the forested Severn catchment, the first process studies were concentrated in the Wye catchment, which had remained largely grassland. The mysterious absence of surface runoff, which was generally assumed to be one of the main pathways for runoff from upland catchments, led to speculation about the ways in which runoff water generated on the permanently wet upper parts of the catchment might reach the stream channels. Natural pipes in the peat deposits and peaty horizons of soil profiles on the slopes were identified quite early on as a possibly significant runoff process. The processes involved with the origin and flow properties of soil pipes were first noted in the Wye headwaters by Knapp (1974) and subsequently by Gilman and Newson (1980) and others. Studies of preferential flows in shallow soil pipes at on the western flank of the Plynlimon mountain were undertaken by Jones (1990, 2010).

Peat has unusual hydrological properties; its tendency to oxidise and crack irreversibly creates a special form of secondary porosity which conducts sufficient volume and velocity of flow to erode conduits. Some of these pipes were large, effectively bedrock channels that had become covered by peat growth. Others were shallow, often developed in the Ao organic horizon of the peaty podsols, or at the boundary between the organic and mineral horizons. The networks tended to be complex and polygonal in nature with evidence of shaping of the channels by the flow, and frequent collapse and occasional upwelling of water caused by obstruction of flows modifying the network. There are two forms of pipes in peaty soils; in deeper 'blanket' peats of plateaus and in the deeper pockets in 'flush' features, quite large, perennial pipes are found and can be the trigger for peat slides when they become surcharged during extreme rainfall events. Ephemeral pipe networks carry flow when fed by surface runoff or when the water table in a peaty podsol rises into the cracked surface peat horizon.

The conclusion from these studies, strengthened after the dry summer of 1976, was that these networks originated as crack systems in the surface peats, with some channels later developed by the flow of water after wetting by autumn and winter storms. Some of the flow in these pipes was shown to be surface runoff generated on low conductivity areas on the upper parts of the hillslopes where the peat-hagg had been eroded down to the mineral soil. This work in Wales has led to a wider recognition of the significance of piping on upland hydrology (e.g. Jones, 2010). Another interesting perspective on these pipe flows was provided by the isotope study of waters flowing in an ephemeral pipe in the Nant Gerig sub-catchment of the Wye by Sklash *et al.* (1996). Although only limited to sampling of a few storms, the indication was that flow in the pipe was dominated by pre-event (or 'old') water, even when the time to peak discharge in this pipe was of the order of only 15 minutes. That water was presumably being displaced from storage in the peat by the input of new event water.

Thanks to innovative recording equipment, the Plynlimon study was able to add to its extensive mapping of both forms of pipes in peaty soils, flow recordings for periods during the 1970s, a time of both flood and drought in mid-Wales. Gilman and Newson (1980) detail the importance of rainfall volume and intensity in initiating pipe flow, in a way similar to the dynamics of the partial/dynamic contributing area concept but with a spatial pattern depending on soil type and slope length (see also Kirby *et al.*, 1991).

Pipes act as a link between shallow groundwater in the soils of the Plynlimon catchments and the stream network. Once the natural pipe networks had been mapped, and their typicality in the context of the wider Wye catchment had been established, instruments were installed to measure flows in individual components of the network and used with local raingauges to develop a simple model of the rainfall–runoff relationship. Flows in pipes typically range from 0.1 to 1 L/s, which is below the range of flows usually measured by V-notch weirs and propeller meters, so hybrid devices were developed expressly for the work in Nant Gerig, consisting of ducted propellers for high flows and tanks that emptied automatically through a siphon. Initially, the counts from the propeller meters and the siphoning tanks were registered mechanically and operators had to be present on site to record the values, but measurements were automated for the 1975–1976 and 1976–1977 winters. As natural pipes mostly flow ephemerally, the runoff record consisted of a series of hydrographs of discrete events, reflecting the response of the pipes to individual rainstorms. Between 1971 and 1977, 15 storm hydrographs were acquired from an average of 5 pipes in the Nant Gerig catchment, for rainfall events ranging from 9 to 164.5 mm. As a rule, the smaller events did not generate pipe flow. A maximum of 13 instrument stations was maintained at any one time, but it was impossible to maintain a complete coverage owing to blockages by sediment, disturbance by animals and mechanical problems.

The rainfall–runoff data from individual pipes were used to build a model of runoff generation through pipe systems. Examination of the hydrographs showed that:

(i) Pipes begin to respond to rainfall after a period of several hours, implying that there is a delay in transmission of water from the surface into the pipe, probably caused by the need to satisfy a water deficit in the surrounding soil.

(ii) During a rainfall event, flows respond quickly to changes in rainfall intensity, indicating rapid movement of water from the surface into the pipe once flow has started.

(iii) The recession limb of the hydrograph is not usually exponential, as with a perennial stream, but the recession rate increases as the discharge falls, perhaps because of leakage out of the pipe.

It was possible to simulate pipe flows using a simple hydrological model, whose parameters, after optimisation, confirmed that pipe flow occurs after a deficit ranging from a few millimetres to 25 mm is satisfied. The flow is sensitive to variations in rainfall intensity, and the steepening of the recession can be accounted for by a steady decline of 1–4 mm/h.

Investigations at Cerrig yr Ŵyn in the Wye headwaters concentrated on the hydraulics of flow in pipes (Newson and Harrison, 1978). Leakage from pipes was measured by diverting water from perennial pipes to create an artificial, controllable flow through an ephemeral pipe system, and measuring the loss from the pipe and the mean flow velocity. The parameters obtained for a typical pipe in the network were used to estimate the velocity of a flood wave passing down a piped slope. This is a first step towards extending the pipe model to cover the propagation of runoff down a piped slope.

The key to understanding the significance of natural pipes to catchment hydrology is the contributing area concept, which relates the generation of flood flows to limited areas within the catchment, for instance valley bottoms or interfluvial wetlands. In the Wye catchment, these contributing areas appear to be the blanket bogs on the interfluves, flushed areas, sometimes with perennial pipes but without strongly developed open stream channels, and the valley bottom wetlands along the stream banks.

Pipe networks offer a short circuit between isolated contributing areas, the blanket bogs in the upper parts of the catchment, and the saturated areas along the streams. However, the pipes on the grassy slopes do not conduct flow except during and soon after rainstorms, and there are limits to the quantity of water that can be carried without efflux to the ground surface. Thus, pipe flow may be significant only for intermediate rainfall events, with surface runoff taking a more important part in the larger events leading to larger flood volumes.

More recent work on a large number of catchments in the UK has widened the interest in natural pipes as a significant hydrological and geomorphological process (Holden and Burt, 2002; Jones, 1990; Jones, 2004). Mapping of ephemeral pipes has been extended to deep blanket peat in the Pennines, using geophysical techniques to locate and trace flow through cavities deep below the surface (Holden et al., 2002). The Pennines studies pointed to an integrated network comprising both ephemeral and perennial pipes, carrying a substantial proportion (10 – 30%) of the runoff from the catchment, with deep-seated perennial pipes contributing significantly to base flow (Smart et al., 2013). Investigations are also continuing into the hydraulic properties that determine the hydrological functioning of natural pipes in peaty soils.

A recent study found a wedge of relatively high conductivity immediately above a pipe, forming a more permeable conduit between surface soil and the pipe (Cunliffe et al., 2013).

4.4.4 Artificial drainage

Whilst 'natural pipes' may be limited in extent to a few tens of metres, and in space to particular soil types and topographic locations, huge areas of Britain and elsewhere have artificial field drainage schemes (Robinson and Armstrong, 1988). They have been installed both on poorly drained soils in high rainfall upland areas to improve grazing, as well as on lower rainfall lowland areas with more permeable soils as an insurance for high value arable and horticultural crops.

Extensive networks of open ditches were cut in upland areas during the early and mid-twentieth century, and it is generally accepted that these, and ditches for forestry, have only a limited impact on soil water contents and water tables in the low conductivity peat soils. Water moving on the surface and through the upper more permeable layers of peaty soils quickly reaches the open channels and can then flow rapidly through the ditch network. As a result, they generally lead to increases in flood peaks (Conway and Millar, 1960; Nicholson et al., 1988). Currently, there are a number of schemes underway in the Pennines and Exmoor to block ditches, in part to reduce the risk of downstream flooding.

The situation is less clear for subsurface pipe drainage adopted in more productive lowland soils. Would it exacerbate flooding by providing faster flow pathways for water to ditches and streams, or would it reduce runoff coefficients by increasing soil deficits between storms? While other work was conducted by the Ministry of Agriculture, Fisheries and Food (MAFF) into the effectiveness of drainage systems for controlling water tables under farmland, a number of hydrological field drainage experiments were established around the country to compare the outflows from undrained and subsurface drained land in order to understand what effect this might have on river flow regimes and flooding (Robinson and Rycroft, 1999). Following WW2 there were generous government subsidies for field drainage of agricultural land to increase food production, yet it was not properly understood if there would be adverse impacts further downstream.

One of these drainage studies was on heavy clay soils at Grendon Underwood in Buckinghamshire (Beven, 1980) within the Institute's River Ray experimental catchment (see Chapter 2). This small lowland catchment is mostly used as permanent pasture. The plots were bounded by field drains (Fig. 4.13). This field had a ridge and furrow topography from ancient ploughing patterns that also served to improve the drainage of the soil between the furrows. On the drained plot, mole drains had been drawn in the furrows, with a tile drain collector at the downslope edge of the plot. A similar tile drain was used to collect subsurface drainage from a second plot without mole drains. Discharge from the tile drains was measured in V-notch weir boxes, and transects of mercury manometer tensiometers and neutron probe access tubes were set up between the ridges and furrows.

Keeping the tensiometers working was difficult, when some of the deeper tensiometers were clearly installed in peds of very low conductivity clay and might take several days to come to equilibrium (by which time there might be air bubbles in the manometers again). This slow response was in contrast to the discharge response when both plots showed rapid response to rainfall. In the case of the undrained plot, once wetted there would often be saturation in the furrows while in the drained plot, the mole drains responded rapidly, and no surface saturation was observed.

Installation of mole drains involves dragging a steel ball through the soil at 40–50 cm depth creating a continuous tunnel. When soil conditions are right, this will also have the effect of producing additional cracks to the surface which facilitate water moving down to the drain. The soil water profiles suggested, however, that the range of this effect was rather limited. To investigate this further, a tracing experiment was carried out after winter wetting in March 1979. A 2 m by 1 m plot orthogonal to one of the mole drains was watered by hand using water containing a solution of red

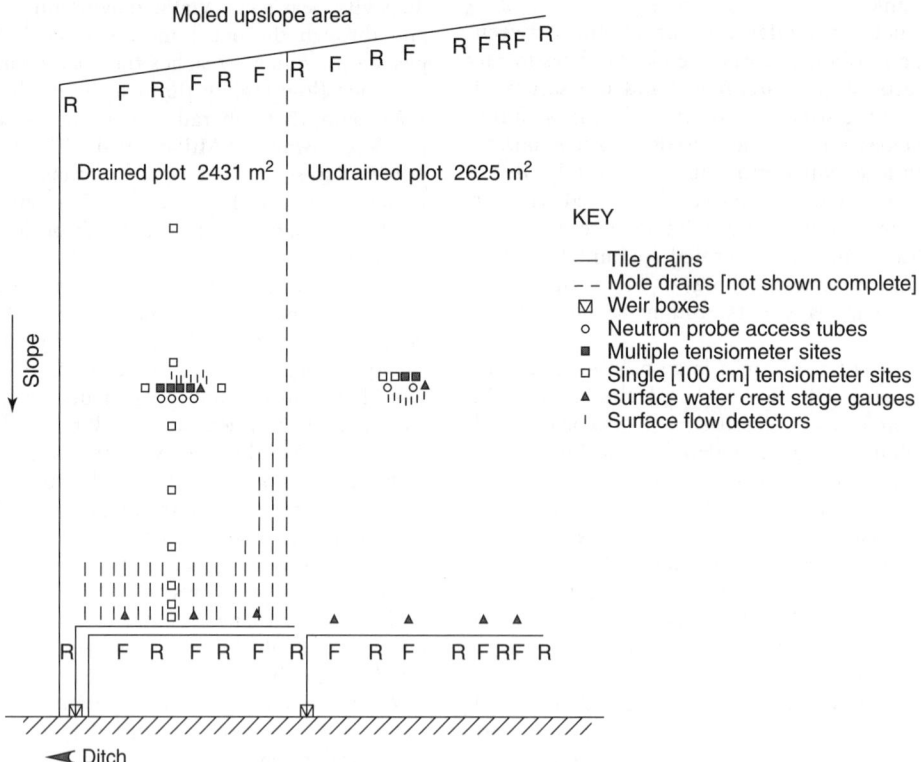

Fig. 4.13 Grendon drainage study. (Source: Reproduced with permission of CEH.)

Rhodamine dye. No dye was added immediately over the mole drain. To one side of the plot, a trench was dug down to the level of the mole drain so that arrival at the drain could be observed. The dye arrived in less than one minute. Observation of the dye flow pathways in the trench suggested that there was some general infiltration in the more organic A horizon, but that as soon as the structured clay peds were encountered there was distinct and rapid preferential flow through macropores between the clay peds (see photograph of this experiment in Beven and Germann, 2013). Thus, in this cracking clay soil, there were still effective preferential flow pathways even at the end of a winter of wetting and swelling. The excavations also revealed that the roots of the pasture grasses were mostly on the outside of the clay peds with very little penetration. This might also have kept macropore pathways open. Detailed studies of the outflow hydrographs showed a seasonal change in their response, with a more rapid response to rainfall from the drained plot in summer when soil water deficits were high (and soil cracks had opened) and higher peaks from the undrained plot in winter when the cracks had closed and surface saturation occurred (Robinson and Beven, 1983).

Further work at different sites across Britain has indicated that the effect of subsurface field drainage on streamflow is quite complex and may increase or reduce storm runoff compared with undrained land, but this can be broadly characterised (and predicted) from the site characteristics, particularly the soil type and rainfall (Robinson and Rycroft, 1999).

4.5 Hillslope-Streamflow Linkages

4.5.1 Betwa Project, India

Following the success of the soil physics based groundwater recharge studies at Thetford, the opportunity arose to apply the techniques in an overseas context, during the Indo-British Betwa Groundwater Project, in Central India. This was an ODA-funded collaborative project, involving BGS, IH and the Central Groundwater Board of India, which ran from 1977 to 1980. A component of the Betwa catchment study was to elucidate the soil water processes in relation to the varied soil types within the catchment (Hodnett and Bell, 1981, 1986).

The Betwa is a south bank tributary of the Jamuna River with a catchment area of 20,000 km^2.

There are three main geological zones in the catchment, and the aim was to evaluate the groundwater resources. This work involved hydrological and soil physical studies to measure groundwater recharge, a key aspect of a groundwater resource. The most extensive geological unit is the Deccan Trap Basalts and the recharge studies were focussed on this unit.

The soils are swelling clay vertisols (black cotton soils). Three soil areas were defined, one with shallow soils (<2 m to weathered basalt), and two with deep soils (>2 m, underlain by up to 10 m of yellow silty clay) overlying the basalt. Each area was instrumented with a network of neutron probe access tubes, in cropped and uncropped areas. Later, detailed measurements of hydraulic conductivity were carried out at 3 representative sites. The soils were found to be very uniform in water holding capacity with a typical seasonal range of storage of about 230 mm under cropped areas with most of the uptake from <1.5 m and none from >2.5 m. Under grass/scrub areas, seasonal soil water depletion was >500 mm and to >3 m depth. Actual evaporation was far less than potential evaporation, except in the wet season and at the peak of crop growth in late January. After harvest in March, when the soil was mainly bare, actual evaporation fell to less than 0.5 mm/d until the start of the monsoon in June.

On the deep soils, the soil physical studies showed that the structured soils in the upper profile form an upper aquifer system, which is virtually isolated by poorly conductive yellow clay from the weathered basalt aquifer beneath. This means that the soils can be modelled as a tank with a capacity of about 230 mm (depending on vegetation). Once full, the tank overflows to produce surface runoff. In grass/scrub areas, the deeper soil water uptake creates cracking to greater depth, and recharge may occur where this zone reaches down to the weathered basalt. In the shallow soils, the poorly conductive yellow clay layer is absent, so the structured black cotton soil and the permeable weathered basalt form a composite aquifer and recharge is not limited.

4.5.2 Hillslope hydrology – Amazon

Hodnett *et al.*, (1997a,b) examined 3 years of soil water data from the Manaus forest site on the forest plateau, slope and valley floor to identify differences in subsurface hydrological response. Under the plateau and slope, changes of water storage were very similar with no indication of surface runoff on the slope. Soil water storage changes beneath the flat valley floor (floodplain) were controlled by the behaviour of a shallow water table whose level was largely determined by the discharge of groundwater which maintained dry season streamflow. In dry periods, soil water changes were far smaller than on the plateau because root uptake was largely replenished by the groundwater flow towards the stream.

The groundwater was recharged by vertical drainage through the deep unsaturated zone (up to 25 m thick) beneath the plateau and slope. The groundwater gradient at the foot of the slope increased (showing the arrival of recharge) more than a month after the measured soil water deficits from the previous dry season had been replenished, indicating a significant lag caused by travel time through the deep unsaturated zone. Valley ground-water levels peaked more than a month after the start of the dry season.

Following a wet season with little recharge, the reduced groundwater flow towards the stream provided an ever smaller contribution to the evaporative demand of the valley floor vegetation and the stream eventually dried up. The valley soil water storage then behaved similarly to that on the plateau and slope. Early in the wet season, the water table responded to local vertical recharge through the floodplain deposits, creating a temporary gradient towards the hillslope. This did not occur after deep recharge from the plateau and slope arrived. In the late wet season, the water level is almost at the floodplain surface and may create seeps on the lower slopes in very wet years. For the period 1966–1989, the recharge was estimated to range from 290 to 1601 mm with a mean of 1087 mm. Published data show that base flow is 91% of annual runoff. Stormflow is generated as saturation overland flow on the floodplain, and water table recessions after rainfall events show that the runoff response depends on the depth to the water table.

On the plateau and slope, the highly conductive zone of very high porosity between 0.4 and 1.1 m, underlain by a less conductive layer, is a possible route for interflow during, and for a few hours after, heavy and prolonged rainfall. It was concluded that the depth of the valley floor water table would have a marked impact on the generation of flood runoff. These results are from areas with deeply weathered and permeable soils; in areas of Amazonia with shallower soils, the predominant flow generation processes will differ.

The soil water data from the ABRACOS sites has been widely used in other studies, including investigations into controls on forest evapotranspiration (Juarez *et al.*, 2007), and the calibration of land surface models (Harris *et al.*, 2004).

4.5.3 Hydrology of small catchments – Amazon

During the ABRACOS project, analysis of soil water content and water level data in the forest valley floor clearly indicated that this zone was important in generating stormflow, and that the water levels in the valley which influenced stormflow generation were controlled by deep recharge beneath the plateau and slope. In 2000, a study began of the 6.58 km^2 Asu catchment, near Manaus in central Amazonia. It is about 40 km SW of the Fazenda Dimona pasture/forest site and one of very few microcatchments in the world that have a flux tower. Soil water and groundwater levels were measured along a transect from the plateau to the valley floor adjacent to the stream. Rainfall was measured at 5 sites (4 above the canopy). Throughfall was also measured to determine interception. Runoff, EC and dissolved organic carbon (DOC) were measured on a continuous basis. Landforms and soils were very similar to those at the ABRACOS site at Fazenda Dimona. The data from the Asu catchment were analysed to investigate the flow generation processes and role of soil, unsaturated zone and groundwater storage in the overall catchment water balance.

The discharge totals from the Asu catchment in 2002 and 2003 were 1433 and 822 mm for rainfall totals of 2976 and 2054 mm, respectively. Stormflow was 23% and 13% of annual rainfall, much higher than in previous studies (in smaller catchments) in the same area. The valley floor occupied 38% of the Asu catchment and was the main source of stormflow, generated as saturation excess overland flow. However, stormflow discharge exceeded 38% of the rainfall following seven large rainfall events, implying a further source of stormflow. This could be explained by rapid throughflow through the macroporous zone (described above) beneath the slope, feeding water to the valley floor groundwater at the foot of the slope during large events.

Increases in soil water storage beneath the slopes exceeded throughfall inputs in large rainfall events, also suggesting a transfer of water from the plateau. Comparisons with other sites show that the amount of stormflow depends on the proportion of valley floor in the catchment, which depends on catchment size and morphology. This must be taken into account when scaling up process studies to larger catchments.

Tomasella *et al.* (2008) studied 3 years of data from the Asu catchment in Central Amazonia, particularly the water balance and the role of different storage zones within the catchment. Variations in groundwater storage cause a strong 'memory effect', which can result in the carryover of seasonal rainfall anomalies between years. This can affect the hydrological response for some time after the anomaly. Storage within, and transit times through, the deep unsaturated zone affected the groundwater recharge and played a key role in reducing most of the intra-seasonal variability. The storage/memory effect is crucial for sustaining streamflow and evaporation in years with low rainfall. The memory effect caused by storage in the groundwater and unsaturated systems calls into question the often used assumption that storage in a catchment returns to a standard state each year and may prevent the closure of annual large-scale water balances. Other work using data from the Asu catchment showed marked interannual variation in interception losses.

4.5.4 Bedrock groundwater controls on runoff generation

In early 2000, a joint initiative between the CEH and BGS was established to investigate the role of bedrock groundwaters in generating runoff in the upland Plynlimon catchments, Wales. Streamflow generation in upland headwater regions has traditionally been modelled as a simple rainfall–runoff process where the bedrock geology is considered impermeable. Rain landing on the catchment was thought to travel along near-surface flow pathways – over and through the soils – into the stream channel. Hydrological and hydrochemical data collected at Plynlimon appeared largely consistent with this hypothesis; the stream responses to rainfall were rapid – indicating a fast travel time – and the 'spikes' in streamflow were associated with a corresponding increase in soil-derived components such as dissolved organic carbon (Neal *et al.*, 1997b). However, not all the hydrochemical data could be explained. For example, a highly variable chloride and stable isotope signal in rainfall was damped in the stream (Neal *et al.*, 1988), suggesting

the presence of a reservoir of water or 'mixing-pot' within the catchment, where the signal is buffered before reaching the stream channel. Sklash *et al.* (1996) also showed that the water flowing through soil pipes during storm events in the Nant Gerig sub-catchment of the Wye catchment at Plynlimon was overwhelmingly 'old' pre-event water. Subsequent work in the catchment identified a significant groundwater reservoir in the bedrock geology (Neal *et al.*, 1997a). However, the role of this reservoir and its contribution to streamflow generation processes, if any, remained unclear.

To address these questions, a detailed physico-chemical process study was established in the headwaters of the Afon Hafren at Plynlimon. The geology (Ordovician/Silurian mudstones and shales) and surficial deposits have been well described by Bell (2005). Flow pathways to the river were determined by detailed monitoring of soil water and groundwater processes adjacent to the stream channel. Nested boreholes, along a 50 m transect, were drilled into the river bank and lower hillslope. The boreholes were sealed, for depth-specific sampling, using a novel resin-injection method developed in-house by the IH/CEH drilling geologist and the IH workshops. In addition to the existing hydrometric infrastructure at Plynlimon, supplementary monitoring instrumentation, comprised logging tensiometers to measure soil water potentials and measurements of groundwater head responses to rainfall were installed. Water quality sampling was carried out for a range of determinands including major and minor ion and stable isotope chemistry. Details can be found in a number of published papers (Haria and Shand, 2004, 2006; Haria *et al.*, 2013; Shand *et al.*, 2004, 2005).

Hydrograph separation of isotope data from waters sampled over a storm event identified how the storm response in the stream was dominated by 'old' pre-event water. Subsurface hydrological monitoring identified a source of 'old' water in the bedrock geology to a depth of at least 30 m. This groundwater was shown to respond rapidly to rainfall (by pressure wave propagation) resulting in flow to the stream channel. The groundwater was stratified, showing extremes in chemistry, but mixing between the different groundwater compartments also occurred and was likely to explain attenuation of the chloride and isotope rainfall signal observed in stream waters. In addition, bedrock groundwater was observed to rise up into the lower soil horizons, where groundwater upwelling increased with proximity to the stream channel; the upper soil horizons remained unsaturated. A fast lateral flow path at the soil–bedrock interface channelled this water quickly downslope, such that it entered the stream channel through the river bank. Rapid displacement of groundwater into the soils and then into the stream explains the soil chemical signal associated with an 'old' pre-event water signal during storm response. This study shows that a far deeper component of the water cycle has the potential to reach the stream channel quickly via the soil horizons. Determining the sources of soil water saturation is important when attempting to understand the processes controlling surface water quality. However, the groundwater contributions to streamflow observed at this site could not account for the main source of pre-event water during runoff response to rainfall. Detailed in-stream hydrochemical profiling highlighted a potentially large discrete groundwater discharge directly into the stream channel upstream of the study site (Haria *et al.*, 2013). Further work is required to determine the spatial occurrence and importance of such discrete groundwater inputs for both stream discharge and surface water quality.

The collaboration between CEH and BGS described here has been instrumental in challenging the existing paradigm of streamflow generation processes in headwater catchments in the United Kingdom and globally. The previously accepted top-down driven mechanism of rainfall–runoff has been replaced by a bottom-up model where 'old' pre-event waters have been shown to dominate the storm hydrograph. These groundwaters exhibit a range of residence times and therefore a spectrum of hydrochemical signatures. The mixing of these waters before they discharge into the stream channel is likely to determine the 'memory' of these catchments to diffuse and non-diffuse pollution events and are therefore instrumental in controlling surface water quality. Models to predict surface water quality in upland catchments, which fail to incorporate groundwater processes in rainfall–runoff simulations may be seriously flawed.

4.6   Looking to the Future

Topics highlighted here in the discussion of likely future developments include further advances in instrumentation, and in conceptual/modelling developments in the representation of preferential flows. The importance of soils in relation to land management issues is also noted, especially for floods.

4.6.1   Real-time soil water monitoring

The Cosmic-ray Soil Moisture Observing System (COSMOS) offers new possibilities for routine monitoring of soil water at the field scale (Zreda *et al.*, 2008). The measurement principle is similar to the neutron probe, but instead of an artificially created radioactive source, it relies on naturally occurring neutrons generated by cosmic rays. In this way, the need for an artificial radioactive source is removed, which brings both practical and logistical advantages and greatly increases the potential of soil water monitoring.

The principles behind the cosmic ray neutron technique and its strengths and weaknesses are discussed in more detail in Chapter 2. COSMOS sensors can be deployed unattended in the field and so provide long-term continuous records of soil water storage. Potentially one of the biggest advantages is the scale of the measurement: each probe is representative of near-surface conditions across an area of approximately 34 hectares with a radius of 350 m radius and to a soil depth of about 0.5 m. Observations of soil water at this scale are hugely valuable, yet impractical to obtain routinely with point measurements, due to the vast number of point sensors that would be required to average out the spatial heterogeneity.

The data need to be processed to adjust for factors such as variations in the incoming neutron flux due to changes in solar activity, atmospheric humidity and the elevation of the site. Once a corrected count rate is obtained, this is related empirically to the soil water content by field calibration (Zreda *et al.*, 2012). The COSMOS probe represents the combined effects of all water sources. For certain environments, there may be other significant sources of near-surface water, such as lying snow, in which case the instrument can potentially usefully measure the water equivalent of the snow. Measurements will also be influenced at sites where there is an appreciable amount of water contained within the biomass and significant seasonal variation in the water stored (Bogena *et al.*, 2013). Figure 4.14 shows a COSMOS time series of soil moisture for a site in southern England.

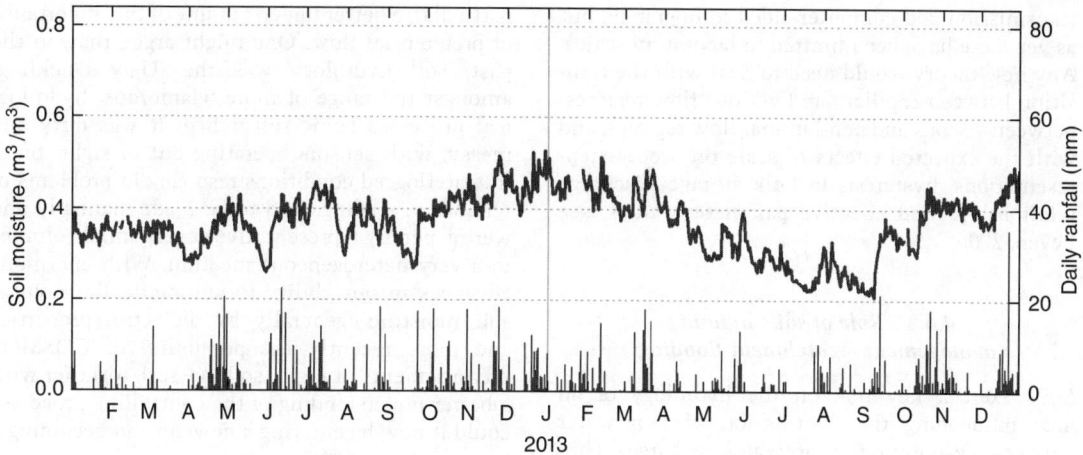

Fig. 4.14 COSMOS data processed and calibrated for Chimney Meadows. (Source: Reproduced with permission of CEH.)

Such data could be invaluable as inputs to aid streamflow prediction (e.g. Robinson and Stam, 1993).

Currently, there are about 70 operational COSMOS sites using cosmic ray neutron detectors around the world: mainly in the USA (http://www.cosmos.hwr.arizona.edu), Europe, southern Africa and Australia. The CSIRO in Australia has 10 sites operating (http://www.ermt.csiro.au/html/cosmoz.html) and in the UK CEH about 20 sites operational in 2015 (http://www.ceh.ac.uk/cosmos/). Stations in the UK network comprise both soil water and meteorological monitoring to facilitate incorporation of the data into hydro-meteorological models.

In addition to fixed installations at these long-term sites, roving studies have been carried out to explore soil water variation along transects using probes installed within vehicles, for example over 37 km in Hawaii (Desilets *et al.*, 2010). The majority of the literature currently originates from the COSMOS group in the United States, who have developed the method over the past few years. There are valuable contributions from other groups around the world reporting on the performance of the technique under various conditions. There clearly remains considerable research to be done to understand the full capabilities and indeed shortcomings of this still emerging cosmic ray technique better. However, the results presented so far suggest promising progress is being made in the challenging task of obtaining representative soil water measurements at the field scale.

4.6.2 Macropores and water flow in soils - where to from here?

It is now more than thirty years since the review paper of Beven and Germann (1982) and even longer since others started to draw attention to the importance of macropores and preferential flows in soils (e.g. Thomas and Phillips, 1979; Bouma and Dekker, 1978). The conclusions of the 1982 paper suggested that more experimental data were required to support the representation of preferential flows in hydrological models and that this would be a far more important limitation on progress than any limitations of theoretical development. Reviewing progress after 30 years, Beven and Germann (2013) conclude that this has been largely borne out by events, but express surprise that the Darcy–Richards equation still exerts such a hold on soil physics, even at the grid scales of climate models, and despite the fact that Richards equilibrium flow experiments are quite the wrong basis for a physics of real soils. Non-equilibrium and preferential flows are important at profile and hillslope scales, but there remains a real lack of combined flow and tracer experiments to properly test theoretical developments. There have been huge improvements in the visualisation of flow processes using light transmission and computer-aided tomography, but as yet these have been limited to laboratory scales. Any new theory would need to deal with the transition between capillary and viscous flow regimes, between viscous and non-laminar flow regimes, and with the expected effects of scale on process representations, hysteresis in bulk storage–discharge relationships and effective parameter values (see Beven, 2006).

4.6.3 *Role of soils in land management – catchment flooding*

Soils exert a key role on the hydrology of an area influencing the distribution of rain water between evaporation, runoff and recharge. Thin soils on chalk may be more able to sustain water to plant roots enabling short vegetation to continue evaporating for longer in summer droughts. Water movement through soils may be through the soil matrix, or may occur as bypass flow through cracks and fissures reaching depths much quicker than would otherwise be expected and potentially moving pollutants. Catchments underlain by clay soils are more prone to flooding (see Chapter 3).

Following the 2013–2014 winter floods in the United Kingdom, when extensive areas of southern Britain areas suffered extensive and prolonged flooding there was much discussion about the potential role of vegetation in reducing rapid runoff. Just as important – if not more important – is likely to be the role of soil. The primary flood reducing role claimed for planting hedge lines and trees may be more subtle than thought. Rather than a direct vegetation impact, changes in the quantities of runoff may be due more to associated changes in farming practice, such as a lowering in the stocking density of animals, thereby reducing soil compaction, increasing infiltration and mitigating flash flooding.

4.7   End Piece

There have been tremendous developments in soil hydrology over the last 50 years. Measurement techniques have been revolutionised and soil flow processes can be observed in the field rather than in artificial sand boxes in the laboratory. There have been major advances in conceptual understanding and modelling of water storage and flows, and, in particular, a better understanding of the importance of preferential flow. One might argue that, in the past, soil hydrology was the 'Ugly Duckling' amongst the range of more 'glamorous' hydrological processes being researched. It was dirty and messy, with sensors operating out of sight, often in waterlogged conditions resulting in problems of reliability; readings had to be made manually and were generally representative of very small volumes in a very heterogeneous medium. With enormous advances in our ability to automatically monitor soil moisture (generally by dielectric properties and more recently the possibility of COSMOS measurements of larger soil masses), together with a better understanding of the controlling processes could it now be entering a new era and becoming a 'Swan' amongst them?

4.8   References

Ah Koon, P.D., Gregory, P.J. & Bell, J.P. (1989) Influence of drip irrigation emission rate on distribution and drainage of water beneath a sugarcane and fallow plot. *Agricultural Water Management*, **17**, 267–282.

Barraclough, D., Gardner, C.M.K., Wellings, S.R. & Cooper, J.D. (1994) A tracer investigation into the importance of fissure flow in the unsaturated zone of the British Chalk. *Journal of Hydrology*, **156**, 459–469.

Batchelor, C.H., Soopramanien, G.C., Bell, J.P., Nayamuth, R. & Hodnett, M.G. (1989) Importance of irrigation regime, drip line placement and row spacing in the drip irrigation of sugarcane. *Agricultural Water Management*, **17**, 75–94.

Bell, J.P. (2005) *Soil hydrology of the Plynlimon catchments – Institute of Hydrology Report 8*. www.ceh.ac .uk/products/publications/documents/plynsoilsreport8 .pdf ISBN: 1903741130.

Bell, J.P., Wellings, S.R., Hodnett, M.G. & Ah Koon, P.D. (1989) Soil water status: a concept for characterizing soil water conditions beneath a drip irrigated row crop. *Agricultural Water Management*, **17**, 171–187.

Beven, K.J. (1980) *The Grendon Underwood Field Drainage Experiment*. Institute of Hydrology Report 65, Wallingford, UK, 30 pp.

Beven, K.J. (2001) Dalton Medal Lecture: How far can we go in distributed hydrological modelling? *Hydrology and Earth System Sciences*, **5**, 1–12.

Beven, K.J. (2006) The Holy Grail of Scientific Hydrology: $Q_t = H(S, R, \Delta t)A$ as closure. *Hydrology and Earth Systems Science*, **10**, 609–618.

Beven, K.J. (2014) BHS Penman Lecture: "Here we have a system in which liquid water is moving; let's just get at the physics of it". *Hydrology Research*, **45** (**6**), 727–736.

Beven, K.J. & Clarke, R.T. (1986) On the variation of infiltration into a homogeneous soil matrix containing a population of macropores. *Water Resources Research*, **22**, 383–388.

Beven, K.J. & Germann, P.F. (1981) Water flow in soil macropores II. A combined flow model. *Journal of Soil Science*, **32**, 15–29.

Beven, K.J. & Germann, P. (1982) Macropores and water flow in soils. *Water Resources Research*, **18**, 1311–1325.

Beven, K.J. & Germann, P.F. (1985) *A distribution function model of channelling flow in soils based on kinematic wave theory*. Proceedings of the International Symposium on Water and Solute Movement in Heavy Clay Soils, Wageningen, Neth. ILR Pubn. 37, 89–100.

Beven, K.J. & Germann, P.F. (2013) Macropores and water flow in soils revisited. *Water Resources Research*, **49** (6), 3071–3092.

Bogena, H.R., Huisman, J.A., Baatz, R., Hendricks Franssen, H.-J. & Vereecken, H. (2013) Accuracy of the cosmic-ray soil water content probe in humid forest ecosystems: The worst case scenario. *Water Resources Research*, **49**, 5778–5791.

Bouma, J. & Dekker, L.W. (1978) A Case Study on Infiltration into Dry Clay Soil, I.Morphological Observations. *Geoderma*, **20**, 27–40.

Calder, I.R., Rosier, P.T.W., Prasanna, K.T. & Parameswarappa, S. (1997) Eucalyptus water use greater than rainfall input – possible explanation from southern India. *Hydrology and Earth System Sciences*, **1**, 249–256.

Conway, V.M. & Millar, A. (1960) The hydrology of some small feat covered catchments in the N Pennines. *Journal of the Institution of Water Engineers*, **14**, 415–424.

Cooper, J.D. (1979) Water use of a tea estate from soil moisture measurements. *East African Agricultural and Forestry Journal*, **43**, 102–121.

Cooper, J.D. (1980) *Measurement of moisture fluxes in unsaturated soil in Thetford Forest*. Report 66, Institute of Hydrology, Wallingford, 97pp.

Cooper, J.D., Gardner, C.M.K. & Mackenzie, N. (1990) Soil controls on recharge to aquifers. *Journal of Soil Science*, **41**, 613–630.

Cunliffe, A.M., Baird, A.J. & Holden, J. (2013) Hydrological hotspots in blanket peatlands: spatial variation in peat permeability around a natural soil pipe. *Water Resources Research*, **49**, 5342–5354.

Davies, J., Beven, K., Nyberg, L. & Rodhe, A. (2011) A discrete particle representation of hillslope hydrology: hypothesis testing in reproducing a tracer experiment at Gardsjon, Sweden. *Hydrological Processes*, **25**, 3602–3612.

Davies, J., Beven, K., Rodhe, A., Nyberg, L. & Bishop, K. (2013) Integrated modelling of flow and residence times at the catchment scale with multiple interacting pathways. *Water Resources Research.*, **49**, 4738–4750.

Desilets, D., Zreda, M. & Ferré, T.P.A. (2010) Nature's neutron probe: land surface hydrology at an elusive scale with cosmic rays. *Water Resources Research*, **46**, W11505 DOI: 10.1029/2009wr008726.

Edmunds, W.M., Lovelock, P.E.R. & Gray, D.A. (1973) Interstitial water chemistry and aquifer properties in the Upper and Middle Chalk of Berkshire, England. *Journal of Hydrology*, **19**, 21–31.

Gardner, C.M.K. (1981a) *The soil moisture databank: moisture content data from some British soils*. Institute of Hydrology Report Series no. 76, 156 pp. Wallingford, UK.

Gardner, C.M.K. (ed) (1981b) *The MORECS Discussion Meeting*, April 1981. Institute of Hydrology Report Series no. 78, 58 pp. Wallingford, UK.

Gardner, C.M.K. (1983) An evaluation of the success of MORECS, a meteorological model, in estimating soil moisture deficits. *Agricultural Meteorology*, **29**, 269–284.

Gardner, C.M.K. & Field, M. (1983) An evaluation of the success of MORECS, a meteorological model, in estimating soil moisture deficits. *Agricultural Meteorology*, **29**, 269–284.

Gardner, C.M.K., Cooper, J.D., Wellings, S.R. *et al.* (1990) Hydrology of the unsaturated zone of the Chalk of south east England. In: Burland, J.B., Mortimore, R.N., Roberts, L.D., Jones, D.L. & Corbett, B.O. (eds), *Chalk*. Thomas Telford, London, pp. 611–618.

Gash, J.H.C. & Stewart, J.B. (1977) The evaporation from Thetford Forest during 1975. *Journal of Hydrology*, **35**, 385–396.

Germann, P.F. & Beven, K.J. (1981a) Water flow in soil macropores I. An experimental approach. *Journal of Soil Science*, **32**, 1–13.

Germann, P.F. & Beven, K.J. (1981b) Water flow in soil macropores III. A statistical approach. *Journal of Soil Science*, **32**, 29–31.

Germann, P.F. & Beven, K.J. (1985) Kinematic wave approximation to infiltration into soils with sorbing macropores. *Water Resources Research*, **21**, 990–996.

Gilman, K. & Newson, M.D. (1980) Soil pipes and pipeflow – a hydrological study in upland Wales. In: *BGRG Research Monograph*, 1. Geo Books, Norwich.

Graham, C.B. & McDonnell, J.J. (2010) Hillslope threshold response to rainfall: (2) Development and use of a macroscale model. *Journal of Hydrology*, **393**, 77–93.

Grindley, J. (1967) The estimation of soil moisture deficits. *Meteorological Magazine*, **96**, 97–108.

Haria, A.H. & Shand, P. (2004) Evidence for deep sub-surface flow routing in forested upland Wales: implications for contaminant transport and stream flow generation. *Hydrology and Earth System Sciences*, **8**, 334–344.

Haria, A.H. & Shand, P. (2006) Near stream soil water-groundwater coupling in the headwaters of the Afon Hafren, Wales: Implications for surface water quality. *Journal of Hydrology*, **331**, 567–579.

Haria, A.H., Shand, P., Soulsby, C. & Noorduijn, S. (2013) Spatial delineation of groundwater-surface water interactions through intensive in-stream profiling. *Hydrological Processes*, **27**, 628–634.

Harris, P.P., Huntingford, C., Gash, J.H.C. *et al.* (2004) Calibration of a land-surface model using data from primary forest sites in Amazonia. *Theoretical and Applied Climatology*, **78**, 27–45.

Hincapié, I. & Germann, P. (2009a) Abstraction from infiltrating water content waves during weak viscous flows. *Vadose Zone Journal*, **8**, 996–1003. doi:10.2136/vzj2009.0012

Hincapié, I. & Germann, P. (2009b) Impact of initial and boundary conditions on preferential flow. *Journal of Contaminant Hydrology*, **104**, 67–73.

Hodnett, M.G. & Bell, J.P. (1981) *Soil physical processes of groundwater recharge through Indian black cotton soils*. Institute of Hydrology Report No 77.

Hodnett, M.G. & Bell, J.P. (1986) Soil moisture investigations of groundwater recharge through black cotton soils in Madhya Pradesh, India. *Hydrological Sciences Journal*, **31**, 361–381.

Hodnett, M.G. & Bell, J.P. (1990) Processes of water movement through a Chalk Coombe deposit in southeast England. *Hydrological Processes*, **4**, 361–372.

Hodnett, M.G., Bell, J.P., AhKoon, P.D. & Batchelor, C.H. (1989a) The control of drip irrigation of sugarcane using 'index' tensiometers: some comparisons with control by the water budget method. *Agricultural Water Management*, **17**, 189–207.

Hodnett, M.G., Bell J.P., Batchelor C.H. & AhKoon P.D. (1989b). *Observations on the wetted zone beneath drip irrigated sugar cane in Mauritius*. Proc. Int. Conf. on Irrigation theory and practice, Inst. Irrign. Studies, Univ. Southampton UK.: Pentech Press, London.

Hodnett, M.G., Da Silva, L.P., Da Rocha, H.R. & Cruz Senna, R. (1995) Seasonal soil water storage changes beneath central Amazonian rainforest and pasture. *Journal of Hydrology*, **170**, 223–254.

Hodnett, M.G., Tomasella, J., Marques Filho, A. De O. & Oyama, M.D. (1996) Deep soil water uptake by forest and pasture in central Amazonia: predictions from long-term daily rainfall using a simple water balance model. In: *Amazonian Deforestation and Climate*, Gash, J., Nobre, C.A., Roberts, J.M., and Victoria, R.L., John Wiley, Chichester, UK, 79–99.

Hodnett, M.G., Vendrame, I. & Tomasella, J. (1997a) Processes of runoff generation in Amazonia. *Annales Geophysicae* Suppl II to Vol, **15**, C319.

Hodnett, M.G., Vendrame, I., Oyama, M.D., Marques Filho, A.D.O. & Tomasella, J. (1997b) Soil water storage and groundwater behaviour in a catenary sequence beneath forest in Central Amazonia. I. Comparisons between plateau, slope and valley floor. *Hydrology and Earth System Sciences*, **1**, 265–277.

Holden, J. & Burt, T.P. (2002) Piping and pipeflow in a deep peat catchment. *Catena*, **48**, 163–199.

Holden, J., Burt, T.P. & Vilas, M. (2002) Application of ground penetrating radar to the identification of subsurface piping in blanket peat. *Earth Surface Processes and Landforms*, **27**, 235–249.

Hough, M.N. & Jones, R.J.A. (1997) The UK Meteorological Office Rainfall and Evaporation Calculation System: MORECS version 2.0 – an overview. *Hydrology and Earth System Sciences*, **1**, 227–239.

Howse, K.R. (1981) A technique for using permanent neutron meter access tubes in cultivated soils. *Experimental Agriculture*, **17**, 265–269.

Hudson, J. (1988) The contribution of soil moisture storage to the water balances of upland forested and grassland catchments. *Hydrological Science Journal*, **33**, 289–309.

Jarvis, N.J., Hollis, J.M., Nicholls, P.H., Mayr, T. & Evans, S.P. (1997) MACRO_DB: a decision-support tool to assess the fate and mobility of pesticides in soils. *Environmental Modelling and Software*, **12**, 251–265.

Jones, J.A.A. (1990) Piping effects in humid lands. In: Higgins, C.G. & Coates, D.R. (eds), *Groundwater Geomorphology: The Role of Subsurface Water in Earth-surface Processes & Landforms* Special Paper No. 252. Geological Society of America, Boulder, CO, pp. 111–138.

Jones, J.A.A. (2004) Implications of natural soil piping for basin management in upland Britain. *Land Degradation and Development*, **15**, 325–349.

Jones, J.A.A. (2010) Soil piping and catchment responses. *Hydrological Processes*, **24**, 1548–1566.

Jones, H.K. & Cooper, J.D. (1998) Water transport through the unsaturated zone of the Middle Chalk: a case study from Fleam Dyke lysimeter. In: Robins, N.S. (ed), *Groundwater Pollution, Aquifer Recharge and Vulnerability, Special Publication 130*. Geological Society, London, pp. 117–128.

Juarez, R.I.N., Hodnett, M.G., Fu, R., Goulden, M.L. & Von Randow, C. (2007) Control of dry season evapotranspiration over the Amazonian forest as inferred from observations at a Southern Amazon Forest Site. *Journal of Climate*, **20**, 2827–2839.

Kinniburgh, D.G. & Miles, D.L. (1983) Extraction and chemical analysis of interstitial water from soils and rocks. *Environmental Science and Technology*, **17**, 362–368.

Kirby, C., Newson, M.D. & Gilman, K. (1991) *Plynlimon research: the first two decades*. Institute of Hydrology Report 109. Wallingford.

Kitching, R. & Shearer, T.R. (1982) Construction and operation of a large undisturbed lysimeter to measure recharge to the Chalk aquifer, England. *Journal of Hydrology*, **58**, 267–277.

Knapp, B.J. (1974) Hillslope throughflow observation and the problem of modelling. In Gregory K.J. & Walling D.E., Fluvial Processes in Instrumented Watersheds. *Institute of British Geographers Special Publication* **6**: 23–31

Marc, V. & Robinson, M. (2007) The long-term water balance (1972 – 2004) of upland forestry and grass at Plynlimon, mid-Wales. *Hydrology and Earth System Sciences*, **11**, 44–60.

Neal, C., Christophersen, N., Neale, R., Smith, C.J., Whitehead, P.G. & Reynolds, B. (1988) Chloride in precipitation and streamwater for the upland catchment of River Severn, Mid-Wales – some consequences for hydrochemical models. *Hydrological Processes*, **2**, 155–165.

Neal, C., Robson, A.J., Shand, P. *et al.* (1997a) The occurrence of groundwater in the Lower Palaeozoic rocks of upland central Wales. *Hydrology and Earth System Sciences*, **1**, 3–18.

Neal, C., Wilkinson, J., Neal, M. *et al.* (1997b) The hydrochemistry of the headwaters of the River Severn, Plynlimon. *Hydrology and Earth System Sciences*, **1**, 583–617.

Nepstad, D.C., de Carvalho, C.R., Davidson, E.A. *et al.* (1994) The role of deep roots in the hydrological and carbon cycle of Amazonian forests and pastures. *Nature (Lond)*, **372**, 666–669.

Newson, M.D. & Harrison, J.G. (1978) *Channel studies in the Plynlimon experimental catchments*. Institute of Hydrology Report 47, Wallingford.

Nicholson, I.A., Robertson, R.A. & Robinson, M. (1988) The effects of drainage on the hydrology of a peat bog. *International Peat Journal*, **3**, 59–83.

Poulovassilis, A., Krentos, V.D., Stylianou, Y. & Metochis, Ch. (1974) *Soil-water properties of layered soils determined* in situ. In: "Isotope and radiation techniques in soil physics and irrigation studies", Proc. Symp. Vienna, 1973. IAEA, Vienna, 205–224.

Price, M., Bird, M.J. & Foster, S.S.D. (1976) Chalk pore-size measurements and their significance. *Water Services*, **80**, 596–600.

Renck, A. & Lehmann, J. (2004) Rapid water flow and transport of inorganic and organic nitrogen in a highly aggregated tropical soil. *Soil Science*, **169**, 330–341.

Richards, L.A. (1931) Capillary conduction of liquids through porous mediums. *Journal of Applied Physics*, **1**, 318–333.

Roberts, J. & Rosier, P. (2005a) The impact of broadleaf afforestation on water resources in lowland UK. I. A comparison of soil water changes below beach woodland and grass on chalk sites in Hampshire. *Hydrology and Earth System Sciences*, **9**, 586–596.

Roberts, J. & Rosier, P. (2005b) The impact of broadleaf afforestation on water resources in lowland UK. I I I. The results from Black Wood and Bridgets Farm compared with those from other woodland and grassland sites. *Hydrology and Earth System Sciences*, **9**, 614–620.

Robinson, M. & Armstrong, A.C. (1988) The extent of agricultural field drainage in England and Wales, 1971 – 80. *Transactions of the Institution of British Geographers*, **13**, 19–28.

Robinson, M. & Beven, K.J. (1983) The effects of mole drainage on the hydrological response of a swelling clay soil. *Journal of Hydrology*, **64**, 205–223.

Robinson, M. & Rycroft, D.W. (1999) The impact of drainage on streamflow Chapter 23. In: Skaggs, R.W. & van Schilfgaarde, J. (eds), *Agricultural Drainage*. American Society of Agronomy / Soil Science Society of America Agronomy Monograph 38, pp. 767–799.

Robinson, M. & Stam, M.H. (1993) A study of soil moisture controls on streamflow behaviour: results for the Ock basin, United Kingdom. *Acta Geologica Hispanica*, **25**, 75–84.

Shand, P., Darbyshire, D.P.F., Haria, A.H., Davies, J. & Griffiths, K. (2004) *Streamflow generation in upland impermeable catchments – is there a need to think deeper?* Proceedings of the Eleventh International Symposium on Water-Rock Interaction, Wanty and Seal II (eds.), Vol. 2, Saratoga Springs, New York, USA; A. A. Balkema Publishers.

Shand, P., Haria, A.H., Neal, C. *et al.* (2005) Hydrochemical heterogeneity in an upland catchment; further characterisation of the spatial, temporal and depth variations in soils, streams and groundwaters of the Plynlimon forested catchment, Wales. *Hydrology and Earth System Sciences*, **9**, 611–634.

Sidle, R.C., Kitahara, H., Terajima, T. & Nakai, Y. (1995) Experimental studies on the effects of pipeflow on throughflow partitioning. *Journal of Hydrology*, **165**, 207–219.

Sklash, M.G., Beven, K.J., Gilman, K. & Darling, W.G. (1996) Isotope studies of pipeflow at Plynlimon, Wales, UK. *Hydrological Processes*, **10**, 921–944.

Smart, R.P. *et al.* (2013) The dynamics of natural pipe hydrological behaviour in blanket peat. *Hydrological Processes*, **27**, 1523–1534.

Thom, A.S. & Oliver, H.R. (1977) On Penman's equation for estimating regional evaporation. *Quarterly Journal of the Royal Meteorological Society*, **103**, 345–357.

Thomas, G.W. & Phillips, R.E. (1979) Consequences of Water Movement in Macropores. *Journal of Environmental Quality*, **8**, 149–156.

Thompson, N., Barrie, I.A. & Ayles, M. (1981) *The Meteorological Office Rainfall and Evaporation Calculation System: MORECS*. Meteorological Office Hydrological Memorandum no. 45, 72 pp. Bracknell, UK.

Tomasella, J., Hodnett, M.G., Cuartas, L.A., Nobre, A.D., Waterloo, M.J. & Oliveira, S.M. (2008) The water balance of an Amazonian micro-catchment: the effect of inter-annual variability of rainfall on hydrological behaviour. *Hydrological Processes*, **22**, 2133–2147.

Watson, K.K. (1966) An instantaneous profile method for determining the hydraulic conductivity of unsaturated porous materials. *Water Resources Research*, **2**, 709–715.

Wellings, S.R. (1984a) Recharge of the Upper Chalk aquifer at a site in Hampshire, England 1. Water balance and unsaturated flow. *Journal of Hydrology*, **69**, 259–273.

Wellings, S.R. (1984b) Recharge of the Upper Chalk aquifer at a site in Hampshire, England 2. Solute movement. *Journal of Hydrology*, **69**, 275–285.

Wellings, S.R. & Bell, J.P. (1980) Movement of water and nitrate in the unsaturated zone of Upper Chalk near Winchester, Hants, England. *Journal of Hydrology*, **48**, 119–136.

Wellings, S.R. & Cooper, J.D. (1983) The variability of recharge of the English Chalk aquifer. *Agricultural Water Management*, **6**, 243–253.

Weyman, D.R. (1973) Measurement of downslope flow of water in a soil. *Journal of Hydrology*, **20**, 267–288.

Williams, R.J., Brooke, D.N., Clare, R.W., Matthiessen, P. & Mitchell, R.D.J. (1996) *Rosemaund pesticide transport study 1987 – 1993*. Institute of Hydrology Report 129, Wallingford 68pp.

Zreda, M., Desilets, D., Ferré, T.P.A. & Scott, R.L. (2008) Measuring soil moisture content non-invasively at intermediate spatial scale using cosmic-ray neutrons. *Geophysical Research Letters*, **35**, L21402. doi:10.1029/2008gl035655

Zreda, M., Shuttleworth, W.J., Zeng, X. *et al.* (2012) COS-MOS: the cosmic-ray soil moisture observing system. *Hydrology and Earth System Sciences*, **16**, 4079–4099. doi:10.5194/hess-16-4079-2012

5 The Physics of Atmospheric Interaction

COLIN LLOYD[1,2], AND SYLVIA OLIVER[2]

[1]Centre for Ecology and Hydrology, Wallingford, Oxfordshire, UK
[2]Ex-Institute of Hydrology

5.1 Introduction

Studies of the physics of atmospheric interactions at the Wallingford span almost 50 years. Research in the period from the mid-1960s through to the early 1980s, described in Section 5.2, is focussed on developing novel instrumentation and experimental methods, and creating new theory and understanding and associated numerical models. From the early 1980s to the mid-1990s, much of the research was concerned with applying these new methods and understanding in the context of improving knowledge of hydrometeorology at the regional and global scale, with a focus on the surface–atmosphere exchanges of energy and water. These are described in Section 5.3. Sections 5.4 and 5.5 consider activity since the mid-1990s, when the scope of studies diversified to include documentation of and modelling of additional atmospheric exchanges, with particular emphasis on carbon dioxide. In Section 6, the first-named author provides reflections drawn upon experience during a long career dedicated to this branch of hydrology defines what he feels to be some of the major contributions of research of the physics of atmospheric interactions at Wallingford.

5.2   Creating New Measurement Methods and Understanding

5.2.1   The early years

Research into atmospheric interactions began at Wallingford in 1966 with the creation of the Hydrometeorology (Hydromet) Section housed in a prefabricated cedar-clad building adjacent to the more permanent office and laboratory accommodation of the then new Hydrological Research Unit (HRU and the launch of the Thetford Project. The Hydrometeorology Section was motivated by the need to answer the question 'Do forests use more water than grassland?' Answering this was important because sheep grazing in upland river catchments was increasingly being replaced by forest and there was anxiety that water resources might be affected. This issue was well recognised and was indeed the initial purpose behind establishing HRU. At that time, it was being addressed by traditional hydrological catchment methods, but although these can define 'what' happens when trees are planted, they do not answer the question 'why' it happens. It was believed that the potential maximum amount of water lost from forest by evaporation could be estimated using the well-established Penman equation (Penman, 1948), but how close this estimate was to reality and what controlled transpiration and evaporation needed closer examination. The Penman-Monteith equation (Monteith, 1965) had developed the Penman equation to explicitly represent the dependency of transpiration on vegetation, and there was emerging recognition that evaporation

could no longer just be estimated as the residual ('unaccounted for' term) in a water balance study. A major study at Plynlimon in Wales was already comparing two catchments, one mostly grassland and a second predominantly planted with conifers. However, the nature of the terrain and the long time-scale inherent in catchment studies were unsuitable for measuring day-to-day details of the forest processes involved in the hydrological cycle. To overcome this problem, the Head of HRU, Jim McCulloch, recruited John Stewart, initially on secondment from the UK Met Office, to measure as accurately as possible the evaporation from forest. John began by undertaking a tour of related research work worldwide and a review of pertinent literature to inform the design of an experiment. This provided information about the type of site and equipment needed for the precise measurement of water use by forests. As early as 1964, it was recognised that instrumentation advances and the automation of data collection and analysis would be crucial in evaporation research, especially over forests.

To gain further insight, an analysis of 85 years of Oxford records were used to evaluate monthly Penman evaporation totals to investigate the long-term variability of evaporation. Although the Penman equation (Penman 1948, 1963) shown below in SI units, sought to adequately describe evaporation from surfaces plentifully supplied with water, it was recognised that this equation might be inadequate for predicting evaporation from plants that exert some control on their water use, including forests.

$$\lambda E = \frac{\Delta Rn + \rho_a c_p (\delta e) g_a}{\lambda(\Delta + \gamma)} \text{ kg m}^{-2}\text{s}^{-1}$$

where λ is the latent heat of vaporisation of water (J kg^{-1}), E is evaporation (kg m^{-2}s^{-1}), Δ is the rate of change of saturated vapour pressure (SVP) (Pa K^{-1}) at air temperature, Rn is net radiation (W m^{-2}) the measured or estimated energy available for evaporation from the free water surface, ρ_a is the density of air (kg m^{-3}), c_p is the heat capacity of air (J Kg^{-1} K^{-1}), δe is the vapour pressure deficit (VPD) (Pa), g_a is the momentum surface aerodynamic conductance (m s^{-1}), and γ is the psychrometric constant (Pa K^{-1}).

Such consideration led John Monteith to devise the Penman–Monteith equation that explicitly defines a dependence of transpiration on vegetation

In writing this chapter I, Colin Lloyd, have been partially guided by the paper written by Jim Shuttleworth in 2007 which documented 'the origin and development of the science of natural evaporation from land surfaces' during the period 1972–2007 (Shuttleworth, 2007). That paper charts a detailed mathematical and physical voyage through the history of evaporation theory and practice during the last 50 years, whereas this chapter provides a personal view of the interaction of the people involved during the last 50 years and their contribution to the furthering of aspects of evaporation theory, practice and results. This research was developed initially and predominantly by the Hydrometeorology group and the subsequent Hydrological Processes Section from its foundation in 1966 to the present time when evaporation research has expanded to include many of the other complex interactions in atmospheric hydrological processes. Along the way, many other aspects of research at Wallingford became entangled with the flux research, especially soil and individual plant evaporation and carbon dioxide measurements. I have tried to maintain a path that describes the history of the group of scientists at Wallingford that were involved with turbulent eddy correlation flux measurements while bringing in those other aspects that had an influential effect upon the overall atmospheric hydrology research.

The IH/CEH Hydrometeorology research programme has always been driven by the need to respond positively, quickly and effectively to international environmental concerns and developmental needs. It has also from the outset been intimately linked to and guided by the relevant modelling studies that are often needed to address these concerns and needs. The Group has always been prepared to exploit the latest technology and either use or develop it to advance its measurement capability. Initially evaporation and its interaction with the surface energy balance between the soil, vegetation and atmosphere was the main concern of the research team at Wallingford. In later years, as international focus shifted and novel instrumentation became available, the group expanded its research into the interaction between evaporation and the carbon balance.

by creating a so-called surface resistance term (r_s), a key advance in understanding forest transpiration (Monteith, 1965).

$$\lambda E = \frac{\Delta(Rn - G) + \rho_a c_p \left(\dfrac{\delta e}{r_a} \right)}{\Delta + \gamma \left(1 + \dfrac{r_s}{r_a} \right)} \text{ kg m}^{-2}\text{s}^{-1}$$

where G is soil heat flux (W/m^2) and r_a is aerodynamic resistance (s/m) – the reciprocal of g_a.

Another equation that would constantly be referred to when assessing the accuracy and precision of our attempts to measure evaporation is the general energy balance equation:

$$Rn - G = H + \lambda E$$

where H is the sensible heat flux (W/m^2). All of the components could be measured with a variety of different sensors, and the ability of these sensors to measure accurately was always how close we could get to balancing this equation. A major difficulty is that the elements on the left-hand side are generally static measurements over a small area above or below ground respectively, whereas the right-hand side elements are typically the summation of individual sources from a large upwind area, an area that dynamically changes with time in direction, size and shape. This mismatch in measurement samples, while not a huge problem over large areas of homogeneous vegetation, would prove to be a major headache over the more globally common heterogeneous vegetation terrain.

From the experimental standpoint, of particular concern was the degree of accuracy that could be obtained for each of the required measurements and how to adequately house, maintain, measure and record a large number of sensors in natural environments distant from mains power. Thus began the Hydromet Section's love for specially designed, instrumented caravans and sheds.

The combination of the energy balance with the Bowen ratio method was the starting point for measuring evaporation. With this in mind, a theoretical study was initiated of the probable errors in the determination of dry bulb temperature and of wet bulb depression (two measurements crucial to the

Bowen ratio method) using natural ventilation in a small instrument screen.

The aboveground measurement of evaporation rates were to be compared, and contrasted with, the development of a continuous weighing lysimeter based on the design of Forsgate et al. (1965). Perceived problems with this design were the effect of temperature change on the weighing system and the constructional difficulties when enclosing an in situ soil/vegetation monolith in field situations. Also undergoing tests and modifications at this time was the HRU Hydrometeorological Station Mk III, a stand-alone climate measurement array that eventually became commercialised as the Didcot Automatic Weather Station (AWS) then, as now, providing long-term data for evaluating local time series estimates of evaporation from many locations around the world. A pilot study of the feasibility and operation of these AWSs at different altitudes was begun at Plynlimon. From the start, this instrument was designed to be simple, robust, and relatively cheap.

Investigation of the feasibility of both conducting a study of the evaporation from forests and the assessment of the required accuracy of measurement continued during 1967. Nick Cowell joined the group to provide the necessary construction skills needed in the planned forest programme. Use of already established catchment sites was ruled out because of the perceived difficult turbulent regime at these upland sites. Extensive areas of level uniform forest (a condition considered necessary at the time for accurate investigation of turbulent characteristics) in the United Kingdom are rare. Thetford Forest, an area of flat, dry heathland (locally called Brecklands) on the border of Norfolk and Suffolk in East Anglia was one of only two sites that satisfied the criteria. It was also the closest to Wallingford. A suitable experimental site in the middle of a large area of Scots and Corsican pine well away from local habitation and possible interference with the expensive equipment was identified, and with the permission of the Forestry Commission, who controlled the area, the Thetford Project was born.

The project was conceived with fairly limited objectives reflecting the initially small number of staff working on it, and in the first years experimental work was concerned with instrument and systems development and building the infrastructure required when working over a tall forest. The Scots pine trees at the Thetford site planted

between 1928 and 1932 were 15 m high. To enable measurements at various heights above the forest canopy required robust towers of 30 m height to support possibly heavy equipment and allow easy maintenance of the sensors. These towers needed to be easily erected by the Hydromet staff and, if required, dismantled and moved to a different location. Commercial folding section scaffold towers were chosen, the type used for access to ordinary buildings that could be erected by a minimum of four people using just pulleys and rope. The manufacturers provided drawings and specified the number of tubes and guys required for such a height, but single column towers of this type had never been put up to 30 m before. However, the experience gained over the next few years in erecting these towers, using simple pulleys for hauling sections to the top of the tower and crossbows (with a fishing line attached to the crossbow bolts for firing guiding lines for the numerous steel guy wires necessary to stabilise the towers), proved of immense value. This was particularly so when in the 1980s, and after, towers nearly twice as high were erected by the Hydromet Section over tropical rainforest in the Amazon and the Cameroon.

The number of towers at Thetford rapidly grew to two 30 m and two 18 m towers (Fig. 5.1).

The proposed large number of instruments to be operated all at the same time required power, so a Petter 10 kW diesel generator set to provide independent power was positioned some 100 m east of the first 30 m tower. A vehicle, usually a Land Rover, was always kept on site for safety in case of fire. Eventually, the site was coupled to the outside world when a request for a telephone line to be installed was granted. This involved laying a 30-mm diameter cable in a 2-km long-deep trench from the nearest road to the site, all for the (at that time) standard telephone line installation charge of £25.

Thus began the Thetford experiment with the initial scientific objective of determining the difference between actual forest transpiration and the then current estimates that assumed forest leaves behaved as free water surfaces in the Penman formula. Of equal interest was how important in the process of evaporation was advection, in the form of downward transfer of sensible heat from the air to the forest canopy compared to direct solar energy input to the canopy. It was recognised that temperature gradients, wind profiles and radiation measurements would be essential and would need

Fig. 5.1 The three tallest towers with the TIS machines on the middle tower. (Source: Reproduced with permission of CEH.)

to be measured with greater accuracy and precision than those routinely used over other terrestrial surfaces, because efficient mixing of the air creates very small atmospheric gradients above forest. It was also recognised that the tower structure would also cause possible errors in the measurements. Such considerations and investigative anemometer experiments guided the construction and placement of lightweight booms of sufficient length to minimise the effect of the tower. A team from Southampton University's Institute of Sound and Vibration Research collaborated with the Hydromet Section to study the turbulent wind flow structure over Thetford Forest using sophisticated hot-wire anemometry. The results of this study pointed to the possibility of using eddy correlation techniques to measure evaporation – a prediction that came to fruition some years later.

Sylvia Marlow and Howard Oliver began working for the Thetford Project during 1968, both Sylvia and Howard having gained their physics degrees at Oxford University. John Gash then joined the Hydromet Group from Imperial College London in November 1969.

5.2.2 Establishing the Thetford project

Experimental work began at Thetford using newly developed Hewlett-Packard quartz crystal thermometers (Swiontek and Hassan, 1965) capable of precision measurements to hundredths of a °C. Profiles of up to 60 of these sensors, serially and rapidly controlled by a Hewlett Crossbar Scanner were installed on one of the 30 m towers, whereas the other 30 m tower had profiles of sensitive photoelectric anemometers with lightweight polystyrene rotor cups. An unusual form of wet bulb consisting of a porous ceramic tube covering the quartz thermometer and supplied with water from a reservoir provided the humidity measurements.

Above canopy was not the only research performed at Thetford. From the outset, other teams measuring various components of the forest water balance were involved in the Project. Two research assistants, plant physiologists Paul Robins and John Roberts, under the direction of Jack Rutter of Imperial College, began the first of many seasons at Thetford measuring the leaf-atmosphere and soil-root biological controls on transpiration using the shorter towers for access to the canopy. And the IH Soil Physics and Physics Groups began work on the distribution of soil moisture and other characteristics of the sandy soil, while an investigation into the components of rainfall interception was started.

A Hewlett-Packard 2116B computer, called at the time a 'mini-computer', but which weighed the same as a small adult and the size of a family-sized fridge was procured to control the increasing number of sensors and to collect, record and analyse the large amount of data being generated at 10-second intervals. This robust computer, built to US military specifications, and first introduced by Hewlett-Packard in 1967, contained an 8 K 16-bit wide memory consisting of individual ferrite-core memory locations. It was claimed that it could be dropped onto concrete from several feet without damage – occasionally this nearly happened. The computer, which operated at a speed of 300 K adds per second and cost in the region of $26,000 (c.f. a modern Apple iPad which has a memory 125,000 times larger, operates over 3000 times faster and costs 50 times less than the HP2116B). Communication with the computer was via toggle switches on the front or punched paper tape fed through a mechanically driven optical tape reader. Data were stored on punched tape or as printed hard copy from a Teletype teleprinter (Fig. 5.2).

As with all new systems, especially one as novel and complex as this system, which can now be called a true data acquisition system (DAS), there were teething problems and breakdowns. Some were mysterious, until a wayward transistor was found to be shorting out by touching an adjacent electronic board – easily solved by the field worker's solution, insulating tape. The team was still very small for the amount of work required to get the project running, so initially a skilled systems analyst was contracted to write the software for the DAS to run the experiment, hopefully to start during the summer of 1969. But the season was fast disappearing and the software was still not ready and, in any case despite being very elegant, the programme was twice the size of the available memory. However, during this

time, it was used at Wallingford, for the calibration of instruments and analysis of data in its new caravan home, together with the other electronics, and a new digital scanning voltmeter. But faced with the possible loss of a year's data, Sylvia (now married to Howard Oliver) put together a simplified version called STANDBY as a stop gap based on the original software from Hewlett-Packard, so that vital preliminary data could be collected.

Sitting in the forest ride surrounded by an array of reference books and computer coding, Sylvia took over programming the computer, deconstructing the incomplete software and learning both FORTRAN and machine code in the process. By the following spring there was a new modular system consisting of very short programmes which each performed a single task and could be tested individually, with redesigned drivers for the hardware, all under the control of a simple programme, MAIN. The requirements of the experiment were paramount and nothing was sacrificed, but every trick that could be found to keep the coding short and efficient was incorporated. With just 16 K of computer memory this data acquisition system initiated and handled readings from over 100 instruments every 15 or 60 seconds, applied calibrations, checked that each measurement was within the correct range and marked those that were not. It calculated 5-minute, 20-minute and hourly averages and printed out the results as tables complete with headings and units. The immediate access to results meant that problems with the equipment could be quickly identified and sorted out and time was not wasted collecting useless data. It was essential that the whole thing could be run by people with no computer knowledge and it was a simple matter of pressing a couple of buttons and typing in the date and time to set it all going. Even the most unlikely set of circumstances or operator mistakes were accounted for by the software. As more and more instruments and a graph plotter were added in following years, the system was upgraded and the original computer was replaced by a solid-state version with a 32 K memory. The system even controlled the physical movement of instruments up and down the tower at one stage. When Sylvia left full-time employment at IH in 1971, she continued to work re-programming the system each year. MAIN ran successfully for many years in various forms and at different sites, but the core structure always remained the same. The whole system, after being used at Wallingford for

Fig. 5.2 The Thetford Data Acquisition System with the HP 2116B computer on the right. (Source: Reproduced with permission of CEH.)

5.2.3 Results from Thetford

Results began to flow from the Project. Those on the theory of evaporation and from the 1969 season were given by (Stewart and Oliver, 1970) and Stewart (1971) described the diurnal progression of Thetford forest canopy albedo. Forest evaporation on dry days, at least, was shown to be substantially less than from grassland or short crops. The sustained time series of forest micrometeorological measurements recorded at Thetford were only equalled by a few groups around the world, for example, by Tom Denmead in Australia (Denmead, 1969) and Baumgartner in Munich, Germany (Baumgartner, 1969).

Colin Lloyd joined the Group in September 1970 from the Atomic Energy Research Establishment at Harwell, who now takes up the narrative. At the job interview Howard Oliver was satisfied that the ability to climb towers and have a head-for-heights ticked the right boxes, but John Gash questioned whether it was a good idea to employ someone who wore a floral shirt with a tie made from the same material. Anyway the job was mine and not a single human resources person in sight. However, as a first test, I was almost immediately sent to Thetford with Nick Cowell to erect our new 36-ft office and daytime accommodation shed, because by now the Group was six people strong and the two instrument and computer caravans were becoming overcrowded for desk analysis and the necessary tea and dinner breaks. A nice touch was the matching blue (male) and pink (female) chemical toilet huts in the forest. Facilities-wise, the site was a forerunner of the Elveden Centre Parks holiday camp that was established just down the road from our site in 1989. For the next few seasons we all stayed overnight at the Crown Inn,

ancillary experiments, was finally scrapped, still working, in the 1990s. Further details and how wind speed measurement was managed by the computer are given in Oliver and Oliver (1973).

in Mundford, some six miles north of the site – a pub famous for housing several members of the 'Dad's Army' TV series during filming.

While I was 'learning the ropes', several more papers emerged from the Project. For example, Oliver (1971) analysed wind profiles within and above the forest canopy over a wide range of stability conditions. A theoretical estimate of the change in wind direction between the top and bottom of the forest canopy, indicating a wind shear of 30°, was confirmed by experimental measurements (Smith *et al.*, 1972).

The project was gaining recognition when Jim Shuttleworth, Chris Moore and David McNeil joined the Section. Jim arrived with a PhD in nuclear physics from Manchester University, Chris came from Flinders University, Adelaide, Australia where he was still working for his PhD and David was a development engineer with a degree from Reading University.

David almost immediately got to work on designing the thermometer interchange system (Fig. 5.3), a large and heavy instrument system interchanging aspirated wet and dry bulb quartz thermometers over a distance of 10 m on one of the 30-m towers. Despite intensive inter-calibration of quartz thermometers deep in soil bore-holes at Wallingford, there were still systematic offset errors in the sensors outputs. Interchanging the thermometers at frequent regular intervals removed this error. Motors, chain wheels and proximity sensors became common talk alongside counter-gradient flow and wind shear. Alongside the pioneering efforts of Black and McNaughton (1971), adaptations of this method (McNeil and Shuttleworth, 1975) became a standard for measuring the Bowen ratio. However, the technical difficulties and cost of this method in applying the Bowen ratio/energy budget method to measure forest evaporation initiated an investigation of possible alternatives such as the eddy correlation technique. But before this system was operational, Gash and Stewart (1975) showed that profile measurements of temperature and humidity could be used to calculate the Bowen ratio and surface resistance of a forest and they produced an average diurnal surface conductance curve that indicated the degree of control of loss of water that the trees exerted through their stomates. When combined with long-term AWS data, they later estimated that in 1975, the trees transpired 353 mm from a rainfall input of 595 mm.

The work was certainly gaining recognition by this time. An episode of 'Bellamy on Botany' entitled 'You can't see the Wood' was filmed around the experimental site and transmitted on BBC TV in November 1975. Visitors also began to arrive for extended stays. Alasdair Thom, already the co-author of the eventual classic textbook, 'Essentials of Meteorology', arrived on sabbatical from Edinburgh University as did Lloyd Gay, on sabbatical leave from the Department of Forest Engineering, Oregon State University. Alasdair's collaboration with John Stewart produced the very important Stewart and Thom (1973) paper.

Below the canopy, the measurement of the separate components in the rainfall interception process was advancing with John Gash taking the lead with me as assistant. At last – something I could understand; water being collected in troughs and running down trunks and being funnelled into raingauges. This was kitchen sink science to me. In 1974 Ivan Wright had joined IH, then he successfully secured a 2-year post at Plynlimon after showing a willingness to climb towers. Part of his new job was looking after the Interception Project installation at Plynlimon. This was the first expansion of the interception work outside Thetford which would eventually include and compare sites at Thetford, Plynlimon, Kielder Forest in Northumberland and Roseisle near Elgin on the Moray Firth, Scotland (Gash *et al.*, 1980). Ivan returned to Wallingford in 1977 and continued working with John Gash, John Stewart and myself, before transferring to Ian Calder's Section in 1980, where he, like Dave McNeil in the Hydromet Group, operated as an engineer-amongst-scientists. He continued interception studies with Richard Harding, Robin Hall and Paul Rosier developing the plastic sheet net raingauge method. He also helped to develop a prototype gamma ray attenuation system with Barbara Olgiska of Strathclyde University for use in her PhD studies. This novel method of estimating rainfall interception and canopy storage values was subsequently moved to the Queen's Forest near Aviemore where further rainfall storage data were taken – but more impressive was its use to record real-time snow canopy storage measurements and develop some unique snow sublimation modelling.

The link with Jack Rutter, one of the pioneers of rainfall interception measurements in the UK, through his Research assistants, John Roberts and Paul Robins at Thetford and Alan Morton at Imperial College, led to a collaboration which produced the paper by Gash and Morton (1978). The IH version of Jack Rutter's differential equation canopy running-balance model of interception loss (Rutter *et al.*, 1971) that I 'coded up' for Gash *et al.* (1979) was heavily criticised by Sven Halldin of Uppsala University – not a good start to my publishing

Fig. 5.3 Two TIS Systems to account for all wind directions. (Source: Reproduced with permission of CEH.)

career; Sven told me to open the window and throw it as far away as possible. I promptly did this and wrote a finite-difference version that provided more operational options than the original. Meanwhile John Gash was developing his own Analytical Model of interception loss (Gash, 1979) – 'Big Theory One' he called it. This theory has now been adopted in many instances around the world as accepted practice.

Another novel approach to evaporation was John Roberts' classic tree-in-a-bucket experiment whereby a Scots pine tree is suspended from a scaffold support, its trunk close to the ground cut through and rapidly placed in a large 50 gallon plastic bucket full of water. A further slice of trunk is then cut away underwater exposing a fresh surface to the water. Transpiration loss is then just how much water you have to add to the bucket to maintain the same water level over a period of time. Apart from being good fun, the purpose of this experiment was to provide an explanation for the discrepancy between two independent estimates of evaporation from Scots pine trees in Thetford. Soil moisture studies by David Cooper (Cooper, 1975) had shown soil water depletion to be approximately 1 mm d^{-1} less than the estimate of evaporation obtained by John Stewart and Alisdair Thom using the Bowen Ratio method (Stewart and Thom, 1973), a discrepancy that could be explained by depletion of water stored within the tree. Unfortunately, the bucket experiment showed that water storage within the tree was most likely to be an order of magnitude less than that necessary to explain the difference. Despite this, the experiment did highlight that any plant physiological theory on root-to-shoot connectivity would need to explain John Robert's findings in this paper.

5.2.4 Expanding studies

At the same time as this research was underway at Thetford, a different approach was being undertaken by Ian Calder at Plynlimon using a 'natural lysimeter' to separate and quantify the components of the surface water balance. As noted above, rainfall interception studies had also been started by the Hydromet Group at Plynlimon to complement those being undertaken at Thetford. The results of this inter-group collaboration appeared in Shuttleworth and Calder (1979) where simple

models of forest evapotranspiration using the combined data showed that the extent of the impact of forests on water resources was critically dependent on the frequency and nature of precipitation at the forested site.

Jim Shuttleworth began to flex his theoretical muscles with Shuttleworth (1975, 1976a, 1976b,b) that also produced a 'discussion' between him and John Monteith (Monteith, 1977; Shuttleworth, 1977).

At this time, the original Hydromet Group began to diverge as IH expanded. John Stewart formed a new group within the Hydrological Processes Division called Vegetation–Atmosphere Interactions and took Howard Oliver, John Roberts and Nick Cowell with him, later to be joined by Jim Wallace, Ivan Wright and Henry Gunston, leaving Jim Shuttleworth to head up the renamed Evaporation Flux Studies Group with Chris Moore, John Gash, David McNeil and myself (Fig. 5.4), essentially to develop and exploit the eddy correlation method. Little did I know it at the time, but this development would eventually propel the new group to all corners of the Earth and to all biomes.

Arch Dyer and Bruce Hicks at CSIRO in Australia had already developed a system called the 'Fluxatron' to produce eddy correlation measurements of sensible heat using propeller anemometers and analogue electronics (Dyer et al., 1967). Chris Moore arrived at IH with a prototype Hyson–Hicks infrared hygrometer (Hyson and Hicks, 1975). If this worked, not only would the previously used delicate and problematic wet fine-wire hygrometer be a thing of the past, but also with no physical sensor presence within the measurement zone, more accurate measurements would become available.

Between Jim Shuttleworth, Chris Moore and David McNeil, an integrated eddy correlation system began to appear – called the HYDRA after the many-headed mythical monster. Initially, it also meant that every time we solved a problem, several more would seem to appear. Required was an alternative way of measuring the vertical windspeed to replace the inertia-prone and delicate mechanical propeller anemometers – solved when a phase-locked looped ultrasonic anemometer was developed (Shuttleworth et al., 1982). An IH version of the Hyson–Hicks hygrometer needed to be miniaturised to fit into the integral frame that David McNeil and the IH workshop staff were also designing and building, both instances where the

Fig. 5.4 The Evaporation Flux Studies group circa 1978. Clockwise from top left: David McNeil, Colin Lloyd, Jim Shuttleworth, Chris Moore, John Gash and Howard Oliver (to depart to join the VAI section). (Source: Reproduced with permission of Colin Lloyd.)

design and prototype manufacturing skills of the workshop proved indispensable and cost-effective.

Some fundamental study comparisons were initially performed between the individual sensors and commercial eddy correlation equipment, such as the new Kaijo Denki DAT-311 sonic anemometer which included comparison with previous studies (e.g. Kaimal *et al.*, 1972). The operational characteristics of the completed device were then investigated by Chris Moore (Moore, 1983). Aiding and abetting in all this was Bob Frost, a contract

electronic wizard who could tell where electrical circuits were failing just by touch alone – how cool (or hot) is that! A small electronics company (Bjorn Board Electronics) situated in Oxford designed and constructed sensor conditioning and power boards that fed into an analogue-to-digital converter (ADC) board (see below). Chris Moore then produced the paper that allowed frequency response corrections to be applied to the physically separated and disparate sensors of the HYDRA (Moore, 1986). This comprehensive paper is still used to formulate the corrections for modern eddy correlation systems. Now, we just needed a suitable computing system to measure, manipulate, correct and store the data from the HYDRA coming in from the sensors at a rate of 16 times a second. RCA had just created new low-power CMOS microprocessors – the CDP1802 COSMAC (Complementary Symmetry Monolithic Array Computer) chip introduced in 1976 and RCA also produced a microboard computer (CDP18S601) and the ADC board (CDP18S643) designed around the 1802 chip.

I was very surprised when Jim Shuttleworth asked me to program the HYDRA. I knew a bit about FORTRAN – but this computer could only be programmed with Assembler machine code, where it takes half a dozen instructions just to operate on a number. Nevertheless, the energy within the Group at this time was extraordinary with members all working in one lab all the hours available to get the HYDRA operational – so I said

yes. It took six months of solid work to learn how to write assembler efficiently and effectively enough to create a program that would scan the sensors, multiply the correlations, apply moving averages, create flux and variance calculations and update running summations every 100 ms (Lloyd *et al.*, 1984). The assembler code, written and developed on the COSMAC development system eventually ran to 16 pages of A4 fan-folded paper.

The initial design measurement rate of 16 Hz only allowed 62.5 ms per measurement cycle, but this eventually was shown to be impossible to achieve and the measurement sampling rate was dropped to 10 Hz. Even then, the timings were so tight that at the end of every hour, a 100 ms measurement cycle had to be given over to writing the data out to a paperback book-sized solid-state 16 KB CMOS Memory bank (GK Instruments, Milton Keynes, UK). Now having a complete working HYDRA (Fig. 5.5), we took it to Thetford

Fig. 5.5 The Mk1 HYDRA. (Source: Reproduced with permission of John Gash.)

for field tests, set it up and turned it on. Unfortunately, we did not know whether it was working because David McNeil in his continuous pursuit of minimising power consumption had installed an LED on the computer board that initially lit up when the system was powered on, but went out when the system was running. Unfortunately, this meant that when the light went out the system was either working, or if the power had failed, not working. David had also thoughtfully put a fuse on the main power line which proved to be serendipitous, because as the battery became exhausted, the voltage dropped but the amperage increased to alarming levels.

Having designed and created a portable eddy correlation system that could be run unsupervised from a 12-V car battery and solar power and erected on a simple mast, we needed opportunities to exploit it. We had a system that was demonstrably able to make routine eddy correlation measurements of surface exchanges over extended periods with limited supervision. This advance provided other researchers with a new confidence in both the eddy correlation technique and the ability to afford and operate a technique that up to then had been resource and manpower hungry, especially over forests where access is difficult and alternative surface flux measurement methods are less effective.

One of the first IH exploits of the HYDRA system was led by John Gash at Berners Heath a site near to Thetford. John's main role in the development of the HYDRA was to field test the instrument – a sort of in-house independent testing. Eight HYDRA systems had initially been made by the IH workshop and they needed to be tested. Rather than just put them all together in a field for comparison – something that probably would not have worked too well anyway as future research in Niger showed, John decided to use them in an experiment. During his student days at Imperial College, John had seen a presentation describing how Eric Bradley of CSIRO had investigated the change in the wind profile as the wind moved from a smooth flat tarmac surface to one composed of a bed of nails on an old airfield runway. Many people had subsequently used Bradley's results to test the usually complex models of boundary layer (BL) adjustment to a change in roughness. Concluding comments at the end of many of these BL modelling papers often quoted a value for the fetch:height ratio needed for equilibrium with the new surface to be

established as being 200:1 or 100:1. For example if the measurement was 2 m high, this "rule of thumb" stated that a fetch of 200 to 400 m would be required before the measurements were free of influence of the upwind surface discontinuity.

With the future plans for the HYDRA including deployment in places that had not been investigated before, including areas that were not as homogeneous as Thetford forest, the question of fetch needed to be put on a better foundation than just the 100:1 rule. Rather than look at the discontinuity in the profile as Eric Bradley had done, John realised that the HYDRA was ideally suited to look at the development of the equilibrium layer itself, something that no one had done before, and so devised an experiment in which turbulence downwind of a roughness change in the vegetation would be measured. Such an experiment would previously have been too expensive to perform (most research groups could only afford to buy and operate a single eddy correlation system). Berner's Heath was an ideal site for this experiment where a large flat area (150 ha) of heather-covered heath was situated to the south-west of a well-defined 10 m high Scots pine and European Larch border fronting an extensive dense plantation of similar height Corsican Pine. The eight HYDRAs were installed at intervals along a line from well inside the heathland to well inside the forest (Fig. 5.6) and data collection begun – not without difficulties – this was also an instrument testing facility as well. Affordable solar panels of the size to power the HYDRA systems were still some way off, so regular exchange of batteries was required, which during the winter meant at least on one occasion the two of us, like something out of 'Scott of the Antarctic', were seen dragging a sled of 8 car batteries across the snow-covered heath to replace the exhausted ones.

The wind, being fickle, seldom blew from the required directions, either directly from the heath onto the forest or from the forest onto the heath. And we had the problem of debugging the software and attendant processor failure – only solved once we reduced the data sampling time from its initial 16 to 10 Hz – there just was not enough time in the 100 ms to get everything done. John's results (Gash, 1986a) showed that the equilibrium layer developed faster than expected, and faster over smooth to rough than rough to smooth. In other words, even the 100:1 'rule of thumb' was a conservative value. Indirectly, the experimental philosophy of looking at the equilibrium layer also led to the development

Fig. 5.6 The Berners Heath–forest interface experiment showing two of the eight Mk1 HYDRAs on a transect between the heath and the forest. (Source: Reproduced with permission of CEH.)

of flux-footprint theory (Gash, 1986b) – which he called 'Big Theory Two'. In this, the upwind fetch is modelled as an array of emission sources in a two-dimensional flow, to quantify the upwind vegetation being sampled. This result is particularly useful for connecting remote sensing estimates of fluxes to micrometeorological measurements and was taken up by researchers in international experiments, such as FIFE, who expanded the footprint theory, introducing realistic wind profiles, cross-wind diffusion, the effects of atmospheric stability and more realistic diffusion models.

Also working at the Berner's Heath site at this time was Robin Hall, who had joined IH in 1979 from the University of Birmingham. Using a wet-surface weighing lysimeter system (described by Calder *et al.*, 1982, 1984), Robin investigated important interception factors such as drainage and evaporation rates and the throughfall coefficient of

the Berner's Heath heather (*Calluna vulgaris*). One of the important findings (Hall, 1985) was the difference between canopy storage at which drainage ceases and a smaller storage value necessary for a completely wet canopy.

Further seasonal tests of the HYDRA were performed at flat extensive sites at Stadhampton near Wallingford in 1978 and Kings Lynn in Norfolk in 1979.

Despite my poor academic record up to this time, Jim McCulloch and IH, funded my part-time Open University BA degree course begun in 1979. I chose mathematics again (the books were lighter for carrying to fieldwork locations) – the degree was finally awarded in 1988.

About this time, Richard Harding joined IH. For the previous 3 years, he had been based at Wallingford, but working with Frank Green on a couple of grant projects from Oxford University. After working for a short while with Howard Oliver, he transferred across to Ian Calder's section and began IH's interest in snow evaporation by initiating a programme of work at Aviemore, Scotland with plastic net rainfall gauges and gamma ray canopy snow evaporation studies alongside Strathclyde University. This was then expanded to a field site in Finse, Norway (with Chris Moore and Liz Morris), then Austria in 1986 (with Liz Morris and Alan Jenkins), followed by an expedition to Greenland led by Henrik Soegaard of Copenhagen University. This work showed the very large energy and evaporation differences between snow-covered forest and open snow surfaces – a result that was quite influential at the time. Eventually, this snow path would lead to participation in the Boreas programme in Canada and the LAPP work in Arctic Svalbard and Finland.

During this time, Jim Shuttleworth and Jim Wallace were developing a one-dimensional model to extend current descriptions of the energy partitioning over vegetated surfaces to accommodate the difficult subject of sparse crops (Shuttleworth and Wallace, 1985). All previous physically based models of the vegetation–atmosphere interaction (as described in e.g. Shuttleworth, 1976a,b) explicitly treated the vegetation as a closed stable canopy of uniform structure. Thetford forest, wheat and grass fields were all regarded as such surfaces. Extending this model required a set of combination equations that not only described the evaporation from vegetation and bare soil but provided a smooth transition from bare soil to closed canopy. This important model provided a tool to extend research into heterogeneous vegetated terrain – a tool that would prove invaluable in many scientists' future research.

5.3 Regional and Global Hydrometeorology Studies

5.3.1 The Amazon

The deforestation of the Amazon rainforest was beginning to appear in news headlines and its potential effect upon local and global climate was already being discussed. Salati *et al.* (1979) had demonstrated that about 30% of areal-average precipitation in Amazonia originated from evapotranspiration within the basin. Consequently there was international concern that this 'self-sustaining' tropical forest was at risk if deforestation continued

unabated. But there was also an unfulfilled need to properly quantify the energy and water exchange between the Amazon forest and the atmosphere. Very little was known about the evaporation and rainfall interception of the rainforest. The remoteness and height of many areas of pristine rainforest, and the cost, had dissuaded researchers from attempting to establish experiments in these regions, but Brazilian Luiz Molion was determined to try. He came to Wallingford on a one-year sabbatical to learn about forest micrometeorology. At the same time, IH Assistant Director Robin Clarke was in Brazil on a year's sabbatical. Robin also recognised the need for a study of Amazonian rainforest and that with the HYDRA and the Thetford forest experience, the Evaporation Flux Studies Section was in a good position to take such an initiative. It quickly became clear that an Anglo-Brazilian collaborative experiment was the way forward. Collaboration with several Brazilian research Institutions was established, including the Instituto Nacional de Pesquisas da Amazônia (INPA) in Manaus, Amazonia; the Instituto Nacional de Pesquisas Espaciais (INPE) in São Paulo and the regional office of Empresa Brasileira de Pesquisa Agropecuária (EMBRAPA) in Manaus.

Berner's Heath was used as a test bed for all the equipment going to Brazil, including putting up the TIS machine. All parts were labelled and then taken down to join a 52-m scaffold tower and the rest of the scientific equipment squeezed into two containers and sent out to Manaus. Included in the consignment was an ex-Thetford Project 8' × 6' wooden shed, whose inclusion, despite some mirth by the rest of the group, was defended by John Gash. Little did we know how important that shed was going to be in the initial construction of the forest site, for keeping both equipment and people reasonably dry during the frequent downpours before a large brick shed was eventually built at the site.

Jim Shuttleworth and David McNeil made a first reconnaissance trip in May 1982, visiting relevant scientific institutions in São Paulo (Instituto Nacional de Pesquisas Espaciais – INPE): Manaus (Instituto Nacional de Pesquisas da Amazônia – INPA), a UNESCO Project in Belem, and CNPq (Conselho Nacional de Desenvolvimento Cientifico e Tecnologico) in Brasilia, as well as possible sites around Manaus in order to establish the infrastructural and general site selection aspects of the project. Robin Clarke and

Luis Molion later visited and identified a suitable site in the protected Reserva Ducke, 25 km north of Manaus – marking the site with empty beer cans. Jim Shuttleworth, John Gash and John Roberts formed the expeditionary force with Chris Moore and I following a couple of weeks later in September 1983. By the time Chris and I arrived at Manaus on a Brazilian Varig Airlines McDonnell Douglas DC-10 that included a tail-end dance-floor and bar (Brazilians will dance anywhere), the problems of working in tropical rainforest were becoming apparent. John Gash's shed (no longer a laughing matter) and several small tents were the only accommodation on site and rapidly filling up with termites. The lorry (on loan from the army) to take the tower and equipment to the site had become stuck on the slippery forest track (Fig. 5.7).

We were therefore forced to carry everything on our backs some 3 km into the forest, but eventually the tower rose to 52 m, both 20 m higher than we had ever put a tower up and also 20 m above the forest canopy (Fig. 5.8). The tower would have gone up faster except for some initial confusion between the English and Brazilian words for 'Pull' and 'Push'. (The Brazilian Portuguese for pull is 'Puxe', which sounds like 'push', and Push is 'Empurrão', which sounds a little like 'Pull' from a distance.) A large generator was carried to the site and the foundations for the concrete block hut were built. Installing the tower guy wires became increasingly more difficult as the tower rose because the crossbow bolt carrying the guide fishing line would often be deflected by branches requiring retrieval of the line and bolt – which invariably became snagged in other branches. Such random deflections also meant that tree trunks were increasingly used to hide behind when the crossbow was being fired. Radiation instruments

Fig. 5.7 The tower-carrying truck with Jim Shuttleworth and John Roberts wondering what to do next. (Source: Reproduced with permission of John Gash.)

Fig. 5.8 The 52 m Manaus tower emerges over the extensive Amazonian rainforest. (Source: Reproduced with permission of CEH.)

and the HYDRA were installed on the tower, the TIS system was also installed and data collection began using the new Commodore PET desktop computer that had recently come onto the market to control and process the data collection. By this time, the termites had eaten the wooden gunstocks of the crossbows and the floors of the tents, although they did not seem to like the creosote on the British shed. Humidity was so high that nothing dried at ground level and fungus and mould would grow on anything from clothes to humans.

Meanwhile, I had volunteered for the task of creating and operating an interception experiment at the site. Having decided to attempt both a random array of raingauges on the forest floor and a

net rainfall gauge of the type that had been shown to work effectively in Plynlimon, the first task was to clear a 10-m square area of vegetation to lay down the heavy-duty plastic sheet. In temperatures of 35 °C and humidity at the forest floor of nearly 100%, it was not an easy job. At the same time, the enormous canopy area of the individual trees had already indicated to me that the normal 20 m square array of raingauge locations was not going to work. In plantation conifer forests in the UK, placement of a 20-m square array anywhere within the plantation would not only contain a statistically acceptable number of trees, but also the array area was deemed to be representative of the whole forest. In the Amazon forest, such an area may only contain the canopy of one tree, or perhaps 3 fern-like trees, or the canopy edges of several trees – none of whose trunks were actually in the area. Expanding the array area was not possible given the limited number of raingauges available so I devised a novel grid that was 100 m long and 4 m wide with locations marked at 1 m intervals to give 505 sampling points. Such a transect arrangement provided a better sample of the forest canopy. Thirty-six bottle raingauges were randomly relocated within this grid after each storm and provided throughfall storm totals. A further 16 tipping bucket raingauges randomly located, but to fixed positions recorded time series of storm duration and intensity. Nineteen stemflow gauges (13 bottle and 6 tipping bucket) recorded the intercepted rainfall running down the trunks. Making stemflow collars for buttressed and spiky fern palm trees was not easy. Incoming rainfall using a tipping bucket raingauge was measured on top of the tower. A further two tipping bucket raingauges were positioned on a platform at a height of 25 m on a tree that had lost its crown. Myself and Carvalho de Moraes of the Brazilian Universidade Federal do Pará in Belém, using a simple rope loop around our waists and the tree, spent a week with a hand auger, hammer and 30-cm long spikes creating a spiked ladder up to the top of the tree – only to be confronted by a 20-cm long centipede as I pulled myself onto the tree-top.

The results of the spatial variability of throughfall and stemflow and the subsequent modelling of the full interception process using John Gash's new Analytical Interception model proved the value of critically determining a sampling strategy consistent with the scale of the canopy and of randomly relocating raingauges (Lloyd and Marques, 1988). Previously, estimates of interception loss had been overestimated mainly because studies had removed apparently erroneous throughfall gauge catches that were greater than incoming rainfall (Lloyd et al., 1988). Having stood in the forest during storms and witnessed what were random canopy drip pathways that channelled water to particular positions, I knew why the previous values were overestimates. A coda to this interception work appeared in 1990 (Lloyd, 1990) when I showed that GCMs needed to be aware of different temporal distributions of rainfall if they were to correctly model rainfall interception.

The net rainfall gauge experiment was a lot of work with little success and was eventually

In the era before mobile or satellite phones, the only communication we had with the UK was by air mail or by telegram and we devised, in order to keep the price of our rationed telegrams down (being charged by the letter), created a lexicon of three letter abbreviations for words. Unfortunately, the telegram is notorious for getting some transmitted words wrong by one letter (e.g. TLX (=Telex) could become TLK (Talk) or TNX (thanks)), so quite often the telegram became unintelligible by the time it reached the UK.

Brazil also had hyper-inflation at this time with prices changing on a daily basis. Our unit of currency was the US dollar which we exchanged for Brazilian Cruzeiro every few days, generally on the 'parallel market', a national method to avoid the prohibitively high exchange rates at the Banks. Parallel market exchange was operated by nearly all shops – as the stable US dollar was much sought after. My first trip on the back of a local moped to a shop in the inner sanctuary of the local Chinatown area to exchange some money with a person who would not have looked out of place in a Bruce Lee movie was a bit unsettling.

abandoned. To achieve a suitable sampling area, the raingauge needed to be 90 m^2 in area. A net raingauge of this size in a non-plantation setting was not only difficult to erect, sealing the plastic sheet to randomly sited and sized trees, but also demanded daily repairs as insects created many holes in the sheet itself leading to a lack of confidence in the collected throughfall measurement. Equally, with storms occasionally producing rainfall intensities of 90 mm over 10 minutes, it was impossible to create a credible measurement system, even including two 500 litre buffer storage tanks and two large 1.5 litr tipping bucket raingauges to cope with storm volumes that could reach 9000 litres in 10 minutes. The lesson learnt was that there is no general solution to many problems, and methods have to be chosen to fit the application.

The remainder of the Amazon region meteorological experiment (ARME) continued to provide data from its combination of eddy correlation, radiation and anemometer equipment leading to several joint papers that cemented the long-lasting friendship between the UK Hydromet Group and the various Brazilian research groups that were operating alongside (Shuttleworth *et al.*, 1984a, 1984b). For many of these young Brazilian researchers, this was their first opportunity to experience hands-on running of a large micrometeorological experiment and the knowledge and experience they gained during our joint three-year experiment provided them with a platform from which they have progressed to be leading figures both in Brazil and internationally. The results of both the major collaboration and measurements were contained in a series of papers from 1984 onwards (Shuttleworth *et al.*, 1984a,b, 1985, 1987). Figure 5.9 (redrawn from Shuttleworth *et al.*, 1984a) shows daily partitioning of net radiation into the constituent latent and sensible heat fluxes for eight days in September 1983. Daily total evaporation for the transpiring canopy accounts for 70% of the available radiant energy – a figure that was two-thirds of the then conventional estimate of potential evaporation. This result, combined with the interception data reported in Lloyd and Marques (1988) and Lloyd *et al.* (1988), suggested for the first time that evaporation from Amazonian rain forest approached, or possibly exceeded, then currently accepted estimates of potential evaporation in wet months and evaporation would fall to 70% of potential rates, or less, in dry months.

Ann Henderson-Sellers of the University of Liverpool visited the site in early 1985 and left an all-sky camera and Graham Thomas to record three months of black-and-white photographs of the hemispherical sky every 30 minutes. Again this research showed that previous assumptions and measurements about the degree of cloudiness were probably wrong and that average daily total cloud amount was 84% (Henderson-Sellers *et al.*, 1987) .

In the wider context, there were important developments (Dickinson *et al.*, 1986; Sellers *et al.*, 1986) in the representation of land surfaces in general circulation models (GCMs), which meant that the interaction of different vegetation covers could be represented in model prediction experiments. However, what these models lacked were data sets to calibrate the models – without these, model outcomes were effectively possibilities rather than probabilities. Providing such calibration became a priority and the data from the Hydromet team's ARME study provided the first such calibration (Sellers *et al.*, 1989). It would not be the last – the linkage between understanding

physical processes and calibration and testing of the Soil-Vegetation-Atmosphere Transfer (SVAT) schemes at the heart of GCMs was forged positioning IH at the forefront of this combined science in the coming decades.

With the success of the ARME project, summarised in Jim Shuttleworth's 1988 Royal Society paper (Shuttleworth, 1988a), and with methods and instrumentation that could be successfully deployed in tropical rainforest under their belt, the Hydromet team moved on to investigate the local climate changes that deforestation was suspected of causing in the areas of Rondônia, Brazil where extensive and ever-expanding clearance of forest for cattle grazing and crops was occurring (Fig. 5.10). International interest in the potential consequences of this deforestation, not just regionally but also globally, had continued to grow leading to the next major new collaborative study between the Hydromet Group and their now experienced Brazilian co-researchers – the Anglo-Brazilian Amazonian Climate Observation Study – ABRACOS which began serious measurements towards the end of 1990.

Acronyms were the name of the game at that time and we were no different, so Jim Shuttleworth thought up the acronym ABRACOS – a nice collaborative feel to it as it means 'big hugs' in Portuguese. However, it also used to provide some merriment from local suppliers when writing ABRACOS on the order form under 'Nome do Projeto'. ABRACOS was predicated on two time scales (1) continuous measurements over a climatological or hydrological time frame and (2) atmospheric and plant physical processes studied during intensive measurement campaigns of weeks or months duration.

The project involved simultaneous measurements of the surface exchanges of three widely separated paired sites of natural forest and nearby post-deforestation pasture land (at Manaus, Marabá in the north-east State of Para, and near Ji-Paraná in the south-west State of Rondônia) (Fig. 5.11). Ivan Wright of Ian Calder's Physical Processes Group was co-opted into the Hydromet team for this experiment as my professional focus moved elsewhere. Results from the initial trial experiment indicated a rapid response of the grass to water stress with evaporation falling from 3.8 to 2.1 mm d^{-1} during a rain-free three week period (Wright *et al.*, 1992). Martin Hodnett and Anna McWilliam also joined to work with John

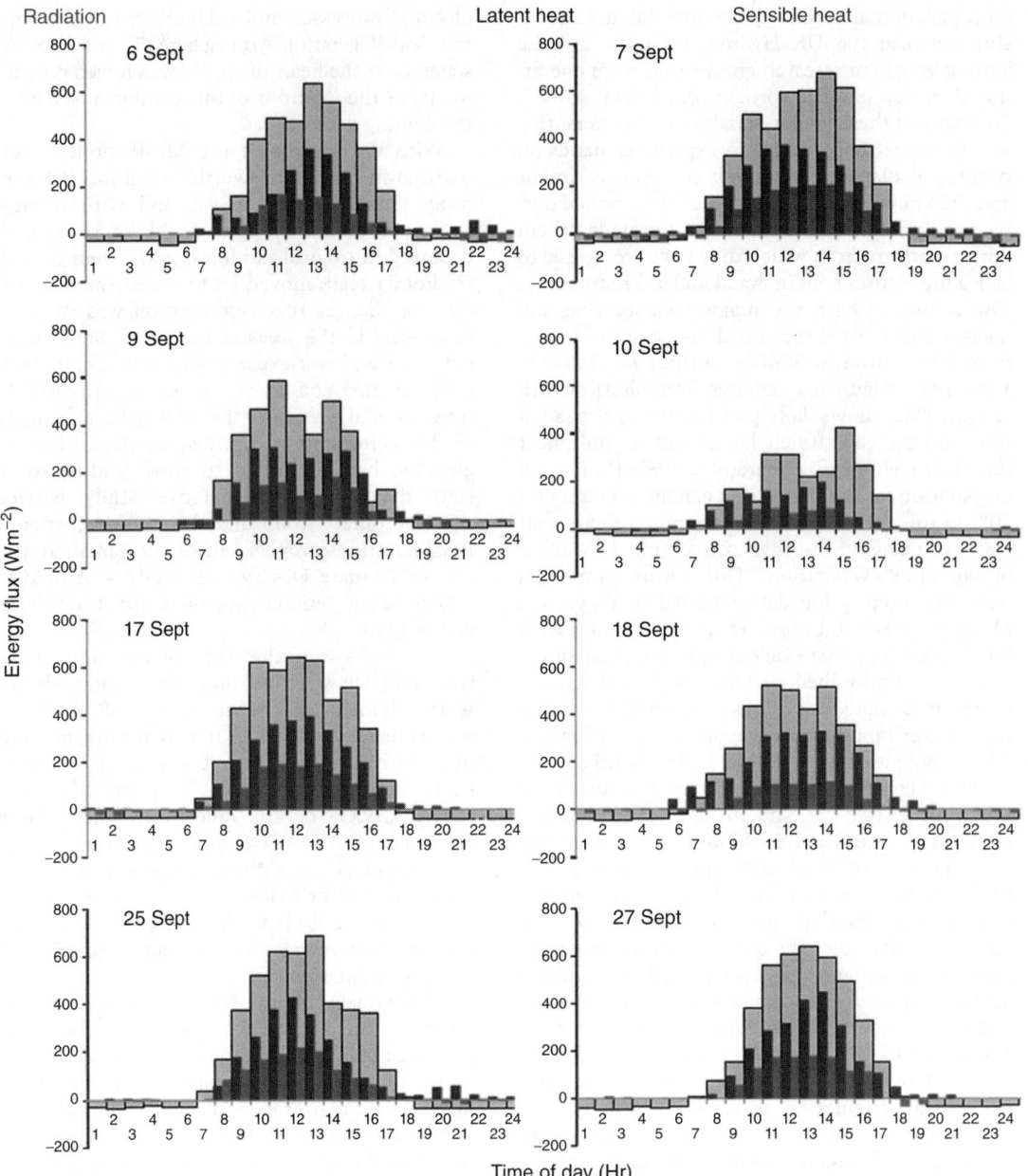

Fig. 5.9 Partitioning of energy at the Reserva Ducke site for eight days in September 1983. (Source: Reproduced with permission of Colin Lloyd.)

Fig. 5.10 Colour modified Landsat satellite image of the herringbone pattern forest clearance (in pink) as areas are cleared to either side of roads. (Source: NASA Open Source.) *(See insert for colour representation of this figure.)*

Roberts on aspects of in-canopy microclimate (see Cabral *et al.*, 1996) and soil moisture and forest sample biomass experiments, and Heidi Bastable joined the micrometeorological team and proceeded to show that there were significant differences in the local climate resulting directly from the clearance operations (Bastable *et al.*, 1993) – a result that had important implications for the general climate of the Amazon basin as a whole. The atmospheric hydrological studies included rainfall interception (Ubarana, 1996), isotopic composition of water vapour (Ribeiro *et al.*, 1996), microclimate of both forest and urban areas (Culf *et al.*, 1995, 1996; Maitelli and Wright, 1996). Many more of the results (Roberts *et al.*, 1990; Grace *et al.*, 1995a,b; Ribeiro *et al.*, 1996) were published elsewhere or contained in a major book (Amazonian Deforestation and Climate) co-edited by John Gash and John Roberts with two Brazilian colleagues (Gash *et al.*, 1996).

Fig. 5.11 One of the ABRACOS experimental sites at the Dimona cattle ranch, 100 km north of Manaus. (Source: Reproduced with permission of CEH.)

5.3.2 *Further international experiments*

While ARME was the start of the Hydromet Group's foray into international collaborative efforts, the general large-scale international experiments could arguably be regarded as starting with the Hydrologic Atmospheric Pilot EXperiment – MOdelisation du BILan HYdrique (HAPEX-MOBILHY) for the study of water budget and evaporation flux at the climatic scale. The work in Brazil on ARME and ABRACOS had become well known internationally. Pierre Morel (then WMO'S Director of the World Climate Research Programme) and Jean-Claude André (then Director of the French Centre National de Recherches Météorologiques) were both interested in establishing a series of mesoscale field experiments with France taking the lead with the pilot experiment, HAPEX-MOBILHY. They organised a meeting in Geneva to which Jim Shuttleworth and interested American scientists were invited. The outcome, for us, was that Jean-Claude André sent Jean-Paul Goutorbe of CNRM to meet Jim McCulloch at IH. Jim agreed to the participation of the Hydromet Group in this experiment and sent Jim Shuttleworth and John Gash to Toulouse to work out the details. In typical Jim McCulloch style (look after the pennies), they were told to take the cheapest route: fly by Gulf Air from London to Paris and then take a train to Toulouse. As Jean-Claude remarked 'You are travelling like students!'.

This was the start of the collaboration with French colleagues that effectively began in 1985 and has continued through to the present time. But the hydromet. capability at IH was beginning to get stretched, both geographically and in terms of the staff numbers. David McNeil and Howard Oliver moved to work elsewhere in IH, and Chris Moore had returned to Australia

HAPEX-MOBILHY was based within a GCM-sized grid square (10^4 km^2) centred in an area of south-west France south of Bordeaux known as Les Landes, an area of mixed coniferous forest and agricultural land (see André *et al.*, 1986 for details about the proposed programme and André *et al.*, 1988 for details of the first results from the experimental Special Observation Period).

The project organisation, comprising a Special Observation Period, where intensive coordinated measurements over both horizontal and vertical scales (by surface, tower, aircraft and satellite), integrated within a longer-term experimental programme, set the pattern for many subsequent international experiments that IH and then CEH were subsequently involved with, for example, Hydrological-Atmospheric Pilot EXperiment in the Sahel (HAPEX-Sahel), ECHIVAL Field Experiment in Desertification Threatened Areas (EFEDA), First ISLSCP Field Experiment (FIFE), Boreas, CarboEurope and African Monsoon Multidisciplinary Analysis (AMMA); ECHIVAL is the European international project on Climate and Hydrological Interactions between Vegetation, Atmosphere and Land surface and ISLSCP is the International Satellite Land Surface Climatology Project.

The IH involvement in HAPEX-MOBILHY comprised a fieldwork team of Jim Shuttleworth, John Gash and myself providing eddy correlation flux measurement and AWS systems located on top of a French tower over an area of Maritime pine (*Pinus pinaster* Ait.) forest 28 km south of Casteljaloux, together with a rainfall interception and soil moisture measurement programme. This was to be the first overseas operation of the new Mk2 HYDRA (Shuttleworth *et al.*, 1988). Much mathematical and wind-tunnel work had been done by Jim Shuttleworth, Chris Moore and David McNeil, together with the IH workshop fabrication expertise, to improve the theoretical, aerodynamic and practical characteristics of the sensor frame and sensors from its Mk1 form. Deployment of the Mk1 HYDRA over semi-arid crops in Syria in 1985 had also highlighted deficiencies in dealing with strong solar heating and large diurnal temperature cycles creating temperature-dependent drifts in the sensors and leading to erroneous downward evaporative fluxes in the real-time calculations. The resultant Mk2 HYDRA (Fig. 5.12) emerged as a sleek easily deployed, battery-powered eddy correlation instrument that would be at the forefront of eddy correlation experiments at remote difficult-to-access sites for the next decade or so until commercial products from Campbell Scientific, Gill Instruments and Licor came to market. Its uniqueness at that time was such that Tony Debney, the then Acting Director of IH, proposed that combinations of Mk2 HYDRAs with an attendant "expert" be marketed as a unit for use around the world (have HYDRA – will travel) – a proposal that was not exactly welcomed by the Hydromet Group – who regarded the HYDRA as just another, albeit highly sophisticated, tool to further IH research interests. In reality, the Mk2 HYDRA (and an 'expert' to install and operate it) did in

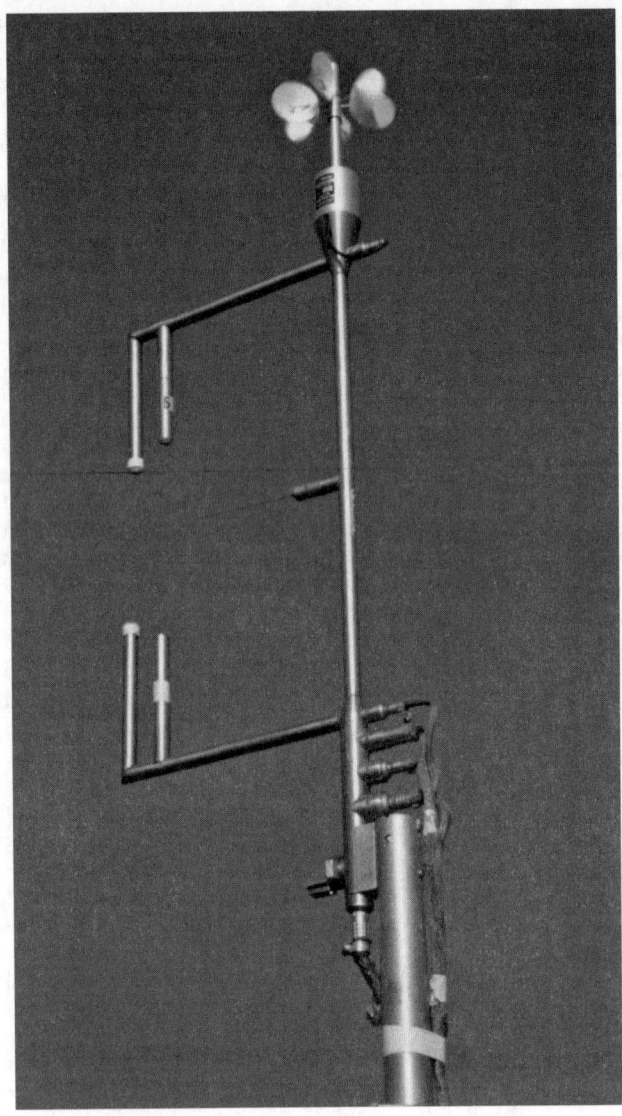

Fig. 5.12 The Mk2 HYDRA eddy correlation instrument. (Source: Reproduced with permission of John Gash.)

fact became the fulcrum instrument around which many research projects revolved right up to the present time.

John Stewart's analysis of the Thetford data had resulted in a seminal paper on modelling surface conductance, and the analysis of the transpiration data from Les Landes was carried out in the framework of the Jarvis–Stewart model, parameterising surface conductance in terms of solar radiation, temperature and specific humidity deficit (Gash *et al.*, 1989).

Compared to UK pine forests where the canopy is near-continuous, the Les Landes forest canopy was sparse with many open spaces between the individual tree canopies. The analysis of the rainfall interception data showed that John Gash's analytical model that had been used with success by IH in the UK and the Amazon and by others over various different forests around the world [e.g. evergreen mixed forest in New Zealand (Pearce and Rowe, 1981); oak forest in the Netherlands (Dolman, 1987); tropical plantation forest (Bruijnzeel and

Wiersum, 1987) and natural rainforest (Hutjes *et al.*, 1990)] did not work well for such a sparse forest. It became clear that, for sparse canopies, the original analytical model was creating both mathematical and practical inconsistencies when attempting to model the HAPEX-MOBILHY data, that could only be corrected by reformulating the model so that the evaporation rate per unit ground area reduced as the canopy cover becomes more sparse (Gash *et al.*, 1995). In other words, the original model was effectively spreading a sparse canopy thinly over unit area beneath a constant rainfall, rather than clumping the canopy in discrete blocks.

In a somewhat similar way, Jim Shuttleworth was encountering problems in attempting to model canopy evaporation in Amazonia. In these attempts to validate GCM models, especially the NCAR Community Climate Model (CCM) which used the Biosphere-Atmosphere Transfer Scheme (BATS) SVAT, the consequences of the average distribution of multiple local storm rainfalls across a grid square became apparent. The models were generally indicating very high wet canopy evaporation when there was insufficient energy to produce it. But in this case individual and localised storm rainfall was being spread across a grid square of complete canopy cover forest. The issue was made explicit as a potential problem for GCMs by Shuttleworth and Dickinson (1989). It was necessary to modify the GCM models so that one or more localised rainstorms within a grid square could be effectively modelled – achieved by clumping individual rainfall areas into a fraction of the grid square – see Appendix 2 of Shuttleworth (1988b) for derivation of a new distribution function, and Dolman and Gregory (1992) for its incorporation into GCMs.

The severe droughts in the Sahelian region of west Africa during the early 1970s and 1980s had become a concern, not only for international governments and aid agencies, but also for climate change scientists who wanted to understand the processes behind these events in a region where there was little information beyond that provided by standard meteorological stations installed and operated following the guidance given by WMO. These were often, for convenience, close to urban areas and kept in a state that was often at variance with local conditions. I have visited Met stations in Africa where instruments were situated in an area of lush grass while the surroundings were definitely desert.

Jim Wallace, then a member of John Stewart's vegetation–atmosphere interactions section, had initiated an ODA-funded experiment in Syria that encountered logistical and operational problems. Jim had joined IH in 1977 to work with John Roberts, while still writing his PhD with the University of Nottingham supervised by John Monteith.

I had already helped Jim with his experiment into the water relations of the heather species *Calluna vulgaris* at a field site on Sneaton Moor, one of the North Yorkshire Moors, near Scarborough, where aerodynamic resistances of the heather estimated from both a weighing lysimeter and a HYDRA eddy correlation system were compared (Wallace *et al.*, 1984). I also used the opportunity to investigate whether eddy correlation instruments should be vertical or normal to the surface at this sloping site for correct or optimum measurements for one-dimensional sonic anemometers. The later adoption of three-dimensional sonic anemometers generally overcame many of the problems associated with sloping sites and mathematically non-normal atmospheric flow, but not all, and papers by Gash and Dolman (2003) and Van der Molen *et al.* (2004) began a series of papers exploring these effects.

5.3.3   *African studies begin at ICRISAT*

The construction of the new International Crops Research Institute for the Semi-Arid Tropics (ICRISAT) research centre at Sadoré, south of Niamey, Niger provided the facilities and collaborative opportunity to relocate the dryland experiment from Syria to the Sahel. ICRISAT was primarily concerned with crop productivity, but it was recognised that precise and frequent measurements of evaporation rate and totals could only benefit the breeding and growing programme. Initial measurements of evaporation flux using the HYDRA Mk2 system and soil evaporation using micro-lysimeters began in September 1985 augmented with plant water relations and photosynthesis measurements beginning in 1986. The initial team consisted of Jim Wallace, John Roberts and John Gash. When I arrived belatedly in July 1986 having just finished my final 10 week field trip to the Amazon, there was just John Roberts there. It was out of the steam cooker of Brazil and into the fire of Niger – temperatures in May at Sadoré could exceed 45 °C. We all stayed at the ICRISAT 'Casa de Passage' in the middle of Niamey. This was a guest house for visiting scientists, with a cook/caretaker and a ferocious-looking Toureg guard complete with spear and daggers. The caretaker, despite our protests that it was not necessary, always ironed our underwear. Eventually, we found out that this killed Tumba Fly eggs that were laid in wet washing. Han Dolman later picked up such a fly egg while playing volleyball and eventually had to have the growing larvae cut out of his back.

When the Casa de Passage closed, we moved into the Hotel Terminus, so-called because it was (hypothetically) situated at the end of the railway from Mali and Burkina Faso – a railway that never materialised. But the swimming pool was very useful for calibrating Infrared thermometers – swimming helped keep the water well mixed. Another local hotel, the Sahel, had an Olympic-sized swimming pool (at least it was until they attached the tiles – then it was 2 cm short of official Olympic size) where the water was always close to 37 °C making it impossible to swim because there was no way to dissipate body heat.

At that time, the ICRISAT centre was still under construction and we operated from within a 20-ft container. Being an old, somewhat battered metal container, it became an oven during the normally sunny weather and then a swimming pool when the roof-leaked torrents of water during the rainy season storms. Luckily power cuts generally also occurred during these storms. We were operating two HYDRA Mk2 systems and two AWS systems at 4 m above a millet field that grew rapidly in its short growing season of about 90 days to reach 2.5 m. Measurement of photosynthesis and respiration using porometers, soil evaporation using micro-lysimeters, soil temperatures, soil tensiometer measurements and Neutron Probe soil moisture measurements completed the experimental plan. The fauna was not quite so prolific as in Brazil, but scorpions liked to hide in the base of the micro-lysimeters, and eventually one stung me. Some of the Neutron probe access tubes went down to 4 m and one presented a particular problem when the probe became stuck at the bottom of a damaged access tube. In trying to get the source up, the signal/support cable broke, and the farm tractor nearly overturned in attempting to pull the access tube out of the ground so a major international radioactive incident was looming. Luckily, the Belgian contractors building

the ICRISAT centre (whose first construction was a 20-m swimming pool for the workers) had a heavy caterpillar-tracked digger that with little effort used its hydraulic bucket to extract the access tube with the probe (luckily) trapped in the end.

The millet work, initially reported in Wallace *et al.* (1989) and further documented in Wallace *et al.* (1990, 1993), indicated that soil evaporation was as important a component of total evaporation as the plant evaporation, and this therefore needed to be explicitly recognised in models of evaporation from millet – an area of theoretical research that had already been studied in Shuttleworth and Wallace (1985). This latter work was subsequently revisited in Shuttleworth and Gurney (1990) where the one-dimensional sparse crop interaction theory was reformulated to allow calculation of canopy resistance using foliage temperature measurements. The measurement and modelling of evaporation in semi-arid land, particularly the Sudano-Sahelian zone, was reviewed in Wallace (1991).

5.3.4 Campaigns across continents

The year 1987 was a busy one for me. Jim Shuttleworth, John Gash and I were at the HAPEX-MOBILHY site in France in March bringing the interception experiment to an end, dismantling all the IH equipment and shipping it back to the UK – in snow! Then following a 3-week fieldwork trip to Niger in May, I was off to Kansas for a 3-week field trip operating the IH contribution (project supervised by John Stewart) to the FIFE programme, taking over the running of the IH experiment from Ken Blyth and Jim Shuttleworth. The FIFE experiment located on the Konza Prairie in Kansas was a large-scale climatology programme designed to improve understanding of surface–atmosphere exchanges through coordinated data collection from satellite, aircraft and ground measurements. During the Intensive Field Campaigns (IFCs), more than 100 science investigators were working at 32 sites within the 15×15 km research area, one of whom was Piers Sellers, later to be an NASA astronaut. John Gash and I had previously met Piers when he was one of John Lockwood's PhD students at Leeds University. He had created a rival interception model (The MANTA Model) to that of Jack Rutter (Sellers and Lockwood, 1981). We both thought that Piers model, while very sophisticated, was overly complicated having multiple canopy layers, and driving parameters that would rarely be routinely measured in applications of the model. The MANTA model produced interception losses that were 20% higher than the Rutter and Gash models – a result that Jim Shuttleworth and John Gash criticised and explained in Shuttleworth and Gash (1981). Jim and Piers Sellers subsequently collaborated on many occasions especially in the context of Piers' more successful simple bio-sphere (SiB) model (Sellers *et al.*, 1986) that is now routinely used in many GCMs

Another rival to John Gash's model was the stochastic model of rainfall interception described in Ian Calder's paper (Calder, 1986), a statistical approach that depended on mean raindrop volume and number of raindrops. One of the outcomes of the model was that, for the same total amount of rainfall, maximum canopy storage will be attained less rapidly for raindrops of larger volume. The differences between Rutter-type running water bal-ances and the statistical variety was examined in Calder *et al.* (1986) and the model was improved in Hall (1992). The difficulty of gaining both raindrop size and number of raindrops without sophisticated equipment meant that the stochastic model was not as easy to implement as John Gash's analytic model which just required raingauge measurements and storm timings.

Then I was back to Niger for a second 3-week field visit to begin the next phase of IH's Sahe-lian work. Millet was just one of four distinct surfaces in that region of Niger, the others being fallow bush (a mixture of sparse woody shrub – *Guiera senegalensis* and bare soil), tiger-bush (a striped terrain of long dense patches of tall shrubs – *Combretum micranthum* and *Guiera senegalensis* – interspersed with completely bare crusted sandy soil) and totally bare laterite soil areas.

Having explored millet evaporation, the Niger experiment expanded to became the Sahelian Energy Balance Experiment (SEBEX) using paired sets of instruments to compare evaporation from two areas of fallow bush and tiger-bush. Not only did the number of sites expand, but also the range of measurements being taken, requiring an increase in the staff numbers to run the experiment.

Alistair Culf joined the Group after gaining a PhD from the University of Lancaster. Alistair arrived on a Tuesday and by Friday he was in Niger to begin measuring and modelling the convective boundary layer. Ivan Wright also re-joined the Group, having explored infrared thermometry theory, calibration and practice for Jim Wallace (Wright, 1990) and proceeded to install many tower-supported and free-standing tripod-mounted infrared thermometers at the SEBEX sites.

For SEBEX, 10-m towers were erected at the fallow bush site at Sadoré and the tiger-bush site at Damari and a range of instruments, including Mk2 HYDRA, terrestrial radiation instruments, AWS and soil temperature and heat flux units were installed to sample the combined and separate energy balances of the bush and soil regimes. Permanent guards were also installed at the more remote Damari site to prevent theft and vandal-ism. However, within weeks of their deployment they complained that bandits had attacked them with bows and arrows during the night (we were shown exhibits), and therefore could they have guards to guard them. Their request was not accepted: we said we would prefer to employ braver guards.

The microclimate of the two sites proved to be markedly different. The sparse leaf canopy of the fallow bushes meant that evaporation from soil and canopy were often very similar, but still required modelling of the separate components to understand the controls on evaporation from the vegetation and the soil (Gash *et al.*, 1991). In contrast, the tiger-bush showed marked differences between the energy balance of the bush strips and the bare encrusted soil areas, where over 70% of the evaporation emanated from the vegetation which covered only 42% of the area (Culf *et al.*, 1993). Evidence in that paper indicated that the bare soil patches channelled rainfall into the bush areas and that the bushes also had access to deep reserves of water. Culf *et al.* (1993) also began to tackle the problem of how to combine the static mea-surements of radiation balance and soil moisture with the dynamic measurement of evaporation flux through use of John Gash's line source fetch model (Gash, 1986a). Applying that model to concentric sectors at varying distances from the measurement point [and results gained from Dolman *et al.* (1992) using a tethered balloon] allowed an estimate of the overall tiger-bush area net radiation to be obtained that was adequate for long-term measurements. However, it was recognised that recent advances in creating two-dimensional source area models that also included the effect of stability in defining fetch distances were needed to successfully model individual days. This further insight into the difficulty of measuring eddy correlation fluxes in heterogeneous terrain, such as tiger-bush, began with Lloyd (1995) investigating the effect of stability and sensor location within the heterogeneous terrain on the flux measurement. I began to write a paper using Hans Peter (HaPe) Schmid's two-dimensional model (Schmid, 1994) to further investigate and map out the upwind source-strengths of water vapour and CO_2 emissions from the extremely different tiger-bush surfaces. I met HaPe at a conference and he was keen to come on board as second author, but my own work pressures stalled the writing and I told HaPe to become lead author in order to get it finished. (Schmid and Lloyd, 1999). The idea of a blending height, already a factor in large-scale numerical models of atmospheric flow, being the distance above the surface where the effect of surface heterogeneity on energy balance measurements is negligible, which was alluded to in Culf et al. (1993), was also graphically identified in Schmid and Lloyd (1999). Han Dolman and Jim Wallace meanwhile were approaching the problem of sparse canopies from the point of view of updating land surface parameterisations in large-scale models (Dolman and Wallace, 1991).

Jim Shuttleworth suggested that I investigate a different approach to estimating sensible heat fluxes. By coupling the variance values of the fine-wire thermocouple of the HYDRA to a model proposed by Tillman (Tillman, 1972) that was based on a similarity prediction by Monin and Yaglom (1971), sensible heat flux could be evaluated. Despite there being a slight area of non-independence because the thermocouple is a common factor, the results could be compared to the HYDRA estimates of sensible heat flux. The resulting paper was written in about a week when myself, Alistair Culf, John Gash and Han Dolman installed ourselves in a single room at IH with four computers and did nothing else except concentrate on the paper (Lloyd et al., 1991). A more robust method of estimating aerodynamic roughness parameters over sparse canopies using eddy correlation was proposed in Lloyd et al. (1992). A combination of HYDRA eddy correlation equipment at the surface and a tethered balloon operating profile measurements up to 250 m over the tiger-bush site showed an increase in value of the roughness length with increasing height

(Dolman et al., 1992), which also pointed to the possibility that using values of roughness length from surface small-scale micrometeorological measurements may not be appropriate for large-scale numerical models such as GCMs.

5.3.5 Field experiment in desertification-threatened areas

Another international collaborative programme with a nested acronym, EFEDA, was also concerned with semi-arid areas – EFEDA was located in the region of Castilla-La Mancha in Spain (Bolle 1993). The ECHIVAL Field Experiment in a Desertification-threatened Area, EFEDA was one of a suite of international field experiments jointly fostered by the World Climate Research Programme (WCRP) and the International Geosphere–Biosphere Programme (IGBP). IH played an important role in this range of related international experiments designed to tackle the inclusion of realistic land surface processes in climate models.

Jim Shuttleworth was involved with the main international committees over-seeing the work and Howard Oliver was the project leader for activities in Spain providing measurements of eddy correlation evaporation, background AWS, soil temperature and soil moisture capacitance probe data from vineyard and vetch agriculture sites designed to investigate the interaction between the soil, vegetation and the atmosphere. Reading University were also providing a separate but complementary micrometeorological project adjacent to our site. About 150 European and American scientists from more than 30 groups also took part. The main aims were typical for these types of experimental programmes: comprising measurement and modelling of the partition of the surface energy balance, the seasonal progression of the vegetation/energy balance interaction across several different vegetation types and comparing aircraft and satellite data to ground-level small area measurements.

EFEDA-I started in 1991 and continued through to 1994. IH's contribution was largely provided by Kevin Sene, who was present on site throughout the 1991 experimental season (Oliver and Sene, 1992). I made a 3-week field trip there in July 1991. Relations between the scientists and the supportive local farmers and farm workers were generally good, helped no doubt by the gift of several half litres of duty-free whisky, although somewhat strained when one group (not us) began to dig deep access trenches to track the vine roots. All the experimental data were transferred under a fixed protocol to an EFEDA land surface information dataset for use in scaling up studies and they proved to be of international importance for climate impact forecasting (Sene, 1994, 1996). This concept of common data base would become the norm for future international programmes. EFEDA-II took place within HAPEX-Sahel during 1992–1993 and John Bromley focused during this time on deep drilling experiments to the underlying water table (see Chapter 4 Terrestrial Hydrological Processes).

5.3.6 *Niger again*

Meanwhile, the presence of IH within Niger and the links formed by IH with French scientists during HAPEX-MOBILHY were a major factor in placing the large-scale international study of Hydrologic Atmospheric Pilot EXperiment in the Sahel (HAPEX-Sahel) within a 1° grid square roughly centred around Niamey. The whole programme was nearly compromised before it even began, when at the end of a reconnaissance visit by the HAPEX-Sahel management and steering group to Niamey, Nigerien rebel fighters staged a minor gun-battle on the road to the airport.

This study, a natural successor to the HAPEX-MOBILHY programme in France, involved 66 separate studies that included ground-based, aircraft and satellite measurements and mesoscale and GCM models and the thoughts of over 200 research scientists and their support staff (Gourtorbe *et al.*, 1994). Of the three intensively investigated Supersite areas within the 1° grid square, the IH team were concerned mainly with the Southern supersite at Sadoré where Jim Wallace was the Supersite coordinator. The SEBEX sites were augmented with the original millet site. However, we now rubbed shoulders with other groups including John Moncrieff's group from Edinburgh University and Giles Harrison's Reading University group. Han Dolman and Alistair Culf operated a mobile tethered balloon that visited all the supersites to sample the boundary layer atmospherics to a height of 200 m. Alistair Culf also released and tracked many radiosonde meteorological balloons during this phase using a simple direction-height recording theodolite. I often helped, and after an hour or so in the heat tracking the balloon, sweat and fatigue would mean I sometimes lost sight of the balloon – but a tug at my leg and an outstretched arm from one of the many local boys, who stood around wondering just what was going on, pointing to where the balloon was, despite being many kilometres away, always impressed me. The spinning radiosondes had a label requesting return to us – and they often were, but some turned up in the local markets, being sold as household spinning ornaments.

The whole HAPEX-Sahel period of little more than 6 weeks long was a very intense period for all the international teams involved. Some people returned home because of accidents or stress inflicted by combinations of intense heat, cultural shock and the logistical and operational difficulties of maintaining continuous measurements. Coordinating and maintaining surface measurements at the three supersites in line with aircraft flights and satellite overpasses, with sensor and logging equipment operating at their specification limits meant many long hours were spent coaxing and repairing measurement systems during the aptly named Intensive Observation Period. But it was not without it's more comical episodes. Project vehicles were frequently stopped by Niamey police, generally for some minor indiscretion. Eventually, the Project management came to an arrangement with the Police whereby vehicles with HAPEX-Sahel logos on the doors would be treated with some leniency – followed very quickly by requests (refused) from local taxi drivers for HAPEX-Sahel stickers.

Simon Allen, Ken Blyth and Cathy Holwill were also part of the Southern Supersite IH team. Simon, having obtained a PhD from the University of Edinburgh, joined John Roberts Vegetation–Atmosphere Interactions section in 1988, He was involved both with SEBEX (Verhoef *et al.*, 1999) and HAPEX-Sahel where he investigated the water relations of the fallow bush (*Guiera Senegalensis*) using sap flow gauges surrounding the stems of the bushes – the results of which appeared in a combined paper with the University of Edinburgh (Levy *et al.*, 1999).

5.3.7 *Extending international collaboration*

International collaboration was the name of the game, partly brought about by the effect of European Community Science Frameworks which were channelling much of the individual European national science budgets into large programmes requiring international collaboration, both as a requirement for funding and as a necessity, because no individual national group could supply the necessary finance and manpower. Such collaboration was evident in HAPEX-Sahel through the many journal papers involving Wallingford staff that emanated from this programme.

Jim Shuttleworth was part of the overarching Scientific Steering Committee for HAPEX-Sahel and Jim Wallace and John Gash were highly involved in the overall HAPEX-Sahel operational project management. In terms of individual contributions to the programme, of the fifty papers in the *Journal of Hydrology* HAPEX-Sahel special issue (Goutorbe *et al.*, 1997a) fifteen papers have one or more Wallingford staff in the authorship list with eight of those havingWallingford staff as main authors (Blyth, 1997; Bromley *et al.*, 1997; Dolman *et al.*, 1997a,b, Gash *et al.*, 1997; Lloyd *et al.*, 1997; Taylor *et al.*, 1997; Wallace and Holwill 1997). The others relevant to this chapter are Goutorbe *et al.* (1997b), Kabat *et al.* (1997), Monteny *et al.* (1997) and Wai *et al.* (1997).

5.4   Eco-Hydrology Studies

Towards the end of the HAPEX-Sahel project, the NERC Terrestrial Initiative in Global Environmental Research (TIGER) programme, designed to understand the terrestrial carbon dynamics between climate and the land surface and planned since 1988, provided funds for further research at the Niger sites within TIGER 3. Howard Oliver began to take an increasingly important role in this programme from 1990. The programme provided funds for the surface and boundary layer measurements and upscale modelling that Alistair Culf and Han Dolman were involved in. The TIGER programme also funded work on combined carbon, water and energy balances of tropical biomes in Brazil, Cameroon and Niger. In the Cameroon, Sam Boyle, Chris Taylor and I erected another 52-m forest tower at a site near Mbalmayo, some 50 km south of the capital Yaounde. This tower was primarily for the use of a group of Edinburgh University scientists, led by John Grace, to conduct their CO_2, energy balance and soil physics work. At the end of the programme of work, John Roberts and I went out to take down the tower and ship it back to the UK. Unfortunately, in the interim, much of the aluminium pole tower structure had 'disappeared' providing useful carrying poles for local palm wine production. The tower was very unstable, and we did think about just pushing it over and collecting the debris but instead we carefully disassembled the remaining tower parts and shipped them to the UK.

Similarly, John Roberts, Heidi Bastable, Alistair Culf and I went out to the Jaru Forest Reserve in the south-western Brazilian State of Rondônia to erect another 52-m tower for Edinburgh University. This site was 7 hours upstream by outboard canoe from the city of Ji-Parana or several hours difficult travel on forest tracks by Land Rover and lorry from the town of Ouro-Preto do Oeste. The tower eventually arrived by lorry with the help of a monster tree clearing vehicle at the edge of the Ji-Parana river, where two canoes arriving from Ji-Paraná ferried the tower across the river to the eventual site (Fig. 5.13). Also involved in this aspect of the Tiger Programme was Robin Clarke, who was now permanently resident in Brazil at the Instituto de Pesquisas Hidráulicas, Porte Alegre, Rio Grande do Sul. Robin, after his sabbatical year in Brazil, had returned to IH and remained there until August 1983, when he become Director of the Freshwater Biological Association before moving to Brazil.

By this time, a relatively low-cost field operable CO_2 sensor (the LICOR 6262) had become available and was being run by several groups in the HAPEX-Sahel experiment. This was mainly operated alongside sonic anemometers to provide estimates of the CO_2 balance (i.e. assimilation of CO_2 by the vegetation during the day and respiration of CO_2 by the soil and vegetation at night). The closed-path sensor (a sample cell as opposed to the open-path sampling of sonic anemometers) used a pump to rapidly draw air from a region close to the sonic anemometer into the sample cell. The method was not without problems, including accommodating the time difference existing between the sonic sample and the CO_2 sample. Condensation in the pump tube and 'smearing' of the turbulent air stream in the tube both led to underestimates of the CO_2 flux. The LICOR 6262 also needed a relatively large amount of power – not so much a problem in the Sahel, but it would prove more problematic in less sunny climes. But these were early days and the future for carbon balance measurements to be integrated into the logical measurements looked bright.

In Canada, the Boreal Ecosystem-Atmosphere Study (BOREAS) was beginning (Sellers *et al.*, 1995). This was a large-scale international

Fig. 5.13 One of our tower section carrying canoes on the Ji-Parana river near our field site at Jaru, Rondonia. (Source: Reproduced with permission of John Gash.)

field experiment that had the aim of improving the understanding of the micrometeorological exchanges between a boreal forest and the atmosphere. In many ways, it was 'HAPEX-Sahel over boreal forest' with multiple ground measurement sites linked to aircraft and satellite overpass measurements. It was located in a 1000×1000 km study region that covered most of Saskatchewan and Manitoba. Richard Harding, Alistair Culf, Mike Stroud and Doug Clarke were primarily involved in this experiment alongside participants from the United States, Canada, France and Russia.

After the frantic nature of HAPEX-Sahel, 1993 was relatively quiet for me, time being mainly spent in analysing the resultant data and paper writing while also looking forward to the next big challenge. Many people were also moving on or away. Jim Shuttleworth left his position as Division Head to take up a permanent post in the United States – Regents Professor at the University of Arizona. Jim Wallace took his place as Division

Head and then as Director of CEH Wallingford when Brian Wilkinson subsequently left. John Gash moved up and out of fieldwork, firstly as section leader and then as Head of Division which was renamed from Hydrological Processes in 1994 to Bio-Physical Processes, and finally to Process Hydrology in 2001. Han Dolman left to return to the Netherlands in 1993, although IH would again work with Han in the CarboEurope programme. Alistair Culf moved over to work on the ABRACOS programme and concentrated on Amazonian work later in the TIGER-sponsored Rondônia Boundary Layer Experiment, and then LBA-EUSTACH before also leaving in 2001. Chris Taylor began to concentrate on mesoscale modelling and the investigation of the interaction of surface hotspots and atmospheric properties leading to the idea of annual persistence in rainfall patterns in Niger (Taylor and Clark, 2001; Taylor *et al.*, 2005). This left just me and Richard

Harding to continue atmospheric hydrological fieldwork.

5.5 Arctic Studies

Climate change was becoming a major political and scientific challenge with one focus being on the area of the world that was evidently responding most rapidly to climate change, that is, The Arctic. With the threat of disappearing snow cover and permafrost, there was the possibility of positive feedbacks as reflection of solar energy at high latitudes diminished and thus speeding up the warming of the soil and permafrost layers in regions where the permafrost contained the majority of the Earth's carbon deposits as peat – leading to increased production of both CO_2 and methane as the soils melted and warmed.

Richard Harding already had a lot of experience in operating experiments in snow-covered terrain in Greenland, Norway and Austria. I began to sketch out a possible experiment in the Arctic. First of all I had to convince CEH's then Director, Brian Wilkinson that we needed one of the new Licor 6262 CO_2 sensors to place alongside our HYDRA eddy correlation system – thus providing a combined energy balance/evaporation/CO_2 flux machine. At the time, CO_2 was seen as the remit of CEH Edinburgh at Bush where David Fowler and his staff had already begun much work with chambers and the new LICOR 6262. Wallingford was 'Hydrology', but eventually we got a LICOR 6262 sensor which David McNeil and the Wallingford workshop worked their magic in creating an integrated stand-alone CO_2 measurement system that contained pump, chemical scrubbers for removal of water vapour and CO_2 (the latter solely for the reference cell) and ancillary equipment within one portable and waterproof case capable of withstanding extreme climates and could be operated from battery power.

Many estimates of the likely size of the CO_2 and methane production from the Arctic used values pertinent to the deep peat bogs and mires of Canada and Fennoscandia, but 30% of the land mass of the Arctic is actually polar desert with possibly little carbon in the soils. Richard and I thought that we needed to balance the projected high levels of CO_2 production from warming Arctic soils by looking at these polar deserts and tundra regions. These regions were largely absent from North America where much of the earlier Arctic research had been performed.

Terry Callaghan (the Father of UK Arctic Research) was instrumental in establishing the NERC Arctic Research Station at Ny-Alesund on the island of Spitsbergen in the Svalbard Archipelago at 79° North. The Station had opened in 1991 and consisted of a large hut 'Harland Huset' containing both sleeping and laboratory accommodation. The fieldwork infrastructure was therefore in place and NERC and CEH agreed to fund this preliminary project. Richard and I, after a meeting with Terry at CEH Merlewood, flew up to Ny-Alesund, via Oslo, Tromso and then using a small feeder plane from the Svalbard capital, Longyearbyen, in late May 1994.

Armed against Polar bear attack with the NERC Arctic Research Station's first World War Czechoslovakian 0.375 rifles (Richard and I had already been on a gun-training course at a remote, rainy and windswept site somewhere on the moors near Oldham), we identified two contrasting sites, one a moss and lichen floodplain and the other a rocky polar desert site before returning to the UK. We procured and assembled equipment and airfreighted it to Ny-Alesund by mid-July except for the large pump-up masts and heavy equipment (too large and heavy to fit in the Dornier 218 feeder plane) which had to go by ship from Longyearbyen. The experiment started in late July and marked the start of nearly a decade's involvement in Arctic research. Along for the ride initially were Richard Harding and Mark Robinson, Mark having been co-opted at the request of Brian Wilkinson to supply some 'hydrology' to this otherwise micrometeorological programme by installing some rainfall–runoff plots. Certainly the climate was different to HAPEX-Sahel – we had quite a few snowstorms even in August.

The new LICOR 6262 CO_2 system was not ready for the 1994 season so measurements were limited to AWS measurements, evaporation and other energy balance measurements with the HYDRA and novel 'frost gauges' that I developed for measuring profiles of temperature, soil moisture and freezing fronts non-destructively in the fragile and often frozen Arctic soils. Again the skill and expertise of the Wallingford workshop enabled these prototypes to come to fruition rapidly. The design and operation of these profile units were reported in Lloyd (1998). The tube profile idea was evidently useful as the UK instrument manufacturer Delta-T brought out the PR2 Profile Probe some years later which measured profile soil moisture with a more sophisticated method than the one I created. With sensors and data loggers becoming cheaper, I began to maximise the number of sensors and data loggers wherever possible. So Bowen Ratio systems and (in 1995) time-domain reflectometry (TDR) probes for soil moisture were also installed using cable testers normally used to find breaks in underground cables. Just the AWS and frost gauge systems were left running over the polar winter.

At the end of August, before returning to the UK, I joined the Tenth International Northern Research Basins Symposium and Workshop when it visited Ny-Alesund and presented our first season results (Lloyd *et al.*, 1994). The return trip to Longyearbyen was in the 3-masted Schooner Rembrandt via the Russian coal-mining port of Barentsburg. During this trip I was also keen to speak to Knut Sand of SINTEF, Norway about a pan-European Arctic Bid into the EC Fourth Framework Programme beginning in 1996 that I was thinking of creating.

Tim Kyte joined the section as a technician to help me initially with the Svalbard work. The 1995 Arctic season began in late May with attempts to firstly find the overwintered equipment beneath the snow and then repair most of it – most damage was caused by reindeer. An AWS raingauge and cable that a reindeer had got caught in was finally found many months later nearly eight miles away. Polar bear tracks also crossed our sites. By the time the new Licor 6262 system arrived on 22nd June, most other equipment was working again. The whole Licor system was largely new to me. David McNeil and Alistair Culf had prepared an assembly and operational video. A really good idea except that with the pump making a lot of noise, every time David leant in to point out and explain really important items, the explanation became lost in the noise.

Both the 1995 and 1996 season started in May and ended in September and produced a unique body of data for high-latitude tundra and polar desert regions (Harding and Lloyd, 1998; Lloyd, 2001a,b; Essery *et al.*, 2005). In May 1995, we were the focus for an ITV News at Ten special report from Svalbard on climate change. It finally aired on Friday 28th May and no doubt caused much laughter amongst the Scandinavian viewers as I langlauf skied across to my field site before explaining our research (Fig. 5.14). My ten minutes of fame! Ken

Hargreaves (CEH Edinburgh) came out for a short time to install and run some soil surface chambers for syringe measurements of CO_2 and methane. Tim Kyte continued his sterling work taking over from me for periods at Ny-Alesund, as did Mike Stroud and Richard Harding. John Gash managed a break from his managerial duties and helped me for a week.

5.5.1 Land arctic physical processes

An expensive bid (the total requested was 2.6M ECU) into the EC Framework 4 programme was (surprisingly to both me and many others) successful – and the land arctic physical processes (LAPP) project officially began in April 1996. It comprised teams from Norway (Knut Sand leading), Denmark (University of Copenhagen – led by Henrik Soegaard and Thomas Friborg – old friends and colleagues from HAPEX-Sahel), Finland (University of Helsinki – led by Tuomas Laurila) and CEH Edinburgh (led by David Fowler). Sites were at Zackenberg in north-east Greenland (Soegaard, Friborg), Kaamanen, (Laurila, Fowler) and the Arctic Research Station at Kevo (Lloyd), both in Finland (Kevo being close to the border with Norway and Russia, Kaamanen some 100 km further south) and Ny-Alesund (Sand, Lloyd). These sites would provide contrasting energy and carbon balances across a wide latitudinal and longitudinal extent of the European Arctic. The Wallingford team continued the work begun during the initial tundra experiment during 1996 on Svalbard and then relocated the entire experiment to Finland beginning in September 1996 with a short site visit to Kevo with some equipment. The following April (1997), Tim Kyte and I spent four days driving a fully loaded UK-rented transit van via the Harwich–Gothenburg ferry all the way up the east coast of Sweden and then up to the Kevo Research Station in northern Finland. The rental vehicle would remain in Kevo until the end of the experiment in 1999.

The new LAPP sites were predominantly deep arctic peat bogs and mires and were chosen to provide marked contrasts between different arctic surface types (Fig. 5.15). This was very different to the Svalbard experience, with the different climate and the bogs creating instrument support and operational problems. During the winter of 1998, the AWS at the Kevo site on one occasion

Fig. 5.14   The Mk2 HYDRA at the Leirhagen site near Ny-Alesund, Svalbard in early May 1995. (Source: Reproduced with permission of Colin Lloyd.)

recorded $-45\,°C$ and $20\,ms^{-1}$ winds. But the frost gauges proved adaptable, providing temperature profiles in the rising and falling bogs. We were now also augmenting solar panel power for the HYDRA, Licor and the other instruments with wind power from small wind generators. These allowed continuous AWS and HYDRA flux measurements to be maintained during the long 'polar night'.

It was during a meeting in Reading in 1997, that John Moncrieff and Paul Jarvis of Edinburgh University persuaded me (after a few whiskies) that it might be a good idea to do a PhD based around the Svalbard Arctic work which, despite having a year out to concentrate on my managerial role within LAPP, was awarded in 2001 (Lloyd, 2001a).

The LAPP teams taking part were highly professional which made my management of the project relatively easy. Data were analysed on time and papers written, most of which appeared in the 'Land Surface/Atmosphere Exchange in High-Latitude Landscapes' Special Issue of the *Theoretical and Applied Climatology Journal* (Halldin *et al.*, 2001). This was a joint effort between the LAPP and the Winter Experiment (WINTEX) part of the Northern Hemisphere Climate-Processes Land-Surface Experiment (NOPEX) communities. This publication brought together the key results from these two European field programmes in the Boreal/Arctic regions. The programmes and this publication illustrated the very important role that snow plays in these areas, not only in the determination of energy, water and carbon fluxes in the winter, but also in controlling the length of the summer active season, and thus the overall carbon budget (Harding *et al.*, 2001). This latter result was highlighted in Lloyd (2001b and c) where the different snow cover periods in 1995 and 1996 at the Ny-Alesund site produced markedly different active season lengths which produced an overall CO_2 sink period in 1995 and an overall CO_2 source

Fig. 5.15 An aerial view of the LAPP mire site at Kevo, northern Finland. (Source: Reproduced with permission of Colin Lloyd.)

period in 1996, results that were successfully modelled using established soil/vegetation carbon and energy balance models modified for the particular carbon and moisture characteristics of byrophytes. The energy balance and integration papers from this Special Issue involving Wallingford staff were by Lloyd *et al.* (2001) and Laurila *et al.* (2001). Robin Hall subsequently took the data taken at the CEH site at Kevo in Finland and used it to drive and test the Meteorological Office Surface Exchange Scheme (MOSES) model (Cox *et al.*, 1999) to provide an improved model of arctic peat soil hydraulic and thermal properties (Hall *et al.*, 2003).

Alistair Culf was inducted into the LAPP fieldwork programme in August 1997 for a week's fieldwork, and Chris Taylor was my co-driver and helper as we dismantled the wetland site equipment, loaded it into the rented Ford Transit and then began the 4-day drive and ferry trip back to the UK in October 1999.

5.5.2 Responding to diverse demands

At the conclusion of the successful LAPP project, I needed to make a decision. To move further into science management, or stay where I was. The LAPP project had shown me that I could do tactical project management, but combining this with fieldwork data collection, data analysis and paper writing meant that no part, in my mind, was fully successful. I knew I would be no good at the more political strategic science management and planning so I chose to stay with experimental science and associated management. Probably a wrong career move, but one I was happy to take.

In 1997, alongside the LAPP project, Nick Jackson began work on a hillslope close to the LAPP wetland site. For this programme of work he and I installed flux and AWS equipment on the hillslope with runoff being measured with a v-notch weir on the stream confluence at the base of the hillslope. The v-notch weir had been made by the Wallingford workshop, and Alan Warwick (workshop manager) and I installed it in April 1998.

Nick Jackson arrived at Wallingford with a PhD on the 'Water use efficiency of arid-zone agroforestry species' from the University of Newcastle in 1993. He then spent seven years in Kenya studying the water use and fluxes by agroforestry and crop systems – some of which links agroforestry to the subject of this chapter (Ong *et al.*, 2000; Jackson, 2000; Jackson *et al.*, 2000). The hillslope work was described in Harding *et al.* (2002). Nick Jackson then co-authored the book *'Tree roots in the built environment'* with John Roberts and Mark Smith (Roberts *et al.*, 2006) becoming Science Coordinator for the CEH Water Programme in 2005.

5.5.3 *Further research moves*

From 1998 onwards, research moved heavily into monitoring-data collection rather than particular process investigation. It was becoming apparent

to me and others that while CEH Wallingford had generated a worldwide appreciation of our expertise in rapid deployment of successful eddy flux based experiments around the world, we no longer had the fieldwork-based scientists (as opposed to technicians) to fully exploit this situation. For me, the future was to become a frenetic chase from one project to another and this change in staff emphasis is why from now on, the fieldwork story increasingly involves me slowly becoming a technician.

Richard Harding and I were invited to join several international projects. First, an EU funded consortium arctic research programme – CONGAS (biospheric CONtrols on trace GAS fluxes in northern wetlands) took me and Roger Wyatt to a reclaimed forest/wetland site at Plotnokovo, near Tomsk in western Siberia during May–September 1999. Logistically, a difficult experiment (Fig. 5.16), which produced just one paper (Friborg *et al.*, 2003). This paper showed that the wetland site, although

Fig. 5.16   Getting the equipment to the Plotnikovo wetland site in May 1999. (Source: Reproduced with permission of Colin Lloyd.)

nominally a sink for carbon was, in terms of Global Warming Potential, a large source which if extrapolated over the entire Siberian wetland area, would produce emissions that equalled 84% of the Russian emission of CO_2 from fossil fuels (equivalent to 24% of US emissions or 6% of the world total fossil fuel emissions (2001 figures).

During the 1980s, European Carbon Cycle research had begun in response to the problem of acid rain. At the 1991 Florence Symposium on 'Responses of Forest ecosystems to Environmental change' (Teller *et al.*, 1992), the effect of climate change became increasingly important. Forests and climate change were therefore high on the agenda for the third (1993–1995) and fourth (1997–1999) EC Framework Programmes. At this time, FLUXNET, EUROFLUXNET and MEDI-FLUX began, all with an atmospheric focus on forests to begin with. Also, starting at this time were ESCOBA and ESCOBA II studying carbon in the ocean, the biosphere and the atmosphere. Richard Harding and I attended the Orvieto workshop of the ESCOBA II project in June 1998. At this meeting, the interdisciplinary project CAR-BOEUROPE was proposed, in order to combine atmospheric, ecosystem and soils based research. Most of the EUROFLUXNET research to this point had been in forests. Richard and I, with support from our LAPP colleagues, argued that for a complete appreciation of the carbon balance of the European region, both wetlands and agriculture needed to be included (see appendix for acronym expansions).

In consequence, a successful bid into the next EC framework programme saw Richard Harding and I set up a new CarboEurope experimental site to investigate the energy and carbon balance of a managed wetland grassland site at Tadham Moor on the Somerset Levels. Previously Owen Mountford (ITE Monks Wood) had produced plant surveys at Tadham Moor and Mike Acreman and David McNeil had run AWS/HYDRA systems at an inundated reed bed site close by at Ham Wall. Acreman *et al.* (2003) showed that the relationship between water table level and evaporation had important implications for local water resources management.

The experiment began in June 2000 and continued through to January 2004, with a forced hiatus in March through May 2001 due to the foot and mouth epidemic. The immediate surrounding area of the Somerset Levels operated a winter flooding/summer draining management system known as Tier 3 (Fig. 5.17). The combination of measurement and modelling of the energy and carbon balances at the Tadham Moor site showed how failing to maintain the Tier-3 water level management scheme resulted in a significant loss of soil carbon, a situation that could be reversed by strictly adhering to the management scheme (Lloyd, 2006). A companion paper (Harding and Lloyd, 2008) showed that most of the incoming radiation goes into creating evaporation and that standard evaporation formulae, for example, Penman–Monteith, provided a reasonable simulation of total evaporation from this site. Acreman *et al.* (2011) combined this data with other CEH measurements of methane, animal and vegetation species composition and other ecosystem components to provide an overview of the trade-off between often conflicting ecosystem services in the Somerset Levels.

During this time, David McNeil and Jon Evans had continued to develop the Mk2 HYDRA into a total integrated instrument with open-path CO_2 and water vapour measurements which eventually became known as the Mk4 HYDRA. The development embedded a redesigned and extended HYDRA Mk2 low-power infrared gas analyser that now had dual channels for CO_2 as well as water vapour, within the measurement volume of a commercial Solent R3 sonic anemometer together with a new low-power datalogger specifically designed for the Mk4 HYDRA. Bringing all the measurements together created an elegant solution that obviated the need for sensor separation corrections that a combined Solent sonic anemometer and the new commercial Licor 7500 open-path CO_2/H_2O sensor had to incorporate. Needing only 4 W of 12 V power, the system needed just a solar panel and a battery to run – a huge improvement over commercial flux systems. In 2002, David McNeil and Jon Evans received NERC capital funding to commercialise the Mk4 HYDRA.

While Tadham Moor was progressing, Richard and I were invited in 2002 to provide micrometeorological measurements for a project at Abisko in northern Sweden. This was a programme, funded by NERC and organised by the School of Biological and Biomedical Sciences, University of Durham. It consisted of a suite of measurements, models and upscaling methodologies to understand the impact of global warming on biogeochemical processes in tundra regions – a project called. Snow in Tundra

Fig. 5.17   The field site at Tadham Moor during the flooded phase of Tier 3 management – January 23, 2003. (Source: Reproduced with permission of Colin Lloyd.)

Environments: Patterns, Processes and Scaling (STEPPS). Combined soil physics and energy measurements at six representative patches of the tundra mosaic pattern and an AWS were installed in late September 2002 followed by CO_2/H_2O flux measurements installed in February 2003 (Fig. 5.18). Initially, the flux measurements were provided by combined Solent R3 sonic Anemometer/Licor 7500 CO_2/H_2O gas analyser – but were augmented by the new Mk4 HYDRA in 2004 after successful field trials both at Tadham Moor in 2002 and in Brazil in 2003. The micrometeorological data collected was primarily used by Durham PhD students and incorporated into their theses and by Andy Fox, a Durham research associate who produced Fox *et al.* (2008), a paper that showed while the flux measurements can adequately represent heterogeneous terrain, significant biases can be seen between chamber upscaling and eddy correlation measurements if the possible errors in both systems are not taken into account. As well as the

propensity for chambers (an alternative CO_2/H_2O measurement methodology) to be sited subjectively in the middle of patches, avoiding damaged areas, rocks, herbivory and possibly operator bias in choosing vigorous and healthy areas increases the upscaled estimates over the flux measurements. Perhaps, a sign of the times was provided by one of the PhD students during our daily 8 km walk back from the field site 'When I complete this PhD I intend to go into science management'.

5.5.4   African monsoon multidisciplinary analysis

Wallingford was also increasingly being involved in a new major international programme of work planned for west Africa – the African Monsoon Multidisciplinary Analysis (AMMA – see http ://www.amma-international.org/about/index) – designed to improve the knowledge and understanding of the west African monsoon (WAM)

Fig. 5.18 Mk4 and Solent R3/Licor 7500 flux systems, wind generators and horizontal solar panels at the upland tundra site near Abisko, northern Sweden. (Source: Reproduced with permission of Colin Lloyd.)

and its variability, with an emphasis on daily to interannual timescales. Chris Taylor and I became joint Task Leaders for the Surface Flux Measurements, with Chris leading on the modelling and satellite observation aspects of the task, whereas I concentrated on the management and provision of a surface flux network – a network that would also contain flux systems run by groups from France and Germany. Experimentally, CEH Wallingford, with our well-known ability to operate stand-alone flux systems in remote regions and harsh climates, were tasked with providing and installing ten flux stations in Mali, Niger and Benin to complement flux stations being installed by the French and German groups. So, with both Tadham Moor and STEPPS still running, a preliminary visit to Mali in August 2003 identified probable sites for the flux systems.

The realisation that data collection rather than fundamental process research was becoming more prevalent at Wallingford was also reflected in Europe where the new CarboEurope programme (CarboEurope IP – Integrated Project)) was being planned at Spoleto, Italy in January 2004. The meeting stated that the next phase of CarboEurope was to be geared towards long-term monitoring of the carbon balance of Europe with creation of large databases with attendant quality control and quality assurance. Investigative research was not ruled out, but it was not the main focus.

Chris Taylor and I then spent a lot of time devising, writing and revising Chapter 3 (EOP Surface Flux Measurements – TT2a) of the AMMA EOP Science Plan – a work scheme for the entire surface flux network, see amma-international .org/library/docs/IIP_v3.0.pdf.

Fig. 5.19 The combined AWS/flux system at the desert margin site at Bamba. (Source: Reproduced with permission of Colin Lloyd.)

The ten CEH self-contained combined AWS/flux systems were designed, constructed and assembled during 2004 and early 2005. The first four were installed in Mali in April 2005 at sites representative of the major terrain types (grassland, acacia forest, bare laterite and desert margin; Fig. 5.19). In May, I went to Benin to identify three representative sites near Djougou in northern Benin before a long drive to neighbouring Niger to identify two sites near Niamey before driving back down to Cotonou in Benin for the flight back to the UK.

Maintaining the STEPPS data collection programme had me in Sweden in June, before returning to Benin in November to install the flux systems at Djougou over forest, crop and grassland sites followed by another long drive to Niger to install the two flux systems over the degraded fallow and bare sand sites at Wankama and Belifoungou near Niamey. The last flux system was installed over a palm tree plantation site near Pobe in Benin close to the Nigerian border in February 2006 followed by a maintenance trip by Roger Wyatt and myself to Mali in early March. I was back in Mali in May, but Tuareg unrest in the northern region of Mali stopped us visiting the northernmost desert margin site. This was to be my last visit to west Africa for CEH. The logistical and instrumental difficulties of maintaining data collection from ten highly sophisticated flux measurement systems, analogous to keeping ten plates spinning on top of poles, stretched our small team to the limit. While the experience was not a total success, the rewards were co-authorship of papers from AMMA colleagues who gained valuable information from the flux network (Mougin *et al.*, 2009;

Timouk *et al.*, 2009; Ramier *et al.*, 2009; Guyot *et al.*, 2009; Tanguy *et al.*, 2012).

In Benin, one of the French teams was running a long-path scintillometer heat flux system across a valley. Scintillometry, another method for measuring heat flux through detecting the light scintillation of a laser source by the turbulent changes in the refractive index of air caused by variations in temperature, humidity and pressure, had begun to be investigated by Jon Evans earlier in 2003, following a link with Henk de Bruin (a previous research partner with John Stewart in the 1980s, see Stewart and de Bruin, 1985). Scintillometry research continued at the Sheepdrove farm complex, fifteen miles south-west of Wallingford.

5.5.5 *International polar year*

Another data collection programme of work emerged from the STEPPS programme when the International Polar Year (IPY), a large scientific programme focused on the Arctic and Antarctic began in 2007. An NERC-funded scientific package called the Arctic Biosphere Atmospheric Coupling at Multiple Scales – IPY (ABACUS-IPY), led by Mathew Williams of Edinburgh University, with Eleanor Blyth and myself as part of the Science Steering Committee. On the basis of CEH's experience in the STEPPS and LAPP projects, sites were chosen at Abisko and northern Finland close to Kevo for the intensive programme of work linking scales and disciplines from plants and soils, through chambers and tower flux systems to light aircraft and Earth Observation satellites over wetland, tundra and birch forest. Despite our AMMA experience, September 2006 found Roger Wyatt and myself driving from the UK to Abisko and then across to Kevo in northern Finland to prepare tower sites for the flux systems and install soil physics sensors before the winter freeze, before driving back to the UK. The following April, Roger and I again did the UK to Abisko and Kevo drive to install towers at the Abisko sites and flux systems at both sites. Jon Evans took over the CEH experimental and instrumental aspects of this project, and Eleanor Blyth continued with modelling aspects.

5.5.6 *Moving on*

John Gash retired in early 2007 and I retired in September 2007 alongside Roger Wyatt, Paul Rosier and John White. But I was subsequently contracted by CEH to install and maintain for three years a combined flux and climate station system over two bioenergy crops at Brattleby near Lincoln, a joint CEH Wallingford/Edinburgh project being coordinated by Jon Finch. One of the energy crops was Miscanthus, which is a fast-growing annual grass that can grow to 3 m in height in 6 months, and the other was coppiced willow which was harvested to ground level on a three-year cycle. The now standard Solent R3/Licor 7500 flux system was installed at both the Miscanthus and Willow sites together with solar panel and wind generator 'power' towers, an AWS and soil physics sensors. The systems ran automatically for the 3 years with monthly visits by me, with annual dismantling/reassembly of the Miscanthus flux system during harvest. A comprehensive database of flux and climate data was produced, some of which was used alongside the soil emissions data provided by CEH Edinburgh in Drewer *et al.* (2011).

Jon Evans continues at Wallingford as the sole surviving atmospheric research experimentalist, with his scintillometer work and his involvement with the COSMOS-UK programme, a UK Soil Moisture Monitoring Network being run by CEH. In a completion of the circle, the COSMOS (Cosmic ray Soil Moisture Observing System) was developed by a team involving Jim Shuttleworth's at the University of Arizona (see Shuttleworth *et al.*, 2010). It involves measuring low-energy cosmic ray neutrons above the ground, whose intensity is inversely correlated with soil moisture content – not dissimilar to the method employed by the Neutron Probe developed at the very start of IH in the 1960s. The COSMOS programme is an exciting development that aims to provide area-averaged soil moisture measurements over areas that complement those of surface flux systems, an aspect of the energy balance of terrestrial systems that has been highly problematic in the past in attempting to distribute evaporation between soil and vegetation.

5.6 Reflections on Contributions to Research

The last 50 years have seen marked changes in atmospheric hydrological research, changes that are mirrored both at Wallingford and internationally as the direction of science has moved from local matters to international policy and their funding considerations. The science has progressed from intensive and detailed investigation of specific aspects of the energy balance and evaporation to routine automatic estimation of areal fluxes of evaporation, carbon dioxide and many other atmospheric constituents. Much of this has been the result of the development of low-power solid-state microprocessors. These have allowed the collection of continuous multi-sensor data to move from static, expensive installations such as the Thetford Project, requiring mains power, large caravans and intensive maintenance and supervision, to highly mobile automatic measurement systems run from solar-powered batteries, requiring little maintenance or supervision beyond visits for data collection. Even this latter aspect is now routinely performed by mobile phone or satellite transmission. For many years Wallingford was at the forefront of this type of development.

Sensors have also progressed, becoming increasingly sophisticated in their design and operation. Much of this has occurred as the provision of scientific instruments developed from niche manufacturers to a main-stream highly profitable enterprise (e.g. Campbell Scientific Inc. has progressed from its creation by Eric and Evan Campbell in 1974, reaching $1M annual sales in 1978 and then moving on to employ 250 people worldwide with an annual revenue of $42M in 2014; (http://www.insideview.com/directory/campbell -scientific-inc). Again, Wallingford, through the development and deployment of prototype sensors and sensor systems, showed manufacturers where the market lay.

The change from running highly complex climate models on a single mainframe computer to multiple models on desktop computers has, in addition, meant a shift in how research is performed with model outcomes often defining measurement programmes such as collection of data for model verification. For example, the stated aims of CarboEurope IP in 2006 were to apply a single comprehensive experimental strategy to provide integrated data for bottom-up process modelling and top-down inverse modelling (http://www.carboeurope.org. – May 2004 CarboEurope IP Flyer).

There are still elements of atmospheric hydrological research that need investigation, but these are mostly to do with measurement precision and accuracy, rather than devising how to measure a large-scale entity such as evaporation. Much

measurement fieldwork is now about data collection from novel terrestrial surfaces, rather than investigative research into individual processes.

A telling statistic that I compiled just before I retired showed that in 1970, 90% of Wallingford staff were actively involved in fieldwork measurement programmes. In 2007, the number of Wallingford staff either actively or even previously involved in fieldwork measurement had fallen to 15%. This may well be the natural progression of this branch of hydrological science. Towards the end of my career, future thoughts were not so much about which individual micrometeorological process needs to be investigated – it was more from where in the world do we need measurements.

Throughout the past 50 years Wallingford, has been at the forefront of investigative energy balance, evaporation and latterly carbon dioxide measurements over terrestrial surfaces. The first integrated forest micrometeorology experiment at Thetford provided a blueprint for the forest experiments found in CarboEurope, whereas the intensive measurement programme in Amazonia in 1984 showed that remote and hostile terrains could be successfully investigated. The development at Wallingford of relatively low-cost battery-powered microprocessor flux measurement systems when microprocessors were in their infancy allowed researchers at Wallingford to react quickly and efficiently, and provide data rapidly in response to changing scientific requirements. Wallingford's participation in international experiments such as HAPEX-MOBILHY, HAPEX-Sahel, LAPP, and AMMA were the direct result of this ability.

Wallingford has contributed hugely to the insight and understanding of most of the components of the atmospheric hydrological processes, from rainfall interception, to evaporation theory and footprint analysis, process modelling of rainfall interception and plant–atmosphere interactions; and the provision of data from the major biomes of the world from the high arctic to equator, from deserts to wetlands and grass to forest.

Acknowledgments

The following past and present members of HRU, IH and CEH contributed this chapter and to the achievements described. Their assistance is gratefully acknowledged: Simon Allen, Heidi Bastable, Eleanor Blyth, Ken Blyth, Alistair Culf, Nick

Cowell, Han Dolman, Jonathan Evans, Bob Frost, John Gash, Robin Hall, Richard Harding, Cathy Holwill, Nick Jackson, Tim Kyte, David McNeil, Chris Moore, Howard Oliver, John Roberts, Jim Shuttleworth, John Stewart, Mike Stroud, Chris Taylor, Mary Turner, Jim Wallace, Alan Warwick, Geoff Wicks, Ivan Wright, Roger Wyatt.

5.7 References

Acreman, M.C., Harding, R.J., Lloyd, C.R. & McNeil, D.D. (2003) Evaporation characteristics of wetlands: experience from a wet grassland and a reedbed using eddy correlation measurements. *Hydrology and Earth System Sciences*, **7** (1), 11–21.

Acreman, M.C., Harding, R.J., Lloyd, C. *et al.* (2011) Trade-off in ecosystem services of the Somerset Levels and Moors wetlands. *Hydrological Science Journal*, **56** (8Special Issue: Ecosystem Services of Wetlands), 1543–1565.

André, J.-C., Goutorbe, J.-P. & Perrier, A. (1986) HAPEX-MOBILHY: a hydrologic atmospheric experiment for the study of water budget and evaporation flux at the climatic scale. *Bulletin of the American Meteorological Society*, **67** (2), 138–144.

André, J.-C., Goutorbe, J.-P., Perrier, A. *et al.* (1988) Evaporation over land-surfaces: First results from HAPEX-MOBILHY special observation period. *Annales Geophysicae*, **6** (5), 477–492.

Bastable, H.G., Shuttleworth, W.J., Dallarosa, R.L.G., Fisch, G. & Nobre, C.A. (1993) Observations of climate, albedo and surface radiation over cleared and undisturbed Amazonian forest. *International Journal of Climatology*, **13**, 783–796.

Baumgartner, A. (1969) Meteorological Approach to the Exchange of CO_2 Between the Atmsophere and Vegetation, Particularly Forest Stands. *Photosynthetica*, **3**, 127–149.

Black, T.A. & McNaughton, K.G. (1971) Psychrometric apparatus for Bowen-ratio determination over forests. *Boundary-Layer Meteorol.*, **2**, 246–254.

Blyth, E.M. (1997) Representing heterogeneity at the Southern Supersite with average surface parameters. *Journal of Hydrology*, **188–189**, 869–877.

Bolle, H.-J. (1993) EFEDA: European field experiment in a desertification-threatened area. *Annales Geophysicae*, **11**, 173–189.

Bromley, J., Edmunds, W.M., Fellman, E. *et al.* (1997) Estimation of rainfall inputs and direct recharge to the deep unsaturated zone of southern Niger using the chloride profile method. *Journal of Hydrology*, **188–189**, 139–152.

Bruijnzeel, L.A. & Weirsum, K.F. (1987) Rainfall interception by a young *Acacia auriculiformis* (A. Cunn) plantation forest in West Java, Indonesia: application of Gash's analytical model. *Hydrological Processes*, **1**, 309–319.

Cabral, O.M.R., McWilliam, A.L.C. & Roberts, J.M. (1996) In-canopy microclimate of Amazonian forest and estimates of transpiration. In: Gash, J.H.C., Nobre, C.A., Roberts, J.M. & Victoria, R.L. (eds), *Amazonian Deforestation and Climate*. Wiley-Blackwell, Michigan, USA, pp. 611.

Calder, I.R. (1986) A stochastic model of rainfall interception. *Journal of Hydrology*, **89**, 65–71.

Calder, I.R., Hall, R.L., Harding, R.J. & Wright, I.R. (1982) *An interim report for the Scottish Upland Afforestation Consortium on the use of a wet-surface weighing lysimeter system in interception studies of heathland vegetation*. 24pp, Institute of Hydrology.

Calder, I.R., Hall, R.L., Harding, R.J. & Wright, I.R. (1984) The use of a wet-surface weighing lysimeter in rainfall interception studies of heather (*Calluna vulgaris*). *Journal of Climate and Applied Meteorology*, **23**, 461–474.

Calder, I.R., Wright, I.R. & Murdiyarso, D. (1986) A study of evaporation from tropical rain forest – West Java. *Journal of Hydrology*, **89**, 13–31.

Cooper, J.D. (1975) *Determination of evaporation and drainage from a Scots pine forest at Thetford*. England by soil moisture measurements.

Cox, P.M., Betts, R.A., Bunton, C.B., Essery, R.L.H., Rowntree, P.R. & Smith, J. (1999) The impact of new land surface physics on the GCM simulation of climate and climate sensitivity. *Climate Dynamics*, **15**, 183–203.

Culf, A.D., Allen, S.J., Gash, J.H.C., Lloyd, C.R. & Wallace, J.S. (1993) Energy and water budgets of an area of patterned woodland in the Sahel. *Agricultural and Forest Meteorology*, **66**, 65–80.

Culf, A.D., Fisch, G. & Hodnett, M.G. (1995) The albedo of Amazonian forest and ranchland. *Journal of Climatology*, **8**, 1544–1554.

Culf, A.D., Esteves, J.L., Marques, A.d.O. & da Rocha, H.R. (1996) Radiation, temperature and humidity over forest and pasture in Amazonia. In: Gash, J.H.C., Nobre, C.A., Roberts, J.M. & Victoria, R.L. (eds), *Amazonian deforestation and climate*. Wiley-Blackwell, Michigan, USA, pp. 611.

Denmead, O.T. (1969) Comparative micrometeorology of a wheta field and a forest of *Pinus radiata*. *Agricultural Meteorology*, **6** (5), 357–371.

Dickinson, R.E., Henderson-Sellers, A., Kennedy, P.J. and Wilson, & M.F. (1986) *Biosphere-atmosphere transfer scheme (BATS) for the NCAR Community climate model*. NCAR Tech. Note, TN-275+STR, 72pp.

Dolman, A.J. (1987) Summer and winter rainfall interception in an oak forest. Predictions with an analytical and a numerical simulation model. *Journal of Hydrology*, **90**, 1–9.

Dolman, A.J. & Wallace, J.S. (1991) Lagrangian and K-theory approaches in modelling evaporation from sparse canopies. *Quarterly Journal of the Royal Meteorological Society*, **117**, 1325–1340.

Dolman, A.J. & Gregory, D. (1992) The parametrization of rainfall interception in GCMs. *Quarterly Journal of the Royal Meteorological Society*, **118**, 455–467.

Dolman, A.J., Lloyd, C.R. & Culf, A.D. (1992) Aerodynamic roughness of an area of natural open forest in the Sahel. *Annales Geophysicae*, **10**, 930–934.

Dolman, A.J., Culf, A.D. & Bessemoulin, P. (1997a) Observations of boundary layer development during the HAPEX-Sahel Intensive Observation Period. *Journal of Hydrology*, **188–189**, 998–1016.

Dolman, A.J., Gash, J.H.C., Goutorbe, J.-P. et al. (1997b) The role of the land surface in Sahelian climate: HAPEX-Sahel results and future research needs. *Journal of Hydrology*, **188–189**, 1067–1079.

Drewer, J., Finch, J.W., Lloyd, C.R., Baggs, E.M. & Skiba, U. (2011) How do soil emissions of N_2O, CH_4 and CO_2 from perennial bioenergy crops differ from arable annual crops? *GCB Bioenergy* **1** (4), 408–419. doi:10.1111/j.1757-1707.2011.01136.x

Dyer, A.J., Hicks, B.B.A. & King, K.M. (1967) The Fluxatron – a revised approach to the measurement of eddy fluxes in the lower atmosphere. *J. Appl. Meteorol.*, **6**, 408–413.

Essery, R., Blyth, E., Harding, R. & Lloyd, C. (2005) Modelling albedo and distributed snowmelt across a low hill in Svalbard. *Nordic Hydrology*, **36** (3), 207–218.

Forsgate, J.A., Hosegood, P.H. & McCulloch, J.S.G. (1965) Design and Installation of semi-enclosed hydraulic lysimeters. *Agricultural Meteorology*, **2** (1), 43–52.

Fox, A.M., Huntley, B., Lloyd, C.R., Williams, M. & Baxter, R. (2008) Net ecosystem exchange over heterogeneous Arctic tundra: scaling between chamber and eddy covariance measurements. *Global Biogeochemical Cycles*, **22**, GB2027 doi: 10.1029/2007GB003027.

Friborg, T., Soegaard, H., Christensen, T.R., Lloyd, C.R. & Panikov, N.S. (2003) Siberian wetlands: where a sink is a source. *Geophysical Research Letters*, **30** (21), 2129 doi: 10.1029/2003GL017797.

Gash, J.H.C. (1979) An analytical model of rainfall interception in forests. *Quarterly Journal of the Royal Meteorological Society*, **105**, 43–55.

Gash, J.H.C. & Stewart, J.B. (1975) The average surface resistance of a pine forest derived from Bowen ratio measurements. *Boundary-Layer Meteorology*, **8**, 453–464.

Gash, J.H.C. & Morton, A.J. (1978) An application of the Rutter model to the estimation of the interception loss from Thetford Forest. *Journal of Hydrology*, **38**, 49–58.

Gash, J.H.C., Lloyd, C.R. & Stewart, J.B. (1979) SIM5T/12 – a model of forest transpiration and interception, using data from an Automatic Weather Station. *Developments in Agricultural and Managed Forest Ecology*, **9**, 173–184.

Gash, J.H.C., Wright, I.R. & Lloyd, C.R. (1980) Comparative estimates of interception loss from three coniferous forests in Great Britain. *Journal of Hydrology*, **48**, 89–105.

Gash, J.H.C. (1986a) A note on estimating the effect of a limited fetch on micrometeorological evaporation measurements. *Boundary-Layer Meteorology*, **35**, 409–413.

Gash, J.H.C. (1986b) Observations of turbulence downwind of a forest-heath interface. *Boundary-Layer Meteorology*, **36**, 227–237.

Gash, J.H.C., Shuttleworth, W.J., Lloyd, C.R., Andre, J.-C., Goutorbe, J.-P. & Gelpe, J. (1989) Micrometeorological measurements in Les Landes forest during HAPEX-MOBILHY. *Agricultural and Forest Meteorology*, **46**, 131–147.

Gash, J.H.C., Wallace, J.S., Lloyd, C.R., Dolman, A.J., Sivakumar, M.V.K. & Renard, C. (1991) Measurements of evaporation from fallow Sahelian savannah at the start of the dry season. *Quarterly Journal of the Royal Meteorological Society*, **117**, 749–760.

Gash, J.H.C., Lloyd, C.R. & Lachaud, G. (1995) Estimating sparse forest rainfall interception with an analytical model. *Journal of Hydrology*, **170**, 79–86.

Gash, J.H.C., Nobre, C.A., Roberts, J.M. & Victoria, R.L. (eds) (1996) *Amazonian deforestation and Climate*. Wiley-Blackwell Michigan, USA, pp. 611.

Gash, J.H.C., Kabat, P., Monteny, B.A. et al. (1997) The variability of evaporation during the HAPEX-Sahel Intensive Observation Period. *Journal of Hydrology*, **188–189**, 385–399.

Gash, J.H.C. & Dolman, A.J. (2003) Sonic anemometer (co)sine responnse and flux measurement. I. The potential for co(sine) error to affect sonic anemometer-based flux measurements. *Agricultural and Forest Meteorology*, **119**, 195–207.

Gourtorbe, J.-P., Lebel, T., Tinga, A. et al. (1994) HAPEX-Sahel: a large scale study of land-atmosphere interactions in the semi-arid tropics. *Annales Geophysicae*, **12**, 53–64.

Goutorbe, J.P., Dolman, A.J., Gash, J.H.C. et al. (1997a) HAPEX-Sahel. *Journal of Hydrology Special Issue*, **188–189**, 400–423.

Goutorbe, J.P., Lebel, T., Dolman, A.J. et al. (1997b) An overview of HAPEX-Sahel: a study in climate and desertification. *Journal of Hydrology*, **188–189**, 4–12.

Grace, J., Lloyd, J., McIntyre, J. et al. (1995a) Fluxes of carbon dioxide and water vapour over an undisturbed tropical forest in south-west Amazonia. *Global Change Biology*, **1**, 1–12.

Grace, J., Lloyd, J., McIntyre, J. et al. (1995b) Carbon dioxide uptake by an undisturbed tropical rain forest in south-west Amazonia, 1992 to 1993. *Science*, **270** (5237), 778–780.

Guyot, A., Cohard, J.-M., Anquetin, S., Galle, S. & Lloyd, C.R. (2009) Combined analysis of energy and water balances to estimate latent heat flux of a sudanian small catchment. *Journal of Hydrology*, 375, 227–240.

Hall, R.L. (1985) Further interception studies of heather using a wet-surface weighing lysimeter system. *Journal of Hydrology*, 81, 193–210.

Hall, R.L. (1992) An improved numerical implementation of Calder's stochastic model of rainfall interception – a note. *Journal of Hydrology*, 140, 389–392.

Hall, R.L., Huntingford, C., Harding, R.J., Lloyd, C.R. & Cox, P.M. (2003) An improved description of soil hydraulic and thermal properties of arctic peatland for use in a GCM. *Hydrological Processes*, 17, 2611–2628.

Halldin, S., Gryning, S.-E. & Lloyd, C.R. (2001) Land-surface/atmosphere exchange in high-latitude landscapes. *Theoretical and Applied Climatology*, 70, 1–3.

Harding, R.J. & Lloyd, C.R. (1998) Fluxes of water and energy from three high latitude tundra sites in Svalbard. *Nordic Hydrology*, 29 (4/5), 267–284.

Harding, R.J., Jackson, N.A., Blyth, E.M. & Culf, A. (2002) Evaporation and energy balance of a sub-Arctic hillslope in northern Finland. *Hydrological Processes*, 16, 1419–1436.

Harding, R.J., Gryning, S.-E., Halldin, S. & Lloyd, C.R. (2001) Progress in understanding of land surface/atmosphere exchanges at high latitudes. *Theoretical and Applied Climatology*, 70, 5–18.

Harding, R.J. & Lloyd, C.R. (2008) Evaporation and energy balance of a wet grassland at Tadham Moor on the Somerset Levels. *Hydrological Processes*, 22, 2346–2357.

Henderson-Sellers, A., Drake, F., McGuffie, K. *et al.* (1987) Observations of day-time cloudiness over the Amazon forest using an all-sky camera. *Weather*, 42 (7), 209–218.

Hutjes, R.W.A., Wierda, A. & Veen, A.W.L. (1990) Rainfall interception in the Tai Forest, Ivory Coast: Application of two simulation models to a humid tropical system. *Journal of Hydrology*, 114, 259–275.

Hyson, P. & Hicks, B.B. (1975) A single-beam hygrometer for evaporation measurement. *Journal of Applied Meteorology*, 14, 301–307.

Jackson, N.A. (2000) Measured and modelled rainfall interception loss from an agroforestry system in Kenya. *Agricultural and Forest Meteorology*, 100 (4), 323–336.

Jackson, N.A., Wallace, J.S. & Ong, C.K. (2000) Tree pruning as ameans of controlling water use in an agroforestry system in Kenya. *Forest Ecology and Management*, 126 (2), 133–148.

Kabat, P., Dolman, A.J. & Elbers, J.A. (1997) Evaporation, sensible heat and canopy conductance of fallow savannah and patterned woodland in the Sahel. *Journal of Hydrology*, 188–189, 494–515.

Kaimal, J.C., Wyngaard, J.C., Izumi, I. & Cote, O.R. (1972) Spectral Characteristics of Surface Layer Turbulence. *Quarterly Journal of the Royal Meteorological Society*, 98, 563–589.

Laurila, T., Soegaard, H., Lloyd, C.R., Aurela, M., Tuovinen, J.-P. & Nordstrøm, C. (2001) Seasonal variations of net CO_2 exchange in European Arctic ecosystems. *Theoretical and Applied Climatology*, 70, 183–201.

Levy, P.E., Meir, P., Allen, S.J. & Jarvis, P.G. (1999) The effect of aqueous transport of CO2 in xylem sap on gas exchange in woody plants. *Tree Physiology*, 19 (1), 53–58.

Lloyd, C.R. (1990) The temporal distribution of Amazonian rainfall and its implications for forest interception. *Quarterly Journal of the Royal Meteorological Society*, 116, 1487–1494.

Lloyd, C.R. (1995) The effect of heterogeneous terrain on micrometeorological measurements: a case study from HAPEX-SAHEL. *Agricultural and Forest Meteorology*, 73, 209–216.

Lloyd, C.R. (1998) The application of an instrument for non-destructive measurements of soil temperature and resistance profiles at a high Arctic field site. *Hydrological and Earth System Science*, 2 (1), 121–128.

Lloyd, C.R. (2001a) *PhD Thesis: The Micrometeorology of a high Arctic site*, University of Edinburgh.

Lloyd, C.R. (2001b) On the physical controls of the carbon dioxide balance at a high Arctic site in Svalbard. *Theoretical and Applied Climatology*, 70, 167.

Lloyd, C.R. (2001c) The measurement and modelling of the carbon dioxide exchange at a high Arctic site in Svalbard. *Global Change Biology*, 7, 405–426.

Lloyd, C.R. (2006) Annual carbon balance of a managed wetland meadow in the Somerset Levels, UK. *Agricultural and Forest Meteorology*, 138, 168–179.

Lloyd, C.R., Shuttleworth, W.J., Gash, J.H.C. & Turner, M. (1984) A microprocessor system for Eddy-correlation. *Agricultural and Forest Meteorology*, 33, 67–80.

Lloyd, C.R. & Marques Filho, A.d.O. (1988) Spatial variability of throughfall and stemflow measurements in Amazonian rainforest. *Agricultural and Forest Meteorology*, 42, 63–73.

Lloyd, C.R., Gash, J.H.C., Shuttleworth, W.J. & Marques Filho, A.d.O. (1988) The measurement and modelling of rainfall interception by Amazonian rain forest. *Agricultural and Forest Meteorology*, 43, 277–294.

Lloyd, C.R., Culf, A.D., Dolman, A.J. & Gash, J.H.C. (1991) Estimates of sensible heat flux from observations of temperature fluctuations. *Boundary-Layer Meteorology*, 57, 311–322.

Lloyd, C.R., Gash, J.H.C. & Sivakumar, M.V.K. (1992) Derivation of the aerodynamic roughness parameters for a Sahelian savannah site using the eddy correlation technique. *Boundary-Layer Meteorology*, 58, 261–271.

Lloyd, C.R., Harding, R.J. & Robinson, M. (1994) *Preliminary reults of energy and water balance from two tundra sites in Svalbard*, Proc. 10th Int. Northern Res. Basins Symposium and Workshop, Spitsbergen Norway. Aug 28-Sep 3 1994, 609–630.

Lloyd, C.R., Bessemoulin, P., Cropley, F.D. *et al.* (1997) A comparison of surface fluxes at the HAPEX-Sahel fallow bush sites. *Journal of Hydrology*, **188–189**, 400–425.

Lloyd, C.R., Harding, R.J., Friborg, T. & Aurela, M. (2001) Surface fluxes of heat and watervapour from sites in the European Arctic. *Theoretical and Applied Climatology*, **70**, 19–33.

Maitelli, G.T. & Wright, I.R. (1996) The climate of a riverside city in the Amazon Basin: urban–rural differences in temperature and humidity. In: Gash, J.H.C., Nobre, C.A., Roberts, J.M. & Victoria, R.L. (eds), *Amazonian Deforestation and Climate*. Wiley-Blackwell, Michigan, USA, pp. 611.

McNeil, D.D. & Shuttleworth, W.J. (1975) Comparative measurements of the energy fluxes over a pine forest. *Boundary-Layer Meteorology*, **9**, 297–313.

Monin, A.S. & Yaglom, A.M. (1971) *Statistical Fluid Mechanics: Mechanics of Turbulence*. Vol. **1**. MIT Press, Cambridge, Massachusetts, pp. 769.

Monteith, J.L. (1965) Evaporation and environment. *Symposium of the Society for Experimental Biology*, **19**, 205–234.

Monteith, J.L. (1977) Resistance of a partially wet canopy: whose equation fails. *Boundary-Layer Meteorology*, **12**, 379–383.

Monteny, B.A., Lhomme, J.P., Chehbouni, A. *et al.* (1997) The role of the Sahelian biosphere on the water and the CO_2-cycle during the HAPEX-Sahel Experiment. *Journal of Hydrology*, **188–189**, 516–535.

Moore, C.J. (1983) On the Calibration and temperature behaviour of single-beam infrared hygrometers. *Boundary-Layer Meteorology*, **25**, 245–269.

Moore, C.J. (1986) Frequency response corrections for eddy correlation systems. *Boundary-Layer Meteorology*, **37**, 17–35.

Mougin, E., Hiernaux, P., Kegoat, L. *et al.* (2009) The AMMA-CATCH Gourma observatory site in Mali: Relating climatic variations to changes in vegetation, surface hydrology, fluxes and natural resources. *Journal of Hydrology*, **375**, 14–33.

Oliver, H.R. (1971) Wind profiles in and above a forest canopy. *Quarterly Journal of the Royal Meteorological Society*, **97**, 548–553.

Oliver, S.A. & Oliver, H.R. (1973) A Computer method for automatic acquisition of wind speed data. *Journal of Physics E.*, **6** (**4**), 401–403. doi:10.1088/0022-3735/6/4/025

Oliver, H.R. & Sene, K. (1992) Energy and water balances of developing vines. *Agricultural and Forest Meteorology*, **61**, 167–185.

Ong, C.K., Black, C.R., Wallace, J.S. *et al.* (2000) Productivity, microclimate and water use in Grevillea robusta-based agroforestry systems on hillslopes in semi-arid Kenya. *Agriculture, Ecosystems and Environment*, **80**, 121–141.

Pearce, A.J. & Rowe, L.K. (1981) Rainfall interception in a multi-storied, evergreen mixed forest: estimates using Gash's analytical model. *Journal of Hydrology*, **49**, 341–353.

Penman, H.L. (1948) Natural evaporation from open water, bare soil, and grass. *Proceedings of the Royal Society of London*, **A193**, 120–145.

Penman, H.L. (1963) Vegetation and Hydrology. In: *Technical Communications 53*. Harpenden, England, Commonwealth Bureau of Soils.

Ramier, D., Boulain, N., Cappelaere, B. *et al.* (2009) Towards an understanding of coupled physical and biological processes in the cultivated Sahel – 1. Energy and water. *Journal of Hydrology*, **375**, 204–216.

Ribeiro, A., Victoria, R.L., Martinelli, L.A., Moreira, M.Z. & Roberts, J.M. (1996) The isotopic composition of atmospheric water vapour inside a canopy in the Amazon forest: vertical and diurnal variation. In: Gash, J.H.C., Nobre, C.A., Roberts, J.M. & Victoria, R.L. (eds), *Amazonian deforestation and climate*. Wiley-Blackwell, Michigan, USA, pp. 611.

Roberts, J., Cabral, O.M.R. & de Aguiar, L.F. (1990) Stomatal and boundary-layer conductances measured in a terra firme rain forest, Manaus, Amazonas, Brazil. *Journal of Applied Ecology*, **27**, 336–353.

Roberts, J., Jackson, N.A. & Smith, M. (2006) *Tree Roots in the Built Environment*. The Stationery Office, pp. 506 ISBN 0-11-753620-2.

Rutter, A.J., Kershaw, K.A., Robins, P.C. & Morton, A.J. (1971) A predictive model of rainfall interception in forests, 1. Derivation of the model from observations in a plantation of Corsican Pine. *Agricultural Meteorology*, **9**, 367–384.

Salati, E., Dall'Olio, A., Matsui, E. & Gat, J.R. (1979) Recycling of water in the Amazon Basin: an isotopic study. *Water Resources Research*, **15** (**5**), 1250–1258.

Schmid, H.P. (1994) Source areas for scalars and scalar fluxes. *Boundary-Layer Meteorology*, **67**, 293–318.

Schmid, H.P. & Lloyd, C.R. (1999) Spatial representativeness and the location bias of flux footprints over inhomogeneous areas. *Agricultural and Forest Meteorology*, **93**, 195–209.

Sellers, P.J. & Lockwood, J.G. (1981) A computer simulation of the effects of differing crop types on the water balance of small catchments over long time periods. *Quarterly Journal of the Royal Meteorological Society*, **107**, 395–414.

Sellers, P.J., Mintz, Y., Sud, Y.C. & Dalcher, A. (1986) A simple biosphere model (SiB) for use within general circulation models. *Journal of the Atmospheric Sciences*, **43**, 505–531.

Sellers, P.J., Shuttleworth, W.J. & Dorman, J.L. (1989) Calibrating the Simple Biosphere Model (SiB) for Amazonian tropical forest using field and remote sensing data: part I: average calibration with field data. *Journal of Applied Meteorology*, **28**, 727–759.

Sellers, P., Hall, F., Margolis, H. *et al.* (1995) The Boreal Ecosystem-Atmosphere Study (BOREAS): An overview and early results from the 1994 field year. *Bulletin of the American Meteorological Society*, **76** (**9**), 1549–1577.

Sene, K.J. (1994) Paramerisations for energy transfer from a sparse vine crop. *Agricultural and Forest Meteorology*, **71**, 1–18.

Sene, K.J. (1996) Meteorological estimates for the water balance of a sparse vine crop growing in semi-arid conditions. *The Journal of Hydrology*, **179**, 259–280.

Shuttleworth, W.J. (1975) The concept of intrinsic surface resistance: energy budgets at a partially wet surface. *Boundary-Layer Meteorology*, **8**, 81–99.

Shuttleworth, W.J. (1976a) A one-dimensional theoretical description of the vegetation-atmosphere interaction. *Boundary-Layer Meteorology*, **10**, 271–301.

Shuttleworth, W.J. (1976b) Experimental evidence for the failure of the Penman-Monteith equation in partially wet conditions. *Boundary-Layer Meteorology*, **10** (**1**), 91–94.

Shuttleworth, W.J. (1977) Comments on resistance of a partially wet canopy; whose equation fails. *Boundary-Layer Meteorology*, **12**, 463–489.

Shuttleworth, W.J. (1978) A simplified one-dimensional theoretical description of the vegetation-atmosphere interaction. *Boundary-Layer Meteorology*, **14**, 3–27.

Shuttleworth, W.J. (1988a) Evaporation from Amazonian Rainforest. *Proceedings of the Royal Society of London B*, **233** (**1272**), 321–346.

Shuttleworth, W.J. (1988b) Macrohydrology – The new challenge for process hydrology. *Journal of Hydrology*, **100**, 31–56.

Shuttleworth, W.J. (2007) Putting the 'vap' into evaporation. *Hydrology and Earth System Sciences*, **11** (**1**), 210–244.

Shuttleworth, W.J. & Calder, I.R. (1979) Has the Priestley-Taylor equation any relevance to forest hydrology? *Journal of Applied Meteorology*, **18**, 639–646.

Shuttleworth, W.J. & Gash, J.H.C. (1981) A not on the paper by P.J. Sellers and J.G.Lockwood. 'A computer simulation of the effects of differing crop types on the water balance of small catchments over long time periods'. *Quarterly Journal of the Royal Meteorological Society*, **107**, 395–414; *Quarterly Journal of the Royal Meteorological Society* 108, 464–470.

Shuttleworth, W.J. & Dickinson, R.E. (1989) Comment on "Modelling tropical deforestation: a study of GCM land-surface parameterization" by R.E. Dickinson and A Henderson-Sellers. *Quarterly Journal of the Royal Meteorological Society*, **115**, 1177–1179.

Shuttleworth, W.J., McNeil, D.D. & Moore, C.J. (1982) A switched continuous-wave sonic anemometer for measuring surface heat fluxes. *Boundary-Layer Meteorology*, **23**, 425–448.

Shuttleworth, W.J., Gash, J.H.C., Lloyd, C.R. *et al.* (1984a) Eddy Correlation measurements of energy partition for Amazonian forest. *Quarterly Journal of the Royal Meteorological Society*, **110**, 1143–1162.

Shuttleworth, W.J., Gash, J.H.C., Lloyd, C.R. *et al.* (1984b) Observations of radiation exchange above and below Amazonian forest. *Quarterly Journal of the Royal Meteorological Society*, **110**, 1163–1169.

Shuttleworth, W.J., Gash, J.H.C., Lloyd, C.R. *et al.* (1985) Daily variations of temperature and humidity within and above Amazonian forest. *Weather*, **40** (**4**), 102–108.

Shuttleworth, W.J. & Wallace, J.S. (1985) Evaporation from sparse crops-an energy combination theory. *Quarterly Journal of the Royal Meteorological Society*, **111**, 839–855.

Shuttleworth, W.J. & Gurney, R.J. (1990) The theoretical relationship between foliage temperature and canopy resistance in sparse crops. *Quarterly Journal of the Royal Meteorological Society*, **116**, 497–519.

Shuttleworth, W.J., Gash, J.H.C., Lloyd, C.R. *et al.* (1987) Amazonian evaporation. *Revista Brasiliera de Meteorologie*, **2**, 179–191.

Shuttleworth, W.J., Gash, J.H.C., Lloyd, C.R., McNeil, D.D., Moore, C.J. & Wallace, J.S. (1988) An integrated micrometeorological system for evaporation measurement. *Agricultural and Forest Meteorology*, **43**, 295–317.

Shuttleworth, W.J., Zreda, M., Zeng, X., Zweck, C., & Ferre, P.A. (2010) *The cosmic-ray soil moisture observing system (COSMOS): a non-invasive, intermediate scale soil moisture measurement network*. Proceedings of the British Hydrological Society's Third International Symposium: 'Role of hydrology in managing consequences of a changing global environment', Newcastle University, 19–23 July 2010. ISBN: 1 903741 17 3.

Smith, F.B., Carson, D.J. & Oliver, H.R. (1972) Mean wind-direction shear through a forest canopy. *Boundary-Layer Meteorology*, **3**, 178–190.

Stewart, J.B. (1971) The Albedo of a Pine Forest. *Quarterly Journal of the Royal Meteorological Society*, **97**, 561–564.

Stewart, J.B. & Oliver, S.A. (1970) *Preliminary results of a study of evaporation from forests*. Symp. On World Water Balance, IASH/UNESCO, Reading, UK.

Stewart, J.B. & Thom, A.S. (1973) Energy budgets in pine forest. *Quarterly Journal of the Royal Meteorological Society*, **99**, 154–170.

Stewart, J.B. & de Bruin, H.A.R. (1985) Preliminary study of dependence of surface conductance of Thetford Forest on environmental conditions. In: Hutchinson, B.A. & Hicks, B.B. (eds), *The Forest-Atmosphere Interaction*. D. Reidel, Dordrecht.

Swiontek, M.C. & Hassan, R. (1965) The Linear Quartz Thermometer – a new tool for measuring absolute and difference temperatures. *Hewlett-Packard Journal*, **16** (7), 12pp.

Tanguy, M., Baille, A., Gonzalez-Real, M.M. *et al.* (2012) A new parameterisation scheme of ground heat flux for land surface flux retrieval from remote sensing information. *Journal of Hydrology*, **454–455**, 113–122.

Taylor, C.M., Harding, R.J., Thorpe, A.J. & Bessemoulin, P. (1997) A mesoscale simulation of land surface heterogeneity from HAPEX-Sahel. *Journal of Hydrology*, **188–189**, 1040–1066.

Taylor, C.M. & Clark, D.B. (2001) The diurnal cycle and African easterly waves: a land surface perspective. *Quarterly Journal of the Royal Meteorological Society*, **127**, 845–867.

Taylor, C.M., Parker, D.J., Lloyd, C.R. & Thorncroft, C.D. (2005) Observations of synoptic-scale land surface variability and its coupling with the atmosphere. *Quarterly Journal of the Royal Meteorological Society*, **131**, 913–937.

Teller, A., Mathy, P. & Jeffers, J.N.R. (Eds.) (1992) *Responses of forest ecosystems to environmental changes.* Proc. 1st European Symposium on Terrestrial Ecosystems: Forests and Woodlands. Florence, Italy, 20–24 May. Elsevier Applied Science, London, 1009pp.

Tillman, J.E. (1972) The indirect determination of stability, heat and momentum fluxes in the atmospheric boundary layer from simple scalar variables during dry unstable conditions. *Journal of Applied Meteorology*, **11**, 783–792.

Timouk, F., Kergoat, L., Mougin, E. *et al.* (2009) Response of surface energy balance to water regime and vegetatin development in a Sahelian landscape. *Journal of Hydrology*, **375**, 178–189.

Ubarana, V.N. (1996) Observations and modelling of rainfall interception at two experimental sites in Amazonia. In: Gash, J.H.C., Nobre, C.A., Roberts, J.M. & Victoria, R.L. (eds), *Amazonian Deforestation and Climate.* Wiley and Sons Ltd, Chichester, pp. 151–162.

Van der Molen, M.K., Gash, J.H.C. & Elbers, J.A. (2004) Sonic anemometer (co) sine response and flux measurement: II. The effect of introducing an angle of attack dependent calibration. *Agricultural and Forest Meteorology*, **122**, 95–109.

Verhoef, A., Allen, S.J. & Lloyd, C.R. (1999) Seasonal variation of surface energy balance over two Sahelian surfaces. *International Journal of Climatology*, **19**, 1267–1277.

Wai, M.M.-K., Smith, E.A., Bessemoulin, P., Culf, A.D., Dolman, A.J. & Lebel, T. (1997) *Variability in boundary layer structure during HAPEX-Sahel wet-dry season transition.*

Wallace, J.S. (1991) *The measurement and modelling of evaporation from semiaridland.* In: M.V.K. Sivakumar, J.S. Wallace, C. Renard and C. Giroux (Eds.) Soil Water Balance in the Sudano-Sahelian Zone. Proc. Niamey Workshop, February 1991. IAHS Publ. no. 199: 131–148.

Wallace, J.S., Lloyd, C.R., Roberts, J. & Shuttleworth, W.J. (1984) A comparison of methods for estimating aerodynamic resistance of heather (*Calluna vulgaris* (L.) Hull) in the field. *Agricultural and Forest Meteorology*, **32**, 289–305.

Wallace, J.S., Gash, J.H.C., McNeil, D.D., Lloyd, C.R., Oliver, H.R., Keatinge, J.D.H. & Sivakumar, M.V.K. (1989) *Measurement and prediction of actual evaporation from sparse drlyland crops.* ODA Report no. ODG 89/1, Institute of Hydrology, Wallingford, 39pp.

Wallace, J.S., Roberts, J.M. & Sivakumar, M.V.K. (1990) The estimation of transpiration from sparse dryland millet using stomatal conductance and vegetation area indices. *Agricultural and Forest Meteorology*, **51**, 35–49.

Wallace, J.S., Lloyd, C.R. & Sivakumar, M.V.K. (1993) Measurements of soil, plant and total evaporation from millet in Niger. *Agricultural and Forest Meteorology*, **63**, 149–169.

Wallace, J.S. & Holwill, C.J. (1997) Soil evaporation from tiger-bush in south-west Niger. *Journal of Hydrology*, **188–189**, 426–442.

Wright, I.R. (1990) A laboratory calibration of infrared thermometers. *International Journal of Remote Sensing*, **11**, 181–186.

Wright, I.R., Gash, J.H.C., da Rocha, H.R. *et al.* (1992) Dry season micrometeorology of central Amazonian ranchland. *Quarterly Journal of the Royal Meteorological Society*, **118**, 1083–1099.

6 Water Resources Security

FRANK FARQUHARSON[1], MAX BERAN[2], JOHN BROMLEY[3],
ALAN GUSTARD[2], HELEN HOUGHTON-CARR[4],
GWYN REES[4], JOHN SUTCLIFFE[2], AND ANDY YOUNG[5]

[1]Water Resource Associates, Henley on Thames, UK
[2]Ex-Institute of Hydrology, Wallingford, UK
[3]School of Geography & Environment, University of Oxford, Oxford, UK
[4]Centre for Ecology and Hydrology, Wallingford, Oxfordshire, UK
[5]Wallingford HydroSolutions Ltd., Wallingford, Oxfordshire, UK

Progress in Modern Hydrology: Past, Present and Future, First Edition. Edited by John C. Rodda and Mark Robinson.
© 2015 John Wiley & Sons, Ltd. Published 2015 by John Wiley & Sons, Ltd.

6.1   Introduction

This chapter describes the evolution of water resources research at Wallingford. There are two primary themes. The first describes research aimed at estimating the low flow regimes within gauged and ungauged basins. These were collaborative regional or national scale projects funded by environmental agencies where the research output would often be used by their staff to solve design problems. Specific application of these techniques included abstraction and effluent licensing, setting environmental flows, irrigation, designing small hydropower schemes, navigation and estimating drought frequency. The research often addressed national and international water law and policy requirements most notably the implementation of the Catchment Abstraction Management Strategy (CAMS) in England and Wales and the European Water Framework Directive. A fundamental change over five decades of research has been the method by which design methods have been transferred to the user community. Initially in the form of written reports, they are now presented in a portfolio of software products underpinned by comprehensive digital time series and spatial databases with bespoke applications for specific clients.

The second theme was where external clients posed specific and usually very challenging water resource problems, often requiring innovative solutions. Much of this work was undertaken in overseas countries rather than in the UK, in part because during the late 1960s and 1970s the Water Resources Board were responsible for national water resource assessment and planning studies in the UK. However, another reason why many of the studies were in developing countries was pragmatic: because one of the first senior Wallingford staff joined after working for one of the UK's large firms of consulting engineers. Rather than recruiting a replacement hydrologist, this company utilized the growing pool of high-quality hydrologists at Wallingford to support a number of water resource studies in the Middle East, Africa and Southeast Asia through a partnering arrangement (see Section 6.6).

6.2   Low Flow Studies 1974–1987

6.2.1   *Introduction*

In 1974, the Institute of Hydrology (IH) was commissioned by the Department of the Environment to develop national procedures for estimating the frequency of low flows at both gauged and ungauged sites. The study (Beran and Gustard, 1977; Institute of Hydrology 1980) benefited from and was modelled on the statistical component of the recently completed Flood Studies Report (Chapter 3). There were essentially four main components to the research programme:

1 Assembling a database of river flows and deriving a number of low flow indices for each gauging station record.
2 Deriving the characteristics of upstream catchment areas.
3 Analysing the relationship between low flows and upstream catchments
4 Presenting the methodology in a format which could be readily used by the UK Water Industry.

In 1970, there were no standard methods for estimating low flows although the dry weather flow (often defined by mean of seven day minima) and the 95 percentile discharge were the most frequently used. A review of the literature and discussion of techniques with the UK water industry identified a number of different ways of describing the river flow regime in the context of low flows. These are summarised on Table 6.1 and a key achievement of the study was the development of methods for estimating them at ungauged sites for any location in the UK. The low flow regime of a river was found to depend largely on the geology of the catchment. A major breakthrough was to develop an automated base flow separation procedure from which a Base Flow Index (BFI) could be derived from river flow data and could be estimated at ungauged sites from mapped solid and superficial geology. BFI has found subsequent international use as a standard hydrometric variable and has even been extended for use in regionalizing flood response (Chapter 3) and estimating groundwater recharge.

The study benefited from a computer database of gauging station information and catchment

Table 6.1 Summary of low flow regime measures

Regime measured	Property described	Data employed	Applications
Mean flow	Arithmetic mean of the flow series	Daily or monthly flows	Resource estimation
Coefficient of variation in annual mean flow	Standard deviation of annual mean flow divided by mean flow	Annual mean flow	Resource estimation
Flow–duration curve	Proportion of time a given flow is exceeded	Daily flows or flows averaged over several days, weeks or months	Licensing abstractions or effluents, hydropower design
Annual minimum series	Annual lowest flows (of a given duration)	Annual minimum flows – daily or averaged over several days	Drought return period; initial storage/yield analyses
Recession indices	Rate of decay of hydrograph	Daily flows during dry periods	Short-term forecasting; hydrogeological studies; modelling
Base flow index	Proportion of total flow from stored catchment sources	Daily flows	Regionalisation, hydrogeological studies, recharge estimation
Low flow spells durations	Frequency with which the flow remains below a threshold for a given duration	Periods of low flows extracted from the hydrograph followed by a statistical analysis of durations	Water quality, fisheries amenity, navigation; drought frequency
Low flow spell volumes	Frequency of requirement for a given volume of "make-up" water to maintain a threshold flow	Same as above, except the analysis focuses on the volume below the threshold	Preliminary design of regulating reservoirs, drought frequency
Runoff accumulation time	Time required to accumulate a given volume of runoff	Daily or monthly flows	Estimating probability of reservoir infill
Storage yield	Storage required for given yield and probability of failure	Daily or monthly	Reservoir design

Source: Gustard, Bullock & Dixon 1992. Reproduced with permission of IAHS.

characteristics developed in the Flood Studies Report (Chapter 3) and from a national river flow database then held at the Water Data Unit (Chapter 11). Research at the Institute considerably benefited from the closer institutional and computer links with the National River Flow Archive (Chapter 11) following its transfer to NERC in 1982. The subsequent ease of access to a comprehensive national database of daily river flows (including regularly updated gauging station information and digital thematic data) has considerably enhanced low flow research at Wallingford for more than four decades. Relationships were sought between the low flow regime of a river and the upstream catchment and it was therefore essential that the

catchments used in the study were relatively natural and had accurate measurements of low flows. Over 1400 flow records were considered for the study of which 517 were found suitable. Catchment characteristics were derived manually, including the average annual rainfall, length of the main stream, stream frequency, slope and the proportion of the catchment covered by a lake or urban area. In contrast to current automatic derivation, this was a very labour intensive task (Chapter 11) (Box 6.1).

6.2.2 Low flow analysis

For the first time in the UK, the study developed a suite of objective techniques for analysing the

Box 6.1

The 1974 Low Flow Study was funded by the Department of Environment – a direct result of the implementation of the Rothschild report in 1971. This led to approximately 25% of funds being transferred from all the UK Research Councils and their research institutes to Government Departments. The aim was to ensure that a proportion of research was providing solutions to practical issues defined by government and not decided by the research community. This met with a hostile reception from the research community who valued their independence in identifying research priorities. However, it was beneficial to those working in applied research and in the case of the Institute's low flow studies, initiated an innovative programme of research lasting over 30 years.

time series of daily flows to produce summary descriptions of the overall flow regime of a river with an emphasis on low flows. The proliferation of methods is a result of the following:

1 The different definitions of a low flow event, for example an event can be expressed as annual minima, a threshold discharge, the time during which the discharge is below a threshold or the rate of recession.

2 The different methods of expressing frequency, for example as a proportion of time during which a discharge is exceeded, for example the flow–duration curve, or as a proportion of years during which a given low flow occurs, for example, extreme value analysis.

3 Different durations or averaging periods. Many applications require information over a set period, such as seven or thirty days or for different seasons.

Of the ten methods of analysis (Table 6.1), five are summarised below. In the Low Flow Study, the term "low-flow indices" was used for specific values derived from an analysis of low flows.

6.2.2.1 Base flow index

The most far-reaching achievement of the Low Flow Study was the "invention" of the Base Flow Index (BFI) as the ratio of base flow to total flow. An automatic hydrograph separation procedure was developed and applied to over 500 time series

of daily flows. Values of BFI range from 0.15 for an impermeable catchment, for example clay with a "flashy" flow regime, to more than 0.95 for catchments with a stable flow regime, for example chalk. The BFI (Institute of Hydrology, 1980; WMO, 2008) has been subsequently widely adopted as a general index of catchment response. Following its initial application in regional low flow studies in the United Kingdom (Gustard et al., 1992), it is now routinely calculated for over 1000 gauged records published in the UK Hydrometric Register (Marsh and Hannaford, 2008). The BFI has also been used to classify the hydrological response of soil types for regional flood studies (Boorman et al., 1995). Furthermore, it has been implemented as a general catchment descriptor for hydrological modelling, as a tool for selecting analogue catchments, and for estimating annual and long-term groundwater recharge. The BFI has been used not only in the UK but also in low flow studies in mainland Europe, New Zealand, Southern Africa and the Himalayas. It has also had extensive use in Canada (Piggott et al., 2005), where it has been mapped in support of regional low flow studies and used to map recharge and discharge zones, to investigate the impact of climate change on groundwater resources and also to relate flood response to soil type. A review of alternative automated separation procedures can be found in Nathan and McMahon (1990). International studies have shown that in tropical climates, the index is heavily influenced by the seasonal climate regime. High index values may be observed downstream of glaciers. In snow-dominated catchments, it has been found useful to calculate BFI from the period without snow, that is to calculate annual summer BFI (Hisdal et al., 2004). Examples of base flow separation for two United Kingdom catchments are shown in Figure 6.1.

6.2.2.2 Recession curve

A key low flow characteristic of a river is the rate of decay of river flows during dry weather. This recession behaviour has applications in hydrological modelling, low flow forecasting and is a key property used by hydrogeologists evaluating storage and transmission properties of aquifers. A new automated technique (Beran and Gustard, 1977) was developed for estimating the recession constant based on a scatter diagram of 'today's flow' against 'the flow two days ago' obtained from low flow recession periods. Trials showed that an envelope could be objectively defined by

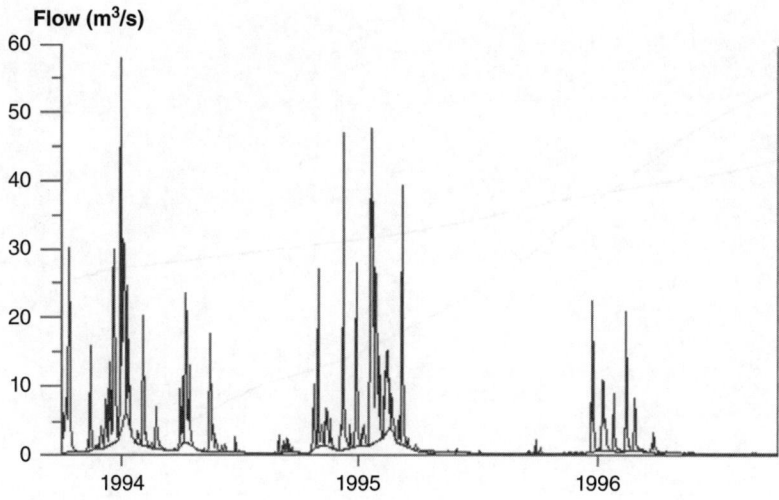

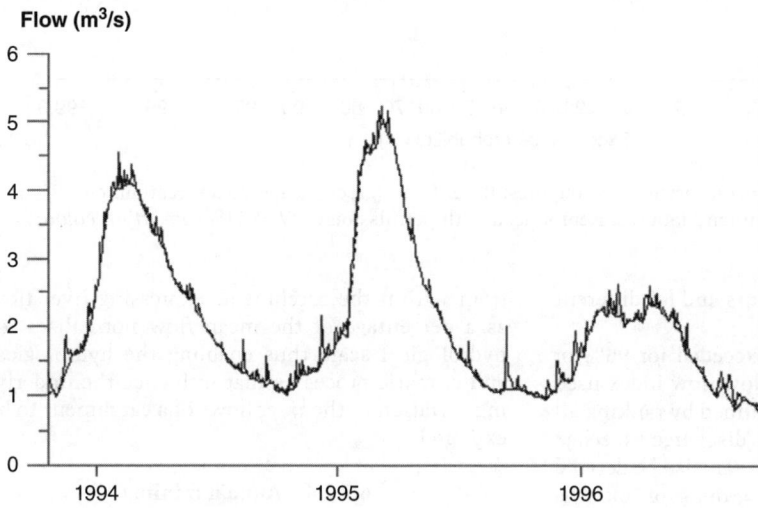

Fig. 6.1 Base flow separations for contrasting flow regimes: River Beult (upper); impermeable catchment, BFI = 0.23 and River Lambourn (lower); permeable catchment, BFI = 0.97. (Source: Reproduced with permission of WHS.)

connecting recession points to produce a master recession curve. The slope of this curve defined the recession constant and this was calculated from over 500 daily river flow records. Values ranged from 0.90 for a very impermeable catchment to 0.99 for a very stable chalk catchment.

6.2.2.3 Flow–duration curve

The flow–duration curve (FDC) is a graph of river flow plotted against the probability of exceedance (Fig. 6.2) and is normally derived from the complete time series of recorded river flows. The construction is based on ranking the data (normally daily discharge) and calculating the proportion of daily values exceeding each value. It thus reorders the observed hydrograph from one ordered by time to one ordered by magnitude. The percentage of time that any particular discharge is likely to be equalled or exceeded is read from the plot. Comprehensive reviews of the FDC are given by Hisdal *et al.* (2004). A normal probability scale was used for the frequency axis and a logarithmic scale for the discharge axis. The plots approximated to straight lines for most UK flow series indicating that the logarithms of discharge were normally distributed. This form of presentation facilitated comparison of

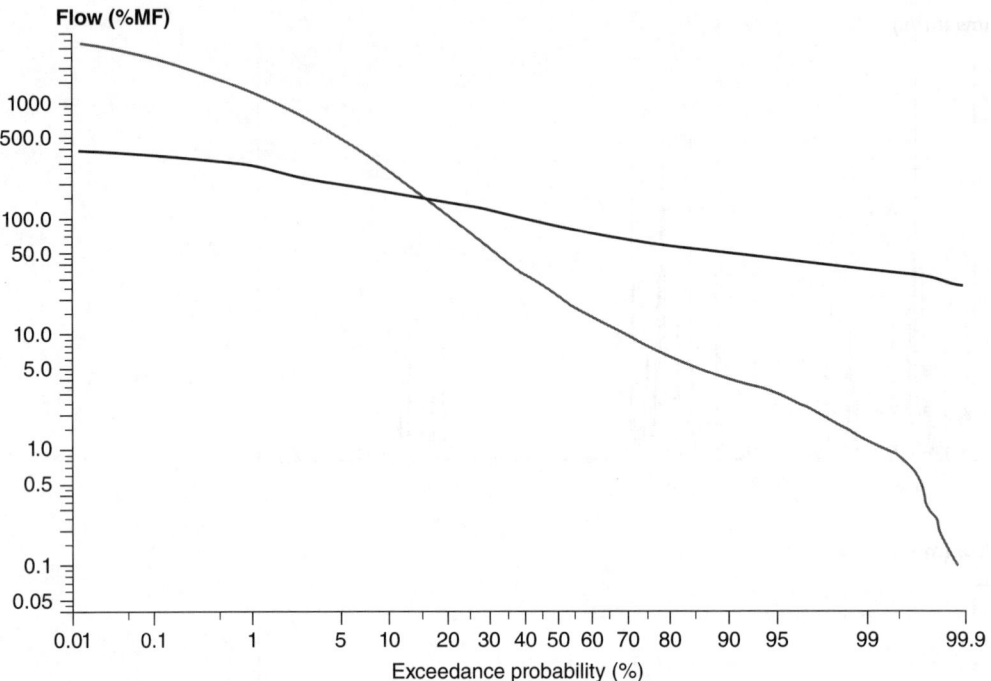

Fig. 6.2 Dimensionless FDCs for contrasting flow regimes: River Beult (green); impermeable catchment, River Lambourn (blue); permeable catchment. (Source: Reproduced with permission of WHS.) *(See insert for colour representation of this figure.)*

distributions between catchments and for different durations and seasons.

Q95 the flow equalled or exceeded for 95% of the time is the most common low flow index used internationally. It can be determined by ranking all daily discharges and finding the discharge exceeded by 95 per cent of all values. Q95 can also be derived from individual months, for groups of months or any specified periods. Other percentiles can similarly be derived from the flow–duration curve. In the Low Flow Study, flow–duration curves were derived from a 10-day moving average of the daily data in order to smooth minor artificial variations in daily discharges. The Q95 (10 day) index was then calculated from over 500 flow records. Monthly and seasonal flow–duration curves were derived from groups of months and methods developed to estimate them at an ungauged site. Discharge was expressed as a percentage of the mean flow to assist in regionalisation. The mean flow of a catchment is predominately influenced by the size of the catchment, how much precipitation is incident on the catchment and the evaporation that takes place

from within the catchment. Expressing river flow as a percentage of the mean flow normalises for hydrological scale thus enabling the hydrological and climatic processes that influence the underlying variation in the river flows of a catchment to be explored.

6.2.2.4   Annual minima

Annual minima can be derived from a daily flow series by selecting the lowest flow every year. In the Low Flow Study, minima over periods of 1, 5, 7, 10, 30, 60, 90 and 180 days were calculated and low flow–frequency curves plotted using a Weibull distribution. Mean annual minima for each of these durations were calculated. These were used to produce a family of frequency plots standardised by the mean annual minima for the range of different durations.

6.2.2.5   Mean flow

The Q95 and MAM (10) were standardised by dividing by the mean flow. As discussed this facilitated comparison of low flows between catchments of different size and contrasting rainfall and furthermore with different record lengths. The mean flow was calculated for all flow records and a procedure was developed for estimating it at the ungauged site.

6.2.3 Regionalisation

The regionalisation of the flow–duration curve and annual minima series was based on an analysis of over 500 flow records and involved four major components:

1 Multivariate regression analysis relating Q95 (10) and MAM (10) to catchment characteristics following the methodology developed in the Flood Studies Report.
2 Deriving relationships between low flow indices of different durations that could then be applied at the ungauged site.
3 Deriving regional frequency relationships for the flow–duration curve and annual minima series.
4 Estimating the mean flow from average annual rainfall and potential evaporation.

Once the variables had been standardised, the dominant predictive variable was the Base Flow Index. At the ungauged site, this was estimated from relationships between BFI and catchment solid and superficial geology. Guidelines were presented of how these could be developed to take account of mixed geologies and examples were provided for some UK geological regions. The importance of using local data from up or downstream locations or from analogue catchments was highlighted. The methodology was summarised as 'comparison with analogue catchments' but could also be described as an "informal region-of-influence approach" (Section 6.4). In addition to estimating low flows from BFI, a number of operational procedures using local data were presented, highlighting that using data from up or downstream, from analogue catchments or from current meterings was often preferable to using a regional model. Tallaksen and van Lanen (2004) and WMO (2008) provide a very comprehensive review of databases, low flow processes, analysis methods, techniques for estimation at the ungauged site and operational case studies (Box 6.2).

6.2.4 Water industry implementation

The results of the Low Flow Study were delivered to the UK water industry in the form of series of

---

Box 6.2

In addition to undertaking hydrological research, the Institute was invited to give evidence to Government select committees. One example (Wilkinson *et al.*, 1996) was the Environment Committee on Water Conservation and Supply ordered by the House of Commons following an extreme drought in 1985 particularly in Yorkshire. For many months between September 1995 and January 1996, reservoirs in the west side of the region ran dry and water had to be taken by 700 tankers from the east side of the region in a convoy of trucks with 3500 daily deliveries along the M62 in a drastic emergency measure which cost £3 million a week. During the drought, Trevor Newton the then Chief Executive of Yorkshire Water said: "I haven't had a bath or a shower now for three months" Unfortunately for him the Daily Express discovered his secret with the headline: "Clever Trevor takes a bath at parents' home". IH staff was not asked to comment on this, but they responded to a wide range of questions including: the impact of afforestation, urbanisation, reservoir construction, climate change and groundwater abstraction.

---

reports. In addition, in excess of 15 residential Low Flow Courses were held for hydrologists from a wide range of UK organisations including water supply, sewerage, environmental protection and consultants. The industry welcomed the formalisation of low flow analysis from gauged data, the suite of techniques for using local data and in some instances embarked on regional low flow investigations often commissioning IH to carry them out. Examples included North West England, South West England and Scotland. The latter was based on a comprehensive study of 232 flow records in Scotland (Gustard *et al.*, 1987a). BFI for each reach were estimated from up or downstream, or nearby analogue-gauged BFI, or from the regional relationships. It was the first large-scale application of the "informal region-of-influence approach". Another innovation was that the results were presented in the form of an easy-to-use map of BFI. Subsequent feedback from the UK water industry also identified the need for regionalisation of seasonal flow–duration curves and runoff accumulation time and these were completed in 1987.

6.3 Low Flow Studies 1988–2000: Capitalizing on Digital Cartography

6.3.1 *Early applications*

Although the Low Flow Study provided useful design procedures for the UK Water Industry it was constrained by relatively short flow records, the absence of digitised databases of thematic data and the need to develop localised BFI estimation methods. A subsequent study (Gustard *et al.*, 1992) capitalised on the rapid expansion of the gauging network between 1960 and 1980 resulting in an increase of 60% in the number of stations used. The second major advance was the availability of digital terrain data, rivers and thematic data and the tools to exploit them. Catchment boundaries, isohyetal and soil association maps were digitised and catchment characteristics calculated automatically by digital overlay. Although the first digitisation step was manual, the overlay was automatic and saved considerable time and improved the accuracy of calculation compared with the 1974 study. Another major advance was the results from the Hydrology of Soil Type (HOST) project. It was based on collaboration between the IH and Soil Survey organisations of the UK. Over 900 soil associations in the UK were grouped into 29 HOST classes based on their physical soil characteristics the catchment response indexed by BFI calculated from over 800 daily flow records (Boorman *et al.*, 1995). The digital HOST map proved very successful both in improving the estimate of low flow parameters and also in subsequent Flood Studies (Reed 1999). In the 1992 Low Flow Study, the 29 HOST classes were grouped into twelve Low Flow Host Groups (LFHG) and used for low flow prediction (Fig. 6.3).

6.3.2 *Micro low flows*

In the late 1980s, the then North West Water and South West Water Authorities requested low flow estimates for every reach of their river systems. The first product to be delivered was a computer file containing several thousand flow estimates. It was made possible by an ongoing project led by IH to digitise all UK rivers at a scale of 1:50 000

(Chapter 11). In the absence of a digital terrain model for automatically estimating catchment boundaries, this digital river network was overlaid on digital rainfall, evaporation and LFHG maps to calculate upstream catchment characteristics and thereby flow estimates (Section 6.3.1) for over two thousand river reaches (Young *et al.*, 2000). This was carried out on a mainframe computer but subsequently implemented on a PC. Following privatisation of the water industry, and working with the South West region of the National Rivers Authority (NRA), this method was extended to automatically estimate catchment areas by assigning 1-km resolution grid cells to river reaches on a closest distance basis. Following publication, this novel approach was replicated in several international studies for the same requirement. By 1995, the PC system known as Micro LOW FLOWS was implemented and operationally used in seven out of the eight NRA regions and equivalent versions were under development for other regions of the UK including the Scottish Environment Protection Agency and the Environment and Heritage Service within Northern Ireland. This provided the water industry with easy-to-use software incorporating the most recent design improvements and marked a turning point in low flow estimation in the UK and the beginning of a period of very constructive cooperation in software which has continued for over two decades (Section 6.4).

6.3.3 *Artificial influences*

6.3.3.1 **Abstractions, Impounding Reservoirs and Discharges**

A key problem in estimating low flows in the UK is the impact of a wide range of human-induced changes. These include surface and groundwater abstractions, effluent discharges and impounding reservoirs. For the purposes of the 1992 study (Gustard *et al.*, 1992), a database of the details for over 43 000 abstraction licences and 96 000 discharge consents was compiled for Great Britain, no mean feat as this entailed harmonizing datasets from nine different regions within England and Wales, all with different data structures, nomenclatures and levels of hydrologically useful records. Of course, within Scotland at this point in time, there was no abstraction licensing and thus the work in Scotland required the compilation of information from many, disparate data sources. Combining

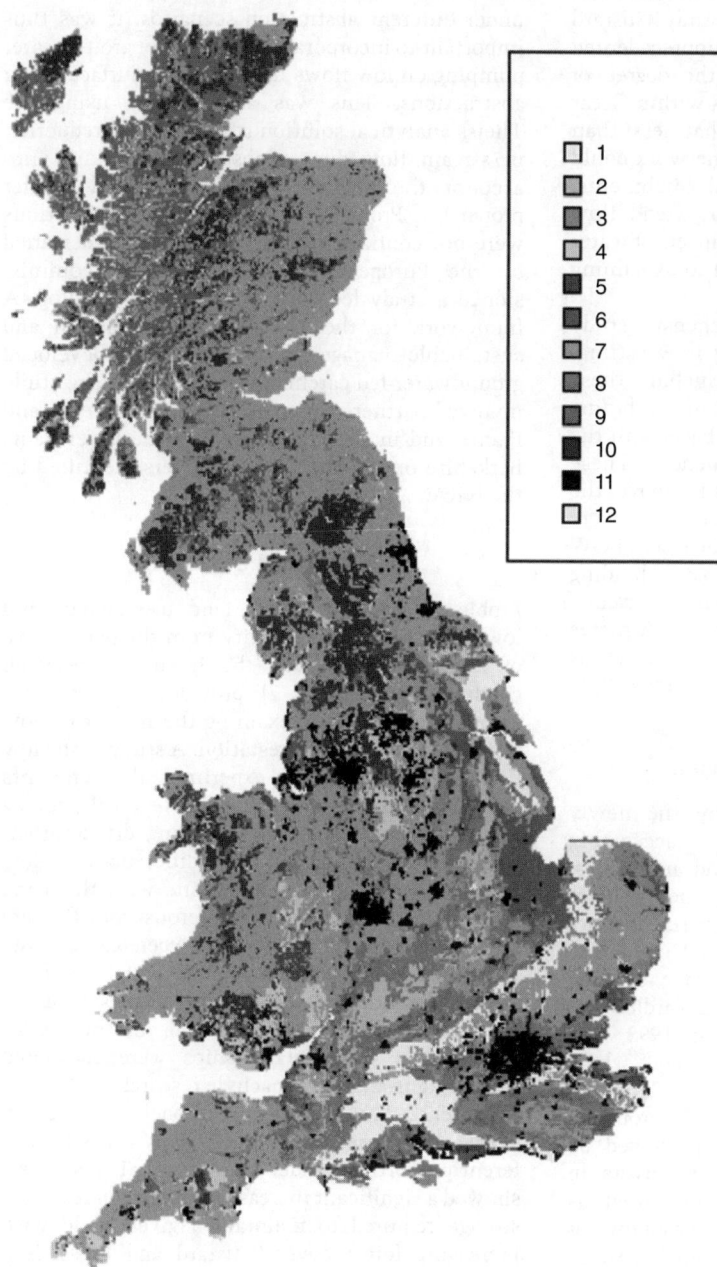

Fig. 6.3 Low Flow HOST groups. (Source: Reproduced with permission of CEH.) *(See insert for colour representation of this figure.)*

these data with the reservoir archives developed as part of the 1987 Compensation Flow Study (Gustard *et al.*, 1987b) gave researchers the unprecedented ability to classify (albeit crudely) the degree of influence on all gauged flows records within Great Britain. This analysis concluded that less than 20% of the primary gauging station network could be considered to be gauging natural catchments. This clearly identified the need to extend flow estimation methods to address the impacts of water use on available resource in addition to explaining the natural variation in river flow.

The NRA commissioned a comprehensive study (Bullock *et al.*, 1994) to widen low flow estimation from natural catchments to include these influences. The approach was based on combining natural flow statistics on a monthly basis with the monthly variation in artificial influences. These were then accumulated at all points above the ungauged location. These methodologies were implemented in a second version of the Micro LOW FLOWS software system (Section 6.3.2) holding details of abstraction licences, discharge consents, reservoir releases and spot gaugings. Estimates could be displayed for a single site or at numerous locations along a river to construct a residual flow diagram.

6.3.3.2 Groundwater Pumping

A key issue addressed in 1989 by the newly formed NRA was the problem of unacceptably low flows in many rivers in England and Wales. Although exacerbated by drought, the problem was primarily due to excessive authorised abstractions many of which were authorised under the 1963 Water Resources Act, which gave existing abstractors the right to abstract regardless of any environmental implications. In 1993, the NRA identified a list of "Top 40 Rivers" where low flow problems were most acute. This list highlighted the key environmental problems associated with water quantity as perceived by the public and the NRA. Thirty one rivers in the 'Top 40' (NRA, 1993) were identified as having low flow problems due to groundwater pumping and nine due to excessive surface water abstraction.

The importance of both reviewing and addressing these issues underpinned the Authority's support for low flow and hydroecological research at the Institute. Micro LOW FLOWS provided a valuable reconnaissance tool where observed

flows could be compared with predicted flows under different abstraction scenarios. It was thus important to incorporate the impact of groundwater pumping on low flows in addition to surface water abstractions. This was achieved by using the Theiss analytical solution to identify the reduction in stream flow due to abstractions, taking into account the location of boreholes and aquifer properties. Problems of groundwater abstractions were not confined to the UK. This was identified by the European Union (2000) who commissioned a study led by the Institute to develop 'A framework for the assessment, remediation and sustainable management of intensively developed groundwater-fed catchments in Europe'. The study involved partners from the UK, Spain, Greece and France and in the UK focused on the River Pang, Berkshire one of the "top 40" rivers identified by the NRA.

6.3.3.3 Land Use Change

Problems associated with land use change and low flows were not a priority from the perspective of the NRA. However, the Institute's research catchments (Chapter 2) provided an excellent database on which to examine the impact on low flows of coniferous afforestation. A study of the low flows of the Plynlimon experimental catchments demonstrated the strong control of local geology in maintaining steam flows during dry weather. The analysis concluded that differences in low flows between the subcatchments with the same vegetation (grassland or coniferous forest) were as large as the differences between catchments with different vegetation (Institute of Hydrology, 1991). The impact of coniferous afforestation on the flow–duration curve, annual minima series and storage yield relationships were estimated using data from the Monachyle research catchment (Chapter2). The IH land use model was used to derive time series of simulated daily flows for different proportions under forest cover. The analysis showed a significant increase in the size of reservoir storage required to maintain a given yield with increasing forest cover (Gustard and Wesselink, 1992). Comparing a fully forested catchment with grassland using the HBVOR model, Tallaksen (1993) found a 20% decrease in Q95 in simulated Monachyle flows and a reduction in the mean flow of 29% and 4% from a Dutch and Norwegian catchment (Box 6.3).

Box 6.3

It is interesting to reflect that one of the driving forces for establishing the Plynlimon research catchments in the 1960s was the concern about the effect of coniferous afforestation on water resources. Additional research catchments (Chapter 2) included the Coalburn, River Ray and River Cam – identified as having very contrasting chalk geology. Unfortunately, research on the Cam concluded in 1985. By the 1990s, the primary concern of the NRA and subsequently the Environment Agency was the impact of groundwater abstraction on low flows and the Cam is now classified by the Environment Agency as "over licensed and over abstracted". With the benefit of hindsight, an active research programme on permeable catchments would have positioned the Institute more favourably to meet the challenges of the 1990s. This was addressed by NERC in 2000 when a thematic research programme LOCAR (The Lowland Catchment Research Programme) was initiated to support, interdisciplinary hydrological, ecological and hydrogeology research. This was based on five groundwater-dominated catchments, the Frome/Piddle in Dorset, the Pang/Lambourn in Berkshire and the Tern in Shropshire.

6.3.4 Instream ecology

A study of compensation releases (Gustard *et al.*, 1987b) in the UK developed a comprehensive database of the historical development of the operating rules and compensation releases of over 500 UK reservoirs. This enabled the regional and historical variation in compensation flow releases to be identified. One of the main conclusions of this work was that the majority of compensation flows released below UK reservoirs in the 1980s were set to satisfy river interests that no longer applied. Guidelines were presented to assist in setting awards on a more rational basis including addressing the resource needs of freshwater ecology. A key recommendation was to explore the use of the Instream Incremental Flow Methodology (IFIM) developed by the US Fish and Wild Life Service in 1976. The computer model (Johnson *et al.*, 1995) estimates the relative amount of physical habitat

available with an incremental change in river flow. At the time the approach was a mandatory requirement in some US states for setting instream flows and was being implemented in Canada and New Zealand. This recommendation was followed by the NRA who supported the implementation of the model. This initiated close cooperation between the then Institute of Ecology and Institute of Hydrology in applying and developing the methodology over more than three decades. By 1996, the model had been applied at over 35 sites in England and Wales (Chapter 9).

6.4 Low Flows 2000 to 2015: Delivering UK and EU Policy Requirements

6.4.1 Introduction

The European Water Framework Directive (WFD) came into force in 2000 and was transposed into UK law in 2003. Its purpose is to enhance the status, and prevent further deterioration, of the ecology of aquatic ecosystems and their associated wetlands and groundwater. The directive required member states to both initially characterise the physico-chemical status of water bodies (as defined with the Directive) and to develop River Basin Management Plans, which set out the management objectives and plans to yield the outcome of good ecological status by 2015. From a surface and groundwater pressure perspective, the primary legislation has been implemented in England and Wales at an operational level through the pre-existing CAMS (Catchment Abstraction Management Strategy) process (Environment Agency, 2002). In contrast, within Scotland the Directive was introduced through wholly new primary legislation, the Water and Environment Services Act, which was implemented in 2006 through the Controlled Activities Regulations (Scotland).

The detail of the scope of these processes is beyond the current text, but a key element of both is to assess the available water resource and ability to meet both current and future licensed abstraction demand whilst ensuring the flow requirements of instream ecology are maintained. This makes it possible to identify river reaches that have the potential for further development, that are over abstracted or over licensed or that have insufficient water for further development. The UK gauging network is dense by international standards; however, over 95 per cent of reaches in England and Wales are located far from a flow-measuring station, many of these are small catchments and the requirement for estimates of river flows within these small catchments moved the low flow science of Wallingford towards the centre of regulation. This was delivered by a major re-development of the Micro LowFlows software. Much of this work was undertaken in collaboration with the regulatory agencies within the UK and as part of a wider international research programme (Section 6.5).

This need for a rapid, nationally consistent approach to estimating natural and artificially influenced FDCs within ungauged catchments led to CEH developing the LowFlows software system (Young *et al.*, 2003). The system is underpinned by regionalised hydrological models that enable the natural, long-term FDCs to be estimated for any river reach in the UK mapped at a 1:50 000 scale. Both long-term "annual" and "monthly" statistics are provided. Within the software the impact of artificial influences is simulated using a geographically referenced database that quantifies seasonal water use associated with individual features. Building on the science of the previous 20 years, the software was a natural evolution that also reflected the massive leap in desktop computing power over the previous two decades. The result was a set of hydrological methods and tools in the right place and at the right time to provide a very significant input in to the implementation of the Water Framework Directive in the UK.

6.4.2 Estimation of FDCs using a region-of-influence model

The 1992 study (Gustard *et al.*, 1992) moved away from the *a priori* definition of hydrological regions and proposed a UK-wide methodology to address the problems of transregion catchments and discontinuities at the boundaries between regions. This was driven by the availability of pan UK catchment descriptor datasets. However, operational experience quickly showed that these national models had limitations in explaining the full-range variability of UK hydrology. Whilst the dispensing of regions was attractive from a research perspective, from an operational perspective it is counter-intuitive that the same model structure and parameterisation should be applicable to a lowland chalk catchment within southern England and a hard rock, high relief catchment

in Snowdonia. The underpinning research for the LowFlows software system sought to address this through the use of a region-of-influence approach, which removes the need for a priori identification of regions and instead develops a "region" of catchments similar to the ungauged catchment. The approach is founded on the dynamic construction of a region, based upon the similarity of the characteristics of the gauged catchments to those of the ungauged catchment. The original application of this approach for estimating "annual" and "monthly" flow–duration statistics was described fully by Holmes and Young (2002) and Holmes *et al.* (2002*a*, 2002*b*). In summary, the similarity between the ungauged catchment and other catchments is assessed based on the distribution of soils and parent geology classes using an Euclidean distance metric. A region of the most similar catchments is then identified from a good quality dataset of gauged catchments with natural flow regimes. Estimates of the flow statistics for the ungauged catchment are then calculated as a weighted combination of the observed (standardised) flow–duration statistics for the catchments in the region. The standardised annual FDC is re-scaled by multiplying by an estimate of annual mean flow from a national runoff grid derived from a daily soil moisture accounting model (Holmes *et al.*, 2002*b*). A similar approach is used to determine the FDC for any month based on a distribution of annual runoff within the year (Holmes and Young, 2002).

6.4.3 The LowFlows software system

The LowFlows2000 software was originally released in the year 2000 within England and Wales. In 2004, the software was exclusively licensed to a new spin out company from CEH called Wallingford HydroSolutions (WHS). The rationale was that the requirements of software support and development did not sit well within the structure of a research council laboratory and were better served by being undertaken within a commercial framework.

The software and methods have been through a series of subsequent revisions, both in terms of extending the methods and software to Scotland and Northern Ireland and regular revision of the underpinning science base to reflect the availability of new records and to address issues within the original methodology identified through operational use. The latest software is available both

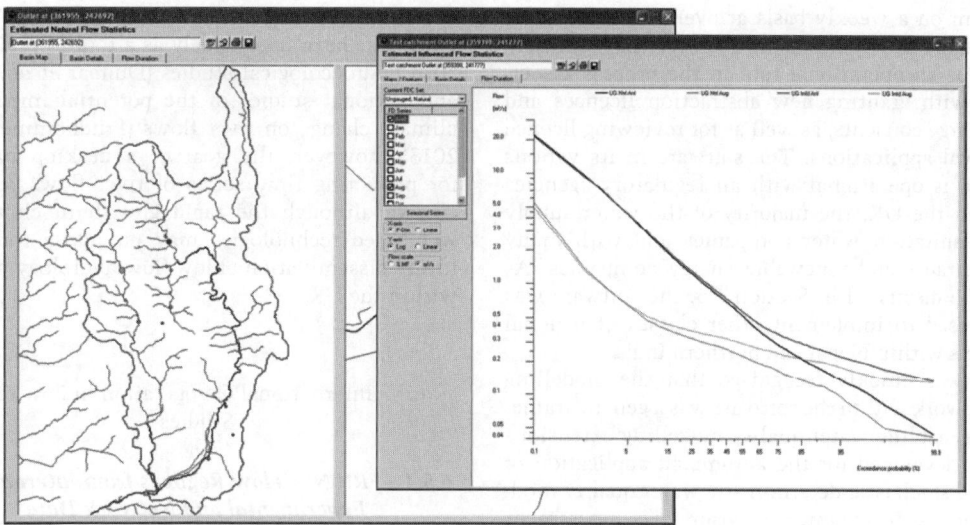

Fig. 6.4 FDCs for natural and influenced flow regimes obtained from the LowFlows 2000 software. (Source: Reproduced with permission of WHS.) *(See insert for colour representation of this figure.)*

as a functionally rich "Enterprise" version and a version focused on just the delivery of the natural flow estimation procedure.

The LowFlows2000 software (Young *et al.*, 2003) and subsequent Enterprise versions incorporate the regionalised hydrological models within a PC-based software framework using contemporary programming tools. A geographic information system-based graphical interface provides access to the spatial datasets of catchment characteristics and the climatic variables required for the application of the regionalised models. These are defined for the entire UK at a 1 km × 1 km resolution. A 1:50 000 scale vectored digital river network and a set of digitised catchment boundaries are used in conjunction with a digital terrain model to define catchment boundaries.

In practice, natural flow duration statistics are obtained by first selecting a point on the digital river network which defines the catchment outlet. A boundary is automatically generated by a digital terrain model. This boundary is overlaid onto the spatial datasets to obtain the catchment characteristics, such as the distributions of soil classes within the catchment, and the other variables required by the underlying hydrological models described above. The required natural flow duration statistics are returned at a "monthly" and "annual" resolution and are displayed in tabular

and graphical form. Examples of output from the software are shown in Figure 6.4.

Water use within the catchment is simulated by utilizing data stored in a flexible database system. Seasonal water use patterns associated with point influences, including abstractions, discharges and impounding reservoirs, are geographically referenced to enable catchment-based data retrievals. The net influences acting within a catchment are calculated by summing the individual influences, where discharges and releases from reservoirs are positive and abstractions are negative. The "influenced" flow–duration statistics are presented, together with the natural statistics, at a "monthly" and "annual" resolution. Hence, practitioners can make a comparison between the natural regime and associated river flow objectives, and the regime as modified by current water use within the catchment. Furthermore, the software can be used for scenario analysis, for example, to investigate the effect of increasing the abstraction rates across a particular catchment for simulating a future water use strategy.

The Environment Agency (EA) in England and Wales and the Scottish Environment Protection Agency (SEPA) have both adopted the LowFlows2000 system as the standard decision-support tool for low flow estimation in ungauged catchments. More than 170 trained staff use the system on a weekly basis at over 95 installations across the United Kingdom. The system is routinely used as an operational tool in the process associated with granting new abstraction licences and discharge consents, as well as for reviewing licence renewal applications. The software in its various guises is operational with all regulatory agencies within the UK, the majority of the water supply and sanitation water companies and with many consultants and renewable energy companies. As will be discussed in Section 6.5, the software was also used to implement other classes of regional models within Nepal and northern India

It was quickly recognised that the modelling framework within the software was a generic framework enabling water quality modelling extensions to be developed for the automated application of hybrid stochastic-deterministic water quality models for use in chemical exposure risk assessment models using the modelling framework defined by Feijtel *et al.* (1998). In 2009, this modelling software was used by CEH and WHS to develop the first England and Wales wide risk assessment tool for oestrogens within rivers (Williams *et al.*, 2009).

6.4.4   Low flow estimation: from FDCs to time series

Throughout the late 1990s and 2000s a parallel programme of research, jointly funded by CEH and the EA, was conducted to develop broadly equivalent models and methods for estimating daily time series of river flows within ungauged catchments. The primary driver for this research was the requirements of the Water Framework Directive and the requirement for an ecological end point to regulation. The objective of the research was to develop a rainfall–runoff modelling framework that could be applied within ungauged catchments without recourse to calibration. The models that have been developed from this work (Young, 2006) were initially lumped conceptual rainfall–runoff models based upon the Probability Distributed Model (Moore, 1985) that were calibrated in many catchments and for which regression models relating the model parameters to catchment descriptors were sought. The success of this work led to the development of the Continuous Simulation of River Flows (CERF) model. This is a semi-distributed hydrological response unit model for which the model structure and the parameterisation are a function of catchment

descriptors (Young *et al.*, 2006 and 2008). To date CERF has been used widely as a tool to undertake both hydroecological studies (Dunbar *et al.*, 2006) and national studies of the potential impacts of climate change on river flows (Prudhomme *et al.* 2013); however, the goal of a desktop package for predicting time series of river flows remains elusive although the rapid growth in cloud and web-based technologies may provide a route for future dissemination of low flow hydrology models within the UK.

6.5   International Co-operation in Low Flow Studies

6.5.1   *FRIEND: Flow Regimes from International Experimental and Network Data*

The low flow research programme at the Institute benefited from and contributed to a number of international initiatives. The first and most important of these was the FRIEND (Flow Regimes from International and Experimental Network Data) project. This was initiated in 1985 in Wallingford as a key component of UNESCO's International Hydrological Programme (IHP). The aim was to study the variability of hydrological regimes in order to improve the management of water resources at basin and regional scales. A small team of hydrologists primarily from The United Kingdom, Germany, The Netherlands and Norway worked for three years on a range of topics including floods, low flows and physically based modelling. Their research was underpinned by a comprehensive database of river flow and thematic data derived from over 3000 catchments operated by research institutes, universities and operational hydrometric agencies. This collaboration enabled low flow analysis to be carried out on a wide range of different hydrological regimes and for research participants to benefit from the broader insights gained by working in an international group. In addition, it facilitated for the first time in Europe, regional scale hydrological research to be carried unconstrained by national political boundaries. Since its inception in Europe the FRIEND project has developed to include 169 countries organised in eight regional groups (Demuth and Gandin, 2010, Table 6.2 and Fig. 6.5). Each group has several research projects. For example, in addition to low flows the EURO FRIEND now has projects as

Table 6.2 International coverage of regional FRIEND groups in 2010

FRIEND regional groups	Number of countries	Year established
European	31	1985
Mediterranean	18	1991
West and Central Africa	18	1992
Southern Africa	12	1992
The Hindu-Kush Himalayas	12	1996
Nile	6	1996
Asian Pacific	31	1997
Latin American & Caribbean	41	1999

Source: Demuth & Gandin 2010. Reproduced with permission of IAHS.

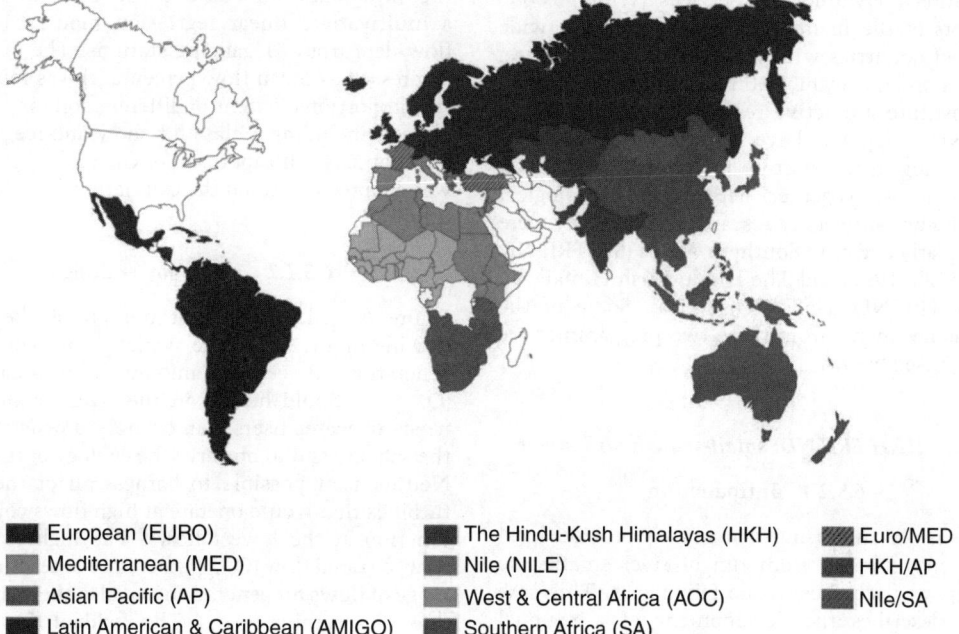

■ European (EURO) ■ The Hindu-Kush Himalayas (HKH) ▨ Euro/MED
■ Mediterranean (MED) ■ Nile (NILE) ■ HKH/AP
■ Asian Pacific (AP) ■ West & Central Africa (AOC) ▨ Nile/SA
■ Latin American & Caribbean (AMIGO) ■ Southern Africa (SA)

Fig. 6.5 FRIEND regional groups. (Source: Reproduced with permission of UNESCO.) *(See insert for colour representation of this figure.)*

diverse as database development, floods, large-scale hydrology and hydrological processes. FRIEND initially focussed on research but now includes capacity building including software and database development, training courses, operational guidance manuals and the provision of facilities for research students (Gustard and Bonell, 2006). Global coordination of FRIEND is achieved by holding an international conference every four years and publishing conference proceedings (e.g. van Lanen *et al.*, 2014) and a report which reviews

the key achievements and future plans of each of the eight FRIEND groups (e.g. Huang and Demuth, 2010).

The FRIEND project has facilitated building hydrological research networks which have participated in a number of international research projects. In Europe, this has included the ARIDE (Assessment of the Regional Impact of Drought in Europe) project (Demuth and Stahl, 2001), funded by the EU, key deliverables were drought analysis at the continental scale, analysis of trends and drought visualisation. This was followed by a second major initiative (Tallaksen and van Lanen, 2004) involving over 25 European hydrologists who compiled a comprehensive textbook and established the European Drought Centre (http://www.geo.uio.no/edc/). The text book and accompanying CD provide an introductory description of drought processes and analysis techniques suitable for undergraduates and a more advanced statistical analysis appropriate for the graduate hydrologist, including a compendium of operational examples of drought applications. In 2008, a manual was prepared on "Low Flow Estimation and Prediction" to meet the needs of National Hydrological Services (WMO, 2008). Authors of the manual were drawn from a wide range of countries with complimentary expertise, many from the eight FRIEND groups. Staff from the Institute are active researchers in the EURO FRIEND group but have also played a key role in assisting in the start-up phase of new FRIEND communities. Together with local hydrological agencies and university research groups, they were particularly active in Southern Africa (SA) FRIEND (UNESCO, 1997) and The Hindu-Kush Himalayan (HKH) FRIEND (UNESCO, 2004a). Some of the key achievements from these two programmes are summarised below.

6.5.2 HKH FRIEND: small-scale hydropower

6.5.2.1 Introduction

There is an increasing demand to provide local electricity supplies from run of river small-scale hydropower schemes in the Himalayas. This case study describes the development of a regional model for predicting power availability within Nepal and the state of Himachal Pradesh in northern India. In contrast to major hydropower schemes with significant reservoir storage, small-scale hydropower schemes frequently have no artificial storage to provide a constant supply of water for power generation. These schemes therefore rely entirely on the natural river flow to generate electricity.

The Himalayas are characterised by rapid changes in elevation and a monsoon-influenced climate but have a relatively sparse network of hydrometric and meteorological stations which compounds the difficulty of hydrological modelling. The methodology for Himachal Pradesh region was confined to a relatively hydrologically

homogeneous rain- and snow-fed region at an elevation between 2000 m and 5000 m. Having favourable topography, relatively easy access and a reliable supply of water throughout the year, this part of the state is the most suitable for small-scale hydropower development. At lower altitudes (below 2000 m), there is little variation in topography and, with runoff derived solely from rainfall, rivers tend to be ephemeral. At higher altitudes (above 5000 m), the remoteness and harsh conditions rule out extensive small-scale hydropower development. The regional model for Himachal Pradesh (Rees *et al.*, 2002) enabled the flow–duration curve to be estimated using a multivariate linear regression model based on flow data from 41 gauging stations. The Q_{95} flow (expressed as mean flow percentage) was related to the proportional extent of different soil (or geology) classes, including a class for snow and ice, within a catchment. The approach of Gustard *et al.* (1992) was adopted to extend the estimation from Q_{95} to a full FDC

6.5.2.2 Hydropower design

Figure 6.6 illustrates that not all of the water flowing in a river will be available for hydropower generation. A certain amount of residual flow ($Q_{residual}$) should be left in the river to meet the needs of water users immediately downstream of the scheme and to preserve the ecology of the river. Neither is it possible to harness all of the flow: turbines that would operate at high flows could not function at the lowest flows. The highest design flow, or rated flow (Q_{rated}), therefore determines the range of flows for generation and also the minimum flow of a turbine (Q_{min}). The residual flow, rated flow and minimum flow, together with the FDC for the site, thus determine the volume of water the scheme will use.

The results of this research have been delivered via HydrA-HP a software package that incorporates the Q_{95} and mean flow model as separate grids at a spatial resolution of 1 km. It provides an easy-to-use, menu-driven interface for rapidly estimating the FDC and, hence, the hydropower potential at any prospective site. The hydropower potential (annual energy output, maximum power and rated capacity) can be calculated for up to eight types of turbine. Once the user has entered the values of hydraulic head and rated flow, the software refers to the operational envelopes of eight standard turbine types to identify which

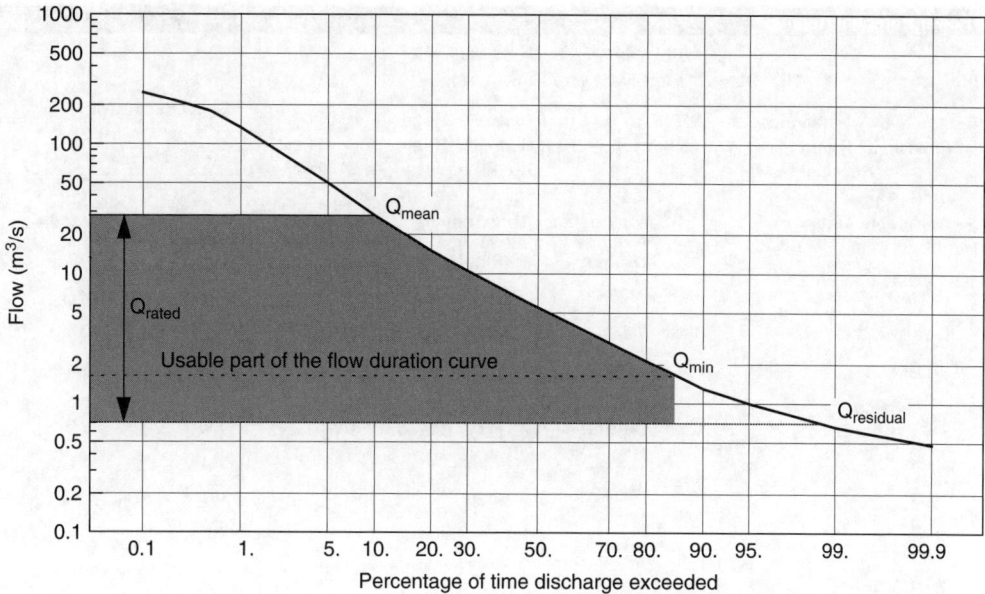

Fig. 6.6 The FDC in hydropower design. (Source: Reproduced with permission of CEH.)

will operate under the stated conditions. For each selected turbine type, the software determines the usable part of the FDC (determined by the rated capacity of the turbine and the residual flow that is to be left in the river). Subsequently, by referring to the relevant flow–efficiency curves, it calculates the average annual energy potential (MWh) and the power generation capability (kW) of the site. The single page report that results from this procedure is shown in Figure 6.7. In addition to the Himalayan region, versions of the software were developed for UK, Spain, the Republic of Ireland, Belgium and Portugal.

6.5.3 HKH FRIEND: deglaciation

The apparent wastage of glaciers has led to warnings of potentially serious water shortages around the world. Although claims that Himalayan glaciers could vanish within 40 years (Pearce, 1999) have largely been dispelled (Cogley *et al.*, 2010), considerable concern remains over how the 500 million people living downstream in South Asia will adapt to inevitable changes in water resources as glaciers recede. In response, the CEH began a project in 2001 "*to assess the seasonal and long-term water resources in snow and glacier fed rivers originating*

in the Hindu Kush – Himalayan region and to determine strategies for coping with the impacts of climate change induced deglaciation on the livelihoods of people in the region." The project, called SAGARMATHA (the Nepali name for Mt. Everest), was funded by the Department for International Development and involved experts from Nepal, India and the UK.

A key part of the project was the development of a new regional hydroglaciological model, which gave forecasts of annual and seasonal river flows for a future 100-year period according to a variety of climate change scenarios. Model results suggested that, for many areas, deglaciation is unlikely to have an adverse effect on freshwater availability for several decades to come. In some catchments, river flows were seen to continually increase over the period considered. However, the results showed that hydrological impacts vary considerably within the region and within catchments: highly glaciated catchments, and catchments where meltwater contributes significantly to the runoff, appear particularly vulnerable to glacier retreat (Rees and Collins, 2006). Indeed, the model identified specific areas, such as in the Upper Indus, where significant reduction of river flows may occur within the next few decades if temperatures continue to

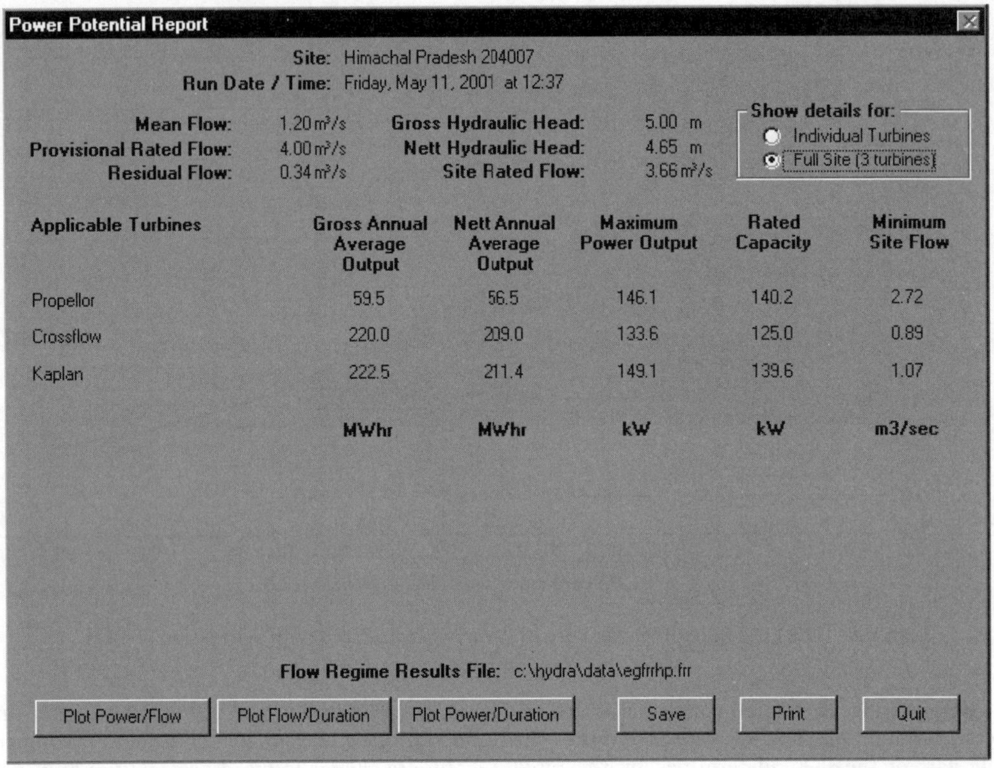

Power Potential Report

Site: Himachal Pradesh 204007
Run Date / Time: Friday, May 11, 2001 at 12:37

Mean Flow:	1.20 m³/s	Gross Hydraulic Head:	5.00 m
Provisional Rated Flow:	4.00 m³/s	Nett Hydraulic Head:	4.65 m
Residual Flow:	0.34 m³/s	Site Rated Flow:	3.66 m³/s

Show details for:
○ Individual Turbines
◉ Full Site (3 turbines)

Applicable Turbines	Gross Annual Average Output	Nett Annual Average Output	Maximum Power Output	Rated Capacity	Minimum Site Flow
Propellor	59.5	56.5	146.1	140.2	2.72
Crossflow	220.0	209.0	133.6	125.0	0.89
Kaplan	222.5	211.4	149.1	139.6	1.07
	MWhr	MWhr	kW	kW	m3/sec

Flow Regime Results File: c:\hydra\data\egfrrhp.frr

| Plot Power/Flow | Plot Flow/Duration | Plot Power/Duration | Save | Print | Quit |

Fig. 6.7 Power potential report. (Source: Hydra software – Gwyn Rees. Reproduced with permission of CEH.)

rise (Fig. 6.8). Such changes would have serious consequences for future water availability in these areas, and urgent measures to mitigate the impacts were recommended.

The project also involved an assessment of the impacts of glacier retreat on livelihoods at community, basin and regional level. This consultation-based activity identified potential threats from deglaciation and identified a variety of adaptation strategies applicable at different scales, including changing cropping patterns; increasing crop diversity; changing animal husbandry techniques; improved water saving; building public awareness of climate change; improving systems of data collection; developing sustainable water management and planning strategies and improving communication and infrastructure in mountain areas.

The project's recommendations were disseminated to policy makers, water managers and planners at two regional stakeholder consultation workshops in Nepal and India in 2004. The project benefitted from the FRIEND philosophy of enabling regional hydrology to be based on extensive databases not constrained by national political boundaries. There were many other projects (UNESCO, 2004a) carried out by research groups and hydrometric agencies in the HKH FRIEND region including the estimation of small-scale hydropower potential described in the previous section.

6.5.4 Southern Africa FRIEND: drought research

For many years, the most widely used methods for low flow estimation in Southern Africa were those developed at a national level, for example for Malawi (Drayton *et al.*, 1980) and South Africa (Water Research Commission, 1994). SA FRIEND, involving 12 countries, was the first initiative to be able to make use of a large international dataset to explain the spatial and temporal variability of

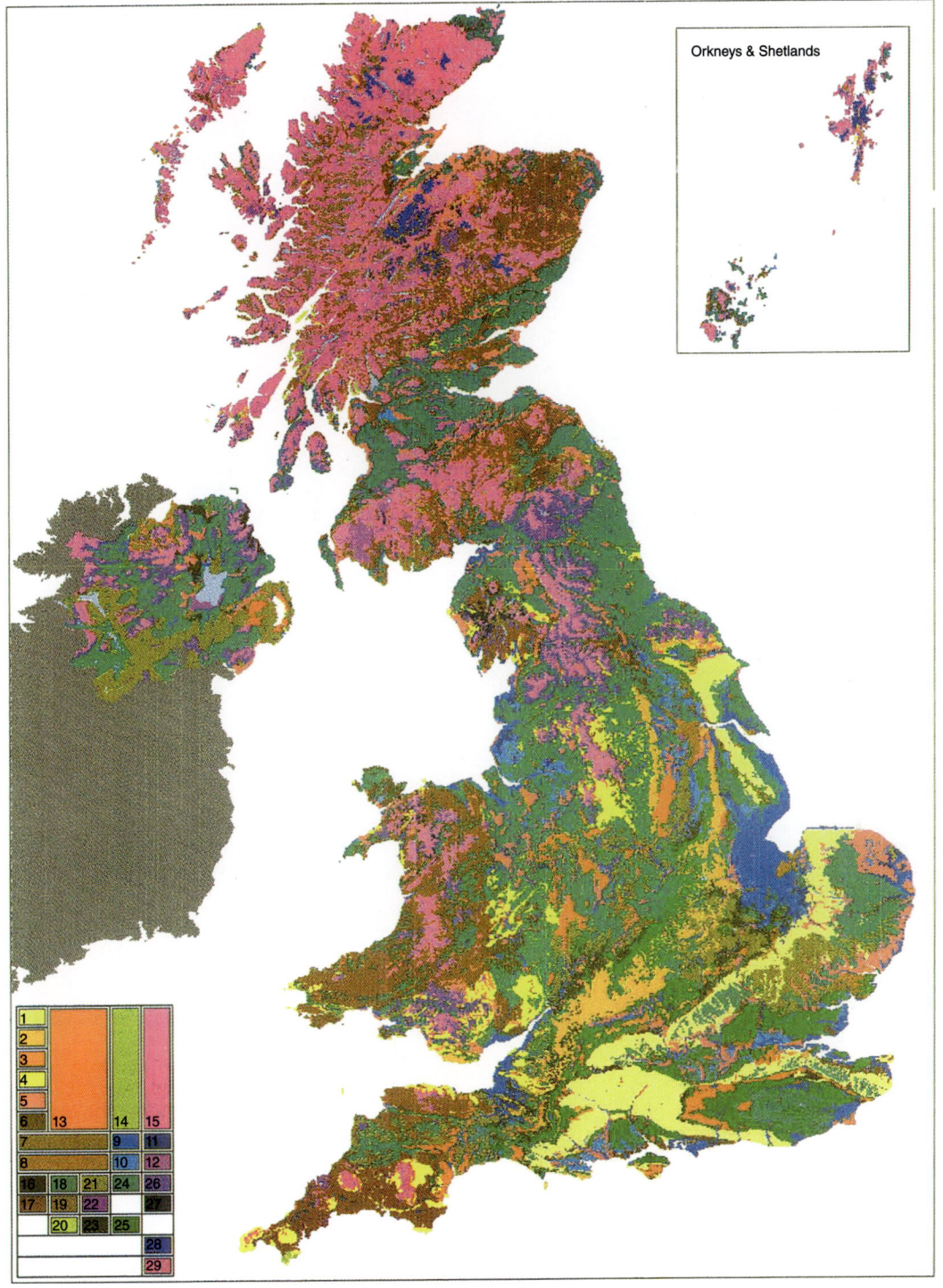

Fig. 3.9 HOST class for each kilometre square. (Source: Reproduced with permission of CEH.)

Progress in Modern Hydrology: Past, Present and Future, First Edition. Edited by John C. Rodda and Mark Robinson.
© 2015 John Wiley & Sons, Ltd. Published 2015 by John Wiley & Sons, Ltd.

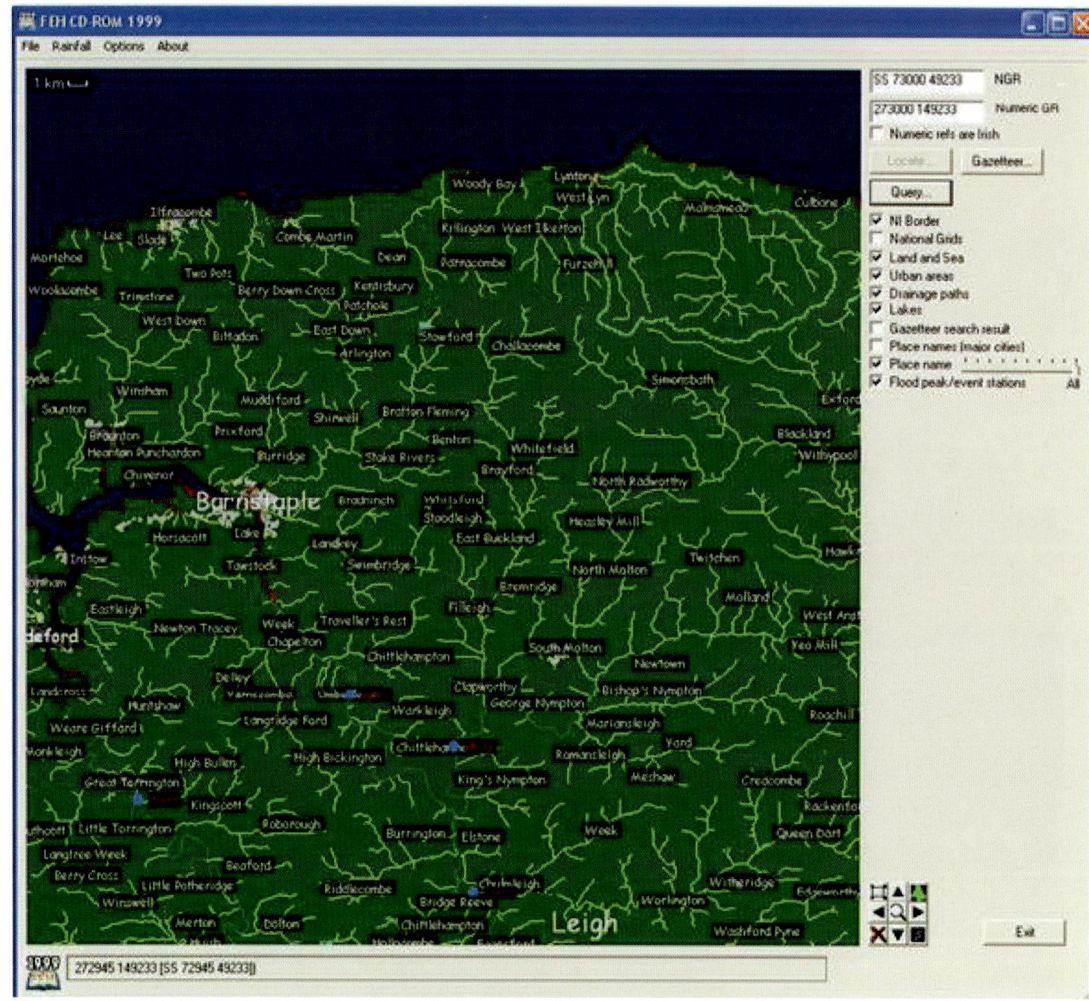

Fig. 3.10 FEH CD-ROM screen shot showing the IHDTM derived river network. (Source: Reproduced with permission of CEH.)

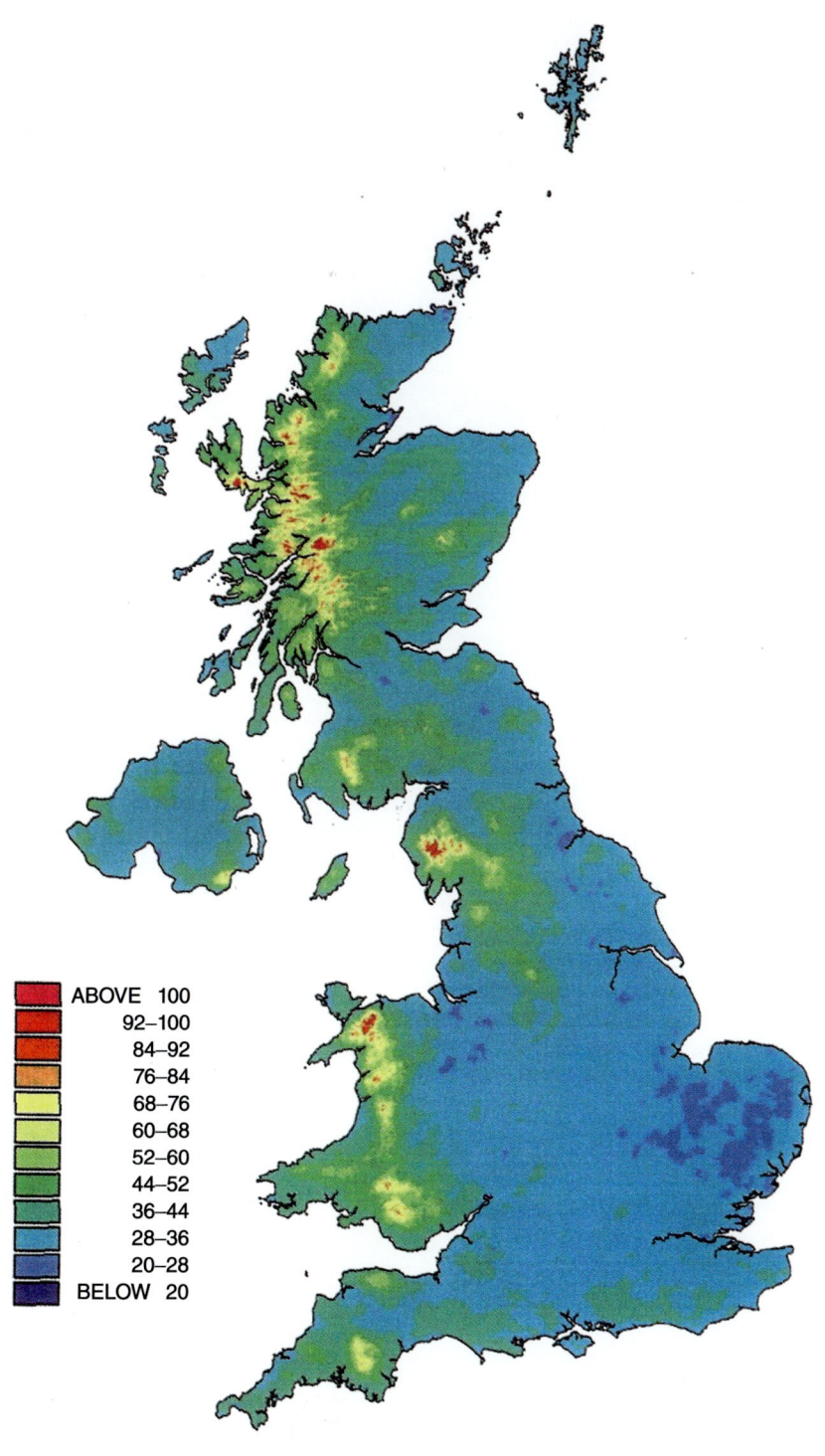

Fig. 3.13 One day Rmed map (mm). (Source: Reproduced with permission of CEH.)

ABOVE 100
92–100
84–92
76–84
68–76
60–68
52–60
44–52
36–44
28–36
20–28
BELOW 20

Fig. 5.10 Colour modified Landsat satellite image of the herringbone pattern forest clearance (in pink) as areas are cleared to either side of roads. (Source: NASA Open Source.)

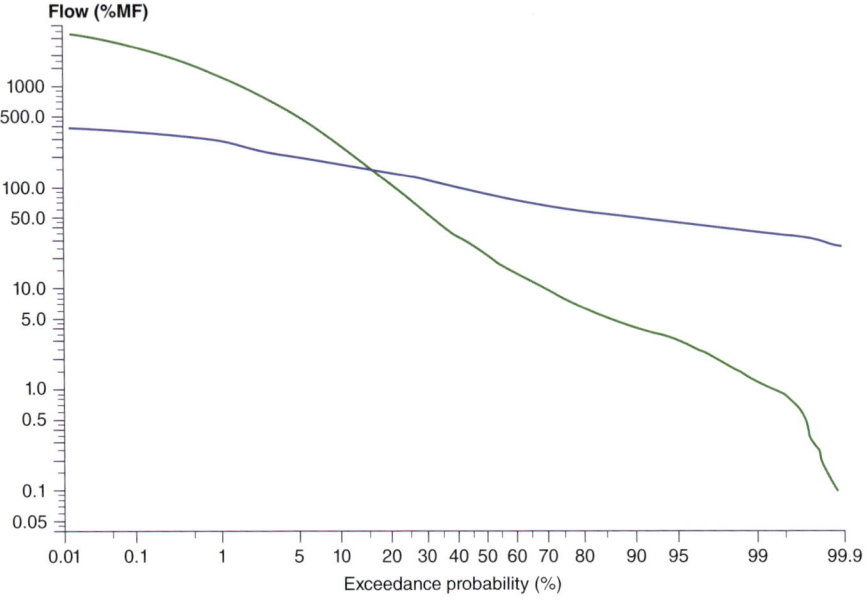

Fig. 6.2 Dimensionless FDCs for contrasting flow regimes: River Beult (green); impermeable catchment, River Lambourn (blue); permeable catchment. (Source: Reproduced with permission of WHS.)

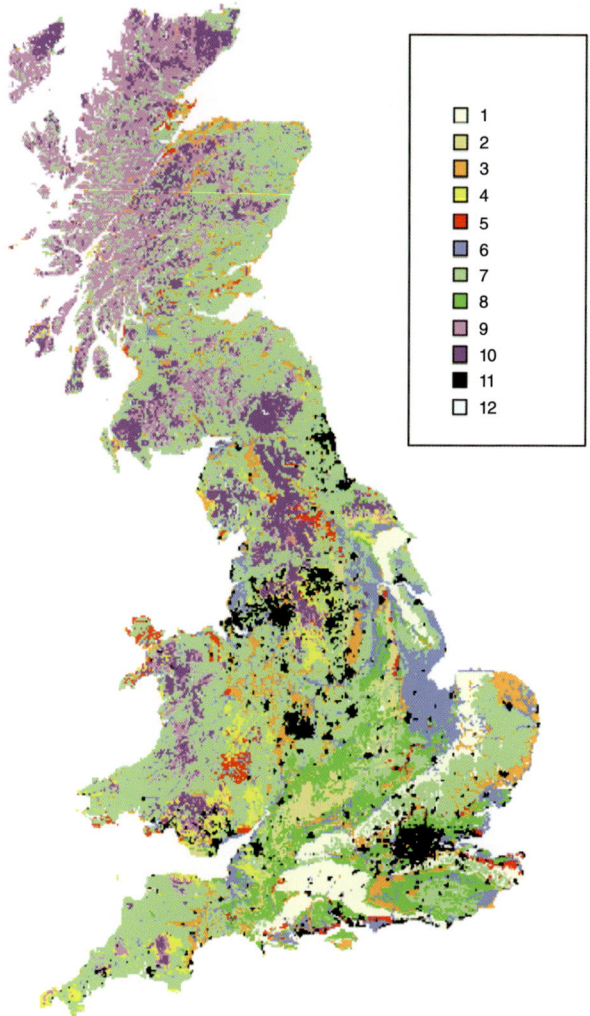

Fig. 6.3 Low Flow HOST groups. (Source: Reproduced with permission of CEH.)

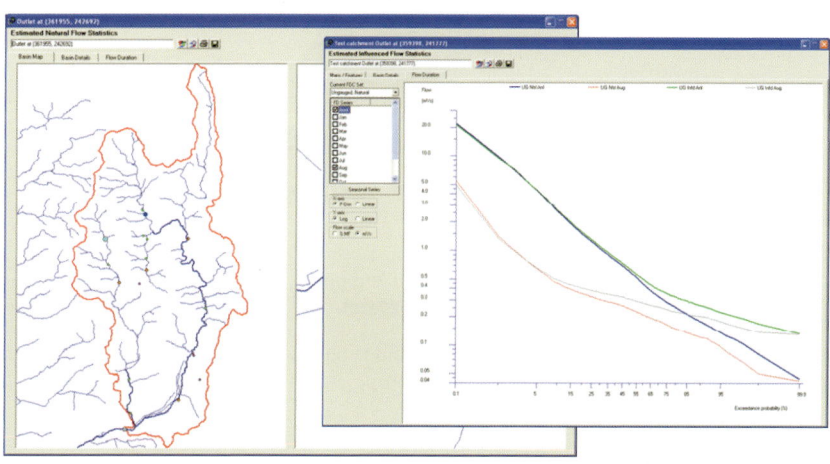

Fig. 6.4 FDCs for natural and influenced flow regimes obtained from the LowFlows 2000 software. (Source: Reproduced with permission of WHS.)

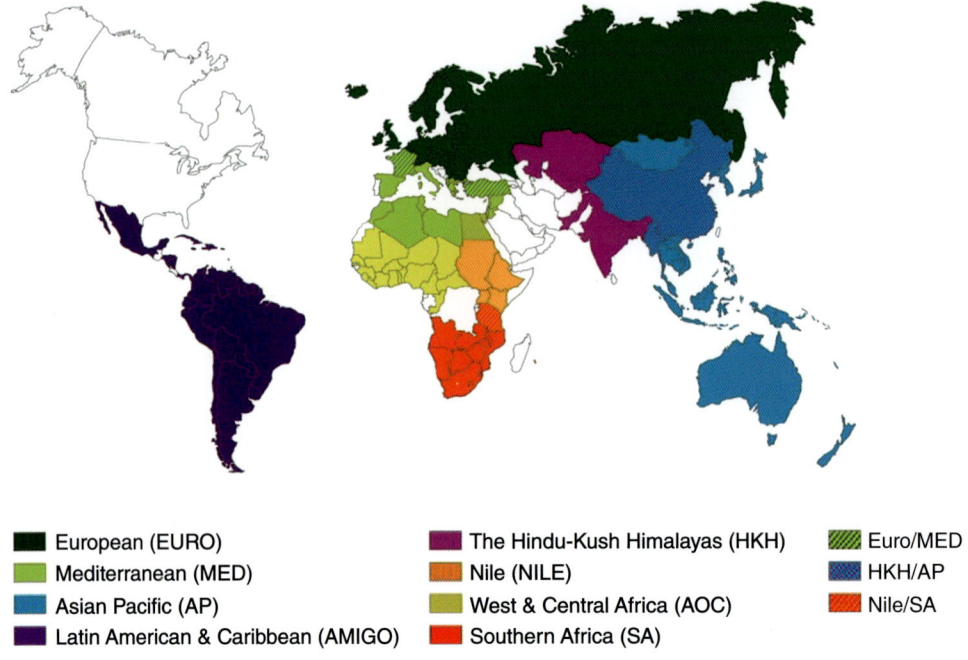

European (EURO)
Mediterranean (MED)
Asian Pacific (AP)
Latin American & Caribbean (AMIGO)

The Hindu-Kush Himalayas (HKH)
Nile (NILE)
West & Central Africa (AOC)
Southern Africa (SA)

Euro/MED
HKH/AP
Nile/SA

Fig. 6.5 FRIEND regional groups. (Source: Reproduced with permission of UNESCO.)

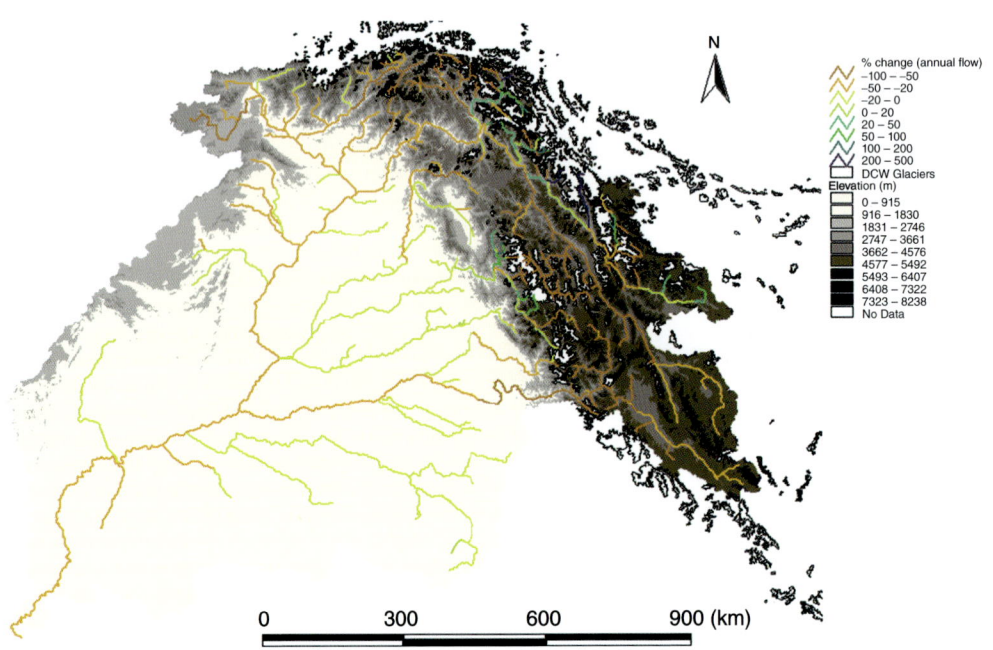

Fig. 6.8 Percentage change in 1961–1990 modelled mean flow for the Indus basin over a 100-year period assuming an average annual temperature increase of 1°C per decade. (Source: Reproduced with permission of CEH.)

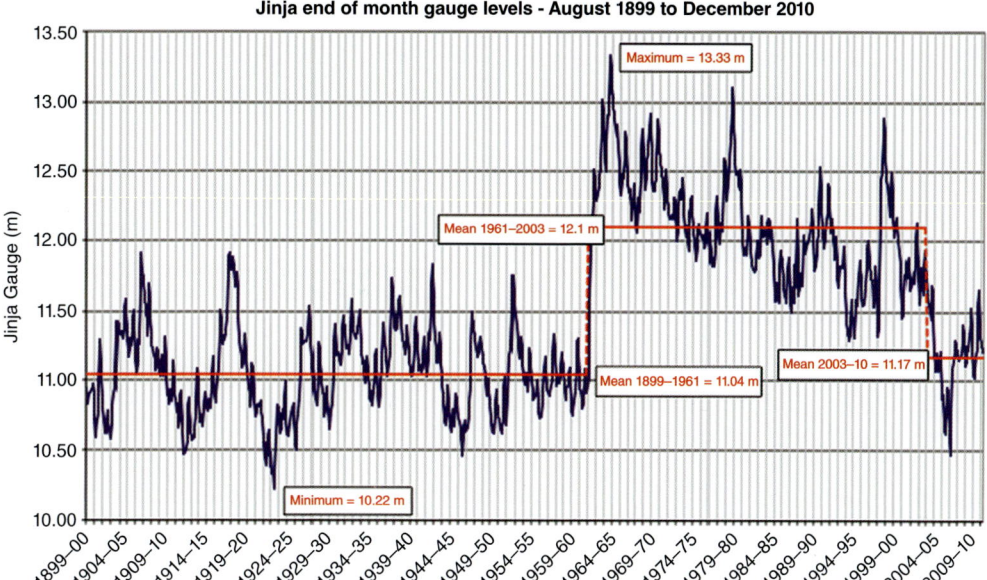

Fig. 6.10 Historical water levels of Lake Victoria measured at Jinja in Uganda. (Source: Water Resource Associates. Reproduced with permission of WHS.)

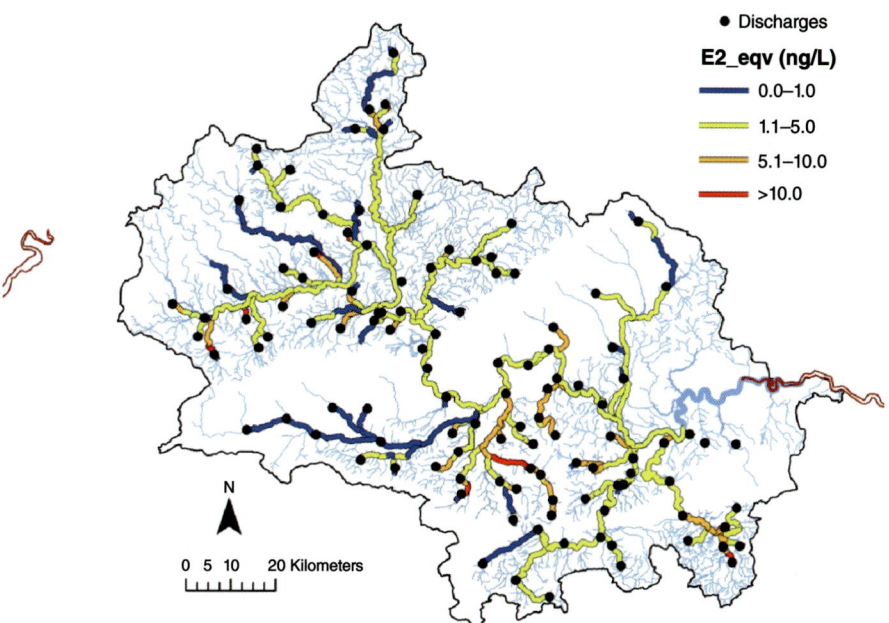

Fig. 8.2 Prediction of combined steroid oestrogen concentrations (estradiol equivalents, E2_eqv) in the Thames catchment originating from the human population and their sewage treatment plants for the month of August. Note concentrations above 1 ng/L suggest some level of endocrine disruption with more significant disruption occurring over 5 ng/L. Thin blue lines denote the river network not impacted by significant discharges. (Source: Reproduced with permission of CEH.)

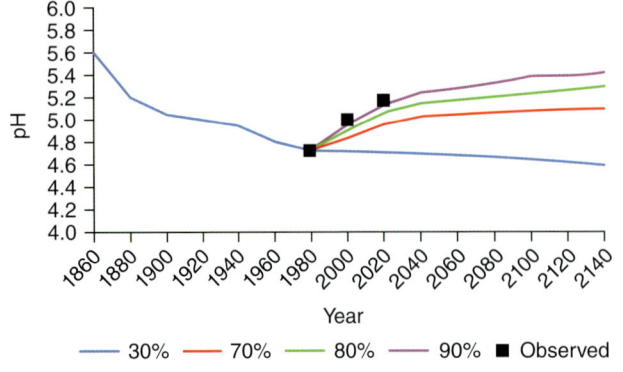

Fig. 8.3 Reconstruction of past pH at Scoat Tarn in the Lake District together with predictions for various sulphur reductions – plus (in red) the observed pH from the Acid Waters Monitoring Network at CEH. (Source: Reproduced with permission of CEH.)

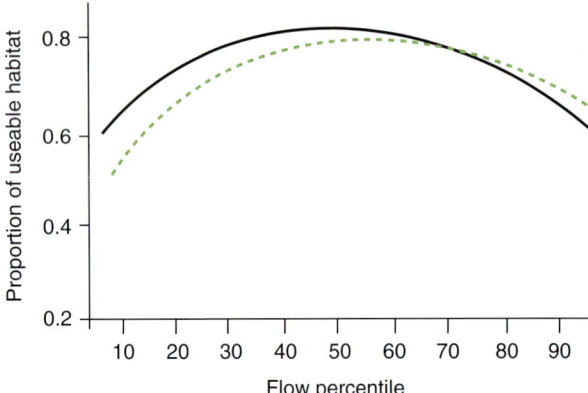

Fig. 9.2 Rapid assessment of flow *v* habitat curves from PHABSIM (black-solid) and RAPHSA (green-dashed). (Source: Reproduced with permission of CEH.)

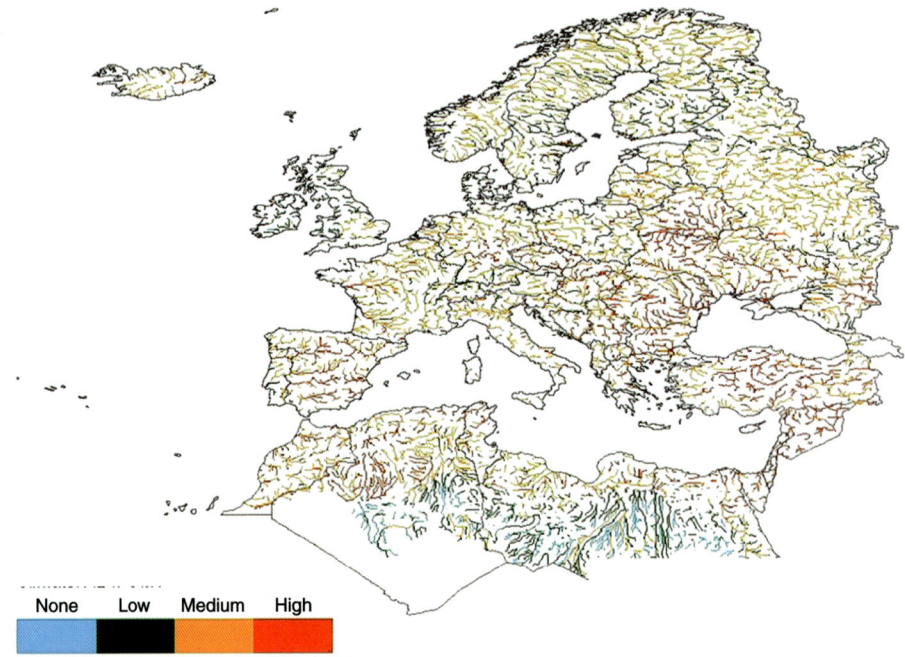

Fig. 9.4 Risk of river ecosystem change in European rivers by 2050. (Source: Adapted from Laizé *et al.* 2014. Reproduced with permission of John Wiley & Sons.)

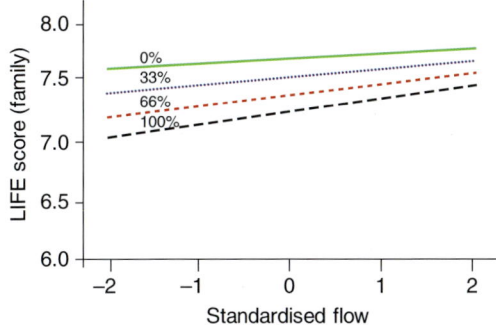

Fig. 9.6 Variations in flow–river ecosystem relationships with channel alteration (given as % modification from natural). (Source: Reproduced with permission of CEH.)

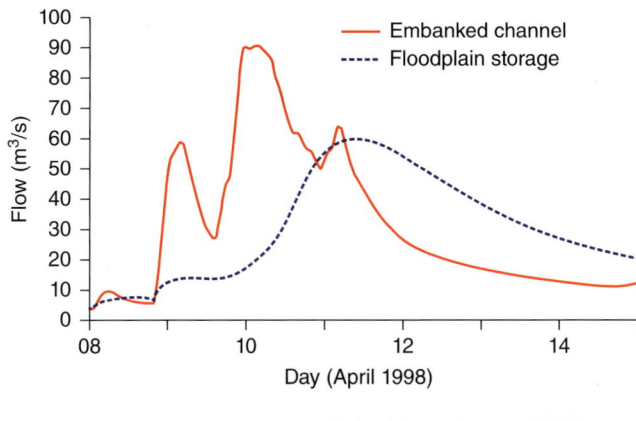

Fig. 9.8 Flood hydrographs at Oxford for April 1998 with and without (embanked) floodplain storage upstream. (after Acreman *et al.*, 2003.) (Source: Reproduced with permission of CEH.)

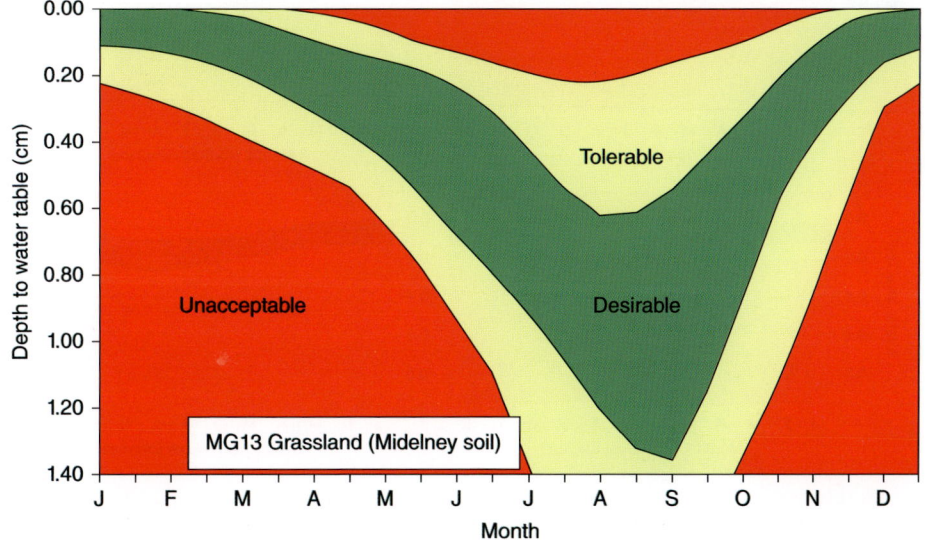

Fig. 9.9 Water-regime requirements of MG13 wetland plant community (Source: Wheeler *et al.* 2004. Reproduced with permission of CEH.)

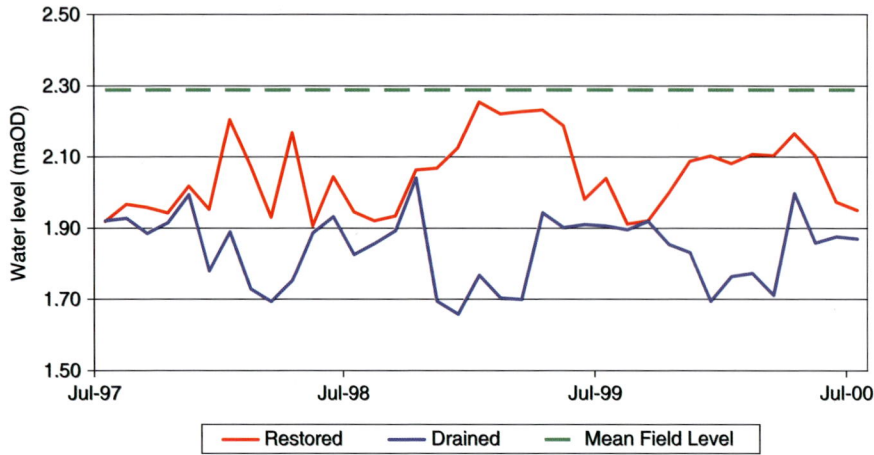

Fig. 9.10 Time series of water table elevations in restored and drained plots at Tadham Moor, from July 1997 to July 2000. (Source: Reproduced with permission of CEH.)

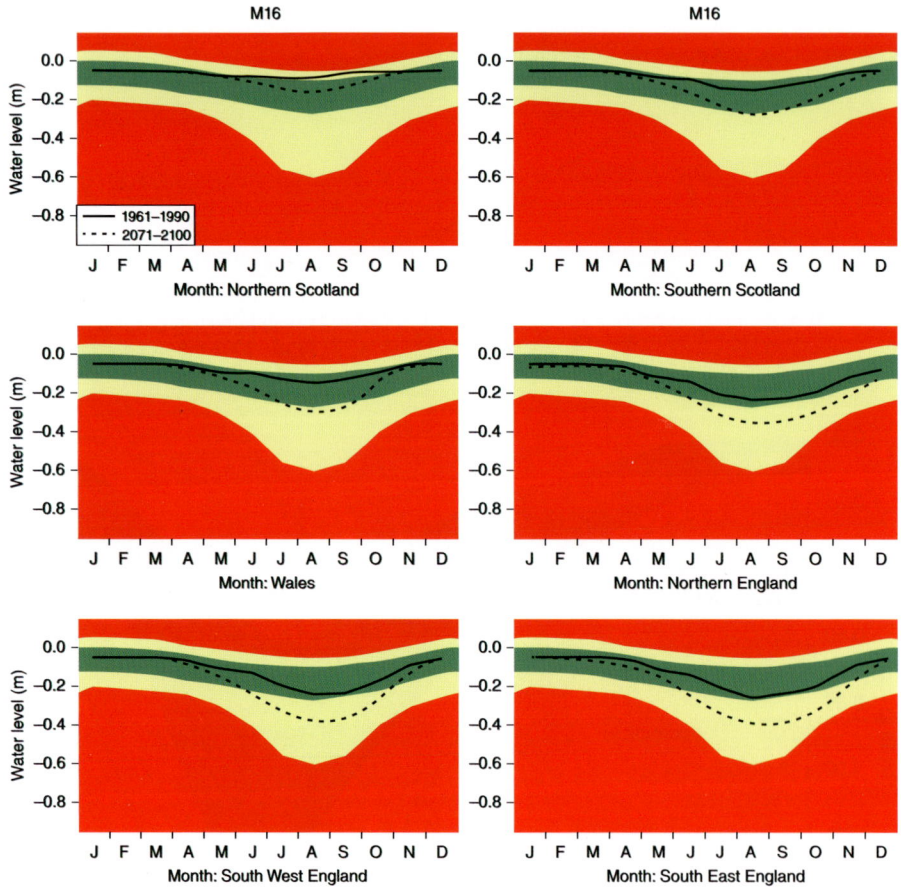

Fig. 9.12 Monthly mean water table levels (*m*) for the rain-fed wetland type, superimposed on the M16 wet heath water level requirements zone graphs, at the six sample sites across GB. The solid lines represent baseline (1961–1990) climate, the dashed lines show the projected values for the 2080s (2071–2100) using UKCIP02 (Hulme *et al.*, 2002) climate change data for the 'Medium–High' emissions scenario. (Source: Reproduced with permission of CEH.)

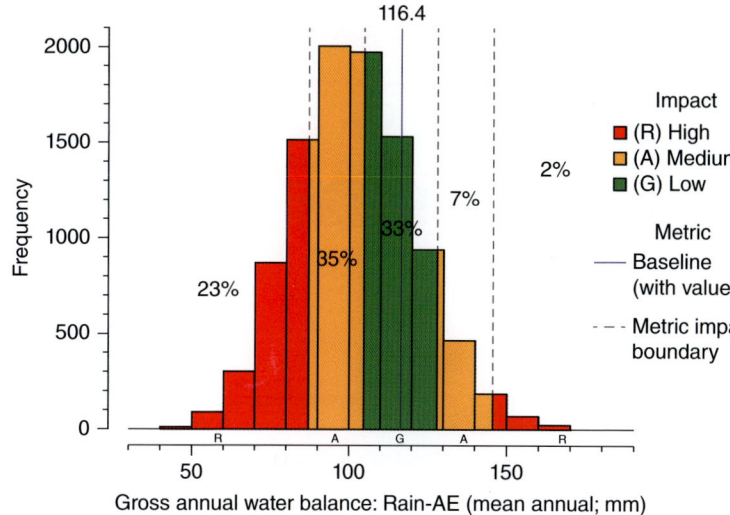

Fig. 9.13 'Wetland Tool for Climate Change' example output showing the potential impact of climate change on wetlands for the 2050s under the 'Medium' emissions scenario – Gross mean annual water balance for a rain-fed M16 wet heath in East Anglia. (Source: Reproduced with permission of the Environment Agency.)

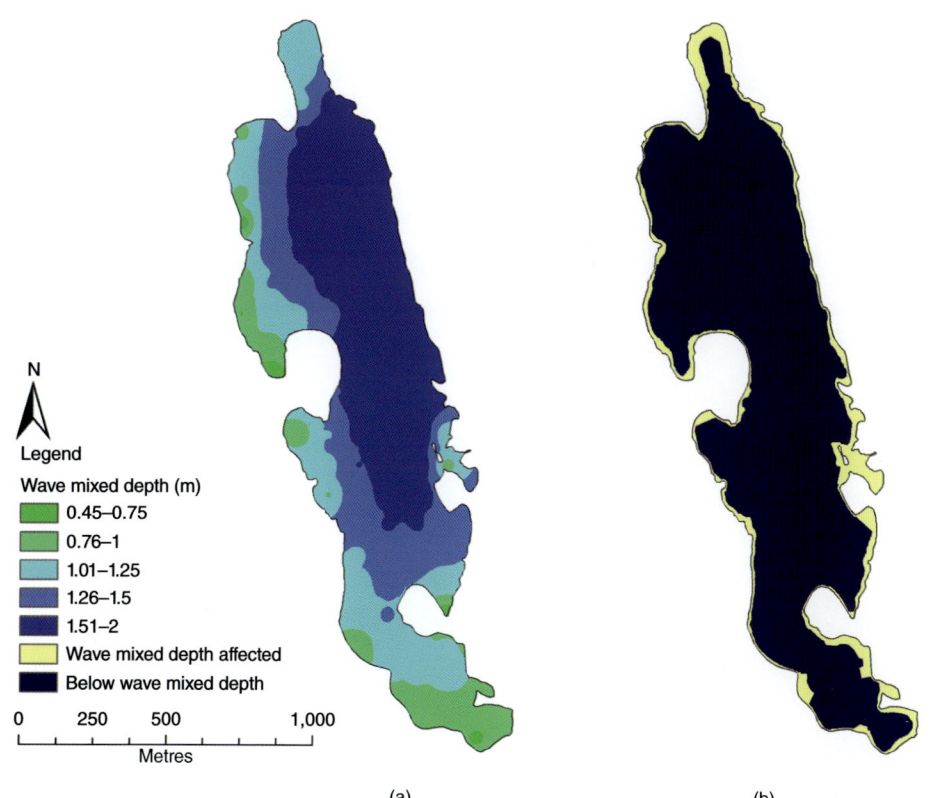

Fig. 9.18 Esthwaite Water (a) average maximum wave mixed depth (m) for 2005–2009 and (b) wave affected area based on this wave mixed depth. (Source: Adapted from Mackay *et al.*, 2012. Reproduced with permission of John Wiley & Sons.)

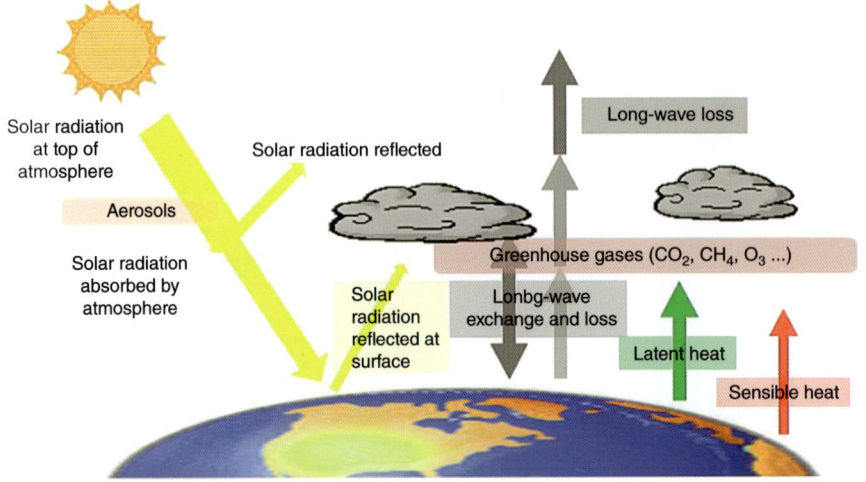

Fig. 10.1 Schematic of main energy flows in the atmosphere and greenhouse effect. (Source: Reproduced with permission of CEH.)

Fig. 10.3 Cloud developing over the Le Landes forest. (Source: NOAA—National Oceanic and Atmospheric Administration.)

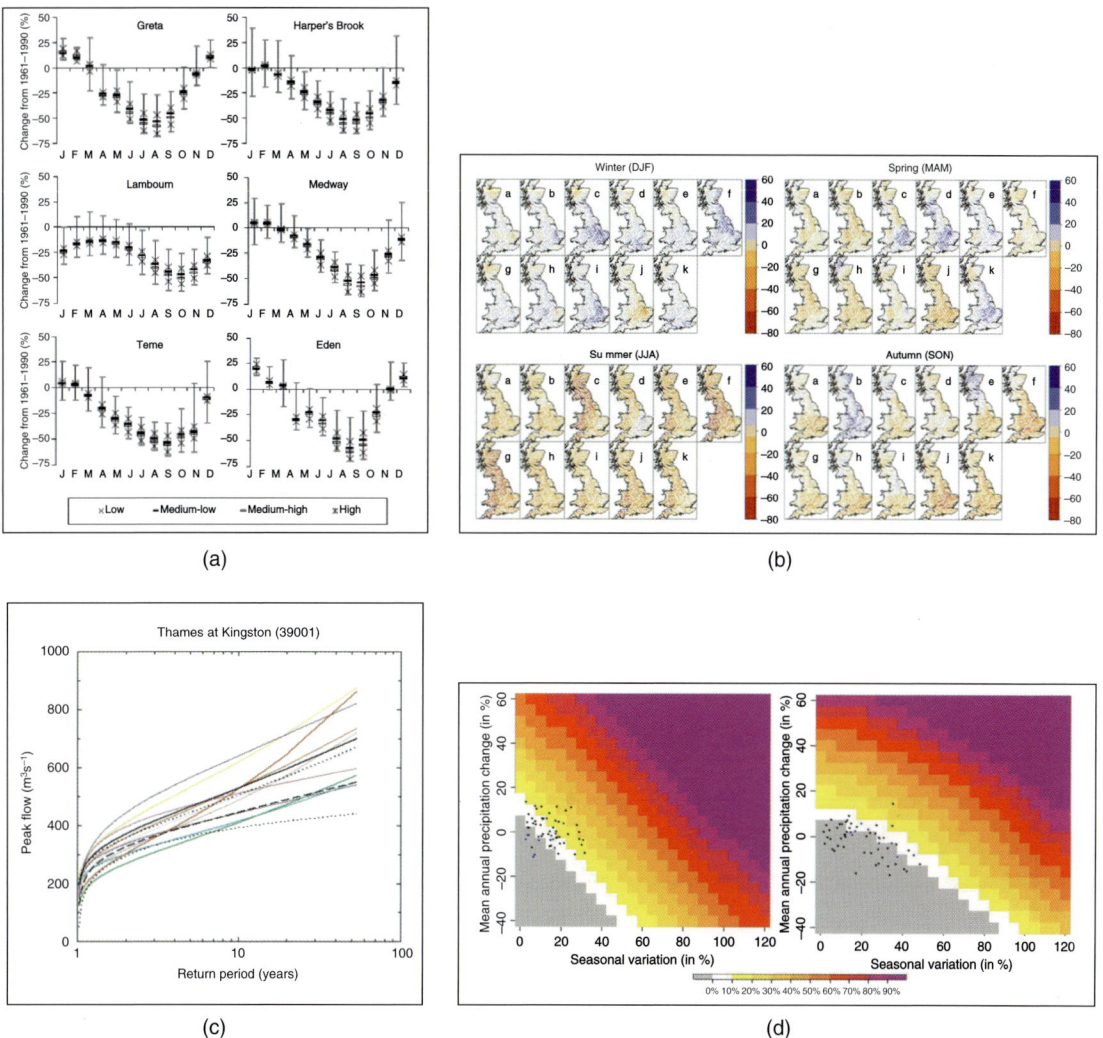

Fig. 10.5 Example impacts on river flows in the UK. (a) Change in mean monthly river flows in the 2050s, with UKCIP02 and other scenarios. (Source: Arnell 2004. Reproduced with permission of John Wiley & Sons.) (b) Change in seasonal runoff in the 2050s, under the 11 UKCP09 regional climate model scenarios (Prudhomme *et al.*, 2012b. Reproduced with permission of John Wiley & Sons.). (c) Flood frequency curves under the current climate (dashed line) and in the 2080s, under the 11 UKCP09 regional climate model scenarios. (Source: Bell *et al.* 2012. Reproduced with permission of Elsevier.) The dotted lines show the uncertainty range for the current frequency distribution. (d) Response surfaces for percentage change in the 20-year return period flood, against magnitude of annual rainfall change (vertical axis) and change in seasonal variation in rainfall (horizontal axis). (Source: Prudhomme *et al.* 2010. Reproduced with permission of Elsevier.) Left: Endrick catchment, North East Scotland and right: Roding catchment, South East England.

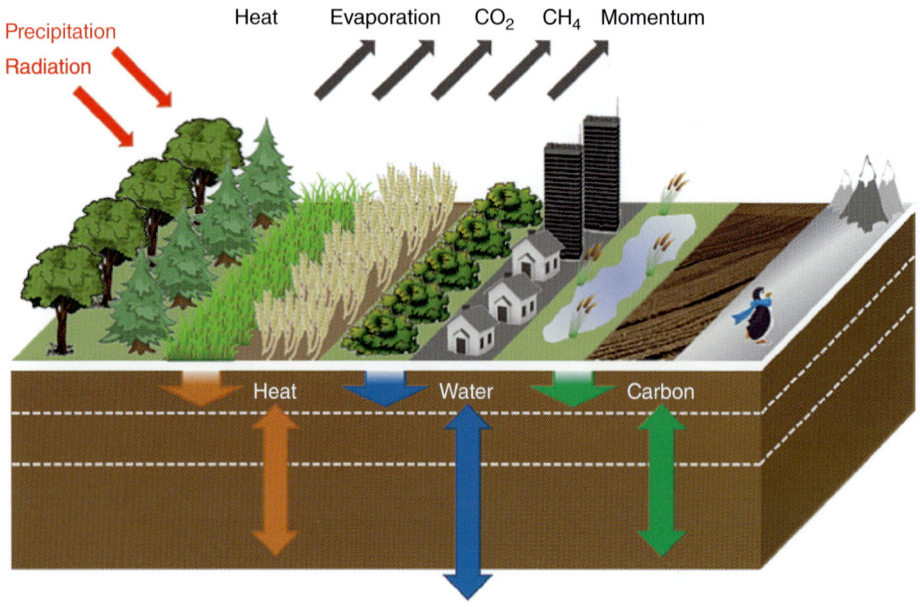

Overview of the science of JULES

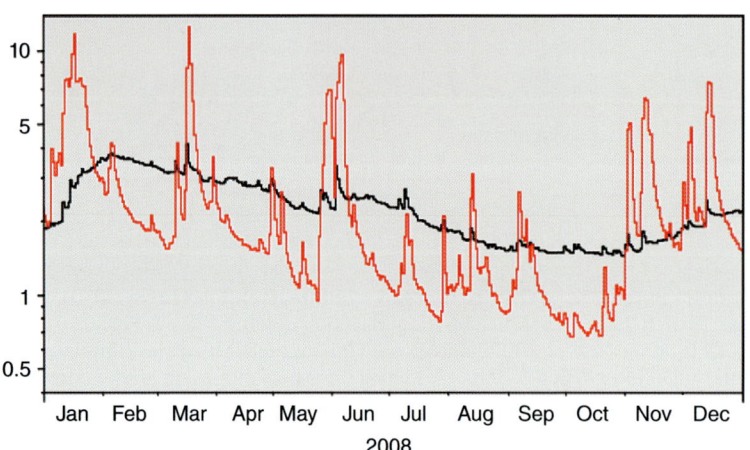

39019 Lambourn at Shaw, Gauged Daily Flow   39081 Ock at Abingdon,
Gauged Daily Flow m^{3}s^{-1}

Fig. 11.2 Contrasting river flow patterns in neighbouring Chalk (black trace) and Clay (red trace) catchments in central southern England. (Source: Reproduced with permission of CEH.)

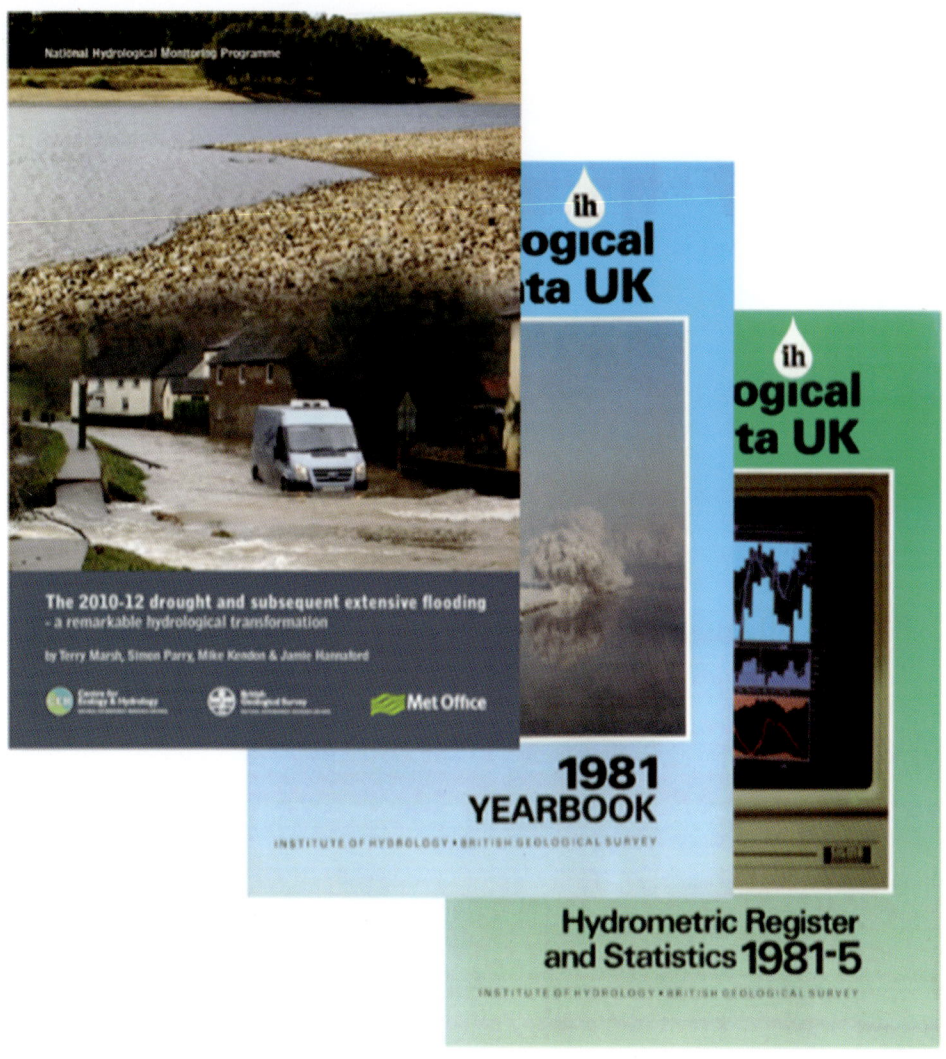

Fig. 11.5 Examples of publications in the National Hydrological Monitoring Programme series. (Source: Reproduced with permission of CEH.)

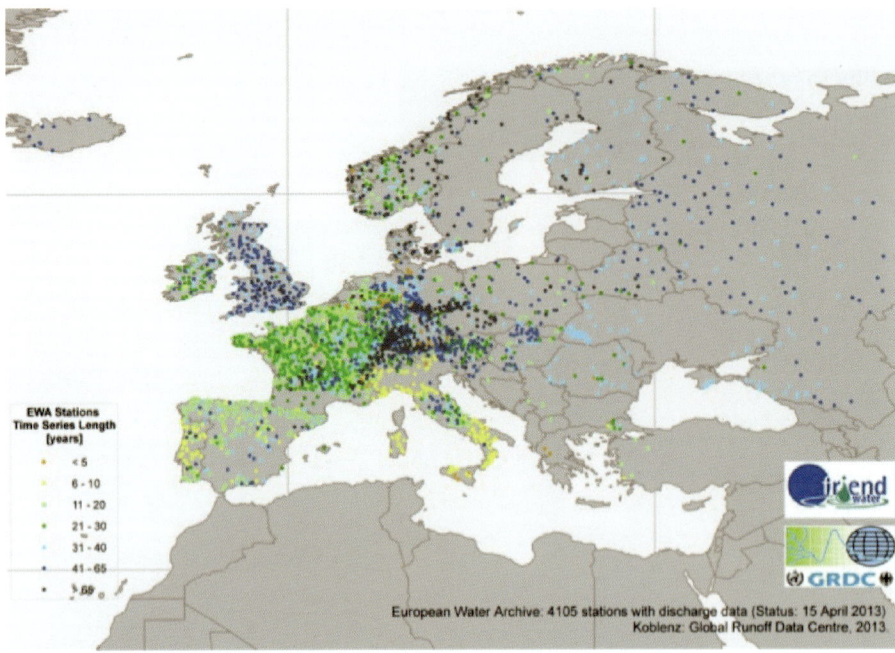

Fig. 11.8 Distribution of gauging stations for which flow records are held in the European Water Archive. (Source: Reproduced by permission of Global Data Runoff Centre.)

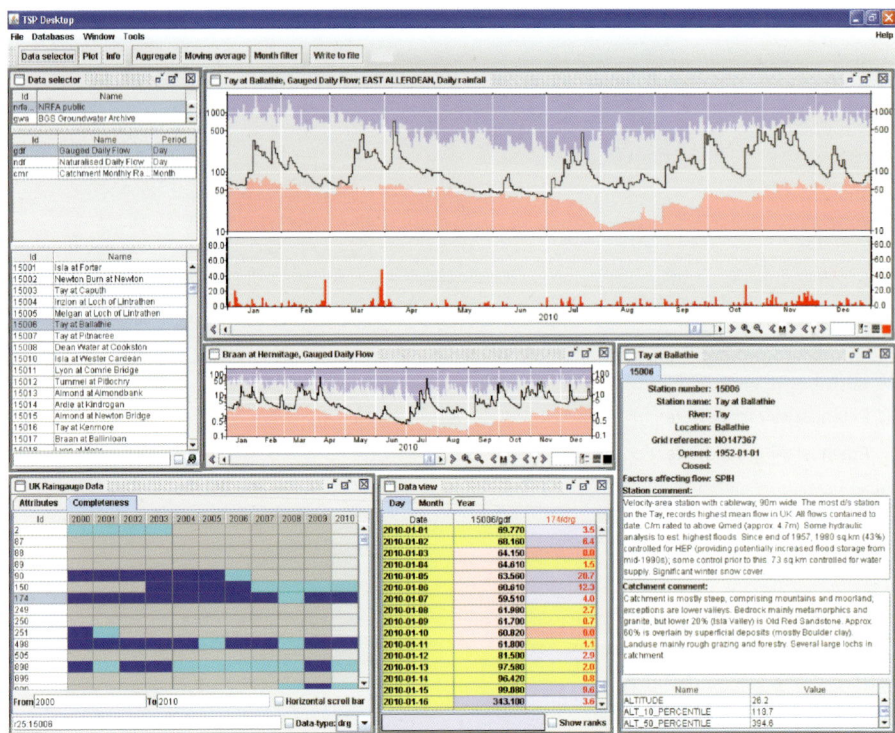

Fig. 11.10 Components used in the Service Level Agreement to identify and rectify errors in river flows. (Source: Reproduced with permission of CEH.)

Susceptibility to groundwater flooding of upper Pang valley, Berkshire

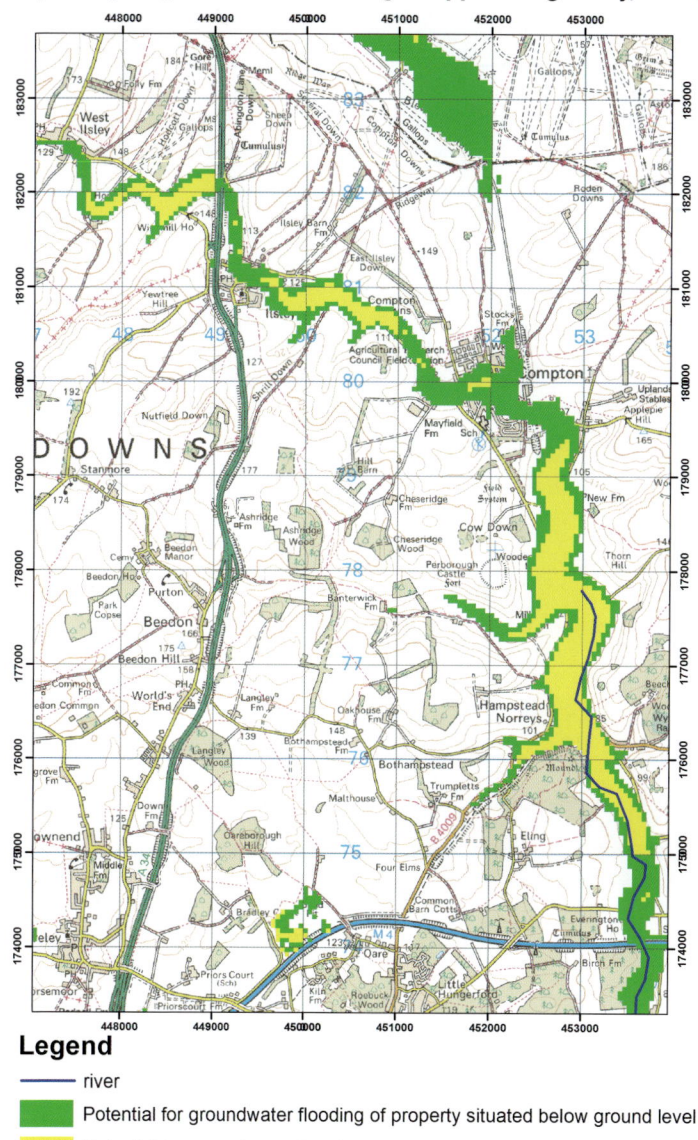

Legend

—— river

Potential for groundwater flooding of property situated below ground level

Potential for groundwater flooding to occur at surface

Fig. 11.12 Susceptibility to groundwater flooding in the upper Pang valley, Berkshire. (Source: Reproduced with permission of British Geological Survey.)

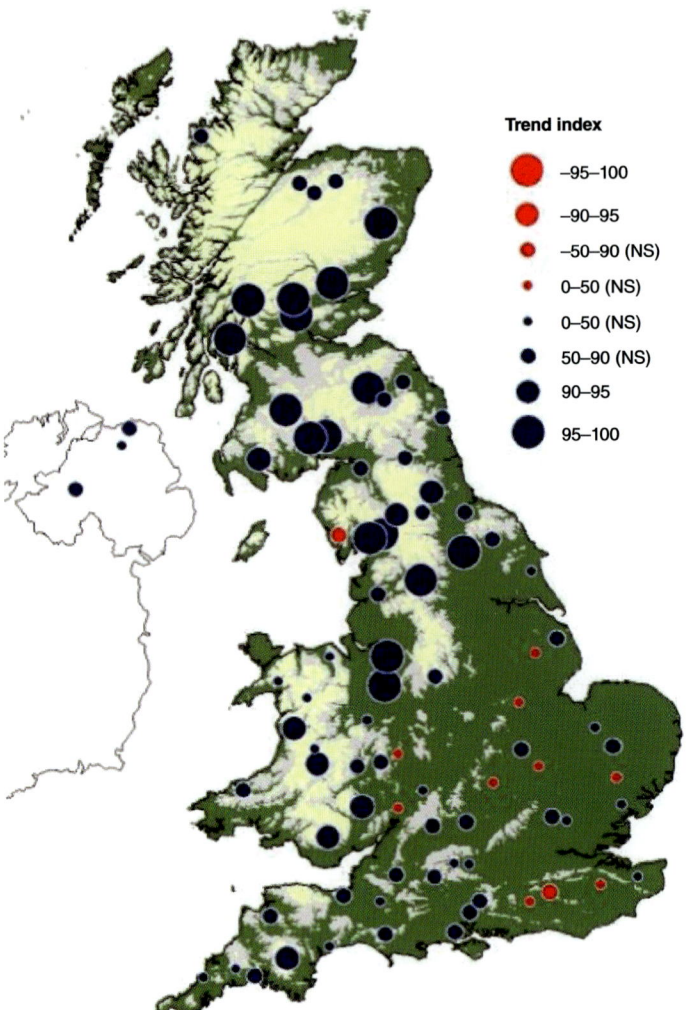

Trend index

- −95–100
- −90–95
- −50–90 (NS)
- 0–50 (NS)
- 0–50 (NS)
- 50–90 (NS)
- 90–95
- 95–100

Fig. 11.13 Trends in high flow duration (number of days above q10 threshold) at gauging stations in the UK Benchmark Network. (Source: Reproduced with permission of CEH.)

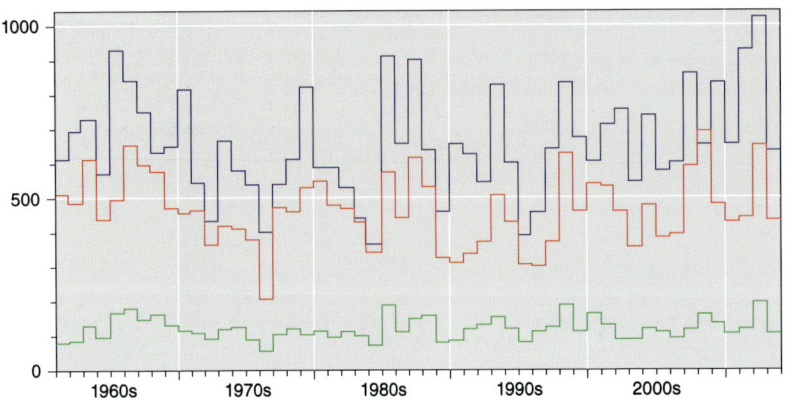

Fig. 11.14 Annual Q_{95} outflows: Scotland (blue), England (red) and Wales (green). (Source: Reproduced with permission of CEH.)

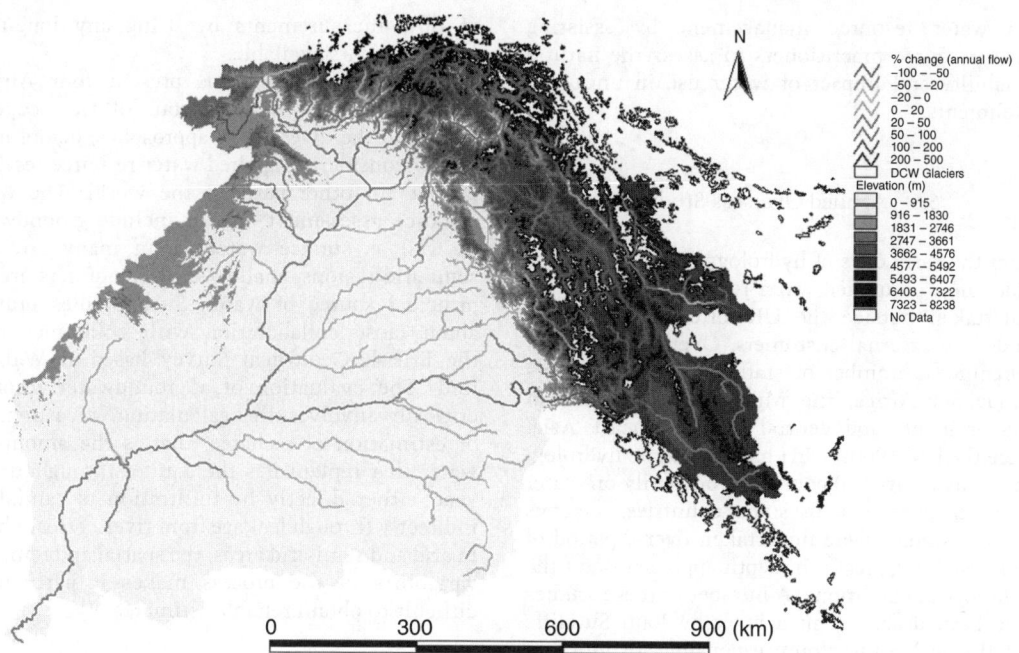

% change (annual flow)
−100 – −50
−50 – −20
−20 – 0
0 – 20
20 – 50
50 – 100
100 – 200
200 – 500
DCW Glaciers
Elevation (m)
0 – 915
916 – 1830
1831 – 2746
2747 – 3661
3662 – 4576
4577 – 5492
5493 – 6407
6408 – 7322
7323 – 8238
No Data

0   300   600   900 (km)

Fig. 6.8 Percentage change in 1961–1990 modelled mean flow for the Indus basin over a 100-year period assuming an average annual temperature increase of 1°C per decade. (Source: Reproduced with permission of CEH.) *(See insert for colour representation of this figure.)*

river flow across Southern Africa. A key objective was to advance the understanding, description and explanation of the mean annual runoff, base flow index and flow duration characteristics building on the work of the original UK Low Flow Studies (6.2). Analyses revealed both broad regional patterns and high local variability (UNESCO, 1997) and laid the foundations for future applied research work in Southern Africa. This included the development of preliminary flow–duration curve estimation procedures (UNESCO, 1997) and regional assessment of rainfall and hydrological drought (Gustard and Cole, 2002).

Drought is a recurrent event in Southern Africa, and being able to accurately forecast droughts and respond effectively to the threat of droughts are priority research and management objectives (Houghton-Carr et al., 2002). Meigh et al. (2002) formulated four principal methods for the analysis of historic droughts and monitoring of current droughts. These methods were delivered by the ARIDA (Assessment of Regional Impact of Drought in Africa) software (UNESCO, 2004b) and released

to national hydrological agencies during regional training workshops. This enabled both objective and consistent analysis methodologies to be used across the southern Africa region.

As is increasingly apparent, several of the activities in the both the SA and HKH FRIEND projects have been concerned with adapting notable achievements from the UK and European low flow and drought research programmes to Southern Africa and the HKH, respectively. One of the most significant examples of technology transfer from the UK was the Southern African (Fry et al., 2004; UNESCO, 2004b) and HKH (Rees et al., 2006; UNESCO, 2004a) pilot implementations of the LowFlows system (6.4.3). A number of modifications to the Low Flows system were required to address the hydrological modelling challenges and different user expectations. These were readily accomplished within the system's flexible modular modelling shell and underlying generic database structure. In both pilot regions, the software tools have the potential to provide a huge leap forward in GIS facilities, flow estimation and water resource management by assisting water resource practitioners to assess the natural variability and impact of water use in ungauged catchments.

6.6   Applied Overseas Studies

From the early days of hydrology at Wallingford, a wide range of applied water resource studies were undertaken outside the UK, often initiated and funded by external customers. Over four decades, a significant number of staff worked on projects throughout Africa, the Middle East, the Indian sub-continent, and central and Southeast Asia. Since the late 1960s, CEH has applied its hydrological skills in over 90 countries, primarily on water resources problems. In some countries, a series of these studies were undertaken over a period of years, enabling greater in-depth appreciation of the hydrological conditions. A number of these studies have been described in a book by John Sutcliffe (2004) who led the group responsible for much of this work at IH.

These overseas studies exposed Wallingford staff to a wider variety of climatic and hydrological conditions than exist in the UK and, thereby often required them to develop innovative approaches and flexible analytical techniques in the applied research problems they encountered. In humid regions such as the UK, the availability of runoff can often be estimated by simple comparison of an isohyetal map of mean rainfall with an estimate of potential transpiration or with regional flow records to deduce basin runoff. In such areas, soil moisture conditions are likely to ensure that basin evaporation is close to the potential rate. In areas where rainfall is concentrated in one season, runoff tends to be the residual after rainfall has eliminated the soil moisture deficit at the beginning of the wet season and transpiration during the wet season. In these areas, water balance methods are also relatively easy to use.

However, this situation is far from common in many areas of the world. For example, in most parts of Africa, the availability of surface water resources is extremely variable, and the concept of assessing climate and hydrological means, even from periods of 30 years, is liable to lead to inaccurate design (Sene and Farquharson, 1998). Therefore, when assessing the water availability of, for example, a dam site, it is essential to extend the availability

of flow measurements by using any long-term measurements available.

The following sections present four African case studies to give a flavour of the scope of work and the diversity of approaches, before more briefly considering applied water resource research projects in other parts of the world. The water resource assessment studies include groundwater as well as surface water as, in many arid and semi-arid regions, shallow or deep aquifers are the principal source of water. Such studies brought about close collaboration with colleagues from the British Geological Survey based at Wallingford. The evaluation of a groundwater resource critically involves the calculation, measurement or estimation of recharge, that is the amount of water that replenishes the aquifer throughout the year, either directly by infiltration of rainfall or indirectly through leakage from river / stream beds. In arid and semi-arid areas, the spatial and temporal variability of the process makes it particularly difficult to obtain reliable estimates.

6.6.1   *Water resource studies in Botswana*

A good example of lengthy involvement by Wallingford is Botswana in southern Africa where CEH undertook 18 studies over two decades. Many of these studies were undertaken in collaboration with consulting engineers, in the case of Botswana Sir Alexander Gibb and Partners, and focused on assessment of surface and groundwater resources for the rapidly developing economy of the country, fuelled in part from diamond mining in the 1980s. Botswana has a semi-arid climate where mean annual rainfall in the populated eastern parts of the country is typically only some 500 to 600 mm, occurring mainly as convective showers between November and March, whereas potential evaporation is over 2000 mm. Annual average runoff varies between 10 and 50 mm, occurring only sporadically following heavy, often localised, convective storms. Topography is flat and, consequently, reservoirs have a high surface area relative to storage capacity resulting in high evaporation losses.

In one of its first projects in 1980, CEH was responsible for investigating how best to meet the increasing demand for water for the capital, Gaborone. The study considered alternative dam sites, potential groundwater sources and increasing the capacity of an existing reservoir that was too small to meet the growing demand for water in the expanding city. Groundwater was also unable to meet the demand and, whilst new dam sites and storage capacities were identified, the preferred option was to raise the existing dam to increase its storage capacity.

Unfortunately, there were no long-term runoff records available to enable accurate assessments of reservoir yield. However, 60 years of monthly rainfall data were available which were used as input to a rainfall–runoff model developed for southern African conditions (Pitman, 1973). This produced a 60-year monthly inflows series for application to a reservoir simulation model where the end of month storage was computed as:

$$S_t = S_{t-1} + Qin_t + P_t - E_t - Abs_t$$

where S_t and S_{t-1} are reservoir storage for the end of month t and month t−1, respectively, Qin_t is the inflow in month t, P_t the rainfall onto the reservoir storage during the month, E_t the reservoir evaporation loss and Abs_t the abstraction for water supply. The simulation was carried out with the existing reservoir capacity and also for a range of potential larger reservoirs.

Although this approach identified the number of months, or years, where reservoir storage was unable to meet the abstraction demand, the approach was too imprecise, with a range of reservoir capacities having similar discrete failure rates of say 1 failure in 60 years. Considering monthly failure rates resulted in only a slight improvement.

Consequently, a new technique based on deficient volumes was developed which related reservoir yield more precisely to failure rates. The approach again used the same reservoir simulation technique but with an infinite reservoir volume. The water balance simulation provided a series of drawdowns, or deficient volumes, which could be analysed statistically by ranking the annual maximum monthly deficient volumes and plotting these using the log-normal distribution (Parks and Gustard, 1982). The method works best where reservoir evaporation is small, but in a semi-arid country such as Botswana the method was adapted to define a series of specific storages from which the associated deficient volumes could be determined. As drought severity increases, the deficient volumes decrease until the optimal net reservoir capacity is reached and a failure occurs after which the deficit is constant (see Fig. 6.9). Interpolation between the trial reservoir capacities simulated enables the optimal capacity able to supply the required yield with the chosen failure return period to be determined. The approach combines the reservoir simulation approach within a probabilistic framework.

A different technique was applied to a 1984 study of the yield of Shashe reservoir which provides water to Francistown in north-east Botswana. For this study, the Gould method, described by McMahon and Mein (1978), was used where a monthly water balance simulation produced a transition matrix of the number of times each end of year state is reached from a range of initial states. The number of occasions the reservoir spilled or failed to meet demand yield was also recorded. The steady-state probability of storage contents could be derived from successive combination of the transition matrix and the starting conditions and the results expressed as probabilities of failure.

In 1986, when estimates of possible reservoir yield were required for significant mining development, only six years of river flow record were available for three gauging stations in the area. No detailed isohyetal map was available, but one was compiled from the rainfall records which existed at the time. Use was made of a regional study by Midgley (1952) who showed that although mean runoff depends on rainfall, slope, soil type and vegetation, these factors are linked with mean rainfall and natural vegetation. He provided a table linking mean runoff with rainfall in zones classified by vegetation and land use. The mean runoff at the gauged sites was estimated by comparing the rainfall during the flow record period to long-term rainfall, and a reservoir site below the most promising gauge was selected for detailed study. This study drew attention to the importance of flow measurement, and gauging was extended to other potential reservoir sites (Parks and Sutcliffe, 1987).

In the 1990s, water resource studies examined much smaller reservoirs whilst investigating the impact of small farm dams within the basins of Gaborone dam and a potential new dam at Bokaa to consider what impact these small dams had on inflows to the capital city's sole source of drinking water (Meigh, 1995). The study also considered Shashe dam, mentioned above. Over 200 small dams of various capacities, used primarily for cattle watering, although some were also used for small-scale irrigation of vegetables, were identified in both the Gaborone and Bokaa catchments using a combination of 1:50 000 maps and remote sensing,

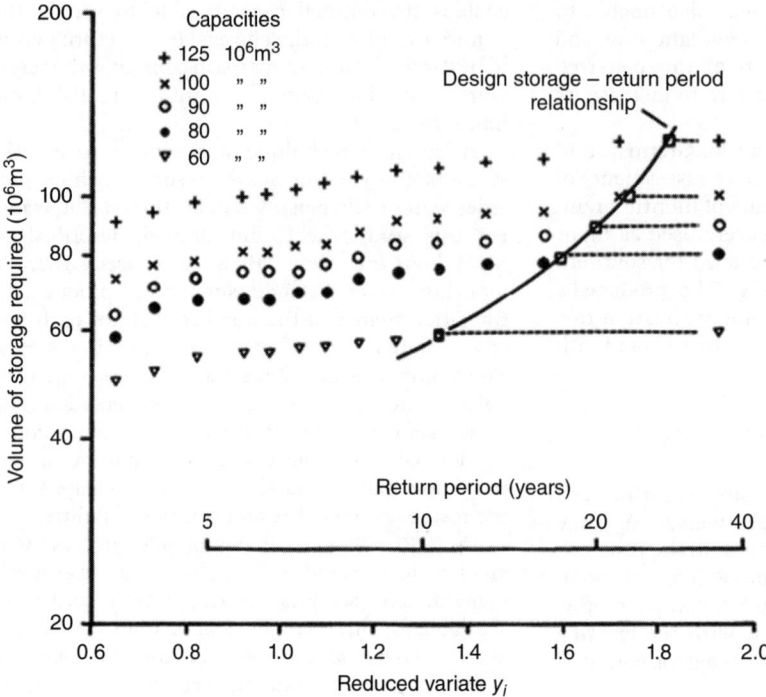

Fig. 6.9 Botswana: Deficient volume analysis. (Source: Parks and Gustard 1982. Reproduced with permission of IAHS.)

whereas there were only 14 dams in the Shashe catchment. Twenty of these reservoirs, thought be representative of the population as a whole, were surveyed to produce area–capacity relationships and a generalised storage capacity relationship was developed:

$$\text{capacity} = 0.7381\,\text{area}^{1.251}\ (R_2 = 93.1\%)$$

where capacity is in thousand m^3 and area is in hectares. This simple relationship enabled the storage capacity of other reservoirs to be estimated, where area was identified from maps or remote sensing.

A model was developed to compute likely evaporation losses and to consider how the water stored in these reservoirs might be used. The modelling showed that the 200 existing small reservoirs within the Gaborone dam catchment were reducing inflows to the main urban water supply reservoir by 25% and also demonstrated that any further development could result in yet greater loss of inflow. Following the study, the Government of Botswana instigated laws to control construction of small dams and used the model to determine

the feasibility of further small farm dams in the catchments of major water supply reservoirs.

This section shows how a series of studies in one country enabled development of a deeper understanding of the country's hydrology and how increasingly sophisticated methodology could be applied allowing greater confidence in outcomes. However, CEH scientists became increasingly interested in regional approaches to problem solving and, in particular, the larger and longer datasets and consequent analytical benefits that could be gained by pooling flow data from different southern African countries together. Much of this work was carried out within the Southern African FRIEND project (Section 6.5.4), but the range of research extended beyond low flows and droughts, also including rainfall–runoff modelling and flood frequency analysis to develop improved approaches for design flood estimation.

Another southern African regional project closely aligned to FRIEND was the SADC (Southern African Development Community) component of the WMO WHYCOS project (World Hydrological Cycle Observation System – http://www .whycos.org/whycos/). Under this, CEH were commissioned by the EU, to work with national hydrological agencies to install 50 new river flow and climate stations throughout 10 SADC countries. The 50 SADC-HYCOS stations used state-of-the-art sensors recording on a datalogger that transmitted data every three hours via data collection platforms (DCPs) through the METEOSAT weather satellite to a regional ground receiver station in Pretoria, South Africa. National hydrological agencies accessed this near real-time data from this regional data centre using the Internet for use in flood forecasting and reservoir operations on shared river basins. This regional data sharing is vitally important as there are many transboundary river basins in southern Africa.

6.6.2 The hydrology of lake Victoria

Transboundary river basins are common throughout the world, but are epitomised by the Nile, the world's longest river, flowing through 10 countries in east and north Africa, and draining one-tenth of the African continent. The water resources of the basin are under pressure from rapid population growth and associated water demand, as well as subject to a highly spatially and temporally variable climate, and exemplify the challenges of transboundary watershed management. Approximately 86% of the Nile water is contributed by the Atbara and the Blue Nile, from Lake Tana in Ethiopia. The other 14% is contributed by the White Nile, from Lake Victoria. The Lake Victoria basin is home to over 30M people and the socioeconomic heart of East Africa. The confluence of the Blue Nile and White Nile is at Khartoum in Sudan.

CEH was involved in studies of the hydrology of Lake Victoria since 1983, starting with an extensive, study funded by the Department for International Development (DFID, formerly the Overseas Development Administration, ODA) which sought to explain the dramatic 2.5 m rise of lake levels seen in 1961/64 compared to a recorded historical range of just over 1 m (Fig. 6.10). The work involved a review of the basic data back to 1900, synthesis of river inflows for ungauged periods, and development of the first water balance model able to explain the rise in lake level (Piper *et al.*, 1986). A follow-on study provided improved models of lake rainfall, an important issue as 85% of the input to Lake Victoria is from rainfall directly onto the lake. This work led to revised estimates of the probabilities associated with high lake levels

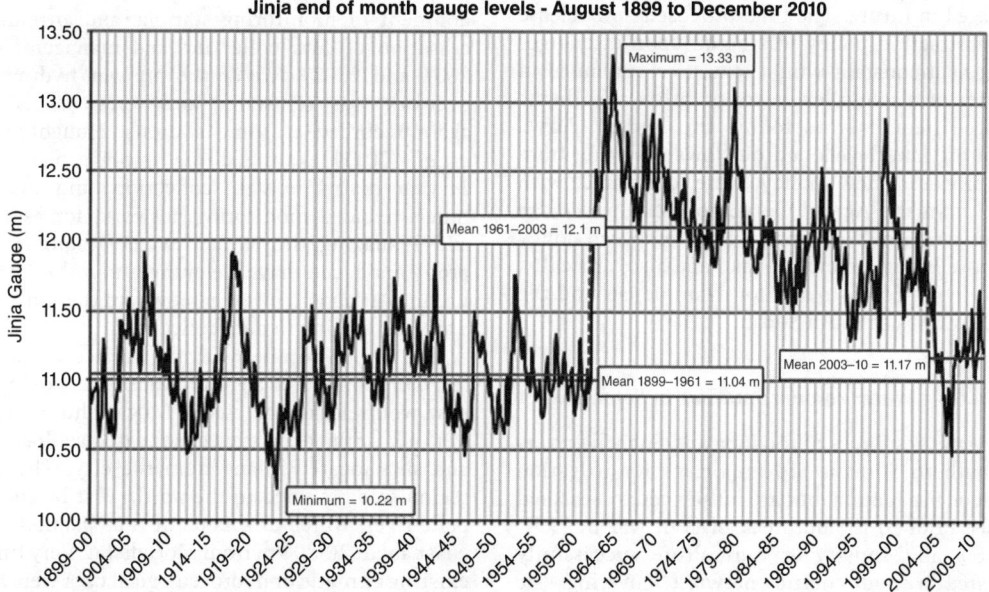

Fig. 6.10 Historical water levels of Lake Victoria measured at Jinja in Uganda. (Source: Water Resource Associates. Reproduced with permission of WHS.) *(See insert for colour representation of this figure.)*

through stochastic modelling of possible future lake behaviour.

In 1992, DFID(ODA) funded additional work to review recent studies of the hydrology of Lake Victoria and outflows into the White Nile and examine the option of installing additional hydropower capacity. The lake water balance was updated and revised estimates of flows available for hydropower development were produced. A second hydropower station at Kiira (Fig. 6.11) was constructed and the increased releases for hydropower contributed to the marked fall in lake levels after 2003 shown in Figure 6.10, although about half this fall was due to a run of below average years of rainfall.

Between 2008 and 2012, CEH was commissioned by the Lake Victoria Basin Commission to develop a new water release and abstraction policy for the Lake Victoria Basin as lake levels had fallen dramatically between 2001 and 2003. Since construction of a dam and hydropower station at Nalubaale (Owen Falls) in 1954, releases had been made according to an "Agreed Curve" that was a representation of the natural rating curve of the sole outflow from the lake into the White Nile at Ripon Falls. CEH undertook a further examination and updating of the basin water balance and a lengthy period seeking views from all stakeholders on how they would wish Lake Victoria to be managed in future. The final proposed new release and abstraction policy was based upon discharges being held constant whilst lake level remained in a series of discrete level ranges rather than being changed frequently to follow the Agreed Curve. Resulting lake levels and outflows resulting from the proposed new policy were compared with those from the Agreed Curve policy as part of an investigation of the impact of the proposed policy on those living around the lake and also, critically, those living downstream, for example, around the Sudd swamps of South Sudan.

6.6.3   Water resource studies in Somalia

The water resources of the Juba and Shabelle rivers in southern Somalia are important for irrigation and food production, but are influenced by seasonal floods. Prior to the outbreak of civil war in 1991, the Somali Ministry of Agriculture successfully operated a hydrometric network covering the Juba and the Shabelle, which provided input data to a flow forecasting model. The network was strengthened in the 1970s and 1980s by support

from CEH and the engineering consultancy Mott MacDonald who provided technical assistance and training as part of the UK government's programme of technical cooperation with developing countries.

One key issue that was identified was the inadequacy of the historical rainfall, river flow and groundwater archiving. Data were held solely in manuscript form and many of the older data sheets had been damaged by moisture or insect attack and in many cases simply lost. In the early 1980s, stand-alone PC computers were becoming readily available which allowed CEH to develop robust database and data processing systems for use on PCs. One such system was designed for climatological and hydrological data, called HYDATA, and a parallel system for groundwater, GRIPS, was also written. HYDATA was a very useful and successful tool that was championed by both the DFID(ODA) and WMO, each of whom recognised the benefits to hydrometeorological services in many developing countries (see Chapter 11). With support from ODA and WMO, Wallingford staff were able to provide PCs loaded with the HYDATA software to over 40 countries and provided training and subsequent support in data storage, quality control and analysis.

The civil war resulted in the neglect and abandonment of monitoring stations and an enforced cessation of data collection and management. In 2001 and 2002, part of the pre-war hydrometric network was reinstated and water levels were again recorded at some stations. Houghton-Carr *et al.* (2011) examined the implications of the 11-year hiatus in data collection, and the now much reduced monitoring network, for assessing and managing the surface water resources. The problems faced have relevance to other basins, within Africa and elsewhere, where there has been a similar decline in data collection.

In 1980, in parallel with the surface water studies, CEH conducted a groundwater study to locate a new wellfield for the city of Mogadishu in an area of thick (150m+) poorly consolidated aeolian sands and silts located inland from the city. The study region was bounded to the north-west by the Shabelle river, which at this point runs parallel to the coast about 20-30 km from Mogadishu. Very limited existing climatic or hydrogeological data were available so a 1-year exploration and drilling study was undertaken to collect as much hydrogeological and climatic data as possible. But to ensure success it

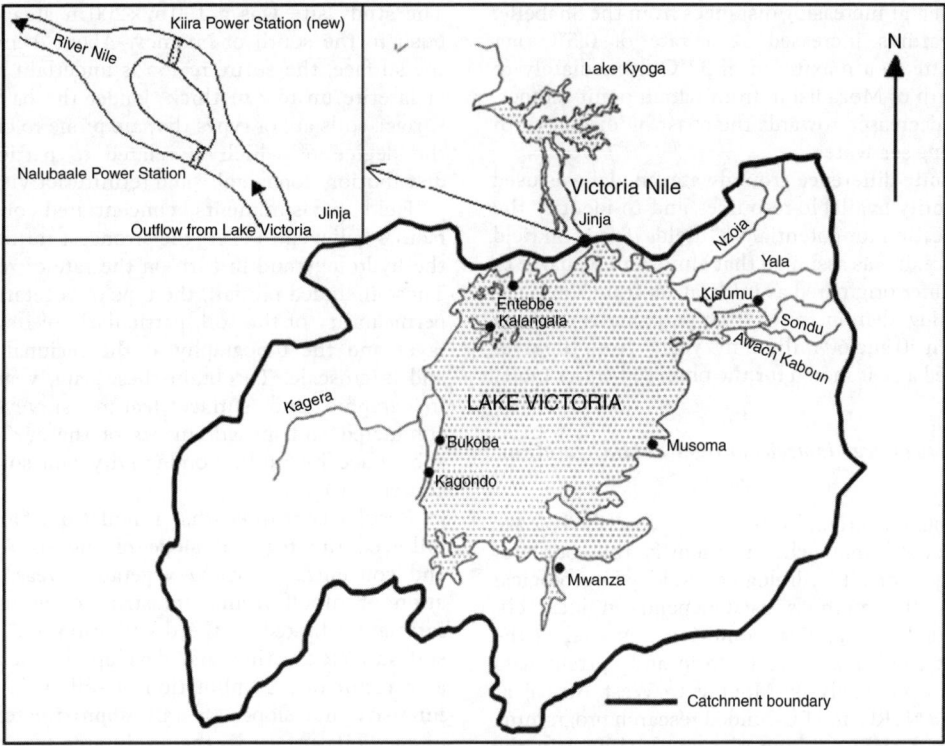

Fig. 6.11 Lake Victoria basin showing the single outlet to the White Nile at Jinja (with schematic of the two hydropower stations shows as an inset). (Source: Water Resource Associates. Reproduced with permission of WHS.)

was necessary first to identify the source or sources of groundwater recharge in the region, and second to quantify the amount.

Water levels taken from project wells and existing sources revealed a pronounced water table mound developed beneath the Shabelle pointing to infiltration from the river channel through up to 50 m of unsaturated sands/silts to the water table.

To identify areas of recent recharge, measurements of tritium concentration, the radioisotope of water, were obtained from groundwater samples throughout the region. The presence of Tritium indicates a post-1952 age, since this was when the radioactive isotope was introduced into the atmosphere by thermonuclear testing. Measurements were also made of deuterium and oxygen-18, the natural stable isotopes contained in water; variations in their concentration depend upon the history of the water when it was in contact with the atmosphere. Tritium levels in the groundwater

pointed to the absence of recharge to the main body of the aquifer since 1952, although this could be partly explained by the presence of 50 m+ unsaturated zone below the river; it is conceivable that tritium-rich water is contained within this zone moving very slowly downward. Results of the natural stable isotope studies were compatible with a river origin for groundwater recharge, because all groundwater values lie between the extremes of river water at all flow stages. The isotope work also confirmed that groundwater movement is dominantly from the river towards the coast.

Temperature profiles from project boreholes demonstrated the presence of a significant temperature gradient away from the river and provided one final piece of evidence to link groundwater in the aquifer with recharge from the Shabelle. In boreholes adjacent to the river groundwater, temperatures range from 27 to 29°C, matching the temperature range of the river itself (24–30°C).

However, at increasing distances from the Shabelle, temperatures increased at a rate of 0.5°C per kilometre to a maximum of 37°C immediately to the north of Mogadishu, from which point temperatures decreased towards the coast by mixing with intruding sea water.

A finite difference groundwater model was used to quantify available resources and to identify the best location for potential wellfields. Based on field evidence, it was assumed that almost all recharge to the aquifer originated as infiltration from the river. Modelling demonstrated that a recharge rate of between 40 and 60 million m³ year⁻¹ from the river provided a best match for the observed water table.

6.6.4   Recharge in areas of tiger bush vegetation, Niger

The Somalia groundwater study described above used several approaches to identify the source of recharge. For all hydrological and hydrogeological studies, the methods used depend on local climatic and geographical conditions, as well as the quantity and quality of historic and current data available for analysis. Moving to West Africa, as part of a NERC and EU-funded research programme (HAPEX-SAHEL – Goutorbe *et al.* (1994)) in the early 1990s, a study was conducted into the process and quantification of groundwater recharge below areas of patterned vegetation in south-west Niger. These distinctive regions display alternating bands of vegetation and bare ground aligned along the contours of gently sloping terrain. It is known locally as 'brousse tigrée' (tiger bush). At the selected study site 45 km to the south of the capital, Niamey, the vegetation bands are 10–30 m wide, separated by 50–100 m wide bands of bare ground.

Groundwater recharge in this region takes place either as indirect (localised) recharge through the beds of ephemeral streams and temporary pools, or as widespread direct (or diffuse) infiltration through the soil. The object of the study was to quantify the level of direct infiltration by rainfall. The mean annual rainfall from an 80-year record at Niamey is 562 mm, much of which falls as high intensity localised storms; intensities exceeding 100 mm h⁻¹ are not uncommon. Such high intensities ensure the generation of runoff and the development of surface crusts even on sandy soils.

The first approach to the issue of recharge was to examine the way rainfall was redistributed over the surface once it falls over this type of vegetation.

The study site was a 190 m × 200 m area of tiger bush to the south of Niamey. About 2 cm below the surface, the entire region is underlain by layer of laterite up to 3 m thick. Under the bare areas, surface soils are of types that are prone to crusting, the degree of which is related to particle size distribution, topography and termite activity.

Field measurements concentrated on those factors believed to exert the strongest influence on the hydrology and in turn on the rate of recharge. These included rainfall, the type of vegetation, the permeability of the soil, particularly of its surface layer and the topography at the regional, meso- and microscale. To obtain these data, vegetation, topographic and surface feature surveys were conducted and measurements of the surface and subsurface hydraulic conductivity and soil water content made.

Results confirmed that runoff from bare areas followed the regional slope of the soil surface, and concentrated below vegetated areas. Generation of runoff from bare strips of ground was intimately linked to the distribution and type of soil surface crusting that develops in these areas as a result of a combination of soil type, rainfall intensity and slope. Overall, approximately 58% of the soil surface in the study area was covered by runoff-generating non-vegetated crusts; the remaining 42% were in areas with thick vegetation cover and considerable surface storage capacity. Such concentration meant that the effective rainfall below vegetation was higher than actual rainfall, a feature which inevitably enhances the ability of the vegetation to survive in an environment where water availability is notoriously erratic from one year to the next. Soil moisture measurements revealed that vegetated areas received between 1.6 and 3.9 times more water than the actual rainfall as a result of runoff from adjacent bare ground. However, the results from subsurface investigations showed that despite this concentration of water, deep infiltration to the water table as recharge was only likely to take place during the most intense rainstorms.

The surface water study highlighted the process by which recharge can take place below tiger bush but could not quantify the amount. To do this a chloride profile method using samples taken from the unsaturated zone within a 77-m deep well dug within the study area was used to provide a long-term average value for infiltration beneath the tiger bush.

During well construction, samples of rock were taken from the unsaturated zone at the following intervals: every 25 cm from 0 to 10 m, every 50 cm from 10 to 62.5 m, then every metre to the bottom of the well. Pore water was extracted from each sample either by centrifugation or elutriation and analysed for chloride; moisture contents of samples were obtained gravimetrically. These data were used to produce depth profiles of pore water chloride concentration and moisture content throughout the unsaturated zone. Low concentrations of chloride indicate higher rates of recharge; high concentrations mean a lower recharge.

From the 77 m profiles it was possible to derive an estimate of historic direct recharge at the site. The chloride concentration of rainfall, which is required to make the estimate, was determined from the analysis of 123 rainfall samples collected from five EPSAT (Estimation des Precipitation par Satellite au Sahel) rain gauges in 1992. A mean recharge rate of 13 mm year (range 10-19 mm) was estimated for the upper 70 m of the profile, with a total residence time of 790 years (range 520–990 years). This is considered to be a representative estimate of the magnitude of direct recharge taking place below tiger bush areas.

6.6.5 Other overseas studies

Previous sections have described a number of water resource studies undertaken by CEH in Africa, but other applied water resource studies were also being undertaken in the Middle East, India, Bangladesh and Pakistan and in various countries throughout Southeast Asia. Typical examples are included in Box 6.4.

6.6.6 Evolution of water resource studies

The discussions above have illustrated how innovative research was applied to challenging problems in countries where data were sparse and climate very different to that of the UK. For the earlier studies undertaken in Botswana, Wallingford staff operated very much as traditional 'consultants' where they visited the region to collect all available data but undertook the bulk of the work back in the UK, writing a final technical report as the project 'output'. There was often a limited legacy of such work and limited opportunity for participation by local staff. The Niger study illustrates how

the working approach changed over time with a conscious effort to involve local government and university staff in the technical work, and with a growing emphasis on technology transfer. This change came about as CEH responded to the evolving strategies of funding agencies such as DFID (Department for International Development) and the European Union (EU). During the last decade of the twentieth century, collaborative research became a fundamental principle of the research programmes of many funding agencies. However, the change also came about through a growing desire to create a more lasting legacy of the research. Certainly much of the effort within the FRIEND programme, described earlier in this chapter, focused on development of improved water resource assessment tools and involved full participation by regional partners.

Over the last decade or so, significant new sets of data have become available such as remote sensed data, digital elevation models (DEMs) and gridded rainfall datasets. Through increased computing power on PCs and use of GIS tools these new datasets allowed CEH to move towards large-scale modelling of water resources. DFID funding supported the development the GWAVA model (Global Water Availability Assessment – see Meigh *et al.*, 1999 which has a gridded water resource modelling capability) developed initially for all of eastern and southern Africa. In GWAVA, water resource components, such as water demands and returns across the three main water use sectors (i.e. domestic, agricultural and industrial), lakes and reservoirs, and any inter-basin transfers, are incorporated into a lumped rainfall–runoff model (CEH's Probability Distribution Function model (Moore, 1985)) to provide an assessment of water availability. GWAVA requires wide ranging inputs of data, increasingly publically available, both to drive the rainfall–runoff model and in order to estimate the anthropogenic water demands at the scale of the grid cell. Modelling on a continental scale may present challenges not usually experienced for smaller-scale applications. Most important is the need to ensure that input data, many of which are produced nationally, are standardised across the continent.

The requirement to manage water resources in a sustainable manner at the basin scale has become a driving force behind the use of hydrological models such as GWAVA in understanding how

Box 6.4   Studies in the Middle East

1973–1976 northern Oman – installation of a new hydrometric scheme and analysis of historical and new rainfall and flood flows in order to estimate long-term runoff and the rates of aquifer recharge. This provided the first scientifically reliable assessment of surface and groundwater resources of the rapidly developing area on the northern Batinah Plains around Muscat.

1979–1980 Jordan – identification of sources of process water for a new potash plant at the southern end of the Dead Sea and of potable water to supply the new township built to house the potash plant. This involved complex modelling of ephemeral wadi flows using limited rainfall data underpinned by a programme of wadi gauging in order to develop a long-term flow series from which recharge into the alluvial aquifer could be estimated and a productive wellfield designed.

1990–1991 Yemen – a rainfall–runoff modelling study to support more reliable surface water and groundwater recharge assessment, with very limited rainfall and flow data available for model input and calibration (Farquharson *et al.*, 1996). A distributed model was developed by sub-dividing catchments into five different runoff-generating zones, each with varying runoff characteristics, and with three runoff-absorbing zones in alluvial areas where surface runoff was lost to shallow groundwater storage to varying extents. Runoff was estimated using the US Soil Conservation Service curve number approach and losses adjusted to allow sufficient groundwater recharge to sustain observed dry-spell flow recession rates. In addition, a simple stochastic rainfall model was developed. This was based upon observations that despite mean annual rainfall varying from less than 150 mm to over 1000 mm the statistical distribution of daily rainfalls was essentially uniform throughout the country, fitting a log-normal distribution. Thus, although wet regions experience a greater number of rain days, a daily rainfall of say 50 mm/day was equally likely to occur in any part of the country. The stochastic daily rainfalls and rainfall–runoff model were used to generate long-term synthetic flow sequences to estimate reliable surface water and groundwater recharge.

1983 Qatar – investigation of rising groundwater levels beneath Doha by identifying and quantifying the main sources of recharge in the city and recommending mitigation strategies to reduce flooding. Results from a detailed engineering inventory of all sources and sinks within the city combined with a field programme of drilling, geophysical logging, pump testing and water level monitoring were used to construct a finite difference model of the region to help quantify regional recharge and to predict the impact of potential mitigation measures.

Studies in the Indian subcontinent

1982–1984 India – collaboration with BGS to assess seasonal soil moisture storage and groundwater recharge in the Betwa basin by water balance techniques. The comparison of rainfall, evaporation and runoff in a monsoon climate provided a means of estimating seasonal recharge and a novel rainfall–runoff model produced (Sutcliffe and Green, 1986).

Early 1980s Bangladesh – hydrological support to ADB-funded multi-disciplinary Small-Scale Irrigation Sector Project in Bangladesh which aimed to reduce the agricultural damage caused by flooding through construction of submersible flood embankments designed to not only keep early season flood waters in April and May from areas growing late-winter boro rice, but also retain some of the receding monsoon flood to provide irrigation for winter crops. This study saw the first use by Institute staff of a portable PC computer which was employed for reservoir simulation and irrigation modelling.

1999 Pakistan – estimation of long-term inflows and sediment loads to Tarbela reservoir and simulation of reservoir behaviour under a range of possible new operating strategies in order to identify how best to prolong the operational life of the reservoir (Tate and Farquharson, 2000). The simulation results were provided to an economist who evaluated the irrigation and hydropower benefits resulting from the various new policies to allow identification of a best option. The study involved extensive collaboration with sediment modellers and also with the group comparing the economic outputs of various management scenarios.

2009–2014 India – collaboration with BGS to provide technical support to the World Bank funded India Hydrology Project which has involved a huge investment in improved instruments, gauging techniques, quality control of data, development of new web-based databases and hydrological information systems, together with significant training. Wallingford staff have been involved in many aspects of the work from project preparation (1993–1994), through both phases of the project to date (see www.hydrology.gov.in). The project involved a large amount of capacity building and in some ways shared similar objectives to the FRIEND project (Section 6.5).

Studies in Southeast Asia

1970–1985 Mekong basin – review of hydrological data and water resource estimates and analysis of the reliability of rain-fed agriculture in Northeast Thailand and Cambodia. Impact of deforestation, changes in cropping pattern, and irrigation and hydropower development in Northeast Thailand on the hydrology of the Mekong. Analysis of the areal coherence of rainfall and the accuracy of estimation of areal rainfall and runoff predicted by basin models used in reservoir analysis.

1984 Thailand – hydrology of the $50\,000\,km^2$ Chi basin in Northeast Thailand, which has relatively low annual rainfall and has sandy soils making it unsuited to rice growing. With improvements in agriculture in Europe reducing the demand for the alternative cassava crop that farmers had been growing, this project examined what alternative crops might be successful in that region and sought potential markets for these new crops.

water resources will be affected by change. Results provide a reference point for long-term strategic planning of water resource development, alert policymakers and stakeholders about emerging problems and allow river basin managers to test regional and local water plans against uncertainties.

Amongst many applications, GWAVA has been used to investigate possible climate change impacts within the Ganges-Brahmaputra and Meghna basins (Fung *et al.*, 2006), to examine how human intervention has affected inflows to and water levels in the Caspian Sea (Erlich *et al.*, 2002) and to assess how agricultural development and dam construction in Angola might affect inflows to the ecologically sensitive Okavango swamp in Botswana (Folwell and Farquharson, 2006). Subsequently, GWAVA has had water quality and environmental flow functionality added. It has been applied globally as part of an Integrated Project Water and Global Change (WATCH, 2007–2011: http://www.eu-watch.org), funded under the EU FP6. This project brought together the hydrological, water resources and climate communities who analysed, quantified and predicted the components of the current and future global water cycles and related water resources states; evaluated their uncertainties and clarified the overall vulnerability of global water resources related to the main societal and economic sectors.

6.7 Conclusions

During the course of this research, there has been a transformation in the availability of digital spatial data including terrain data, river networks, precipitation, evaporation, soil type and geology. Together with advanced data handling tools, this has improved the ease, accuracy and speed of deriving catchment characteristics for use in regional modelling. A second major advance has been delivering these models to the user community via software products. These have been tailored to a specific water sector and have often combined hydrology with other disciplines, for example hydrology and mechanical engineering (hydropower) or hydrology and ecology (environmental flows). The third main development has been software products that integrate natural and artificial data and integrate a range of estimation methods, for example from the ungauged site to the analysis of long flow records, often enhanced by formal or informal data transfer methods. In parallel with software development, there has been a geographical extension of the Institute's water resource research from the UK to mainland Europe, southern Africa, the Himalayas, Indian sub-continent and Southeast Asia and the tropics. Applied research in countries where data are generally scarce has required not only innovation but also judgement in order to extract the most out of limited data, whilst still producing robust, and defensible, results. Research at Wallingford has aimed to improve water management in many countries and provided support to the development and poverty alleviation objectives of DFID, the EU and various UN agencies.

Looking to the future, the pressures on both the quantity and quality of water resources from population growth and climate change will drive the critical need to manage those water resources equitably and sustainably, and the associated demand for appropriate methods and tools. Further improvements in information technology, in the availability and quality of digital datasets, and in analytical software will facilitate this. One of the future challenges will be supporting developing countries, many of whom lack the staff and budgetary resources and institutional capacity to monitor and manage their water resources effectively, whereas experiencing rapid population growth and changing climate that engender accelerated watershed degradation, declining water quality and quantity and, ultimately, increased poverty, to make optimal use of these advances. Indeed, recent European research (Laize *et al.*, 2013) has indicated that socioeconomic factors may have a more significant short-term impact on water resources than climate change. This reiterates the fact that water resources management is not only about water supply, but also about relatively ostracised sanitation which has major implications for water quality, human and ecosystem health and water treatment costs. The continued focus on integrated river basin planning and management will be augmented by increased efforts towards water efficiency and demand management. Perhaps,

formal rainwater harvesting systems for domestic use (e.g. large water tanks under lawns) will become as common as solar panels are in the second decade of the 21st century. Undoubtedly, desalination processes will be forced to become more efficient and more affordable so that this technology can be utilised more widely. Similarly advances in the use of remote sensing to assess water resources will continue which will be of particular benefit to developing countries.

6.8 References

Beran, M.A. & Gustard, A. (1977) A study into the Low Flow Characteristics of British Rivers. *Journal of Hydrology*, **35**, 147–157.

Boorman, D.B., Hollis, J.M., & Lilly, A. (1995) *Hydrology of soil types: a hydrologically based classification of the soils of the United Kingdom*. Institute of Hydrology Report No. 126, Wallingford.

Bullock, A., Gustard, A., Irving, K., & Young, A. (1994) *Low Flow Estimation in Artificially Influenced Catchments*. NRA R&D Project 257.

Cogley, J.G. *et al.* (2010) Tracking the Source of Glacier Misinformation. *Science*, **327** (**5965**), 522.

Demuth, S. & Stahl, K. (2001) *ARIDE: Assessment of the Regional Impacts of Droughts in Europe*. Final Report, EU Contract ENV4-CT-97-0553. Institute of Hydrology, University of Freiburg, Freiburg, Germany.

Demuth, S. & Gandin, J. (2010) *Footprints of international cooperation over a quarter century, Global Change: Facing Risks and Threats to Water Resources*. Proc. of the Sixth World FRIEND Conference, Fez, Morocco, October 2010. *IAHS Publ.* 340: pp. 3–10

Drayton, R.S., Kidd, C.H.R., Mandeville, A.N. & Miller, J.B. (1980) *A regional analysis of river floods and low flows in Malawi*. Institute of Hydrology Report No. 72. Wallingford.

Dunbar, M.J., Young, A.R. & Keller, V. (2006) *Distinguishing the Relative Importance of Environmental Data Underpinning flow Pressure assessment (DRIED-UP)*. Environment Agency, Bristol.

Agency, E. (2002) *Managing Water Abstraction: The Catchment Abstraction Management Strategy Process*. Environment Agency, Bristol.

Erlich, M., Shiklomanov, I., Yezov, V., Georgievsky, V. & Shalygin, A. (2002). *Reasons and predictions of the Caspian Sea water level fluctuations: Impact of climate factors and man's activities*. EGS XXVII General Assembly, Nice, France, April 2002.

European Union (2000) *Technical Report to the European Union ENV4-CT95-0186 –Groundwater and River Resources Programme on a European Scale (GRAPES)*. Institute of Hydrology, Wallingford.

Farquharson, F.A.K., Plinston, D.T. & Sutcliffe, J.V. (1996) Rainfall and runoff in Yemen. *Hydrological Sciences Journal*, **41**, 797–811.

Feijtel, T.C.J., Boeije, G., Matthies, M. *et al.* (1998) Development of a geography-referenced regional exposure assessment tool for European rivers – GREAT-ER. *Journal of Hazardous Materials*, **2076**, 55–60.

Folwell, S.F., & Farquharson, F.A.K. (2006) *The impacts of climate change on water resources in the Okavango Basin. Climate Variability and Change – Hydrological Impacts*. Proceedings of the Fifth FRIEND World Conference held at Havana, Cuba, November 2006, IAHS Publ. 308.

Fry, M.J., Folwell, S.S., Houghton-Carr, H.A. & Uka, Z.B. (2004) An integrated Water Resources Management tool for Southern Africa allowing low flow estimation at ungauged sites. In: Stephenson, D., Shemang, E.M. & Chaoka, T.R. (eds), *Water Resources of Arid Areas*. Taylor & Francis, London, pp. 507–514.

Fung, C.F., Farquharson, F.A.K. & Chowdhury, J. (2006) *Exploring the impacts of climate change on water resources – regional impacts at a regional scale: Bangladesh. Climate Variability and Change – Hydrological Impacts*. Proceedings of the Fifth FRIEND World Conference held at Havana, Cuba, November 2006, IAHS Publ. 308, 2006.

Goutorbe, J.-P., Lebel, T., Tinga, A. *et al.* (1994) HAPEX-Sahel: a large-scale study of land-atmosphere interactions in the semi-arid tropics. *Annales Geophysicae*, **12**, 53–64.

Gustard, A., Marshall, D.C.W & Sutcliffe, M.F. (1987a) Low Flow Estimation in Scotland. Institute of Hydrology Report No. 101, Wallingford.

Gustard, A., Cole, G.A., Marshall, D.W., & Bayliss, A.D. (1987b) *A study of compensation flows in the UK*. Institute of Hydrology, Report No. 99, Wallingford.

Gustard, A., Bullock, A. & Dixon J.M. (1992) *Low Flow estimation in the United Kingdom*. Institute of Hydrology Report No. 108, Wallingford.

Gustard, A. & Wesselink, A.J. (1992) Impact of Land Use Change on Water Resources: Balquhidder catchments. *Journal of Hydrology*, **145**, 389–401.

Gustard, A. & Cole, G.A. (eds) (2002) *FRIEND – A Global Perspective 1998–2002*. Centre for Ecology & Hydrology, Wallingford, UK.

Gustard, A. & Bonell, M. (2006) FRIEND Key achievements and opportunities in international co-operation. In: Demuth, S., Gustard, A., Planos, E., Scatena, F. & Servat, E. (eds), *Climate Variability and Change – Hydrological Impact*. Vol. **308**. IAHS Publ., pp. 3–9.

Hisdal, H., Tallaksen, L.M., Clausen, B., Peters, E. & Gustard, A. (2004) Hydrological drought characteristics. In: Tallaksen, L.M. & van Lanen, H.A.J. (eds), *Developments in Water Sciences 48 Hydrological Drought Processes and Estimation Methods for Streamflow*

and Groundwater. Elsevier Science Publisher, The Netherlands, pp. 139–198.

Holmes, M.G.R. & Young, A.R. (2002) *Estimating seasonal low flow statistics in ungauged catchments*. Proceedings of the British Hydrological Society Eighth National Symposium Birmingham, pp. 97–102.

Holmes, M.G.R., Young, A.R., Gustard, A. & Grew, R. (2002a) A new approach to estimating mean flow in the United Kingdom. *Hydrology and Earth System Sciences*, **6**, 709–720.

Holmes, M.G.R., Young, A.R., Gustard, A. & Grew, R. (2002b) A region of influence approach to predicting flow–duration curves within ungauged catchments. *Hydrology and Earth System Sciences*, **6**, 721–731.

Houghton-Carr, H.A., Fry, M.J., McCartney M.P. & Folwell, S.S. (2002) *Drought and drought management in Southern Africa*. Proceedings of the British Hydrological Society Eighth National Symposium Birmingham, 103–108.

Houghton-Carr, H.A., Print, C.R., Fry, M.J., Gadain, H. & Muchiri, P. (2011) An assessment of the surface water resources of the Juba-Shabelle basin in southern Somalia. *Hydrological Sciences Journal*, **56**, 759–774.

Huang, Y. & Demuth, S. (eds) (2010) *FRIEND a Global Perspective 2006–2010. IHP Non Serial Publications in Hydrology*. German IHP/HWRP Secretariat, Koblenz.

Institute of Hydrology (1980) *Low Flow Studies Report*, Wallingford.

Institute of Hydrology (1991) *Plynlimon Research: the first two decades*. IH Report No109. Wallingford.

Johnson, I.W., Elliot, C.R.N. & Gustard, A. (1995) Modelling the effect of groundwater abstraction on salmonid habitat availability in the River Allen, Dorset, England. *Regulated Rivers: Research and Management*, **10**, 229–238.

Laize, C., Acreman, M., Schneider, C. *et al.* (2013) Projected flow alteration of ecological risk for pan-European rivers. *River Research and Applications*, **30** (3), 299–314.

Marsh, T. J. & Hannaford, J. (eds.). (2008) *UK Hydrometric Register. Hydrological Data UK Series*. Centre for Ecology and Hydrology, Wallingford.

McMahon, T.A. & Mein, R.G. (1978) *Reservoir Capacity and Yield. Developments in Water Science No. 9*. Elsevier, Amsterdam, The Netherlands.

Meigh, J.R. (1995) The impact of small farm reservoirs on urban water supplies in Botswana. *Natural Resources Forum*, **19** (1), 71–83.

Meigh, J.R., McKenzie, A.A. & Sene, K.J. (1999) A grid-based approach to water scarcity. Estimates for Eastern and Southern Africa. *Water Resources Management*, **13**, 85–115.

Meigh, J.R., Tate, E.L. & McCartney, M.P. (2002) Methods for identifying and monitoring river flow drought in Southern Africa. In: van Lanen, H.A.L. & Demuth, S. (eds), *Regional Hydrology: Bridging the Gap between*

Research and Practice. IAHS Publication No. 274, pp. 181–188.

Midgley, D.C. (1952) *A preliminary study of the surface water resources of the Republic of South Africa.* PhD Thesis, University of Natal, South Africa.

Moore, R.J. (1985) The probability-distributed principle and runoff production at point and basin scales. *Hydrological Sciences Journal,* **30**, 273–297.

Nathan, R.J. & McMahon, T.A. (1990) Evaluation of automated techniques for base flow and recession analysis. *Water Resources Research,* **26**, 1465–1473.

National Rivers Authority (1993) *Facts on the top 40 low flow rivers in England and Wales.* National Rivers Authority, Bristol.

Parks, Y.P. & Gustard, A. (1982) A reservoir storage yield analysis for arid and semi-arid climates. In: Lowing, M.J. (ed),Proc. Exeter Symposium, July 1892. IHAS Publ. 135 *Optimal Allocation of Water Resources.* IAHS Press, Wallingford, UK, pp. 49–57.

Parks, Y.P. & Sutcliffe, J.V. (1987) *The development of hydrological yield assessment in NE Botswana,* in: Proc Brit Hydr Soc National Hydrology Symposium, Hull, 12.1–12.11.

Pearce, F. (1999) *Flooded out: retreating glaciers spell disaster for valley communities. New Scientist* (2189), 5 June 1999, 18.

Piggott, A.R., Moin, S. & Southam, C. (2005) A revised approach to the UK IH method for the calculation of baseflow. *Hydrological Sciences Journal,* **50**, 911–920.

Piper, B.S., Plinston, D.T. & Sutcliffe, J.V. (1986) The water balance of Lake Victoria. *Hydrology Science Journal,* **31**, 25–47.

Pitman, W.V. (1973) *A mathematical model for generating river flows from meteorological data in South Africa.* Report No. 2/73, Hydrological Research Unit, University of Witwatersrand, Johannesburg, South Africa.

Prudhomme, C., Haxton, T., Crooks, S. *et al.* (2013) Future Flows Hydrology: an ensemble of daily river flow and monthly groundwater levels for use for climate change impact assessment across Great Britain. *Earth System Science Data,* **5**, 101–107.

Reed, D.W. (1999) *Flood Estimation Handbook.* Wallingford U.K, Institute of Hydrology.

Rees, H.G., Croker, K.M., Singhal, M.K. *et al.* (2002) Flow regime estimation for small-scale hydropower development in Himachal Pradesh. *Journal of Applied Hydrology,* **XV** (2/3), 77–90.

Rees, H.G. & Collins, D.N. (2006) Regional differences in response of flow in glacier-fed Himalayan rivers to climatic warming. *Hydrological Processes,* **20**, 2157–2169.

Rees, H.G., Holmes, M.G.R., Fry, M.J., Young, A.R., Pitson, D.G. & Kansakar, S.R. (2006) An integrated water resource management tool for the Himalayan region. *Environmental Modelling & Software,* **21**, 1001–12.

Sene, K.J. & Farquharson, F.A.K. (1998) Sampling errors for water resource design; the need for improved hydrometry in developing countries. *Journal of Water Resources Planning & Management,* **12**, 121–138.

Sutcliffe, J.V. & Green, C.S. (1986) Water balance investigations of recharge in Madhya Pradesh, India. *Hydrology Science Journal,* **31**, 383–394.

Sutcliffe, J.V. (2004) Hydrology: A question of balance. In: *IAHS Special Publication 7.* IAHS Press, Wallingford, UK.

Tallaksen, L.M. (1993) *Modelling land Use change effects on low flows.* In: Flow Regimes from International Experimental and Network data (FRIEND), Vol. 1 Hydrological Studies Gustard, A.(ed.) Institute of Hydrology, Wallingford UK, pp. 56–68.

Tallaksen, L.M. & van Lanen, H.A.J. (eds) (2004) Hydrological drought – processes and estimation methods for streamflow and groundwater. In: *Developments in Water Science.* Vol. **48**. Elsevier Science B.V., Amsterdam.

Tate, E.L. & Farquharson, F.A.K. (2000) Simulating reservoir management under the threat of sedimentation; The case of Tarbela Dam on the River Indus. *Water Resources Management,* **14**, 191–208.

United Nations Educational, Scientific and Cultural Organization (UNESCO) (1997) *Southern Africa FRIEND.* Technical Documents in Hydrology No. 15, UNESCO, Paris.

United Nations Educational, Scientific and Cultural Organization (UNESCO) (2004) *Hindu Kush – Himalayan FRIEND 2000–2003 Technical Documents in Hydrology No. 68,* UNESCO, Paris.

United Nations Educational, Scientific and Cultural Organisation (UNESCO) (2004) *Southern Africa FRIEND Phase II 2000–2003.* Technical Documents in Hydrology No. 69, UNESCO, Paris.

van Lanen, H.A.J., Demuth, S., Daniell, T. *et al.* (2014) Over 25 years of FRIEND-Water: an overview. In: *Hydrology in a Changing World: Environmental and Human Dimensions Proceedings of FRIEND-Water 2014.* Vol. **363**. IAHS Publications, pp. 1–7.

Water Research Commission (1994) *Surface Water Resources of South Africa 1990.* Vol. **13**. WRC, Pretoria.

Wilkinson, B.W., Law, F.M., Gustard, A. & Armitage, P. (1996) *Environment Committee. First Report: Water Conservation and Supply. House of Commons Session 1996–1997.* The Stationary Office, London, pp. 178–204.

Williams, R.J., Keller, V.D.J., Johnson, A.C. *et al.* (2009) A national risk assessment for Intersex in Fish Arising from Steroid Estrogens. *Environmental Toxicology and Chemistry,* **28** (1), 220–230.

WMO (2008) *Manual on Low Flow Estimation and Prediction.* Operational Hydrology Report NO 50, WMO Geneva.

Young, A.R., Gustard, A., Bullock, A., Sekulin, A.E. & Croker, K.M. (2000) A river network based hydrological model for predicting natural and influenced flow statistics at ungauged sites. *Science of the Total Environment*, **251/252**, 293–304.

Young, A.R., Grew, R. & Holmes, M.G.R. (2003) LowFlows 2000: a national water resources assessment and decision support tool. *Water Science and Technology*, **48**, 119–126.

Young, A.R., Keller, V. & Griffiths, J. (2006) Predicting low flows in ungauged basins; a hydrological response unit approach to continuous simulation. In: *Climate Variability and Change – Hydrological impacts*. Vol. **308**. IAHS Publications, pp. 134–138.

Young, A.R. (2006) Stream flow simulation within UK ungauged catchments using a daily rainfall-runoff model. *Journal of Hydrology*, **320**, 155–172.

Young, A.R., Grew, R, Keller, V., Stannett, J. & Allen, S. (2008) *Estimation of river flow time-series to support water resources management: the CERF model*. In: Sustainable Hydrology for the 21st Century, Proc. 10th BHS National Hydrology Symposium, Exeter, 100–106.

7 Hydrological Modelling

KEITH BEVEN[1], JAMES BATHURST[2], ENDA O'CONNELL[2],
IAN LITTLEWOOD[3], JIM BLACKIE[4], AND
MARK ROBINSON[5]

[1]Lancaster Environment Centre, University of Lancaster, Lancaster, UK
[2]Department Civil Engineering and Geoscience, University of Newcastle upon Tyne, Newcastle upon Tyne, UK
[3]British Hydrological Society, UK
[4]Ex-Institute of Hydrology, Wallingford, UK
[5]Centre for Ecology and Hydrology, Wallingford, Oxfordshire, UK

Never confuse a model with reality

7.1   Introduction

Hydrology is the science concerned with the movement and storage of water, and there have been numerous attempts over many hundreds of years to quantify the available water resource and make predictions of its future behaviour.

The water balance measurements on the Seine by Pierre Perrault in the seventeenth century were the first to prove beyond doubt that rainfall was more than sufficient to account for streamflow. Early popular ideas included the fanciful notion that river flows originated from condensation of waters from air in caverns within mountains. A true appreciation of the link between precipitation and streamflow is the essential basis for a proper understanding of the hydrological cycle, starting with the basic water budget:

Change in soil water storage =

Precipitation – evaporation losses–runoff.

The challenge for hydrology is to represent these terms in ways appropriate for the particular problem at hand. Usually, we observe precipitation to within some accuracy depending upon temporal and spatial scale. Changes in soil water storage may be provided by infiltration equations with various levels of complexity and can be extended to both the unsaturated and saturated zones via Richard's equation for example (Chapter 4). Evaporation generally needs to be estimated although there have been enormous technological advances in its measurement (Chapter 5).

Progress in Modern Hydrology: Past, Present and Future, First Edition. Edited by John C. Rodda and Mark Robinson.
© 2015 John Wiley & Sons, Ltd. Published 2015 by John Wiley & Sons, Ltd.

The earliest hydrological 'models' were simple mass balance accounting schemes that used measurements of rainfall inputs and streamflow outflows to evaluate an overall resource and attempted to estimate evaporation losses indirectly (usually from air temperature) and perhaps changes in catchment storages. In the late nineteenth century, the rate of evaporation was often based on an evaporation pan (see Chapter 2) setting an upper limit to evaporation, any excess being assumed to be recharge to soil or groundwater reserves. In the middle of the twentieth century, the pioneering work of Howard Penman (1948) devised a method for estimating 'potential' evaporation, which when linked to soil water availability provided a much more realistic estimate of actual evaporation. More recent and current water resource studies are discussed in Chapter 6.

This chapter deals primarily with hydrological flood modelling. Its history began in the mid-nineteenth century with the 'Rational Method' for peak flows attributed to Thomas Mulvany (1850). Subsequently, the work of Robert Horton (1933) in estimating surface runoff amounts (essentially his infiltration equation), when combined with the unit hydrograph (UH) methodology of Leroy Sherman (Sherman, 1932) providing information on the shape of the flood hydrograph from a relatively short, intense rain, offered an initial flood prediction methodology framework. Many different models have been developed, varying greatly in scope and level of detail, and are used for different purposes including science-driven testing of ideas to problem-oriented applied studies. But all represent greatly simplified analogies or visions of the real world.

7.1.1 Hydrological modelling at Wallingford

From the earliest days of the Hydrological Research Unit (HRU) at Wallingford, it was recognised as essential to have a sound mathematical basis for the work. Quantities had to be measured (Chapter 2) and by developing an understanding of the behaviour and interactions of water stores and fluxes (Chapters 4 and 5), it would be possible to develop mathematical models to better understand and predict water resources and extreme events (Chapters 3 and 6).

At the simplest level, this implicitly required an assessment of the precision of the measurements – for example the number of gauges required

to estimate the areal rainfall or the number of rivers flow measurements needed to estimate total runoff over a period of time. The first catchment water balance study begun in the late 1960s was for the Ray at Grendon Underwood, a small watertight clay catchment in southern England. Instrumentation included a network of raingauges, a weather station (to estimate potential evaporation), soil moisture measurements using a neutron probe and a flow gauge (see Chapter 2). Annual catchment storage was estimated to be within 5% of the rainfall input although seasonal discrepancies were somewhat higher, which was attributed mainly to the lack of a soil heat flux component in the evaporation formula. To refine the water balance, estimates of heat flux were made from soil temperature measurements and a more intensive soil water network was established. Thus, the use of a simple model was able to inform the data collection program of additional useful measurements.

The IH (Institute of Hydrology) Conceptual Model (see below) grew out of similar methods of water balance analysis that Jim Blackie and Charles Eeles had developed in the late 1960s to analyse and interpret the results from the East African catchments (Chapter 2). One of the first uses of such models was to 'extend' records of runoff, and this use continues up to the present day. Because it is easier to measure rainfall than runoff, the available rainfall records tend to be longer and more numerous than runoff records, a lumped conceptual model could be used to estimate river flows in the earlier days when rainfall was measured, but river flows were not. The reconstructed runoff records can then be used to extend information available for estimating how often extreme flows, floods and droughts will occur (assuming catchment 'stationarity', that is that urbanisation, land use and climate change effects can be ignored).

It was no coincidence that the development of rainfall–runoff modelling, both at IH and more widely, coincided with the growing availability of powerful computation facilities. Until the 1960s, studies of how rainfall gave rise to flow in rivers were generally characterised by: (i) the selection of isolated storms for which the causative rainfall could be clearly related to the flood peak that followed it; (ii) the separation of the baseflow component from the flood hydrograph, usually by graphical and subjective means, the remainder being regarded as 'storm runoff'; and (iii) separation from the total event rainfall, by graphical and largely empirical means, of a proportion ('net rainfall'), equal in volume to that of the estimated storm runoff. Finally, (iv) storm runoff was related to effective rainfall by a linear differential equation describing the behaviour of one or more linear reservoirs, possibly in cascade or in parallel. A linear reservoir can be regarded as a tank with a hole in the bottom, for which output is proportional to storage in the tank. Equating the rate of change in storage to the difference between input $p(t)$ and output $q(t)$ gives an equation $k\ dq/dt + q(t) = p(t)$ where k is a constant of proportionality between the depth of water stored in the tank, and the output from it.

During the early 1970s, Eamonn Nash, the first head of the HRU, used to return to Wallingford as a consultant during part of the summer, and his guidance and experience greatly assisted development of thinking on conceptual modelling at the IH. A series of papers was produced, which outlined the development of more complex models for flow forecasting, describing some of the principles that should influence their structure and pointing to the need for simple models with parameters that are independent in their effects (Nash and Sutcliffe, 1970).

Nash made an important distinction between the two concepts of *forecasting* and *prediction*: forecasting, in the Nash terminology, is concerned with estimating when a particular runoff event (such as a flood peak) will occur and how large it will be; prediction is concerned with how often a particular event will occur in the future (such as how often a flood peak discharge will exceed a particular magnitude). Nash would say that using conceptual models to extend flow records is a procedure to assist flood prediction (and is concerned with the frequency of occurrence of specified runoff events), whilst using them to estimate when an imminent flood peak will occur, and how large it might be, is flow forecasting. This chapter is concerned with non real-time flow forecasting while Chapter 3 deals predominantly with prediction.

From the mid-1970s, IH was in the forefront of hydrological modelling. As part of its programme on flood research (which also included the work that underlay the *Flood Studies Report (FSR)* (NERC, 1975) and *Flood Estimation Handbook (FEH)* (IH, 1999)), a research programme was put in place to incorporate more process understanding into flood predictions and frequency analysis (Chapter 3). The development of modelling required associated computer facilities capable of storing and processing the field data in a form suitable for testing and application.

The basis for the work at IH was a fruitful mix of both inquisitive science-driven basic research and problem-driven applied studies (including overseas and consulting work). The pattern was one in which the former could feed into the latter.

7.2   Different Types of Models

Rainfall–runoff models serve a number of different purposes ranging from infilling gaps in a streamflow record to predicting the likely effect of a change in land use on streamflow regime. The choice of modelling approach will depend on the specific purpose, the nature of the catchment and the resources available (including data, computing power and operator skill). There are many different processes to represent, different ways of doing so, and too many interactions between them for a simple analysis. This has unfortunately led to the proliferation of many different hydrological models, varying greatly in their complexity and data requirements, none of which is suitable for all situations and needs. For convenience, this chapter follows the system suggested by Beck (1991) who proposed a broad classification of models of increasing complexity into metric-, conceptual- and physics-based. The simplest are black-box, data-driven or metric models, which rely solely on observed relationships and have limited or no representation of physical processes. This could, for example be a simple correlation between rainfall and runoff or the derivation of a UH. Conceptual models are a compromise between this empirical approach and one in which all physical processes are represented by mathematical equations. Conceptual models in contrast are structured to explicitly represent the main component hydrological processes that are believed to be of importance, and typically comprise a system of interconnected stores to represent quantities of water in, for example the upper soil layers, or deeper groundwater. These are filled and depleted by hydrological processes such as evaporation or infiltration. Physics-based modelling is an attempt to represent a complex synthesis of all the physical processes of real hydrological behaviour. This approach requires representing multiple interrelated processes operating simultaneously and with great variability in space and time. These models classically employ partial differential equations (PDEs) to represent the water movement and storage in soils, aquifers and channel systems. They are arguably the most attractive to academic hydrologists as they appear to be the closest representations of physical reality. However, this can be illusionary due to the great spatial complexity of catchment systems and the great difficulty of obtaining sufficiently detailed spatial information, particularly of subsurface properties. Excellent reviews of the types and history of development of hydrological rainfall–runoff models can be found in papers including Clarke (1973), O'Connell (1991), Loague (2010), Beven (2012a) and O'Sullivan and Bruen (2014).

In addition, rainfall–runoff models may be classified as '*Lumped*' if they take no account of spatial variability in the catchment processes, or '*Distributed*' if the processes exhibit spatial heterogeneity. Lumped models may be used for a catchment that is believed to be relatively homogeneous, and this approach has the attraction that if the model structure and parameters can be linked to broad physical characteristics of the basin such as soil type, land use and geology, then parameter values may be linked to catchment characteristics. Distributed hydrological models could most simply be the amalgamation of lumped contributions from distinct sub-basins, but are best represented in physically based models.

Some of the main groups of model approaches are summarised below and then examples are given of work undertaken or begun at Wallingford.

7.2.1 Metric models

These models are based primarily on observations (precipitation and runoff) and seek to characterise the resulting rainfall–runoff relationships from those data. For the estimation of design floods, these modelling techniques have been used by engineers and hydrologists for more than a century. During this time, the methodologies have evolved, as a consequence of increased computing power, improvements in available analytical techniques and steadily increasing number and length of records.

The origin of the data-driven metric approach to modelling in hydrology was the development of unit, the 'unit graph' or UH theory by L.K. Sherman (1932), providing the stormflow response to a unit input of excess rainfall. The UH analysis assumes a linear relation between the net

or effective portion of the rainfall that becomes stormflow (i.e. rainfall minus evaporation and contributions to soil and groundwater stores) and the storm hydrograph response (i.e. excluding baseflow). This approach of a UH and losses model forms the core of the UK FSR (NERC, 1975) and requires little/no preconceptions of the underlying hydrological processes and minimal data input, namely, records of rainfall and streamflow. Further details of the FSR are provided in Chapter 3, and only a brief summary is given here. The rainfall loss model adopted in the FSR assumes a declining percentage loss related to soil moisture through a storm to represent changes in catchment contributing area. The variation in the shape of the UH can then be related to differences in catchment characteristics as part of a regional design tool. This model, presented in its simplest form, has three main parameters: time to peak of the UH (T_p), storm percentage runoff (PR) and baseflow (BF). Through the analysis of a large number (1488) of observed flood events, the model parameters were estimated for 143 catchments in the UK. Using multivariate linear regression techniques, the model parameters were linked to mapped catchment characteristics thereby allowing the use of the rainfall–runoff model at ungauged sites throughout the UK.

Subsequently, the FEH was published in 1999 (IH, 1999). A major advance in flood risk estimation in the UK was the release of easily accessible digital catchment descriptors for all catchments draining an area of at least $0.5\,km^2$ (Bayliss, 1999). As part of the FEH research programme, the FSR rainfall–runoff method was updated (Houghton-Carr, 1999), although the model at the core of the design method remained unchanged. The FSR/FEH model adopted multiple linear regression for the development of predictor equations relating model parameters to catchment characteristics and catchment descriptors. The parameter equations were updated, based on the new digital catchment descriptors and, in addition, a new rainfall Depth-Duration-Frequency model was developed (Faulkner, 1999) and incorporated into the FEH rainfall–runoff method. Subsequently, the revitalised flood hydrograph (ReFH) model has been developed to improve the way that observed flood events are modelled and has a number of advantages over the FSR/FEH UH and losses model. Further details of the FIE and ReFH are given in Chapter 3.

Ever since Sherman (1932) introduced the 'unit graph' method, the UH, in several guises, has been described and applied in the literature. Hybrid metric-conceptual models have been developed where a conceptual structure is applied for the rainfall loss. One example of this producing a continuous streamflow simulation is the IHACRES model.

7.2.1.1 IHACRES

This UH-based rainfall–streamflow modelling method emerged over a period of collaboration between researchers at IH/CEH (Centre for Ecology and Hydrology) and the Australian National University (ANU). The method, known as IHACRES (Identification of unit Hydrographs And Component flows from Rainfall, Evaporation and Streamflow data) was published by Jakeman *et al.*, (1990). More technical details are given in

papers in the References outlining how the method came to be developed. Figure 7.1 (Littlewood *et al.*, 1997) illustrates schematically how inputs of daily rainfall and air temperature generate an 'effective rainfall' (the part of rainfall that becomes streamflow) output from the loss module, which then forms the input to the UH module comprising linear stores (two stores in parallel in this case). The UH for total streamflow convolved with the effective rainfall generates modelled streamflow, shown compared with the measured streamflow. A slow-flow UH convolved with effective rainfall generates a slow-flow hydrograph, shown compared with modelled streamflow. Similarly, a quick-flow hydrograph is generated (not shown). The transfer function representing the UH is fitted using the Simplified Recursive Instrumental Variable (SRIV) algorithm published by Young and Jakeman (1979).

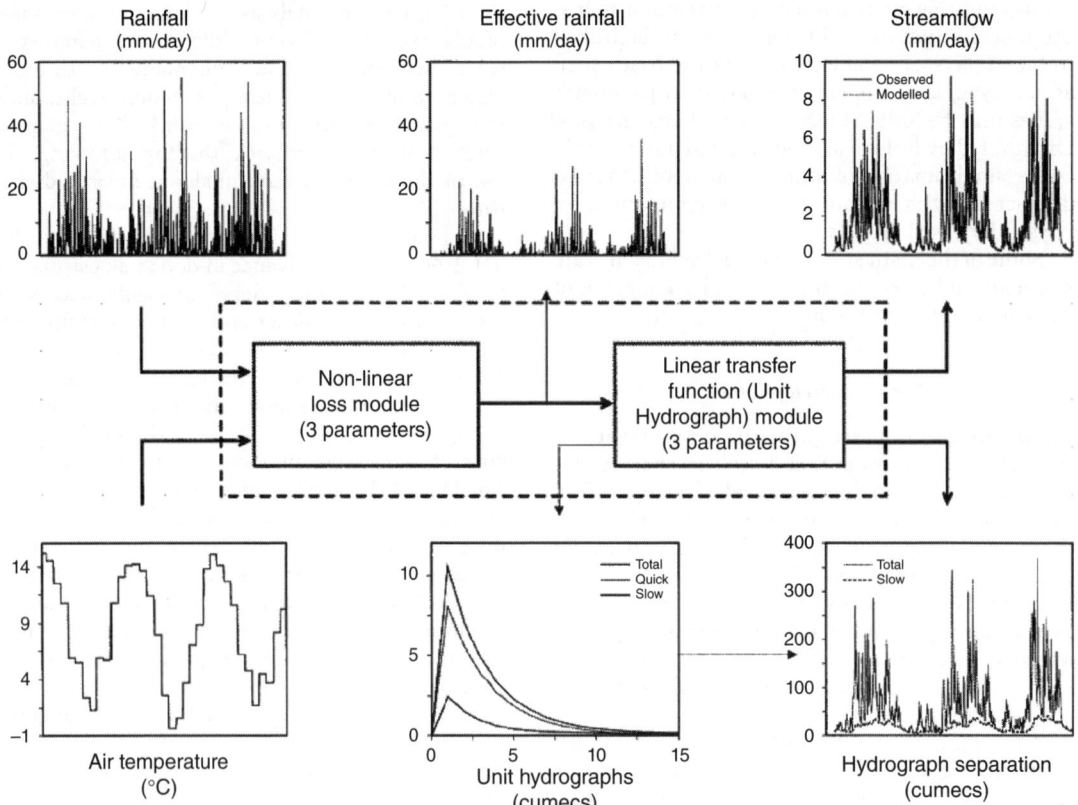

Fig. 7.1 IHACRES schematic: the basic model structure. (Source: Littlewood et al. 1997. Reproduced with permission of CEH.)

Littlewood and Jakeman (1991) argued that IHACRES had four strengths for applied hydrology: Firstly, the UH is for total streamflow, rather than a poorly defined 'direct' component of streamflow. Secondly, it can be calibrated over a single storm or multiple events. Thirdly, the separate quick-flow and slow-flow responses lead to hydrograph separation arguably superior to baseflow indices calculated solely from streamflow. Fourthly, with just five dynamic response characteristics (DRC), IHACRES was well suited for model parameter regionalisation to assist with estimating hydrographs at ungauged sites.

In the mid- to late-1980s, there was a need to characterise rainfall–streamflow and streamflow–stream-pH dynamics for headwater basins in the catchment above Llyn Brianne, Wales. The aim was to gain better understanding of the effects on stream water quality of different land covers, including moorland grass and commercial conifer forest (Bird *et al.*, 1990). This research was undertaken as part of the multidisciplinary Llyn Brianne Acid Waters Project (LBAWP), using the facilities of IH and the Geography Department at Swansea University. Subsequently, Tony Jakeman of ANU Canberra spent a sabbatical at Wallingford working on rainfall–streamflow modelling for the project. Jakeman *et al.* (1990) applied IHACRES to two small moorland pasture basins (denoted CI5 and CI6) of 0.34 and 0.72 km^2, respectively. For each catchment, hourly rainfall and streamflow data were used to calibrate the same IHACRES model structure comprising a loss module (based on air temperature) applied previously by Whitehead *et al.* (1979), followed by a UH module comprising two linear stores acting in parallel, to convert the effective rainfall to modelled streamflow. Other configurations of linear stores in series and/or in parallel were found to be inferior model structures.

The model was calibrated over a 400-hour period containing several streamflow peaks and accounted for 95% of the initial variance in CI6 streamflow and 85% for CI5 (Jakeman *et al.*, 1990). The model was defined by just five DRC: a catchment drying time constant; a factor to ensure that the volume of effective rainfall over a calibration period is equal to the volume of observed streamflow; dominant quick- and slow-flow response decay time constants and a slow-flow index (SFI) being the proportion of modelled streamflow over the calibration period that comprises slow flow). The two decay time constants and SFI completely define a UH for total streamflow, which, when convolved with the effective rainfall, gives separate hydrographs of modelled quick and slow flows that sum to modelled streamflow. For CI5 and CI6, the quick-flow decay time were 8.5 and 4.4 hours, respectively, the slow-flow decay times were 133 and 90 hours; and SFI proportions were 0.74 and 0.61. CI6 is twice the size of CI5, with a lower main stream slope (67 vs 113 m km^{-1}) and lower drainage density (1.16 vs 2.05 km^{-1}). Thus, the IHACRES results suggested that the larger basin, with lower main stream slope and drainage density, has the faster characteristic flow decay times and the smaller SFI, which appears counter-intuitive and suggesting other physical processes, may be controlling streamflow generation in the two basins.

Further work found that the same model structure also worked well for the 894 km^2 Teifi at Glan Teifi in Wales (Littlewood and Jakeman, 1991) and other catchments including Maimai M6 in New Zealand (0.016 km^2), the Kenwyn at Truro (19 km^2) and the Colne at Denham (743 km^2) (Jakeman *et al.*, 1991, 1992a). It appeared that there was potential for finding statistical relationships between IHACRES DRCs and catchment characteristics (Jakeman *et al.*, 1992b; Littlewood and Jakeman, 1993) leading to preliminary IHACRES model parameter regionalisation research using data from the CEH Plynlimon catchments (Sefton *et al.* 1995) and basins in Australia and the USA (Post and Jakeman, 1996; Post *et al.*, 1998).

Jakeman and Hornberger (1993) developed an improved loss module with an extra DRC in the loss module and applied the whole model to seven catchments: the Monachyle (7.7 km^2) and Kirkton (6.9 km^2) at Balquhidder in Scotland; the Orroral Valley (89.6 km^2) and Licking Hole Creek (20.6 km^2) catchments in Australia; Coweeta basins 34 (0.33 km^2) and 36 (0.49 km^2) in the USA; and the constructed Hydrohill catchment (0.00049 km^2) in China. Jakeman and Hornberger (1993) tried other configurations for the UH module but again concluded that two linear stores in parallel was optimal, just as had been found previously. This structure could characterise the rainfall–streamflow dynamic behaviour of many sizes and types of catchment (e.g. Littlewood and Jakeman, 1994).

Sefton and Howarth (1998) calibrated the 6-DRC, two UH stores in parallel, IHACRES model for 60 catchments using daily data and employing a model selection procedure based on Jakeman *et al.* (1990). They investigated statistical relationships between each DRC and suitable catchment descriptors, towards establishing a regionalised rainfall–streamflow model for England and Wales. Boorman and Sefton (1997) employed this work to assist with a regional study of the effects of assumed future climate change on hydrological responses. Later work by Littlewood (2003) employed the match between observed and modelled flow duration curves as an extra modelling objective for 5 of the 60 catchments. It appeared likely that the earlier model calibration was weighted to fitting high flows and had systematically overestimated low flows. This emphasises the need to select appropriate objective functions for a particular investigation.

The data time step employed to calibrate any discrete-time rainfall–streamflow model for regionalisation of its parameters is another factor to be considered. For the flashy $10.6\,km^2$ Wye at Cefn Bryn catchment at Plynlimon, the calibrated IHACRES quick-flow DRC can have good precision ($\pm 2\%$) but very poor accuracy (about 400%) (Littlewood, 2007). If it is assumed that the Cefn Brwyn hydrometric data are error-free, 42% of the absolute inaccuracy in the quick-flow DRC is accounted for by loss of information in the daily effective rainfall (relative to the greater information content of the hourly data), leaving the remaining 58% accounted for by loss of information in the streamflow data (Littlewood and Croke, 2008, 2013; Littlewood *et al.*, 2010). This draws attention to the importance of following a multi-objective approach to IHACRES model calibration that is suitable for the application in question, and to the fact that daily data may be too coarse for accurate calibration of the quick-flow parameters. It is also inherent in the IHACRES methodology that the UH rise time is set equal to one data time step. When a discrete-time IHACRES model was calibrated for the Wye at Cefn Brwyn using hourly data, it was reassuring but not surprising that it gave very nearly the same UH as that yielded by a corresponding continuous-time data-based mechanistic (DBM) model (Littlewood *et al.* 2010; Littlewood and Croke, 2013).

Other applications of IHACRES include the addition of simple snow accumulation–snowmelt modules for modelling of snow-affected catchments in Australia (Schreider *et al.*, 1997) and Scotland (Steel *et al.* 1999); linking soil moisture measurements to the losses module (Robinson and

Stam, 1993) and predicting streamflow from rainfall forecasts for two Brazilian catchments (Littlewood *et al.*, 1997). An alternative, catchment moisture deficit, loss module was developed (Croke and Jakeman, 2004; Croke and Littlewood, 2005) and provided as an option in IHACRES software from ANU (Croke *et al.*, 2006). Furthermore, IHACRES was employed as a focal point for contributions to the Prediction in Ungauged Basins (PUB) Decade (2003–2012) (e.g. Littlewood *et al.*, 2003, 2007a,b), an initiative of the International Association of Hydrological Sciences (IAHS).

7.2.2 Conceptual models

It had long been recognised, of course, that the separation of runoff and rainfall into their 'storm runoff' and 'effective rainfall' components, so as to cast the rainfall–runoff relationship into a form in which it could be analysed using linear theory, was unsatisfactory for a number of reasons. In reality, vegetation, soil and aquifers do not behave as simple linear reservoirs; the separation of rainfall and runoff into components is empirical and subjective and the need to work with clearly identifiable storm events means that much of the available record of a catchment's rainfall and runoff could not be used. However, when better digital computation facilities became available in the 1960s and 1970s, it became possible to estimate changes in the depths of water stored in vegetation cover, in soil and in aquifers with each time step (typically, 1 day), obviating the need to select clearly separated storms from the complete record. Computing the depths of water in the conceptual storages necessarily requires a conceptualised description ('model') of the storages themselves (vegetation, soil layers, aquifers, stream channel, etc.) and of how water is passed between them (and back into the atmosphere as evaporation). The description of the storages, and the flows between them, unavoidably includes non-measurable quantities ('parameters'), which have to be estimated from the available records of rainfall, runoff and potential evaporation. The problem of estimation was greatly facilitated by developments in computation, as described below.

7.2.2.1 Rainfall–runoff models

Given the complexity of the hydrological cycle, a commonly adopted approach is to simulate the hydrological cycle within a catchment by means of a conceptual model based on the argument that models which include descriptions of actual physical processes can be expected to give better short-term forecasts of river flows and water levels than purely statistical models. This involves the introduction of more physical representation of the likely hydrological behaviour by using a number of parameters which may or may not have physical meaning. These parameters of the model are adjusted (whether by eye or by computer minimisation of some specified measure of goodness of fit) to reduce the difference between predicted and observed measurements – usually, but not always streamflow at daily or hourly intervals, depending on catchment size.

For any given set of parameter values, the measured rainfall could be used as input to the model to give a series of outputs in the form of calculated river flows corresponding to that particular set of parameters. The differences between these calculated flows and the flows actually observed gave a measure of how well the model would fit, if this set of parameter values were to be adopted in practice; Nash and Sutcliffe (1970) derived an overall measure of goodness of fit, later termed 'the Nash-Sutcliffe Efficiency R^{2}', such that the closer that R^2 was to unity, the better the model fit. Development of such conceptual models came at a time when computer algorithms for locating the minimum of a given function had been developed by Rosenbrock (1960) and Nelder and Mead (1965). It was then only a short step to use such algorithms to identify that set of model parameters which would maximise the agreement between the river flows actually observed and those calculated using the model (typically by minimising the function $1-R^2$, where R^2 is the Nash-Sutcliffe efficiency). Use of function-minimisation algorithms was not, of course, essential, and some models, particularly those with many parameters, continued to be fitted by trial-and-error. And as discussed more fully later, there may be widely different sets of model parameters which give about the same degree of goodness of fit.

One of the earliest (lumped) conceptual models was the Stanford Watershed Model (SWM), which produced estimates of flow from rainfall inputs using infiltration, UH and recession functions together with soil water budgets evaporation estimates and flow routing techniques (Linsley and Crawford, 1960; Crawford and Linsley, 1966). A parallel development by Sugawara (1961) known

as the 'tank' model simulated the movement of water through the catchment using simple linear reservoirs in series and in parallel. Both these models relied on the adjustment or calibration of parameter values to fit modelled to observed data. Other early models included Nash and Sutcliffe (1970), Dawdy and O'Donnell (1965) and others. The aim of lumped catchment models is to identify the major processes contributing to the response that is of particular interest, usually the streamflow from the catchment, representing those processes and their interactions by simple mathematical functional relationships.

During the 1960s and the 1970s, a large number of lumped conceptual models were developed (see review by Clarke, 1973). Linsley (1982) regretted this development and said that the use of a relatively few models on many catchments would have saved much effort and probably led to the development of a few very good multipurpose models. The models described later are among the more successful and widely used that Wallingford scientists have been involved in developing.

An example of one of the simpler types of lumped conceptual deterministic continuous models is the IH Lumped Model (Nash and Sutcliffe, 1970; Mandeville *et al.*, 1970; Blackie, 1979). This model is based on generally accepted concepts of how precipitation moves through a catchment system and the constraints determining its emergence as evaporation, rapid response runoff or baseflow. The version of the model described here is designed to produce hourly estimates of streamflow from catchment rainfall and potential evaporation derived using the Penman (1948) equation. It consists of four stores notionally representing: interception by vegetation, soil water store, groundwater store and surface runoff reservoir. In this form, the model has 15 parameters, whose values have to be determined from field knowledge or by 'optimisation' of some specified measure of goodness of fit.

The model structure is represented in Figure 7.2. In this form, it would not be appropriate to be applied to catchments in which snowfall accumulation is significant, or those in which soil profiles contain horizons with distinctly differing soil moisture storages and hydraulic conductivities. Yet it can nevertheless provide a sound basis for assessing the most appropriate approach to modelling a particular catchment. Initial hydrograph fitting will, for example throw light on whether a catchment is stormflow or baseflow dominated,

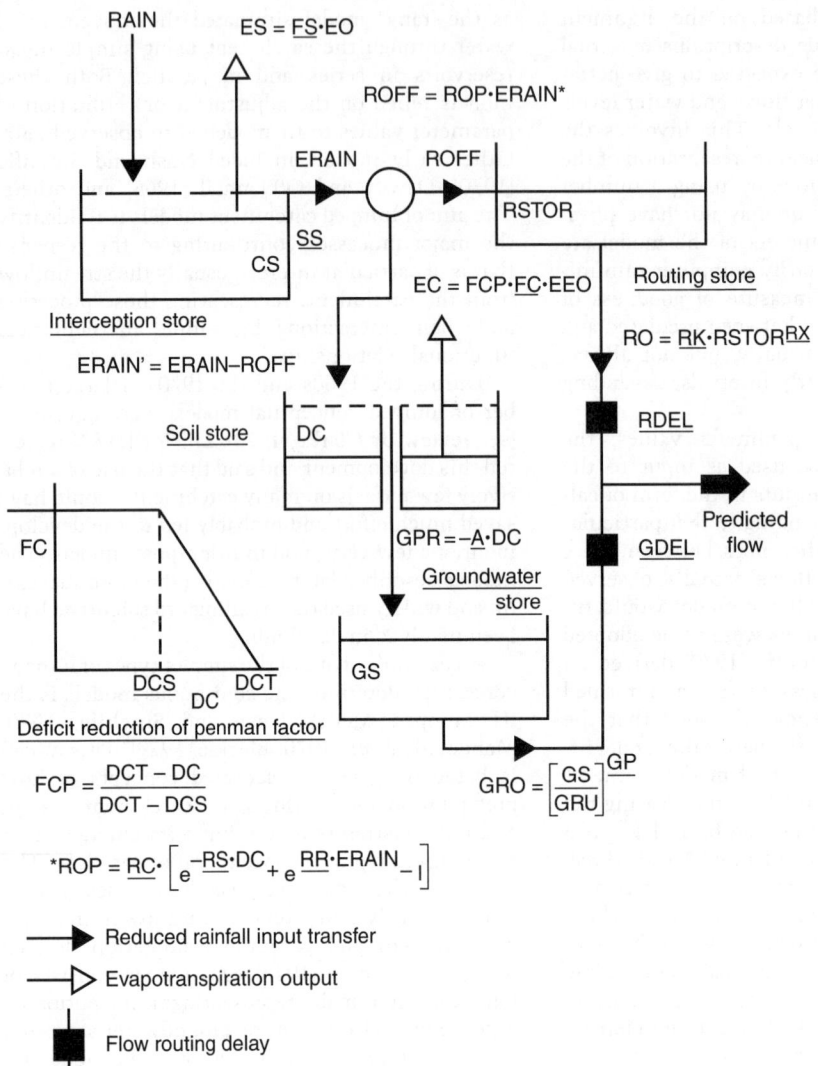

Fig. 7.2 The lumped IH Conceptual model structure. (Source: Blackie & Eeles 1985. Reproduced with permission of John Wiley & Sons.)

while the speed of response of stormflow will indicate the appropriate type of infiltration surface runoff functions to be included. Extra parameters can be included where additional representations of processes are necessary. The fitting process can also provide a valuable form of data quality control, highlighting periods or conditions when the storm recorded response deviates from its modelled 'normal' behaviour. This may represent a period when different runoff processes were operating (perhaps due to a hydrological threshold being exceeded) or, more prosaically, it may be indicative of instrument malfunctions, and provide an objective method for data infilling.

Using values of observed rainfall and estimated potential evaporation as input, the models compute runoff depending on certain parametric values; their values are optimised to minimise the 'difference' (commonly the sum of squared differences) between computed and observed runoff. Optimisation of the model parameters can best be performed by the use of 'automatic' search techniques such as the simplex method and Rosenbrock optimisation. Due to functional dependence between parameters, especially when there are many of them, it is best to limit the number of parameters. Versions of this model have been widely used on catchments varying from the impermeable and highly responsive Plynlimon catchments to the largely chalk and baseflow-dominated Cam catchment. It has also been applied to a data series of approximately 100 years for the River Thames at Teddington.

A series of papers describe the early development at Wallingford of conceptual models for flow forecasting led by Eamonn Nash. The first paper (Nash and Sutcliffe, 1970) outlined a description of basic principles and desirability of a simple model in order to produce stable value of the optimised parameters which are independent. It also proposed a measure of model performance, subsequently widely adopted as the *Nash-Sutcliffe efficiency* measure, to describe the correspondence between an observed and a simulated time series, and whilst there are many other measures of model fit (e.g. Legates and McCabe, 1999), it is now one of the most highly cited papers in hydrology. A second paper (O'Connell *et al.*, 1970) described the application of several simple rainfall–runoff models to an 1186 km^2 (458 ml^2) catchment using a modification of Rosenbrock's optimisation technique to fit the model parameters. This showed the desirability of having a relatively small number of independent parameters if the optimum values have to be well defined.

A third paper (Mandeville *et al.*, 1970) applied simple models to the River Ray catchment to sample drier conditions in which soil water status would be likely to have an important role on runoff. Models incorporating a reduction of potential evaporation at times of high moisture deficits gave a significant improvement. These papers with their convincing argument in favour of robust models, whereby increasing complexity should always be justified by the results, provided a strategy that influenced many UK and overseas hydrologists (e.g. Bergstrom, 2014).

Where catchments have major physical differences (such as vegetation cover), there will clearly be differences in the hydrological processes operating within them, which may then be manifest as differences between the optimum parameter

values. There was a hope that it would be possible to relate such different parameter values to simple measures of catchment properties in an analogous way to the multiple regression equations used in the FSR, linking flood statistics and UH parameters to catchment characteristics such as slope, soil type and drainage density. In practice, however, runoff processes have generally proved to be too complex to be represented in such a simple manner.

Douglas *et al.* (1976) describe a study whereby the streamflow efficiency was optimised for more than one catchment output, namely augmenting sub-daily rainfall data with monthly soil water measurements by neutron probe. Initial results were disappointing largely due to the sparseness of the soil moisture readings, and the authors suggested that the technique would be more appropriate where soil moisture can be measured more frequently. Twenty years later, taking advantage of advances in soil water measurement technology, Robinson and Stam (1993) used the IHACRES model with a losses module informed by continuous soil water measurements by capacitance probe for the 230 km^2 River Ock catchment. They showed that the extra information both improved the model calibration fit and enabled the calibration to be achieved on a shorter period of observation than was possible with rainfall data alone.

A generation of topographically oriented models has evolved, starting from TOPMODEL (Beven and Kirkby, 1979), to more recent models such as TOPKAPI and Grid-to-Grid (G2G) (see below) which achieve a parsimonious representation of the dominant component processes. There is an extensive literature on digital terrain models (DTMs) for surface flow routing.

A common feature of conceptual models is that they contain thresholds, such that the behaviour of a hydrological process above a threshold is different from the behaviour below it. A good example is the interception of rainfall by vegetation; typically, vegetation is conceptualised as a reservoir of depth h (a parameter to be estimated) which might initially be empty. If rain then occurs, the reservoir will fill up until the accumulated rainfall is of depth h; when the interception 'store' becomes full, the threshold h is exceeded, and any further rainfall is passed by the model to the soil surface, where it may infiltrate, or contribute to surface runoff (thus activating another threshold). The model behaviour is similar when rainfall ceases; with a depth h of water stored on the plant canopy, evaporation will continue until the interception store becomes empty, when evaporation from it will cease. A typical rainfall–runoff model will contain many such thresholds.

The presence of thresholds in a model destroys the 'smoothness' of the function that is to be minimised; whether or not derivatives are calculated, it is easier to locate the minima (and there may be several) of a function that varies smoothly throughout the parameter space. This lack of smoothness is a particular problem where a computer algorithm for function minimisation uses derivatives, but is less of a problem where derivatives are not calculated (as in the Nelder-Mead algorithm). At the end of the 1970s, IH modellers also began to question the concept of a single threshold (e.g. for interception storage); the depth h of interception storage varies spatially over a watershed and could be regarded as having a frequency distribution in which the threshold varied with vegetation type from values close to zero and ranging up to a much larger value. If all interception storages are initially empty when rainfall commences, those with very small h will overflow almost immediately, whilst in those storages with large h will overflow much later, if at all. Then, when rainfall ceases, shallow storages will become dry almost immediately, whilst deeper storages will continue to contribute to evaporation.

Thus, replacement of a single store with a single threshold by a population of storages, each with its own threshold, replaces an 'all or nothing' response by a more graduated, smoother response. In the simplest case, the single threshold h in an interception storage is replaced by an exponential probability distribution of thresholds, determined by a single parameter. A threshold that determines the behaviour of soil moisture storage, into which the overflows from interception storages are passed, can also be replaced by a population of thresholds. The random variables for interception storage and for soil moisture storage may or may not be independent, but if they are, their joint probability distribution will factorise to give a product of the two marginal distributions for interception storage thresholds and soil moisture storage thresholds.

At IH, the 'probability-distributed model' (PDM) was originally proposed by Robin Clarke (Moore and Clarke, 1981). After Clarke left IH in 1983, further development of the PDM was continued by Moore who has published extensively on it since (e.g. Moore, 1985, 2007).

The G2G hydrological model that developed from the PDM is a distributed grid-based rainfall–runoff and routing model, for translating rainfall into river flows to predict river flooding. Intensive work on the G2G project began in 2002 at CEH and it is used operationally for countrywide forecasting of river flooding across England, Wales and Scotland. It runs at a 15-minute time step to align with available national hydrometric data and is currently configured at a 1 km resolution. Gridded outputs include estimates of river flow, surface runoff (average water depth over a grid square) and soil moisture deficit. A major driver for developing the area-wide G2G approach was to address the need for forecasting 'everywhere'. The runoff production and routing elements of the model use supporting spatial data sets linked to conceptual formulations of the relevant hydrological processes to represent the spatial and temporal changes in runoff and water flows (Moore *et al.*, 2006). The runoff production is determined by the storm pattern, spatial data sets on landscape properties (land cover, topography, soil and geology) along with the dynamically changing soil moisture as calculated through continuous water accounting. The G2G model can also generate scenarios of how peak river flows may change across the country with climate change (Bell *et al.*, 2007).

7.2.3 *Physics-based models*

Where some extreme or rare event is involved, or some environmental change is envisaged, for example if extensive deforestation occurred, how would a basin's flow regime change? A model based on the physics of the processes may be expected to provide a better chance of providing an adequate forecast. In theory, the application of physics-based approaches to modelling would enable parameters to be derived from field measurements. Such models have also been extended to include processes of sediment release and transport and the transport of chemical compounds.

Most physically based distributed models (PBDMs) are founded on the PDEs, most commonly the Richards equation for subsurface flows and a kinematic wave equation for surface flows. Models of this type include the Institute of Hydrology Distributed Model (IHDM) and subsequently the Système Hydrologique Européen (SHE) models.

From the mid-1970s, IH was in the forefront of distributed hydrological modelling. The first version of the IHDM was developed based on solving equations for overland flow and infiltration and routing through the channel network. The solutions were carried out on a discretisation of the catchment into one-dimensional hillslope planes and channel reaches and, as such, had much in common with the structure of the US Department of Agriculture kinematic wave model KINEROS developed by Roger Smith and Dave Woolhiser; the latter spent a year as a visitor in Wallingford in the late 1970s (Morris and Woolhiser, 1980).

IHDM went through several versions before being completely rewritten to solve both the surface and partially saturated subsurface flow equations on a hillslope plane underlain by an impermeable bedrock. This IHDM4 model used a semi-3D finite element solution of the Darcy-Richards equation in the subsurface, based on a 2D vertical slice of variable depth down the slope, assumed to be homogeneous across a slope that was allowed to vary in width along the slope (Beven *et al.*, 1987). Later, it was improved numerically by Winifred Wood and Ann Calver (Wood and Calver, 1990, 1992) with tests of the model by Calver (1988) and Calver and Cammeraat (1993). The first version of the IHDM was written in 1977 and one of its first applications was to compare forested and grassland catchments at Plynlimon (Morris, 1980).

More recently there has been close cooperation between modellers in CEH and at the Met Office. The Met Office Surface Exchanges Scheme (MOSES)-PDM is an operational system developed at the Met Office for the real-time diagnosis of soil state and surface hydrology. It is based on the MOSES modified to take account of unresolved soil and topographic heterogeneity when calculating surface runoff by incorporating a probability-distributed moisture (PDM) scheme developed by the CEH. High-resolution soil characteristics and land cover data together with analyses of precipitation amount and type, cloud cover and near-surface atmospheric variables are used to drive MOSES-PDM. Hourly values are calculated of snowmelt, runoff, net surface radiation, evaporation, potential evaporation, soil temperature, soil moisture and soil moisture deficit on a 5 km grid. Precipitation amount is derived from radar and surface observations and radiation is calculated from three-dimensional cloud analyses derived from satellite data and surface observations. A fuller description of the implementation of MOSES-PDM can be found in Smith *et al.* (2006).

7.2.3.1 Genesis and evolution of SHE

The genesis of SHE can be traced to the World Meteorological Organisation (WMO) Hydromet Survey of the catchments of Lakes Victoria, Kyoga and Albert in the Upper Nile, initiated in 1967 to gain a better understanding of the region's water balance. In 1975, WMO invited tenders for the development of a mathematical model of the Upper Nile basin in three components: a catchment model, a lake model and a river channel model. The Danish Hydraulic Institute (DHI) was one of the bidders, and Enda O'Connell at IH was invited by Mike Abbott to advise on DHI's hydrological catchment modelling. He took the view that a relatively simple systems modelling approach would be appropriate, given the limited data that would be available. However, the contract was awarded to the Snowy Mountains Corporation, which proposed the use of the Sacramento model, a variant of the SWM IV for predicting the impacts of land use changes on the Upper Nile catchments.

This outcome led to discussions about the need to develop a general physically based modelling system that would supersede the generation of quasi-physically based models that had evolved from the SWM, and that could provide a sounder physical basis for predicting the impacts of increasing human activities on the flow, water quality and sediment regimes of catchments worldwide. IH had already embarked on the development of the IHDM, while O'Connell had been influenced by his contact with Allan Freeze and his pioneering 3D variably saturated flow simulations while working at IBM Research in Yorktown Heights in the early 1970s. Freeze and Harlan (1969) had laid out a blueprint for a general physically based modelling system, while the computational hydraulic surface water modelling systems that had been developed at DHI under the leadership of Abbott led the field at the time, both scientifically and commercially. Abbott was of the view that a competitive future position for Europe in addressing the emerging environmental problems in hydrology depended on a major initiative that would mobilise the complementary strengths of a consortium of European partners with the relevant expertise and build a modelling system that could exploit associated technological developments in computing power, remote sensing/data capture, and so on, in the coming decades. The French consulting organisation SOGREAH, also a leader in computational hydraulic modelling, was invited to join the consortium, drawing in

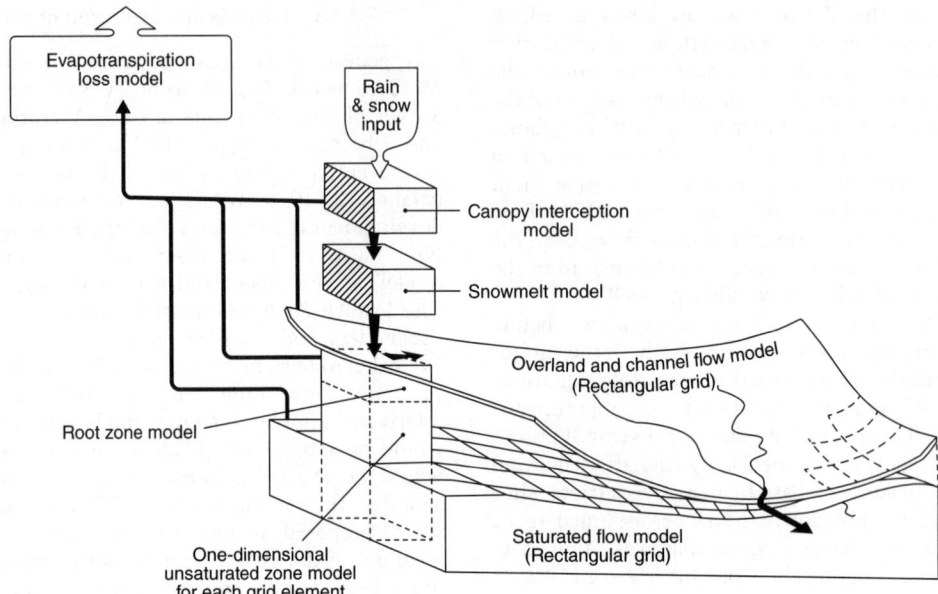

Fig. 7.3 The original schematic representation of the SHE. (Source: Abbott et al. 1986a. Reproduced with permission of Elsevier.)

leading figures such as André Preissmann and Jean Cunge, and the title SHE was conceived to give an identity to the ambitious modelling system that was to be developed.

Funding was procured to develop SHE through a loan from one of the European Commission's IT programmes, there being no other funding source available that could finance such an ambitious undertaking. This implied that SHE would have to be exploited commercially to generate the necessary revenue streams to repay the loan. An association was formed (ASHE, Association pour le SHE) to create the necessary legal entity to manage the development of the SHE and to satisfy the requirements of the EC. A series of meetings of the ASHE partners was held in 1976 during which the structure and numerical architecture of the SHE was conceived. A major challenge at the time was the need to reconcile the structure, and numerical solution schemes chosen with the available computing power. A fully 3D approach was not feasible, so it was decided to link 2D representations of surface flow and of saturated groundwater flow with a 1D representation of flow in the unsaturated zone, with the solutions of the governing physically based equations obtained on a rectangular finite

difference grid (Fig. 7.3). In following the Freeze and Harlan blueprint in integrating both surface and subsurface flow components, SHE went beyond the previous generation of distributed models that generally included only surface runoff and infiltration (Woolhiser, 1996, Beven, 2012b). It also chose a different way of discretising the space from the hillslope segment oriented distributed models of Stephenson and Freeze (1974) and Beven (1977).

7.2.3.2 How the SHE was built

The SHE was constructed in components corresponding to the principal processes within the land phase of the hydrological cycle. These were allocated to the partner organisations for development and coding according to their respective strengths and expertise: snowmelt, interception and evapotranspiration at the IH; overland and channel flow at SOGREAH; and unsaturated and saturated zone flow at DHI. In addition, DHI constructed a central control element, the FRAME, for the parallel running of the process components, which could have different time steps, and for the exchange of information between them.

The individual components were mostly coded during 1977 and 1978, and their respective numerical solution schemes were integrated through the FRAME, to provide a first complete test system in 1978. This version was known by its engineering term as a 'test mule', a name which equally well described the system's stubborn nature and propensity for delivering unexpected kicks to its developers, primarily through problems with numerical coupling! In the next stage of the development, each partner continued to refine its own specific components within its own version of the Mule. At intervals, joint working meetings were held, rotating between the partners but with more extensive meetings at DHI, at which the Mule was upgraded to the next version. These allowed the information exchange between components to be altered as required and enabled the partners to collaborate in joint test runs. Emphasis was placed on ensuring that an overall mass balance was maintained at each time step and this proved especially challenging at the interface between the two-dimensional saturated zone and one-dimensional unsaturated zone components, where unsaturated zone moisture profiles had to adjust to changes in the phreatic surface level. The problem was not fully resolved until the introduction of a fully three-dimensional subsurface model in the 1990s (Ewen *et al.*, 2000).

In those pre-Internet and pre-fax days, communication between the partners was maintained by letter, telephone and telex. Versions of the programs and data sets were carried to meetings manually on large computer printouts or on magnetic tape contained in protective plastic disks around 25 cm in diameter. Computing was carried out on large mainframes, which occupied considerable space in dedicated air-conditioned rooms. For a period in the early 1980s, though, the team used a CRAY-1 supercomputer that was notable for incorporating a bench on which users could rest from their labours! With the computational capabilities at that time, simulations of no more than a week or two could be completed within a reasonable computation time, as numerical stability required short computational time steps (around 5 minutes), limiting the early test applications to the scale of storm events. There were no GIS in those days and graphical outputs were initially rudimentary in nature, but improved rapidly.

The first production version of SHE emerged from this development phase in 1982, although further extensive testing and development continued to increase reliability, running speed and general efficiency. Several test catchments had been proposed to support the system development but most progress had been made using data from the Wye catchment at Plynlimon (10.55 km^2). Demonstration applications were therefore carried out for this catchment, culminating in the publication of a quartet of papers in the Journal of Hydrology in 1986, describing both the SHE and the applications and effectively launching SHE upon the scientific community (Abbott *et al.*, 1986a,b; Bathurst, 1986a,b).

7.2.3.3 Subsequent evolution

From about 1986, following the initial phase of SHE development and its first applications, a certain pattern asserted itself whereby each of the partners began embarking on its own programme of further development based on the then current version of the SHE. In 1984, Enda O'Connell moved to Newcastle University (NU), and it was decided that the SHE should be transferred to Newcastle to facilitate its further development through PhD projects. James Bathurst followed in 1985 and led the extension of SHE to include erosion and sediment transport. At DHI, SHE evolved into MIKE SHE, which now incorporates an extensive erosion, sediment transport, contaminant transport and irrigation modelling capability, and which can be linked with other MIKE modelling systems within ArcViewGIS. It has been used in a wide range of integrated surface and subsurface flow and transport modelling projects worldwide, such as the Everglades restoration project (e.g. Refsgaard *et al.*, 2010). At NU, SHE evolved into SHETRAN, as described below. By 1994, the repayment of the EC loan had been resolved and the partners decided that the ASHE should then be wound up.

7.2.3.4 Early SHE technology transfer and training

The first large-scale SHE application, and the final ASHE collaboration, was the transfer of the SHE technology to the National Institute of Hydrology (NIH) at Roorkee in India in the late 1980s. Important goals were the installation of the SHE software on the NIH computer, the training of six NIH staff and demonstration applications to six sub-basins of the Narmada River in Madhya Pradesh state. The programme was coordinated by DHI and all the work took place at DHI and NIH (Refsgaard *et al.*, 1992; Jain *et al.*, 1992).

User guides had already been produced during the development of SHE but the training programme stimulated their refinement, as well as the production of further support documentation in a number of volumes. NIH staff spent periods of months learning how to use SHE at DHI, taught by staff from DHI, Laboratoire d'Hydraulique de France (taking over from SOGREAH) and NU. The training staff also carried out extended visits to NIH and the field area. Advances in computing capacity meant that simulation lengths had increased from a few weeks to a few years, while the business of interpreting model output was eased by improving computer graphics. It was a particular revelation to be able to see, in shades of colour, the time variation of soil moisture content through the full depth of the soil column on a single plot, highlighting the wetting of the soil by the monsoon rains. Similarly, the size of catchment that was simulated was greatly expanded, up to $5000\,km^2$. However, the necessity of limiting the number of grid squares to maintain reasonable computational times meant large grid resolutions of 1, 2 or 4 km. While allowing a spatially distributed representation of the catchment, this raised the issue of scale and physical basis in parameter evaluation (Beven, 1989, 2001, 2002a), which remains controversial today. The challenge of parameterizing models of large-scale catchments was further emphasised by the eventual need to visit 15 different agencies to obtain the necessary data. Electronic data sets did not exist and all map, time series and other data had to be digitised manually and typed into the computer. The arrival of the fax machine facilitated international communications but the training staff visiting NIH still had to book telephone calls in advance for conferring with far away DHI.

A further such programme was carried out by NU and IH, funded by the UK Department for International Development, to transfer the SHETRAN modelling technology to the Chilean National Forestry Corporation (Corporación Nacional Forestal – CONAF) in the second half of the 1990s. This built on the Indian experience and involved extended training visits to the UK by the CONAF trainees and the instrumentation of several small catchments in Chile to provide experience in data collection to support model operation. By this stage, the Internet had been established and communication by email became feasible.

With hindsight, the technology transfer programmes were probably overambitious for the then status of the software. SHE/SHETRAN did not then have a user-friendly front end and users needed a detailed knowledge of the program code to resolve any simulation problems. The conceptual and technological foundations of physically based modelling were not easily assimilated by agency staff with limited previous modelling experience. Following the transfers, the SHE models that had been implemented received limited use in India and only research use in Chile (at universities associated with the transfer programme). Nevertheless, valuable experience was gained in model application and training, the potential of physically based modelling was established in the consciousness of the institutions involved and continued collaboration enabled further projects to be carried out (e.g. Bathurst *et al.*, 2010; Birkinshaw *et al.*, 2011).

7.2.3.5 SHETRAN development

From the 1980s to the present day, considerable strides have been made in model capability, front-end development, operational understanding and application. At NU, the development of SHETRAN (the 'TRAN' representing a focus on sediment and contaminant transport) took place through a major project in support of the UK nuclear industry's subsurface radioactive waste disposal research programme, a series of European Commission funded projects, the most notable being the flagship Mediterranean desertification project MEDALUS, and a steady stream of PhD studies (Ewen *et al.*, 2000). SHETRAN now incorporates components for sediment transport (including landslide erosion modelling), contaminant transport and the nitrogen cycle, built around a 3D variably saturated subsurface modelling component. It remains recognisably based on the original SHE but has a fully revised architecture and is much more accessible to the new user (http://research.ncl.ac.uk/shetran). The astonishing advances in computational capacity that have taken place since the early 1980s have extended simulation periods from a few weeks on a mainframe computer to a thousand years on a laptop, fully bearing out the expectations of SHE's original developers. Communications between distant users have advanced from airmail letters to almost instant email, web-enabled information exchange and mobile telephone/Skype contact almost anywhere in the world. SHETRAN applications have encompassed land use and climate change impact assessments in a range of countries and environments and have enabled such problems as the relationship between raingauge density and uncertainty in catchment-scale hydrograph simulations and the relationship between catchment scale and sediment yield to be investigated. (A list of publications is given at the above website.) At the physically based end of the modelling spectrum, it has informed the education of students and the training of professionals. Equally significant developments have taken place at DHI, such that, if there can be said to be an industry standard physically based model, it is MIKE SHE. Aspects of physically based modelling remain controversial and the subject of continuing research. However, initial concerns about data provision and parameterization have been greatly alleviated by the widespread availability of remotely sensed data sets, while the uncertainties posed by scale-dependent parameterization can at least be addressed through 'blind testing' and the specification of uncertainty bounds (Ewen and Parkin, 1996; Thorne *et al.*, 2000; Bathurst *et al.*, 2004).

Comparison with other physically based catchment models shows that the SHE approach remains valid and that SHETRAN and MIKE SHE remain leading examples (Bathurst, 2011). Other models have arisen but have not developed the same momentum or visibility, an indication perhaps of the importance of the stability of design, dedicated teams and the maintenance of corporate memory across the decades that have characterised the SHE adventure.

7.2.3.6 Barriers to take up

The introduction of SHE into the scientific hydrological community has had a number of impacts, some positive and some negative. In the early years, SHE was a complex numerical modelling beast that was operating at the limits of numerical stability, owing to the complex coupling of the component surface and subsurface processes. The SHE partners jointly had the expertise and resources to develop and run it, but it was not feasible to provide the training and support that would have enabled wider use in the scientific community. Moreover, there was a commitment at the outset to commercial exploitation to repay the EC loan and to develop a competitive position in

the field for the two commercial partners. This was doubtless viewed as creating a position of potential dominance that was narrowly concentrated in a few large organisations. Moreover, many conventional hydrological problems could be solved using simpler semi-distributed or lumped modelling approaches.

The main justifiable criticism of PBDMs has come from the problem of scale-dependent parameterizations of spatially heterogeneous processes, non-uniqueness in parameter values (equifinality; Freer and Beven, 2001) and the lack of the necessary distributed data sets that would constrain the uncertainty resulting from parameterizations based only on discharge records (e.g. Beven, 2006a,b, 2012b). Indeed, the characterisation of a catchment as a closed system based on a gauging station has been an impediment to the modelling of the rainfall–runoff process as a boundary-value problem. In other fields where the assumption of a closed system has not been possible (e.g. the hydrodynamic modelling of estuaries, numerical weather prediction, oceanographic modelling, groundwater modelling, etc.), there has been a much greater emphasis on the development of field programmes that are run specifically to support model parameterization. Many of the problems that SHE was developed to address still remain, for example the impacts of land use management changes on flooding, and they require field programmes that can constrain the uncertainty in the model predictions. However, this culture has not gained ground in hydrological practice.

7.2.3.7 Whither SHE-type modelling?

There are a number of distinctive strands to the development of a new generation of SHE-type models. Firstly, it has been recognised that PBDMs produce a vast amount of information that has not been properly interrogated, and that methods are needed to abstract this information. In the context of predicting the impacts of land use management on flooding, various methods of tracking information (e.g. water packets labelled with space-time information on entry to the river channel network) have been developed as well as reverse algorithmic differentiation approaches that can be used to derive sensitivity maps of the impact of changes at the grid square scale on the catchment flood hydrograph (Ewen *et al.*, 2013). Such methods have the potential to inform catchment managers on where, within catchments, interventions might most usefully be made to mitigate downstream flooding. Such maps are still subject to uncertainty, conditional on the uncertainty in model parameterizations.

A second strand can be found in the enormous computing power that can now be mobilised to solve the governing PBDM equations on ever finer grids and over large catchment domains. This opens up a new supercomputing frontier for physical-based modelling and employs multi-physics representations (different PDEs/hybrid equations) and multi-solvers with square-grid resolutions typically in the range 5–90 m, thus getting closer to physically representative model parameterizations.

A third strand is the emergence of Hyper-resolution (e.g. 0.1–1 km) Global Land Surface Modelling as a grand challenge for the land surface modelling community (Wood *et al.*, 2011). Six major challenges are foreseen in developing such a modelling system: improved representation of surface–subsurface interactions due to fine-scale topography and vegetation; improved representation of land–atmosphere interactions and resulting spatial information on soil moisture and evapotranspiration; inclusion of water quality as part of the biogeochemical cycle; representation of human impacts from water management; using massively parallel computer systems and recent computational advances in solving hyper-resolution models that will have up to 10^9 unknowns and developing the required in situ and remote sensing global data sets (Wood *et al.*, 2011). There is a degree of resonance here with the rationale for developing SHE (Abbott *et al.*, 1986a).

Some of the problems associated with process representation in this new generation of SHE-type models remain (e.g. Beven and Cloke, 2012) and will need to be tackled with renewed vigour and new thinking by the hydrological science community (Wood *et al.*, 2012), rather than be viewed as interminable obstacles to progress. But SHE can still be viewed as the forerunner of this new generation of distributed hydrological models. The introduction of SHE doubtless influenced the development of the distributed grid-based modelling in the following decades, albeit frequently using simpler representations of the component processes that did not require complex numerical solution schemes. If SHE were to have an epilogue, it would be that it was possibly developed before its time, but it will nonetheless have the distinction of being first!

7.3   Uncertainty in Modelling

In the late 1970s, we were optimistic that incorporating more physics into hydrological models was an important research programme, as set out in the seminal paper of Freeze and Harlan (1969). The reasoning was discussed in relation to the IHDM and SHE models in Beven and O'Connell (1982) and Abbott *et al.* (1986a). This was perhaps a little naïve, but it should be viewed now in the context of its time when the alternative was the optimisation of some conceptual model structure, in the full knowledge that the optimum parameters found would be dependant on the objective function used and the period of data used. Distributed, physics-based modelling was felt to be more objective in its physical foundations, and there was an expectation that, as the available computer power increased and the grid scale could be refined, and as more information became available about the parameters required by such models, then such models would eventually get closer to reality. In addition, the distributed-model predictions would provide useful information for other purposes such as the transport of sediments and pollutants and the triggering of landslides and debris flows. Many hydrologists still hold to this position, and such reasoning is still strong in the development of atmospheric and ocean circulation models.

There was, however, an alternative view that this might be an impossible dream. This was first expressed in the paper on 'Changing ideas in hydrology: the case of physically based models' (Beven, 1989). This paper began as a commentary on the 1986 SHE papers (Abbott *et al.*, 1986a,b; Bathurst, 1986a,b) but had grown to such a length that Jim McCulloch, then Director IH and an editor of *J. Hydrology* where those papers had been published, suggested that it be developed as a separate paper. This took some time, since it was not then sufficient only to provide a critique of physically based models (it was suggested that such models should be considered as lumped conceptual models at the grid scales at which they were being used), but it was also necessary also to suggest a way ahead. The suggestion was that, regardless of what model was used, we had to be much more realistic about the uncertainties generic to both hydrological data and models (see also Beven, 2000, 2001, 2002a, 2002b).

Keith Beven started to explore issues of uncertainty and different processes representations (and has continued to this day, see Beven and Germann, 2013; Beven and Binley, 2013; Davies *et al.*, 2011, 2013). Ways of incorporating macropores and preferential flows into distributed models had been explored (Beven and Germann, 1981; Beven and Clarke, 1986, see Chapter 4 in this volume). Ways of assessing uncertainty had been explored by using the simpler semi-distributed model TOPMODEL (Beven and Kirkby, 1979).

Increasing computer power meant it was possible to run many different model realisations and also generate many stochastic input realisations as a way of producing flood frequency statistics. While TOPMODEL is greatly simplified relative to a fully distributed model (and therefore runs much more quickly), it retains a certain physical basis in generating saturation and infiltration excess overland flows and subsurface stormflows. The continuous simulation of flood frequency was first published as a contribution to the flood research programme (Beven, 1986a,b, 1987) and was later developed at IH and CEH by Calver and Lamb (1996), Lamb (1999) and Lamb and Kay (2004). It was also taken further at Lancaster University by David Cameron, including incorporating the effects of different forms of rainfall generator and of climate change (Cameron *et al.*, 1999, 2000, 2001). Elsewhere the methodology has been used with TOPMODEL in the Czech Republic by Blazkova and Beven (2000, 2004, 2009).

TOPMODEL makes use of a topographic index in defining areas of hydrological similarity in a catchment (see Beven, 2012b). In the original application of the model to Crimple Beck near Leeds and later applications (Beven *et al.*, 1984), this analysis had been done by hand, drawing lines orthogonal to the map contours to define hillslope segments of variable width and then calculating areas and slope angles for each area between the contour lines. Later, the coordinates of the contour intersections were punched onto computer cards and the slope angles and areas calculated by computer. Later still, as digitisers became more widely available, the coordinates could be digitised and stored directly.

Software is now available to calculate numerous geomorphological variables from raster DTM data upstream of any chosen point on the river network at the click of a mouse button, including the topographic index. Some hydrologically relevant information has also been lost, however. In the early days, it was possible to incorporate local

field knowledge into the analysis, for example where road ditches or farm drains were effective in collecting downslope drainage and where drainage might be channelled into culverts. Most DTM analysis ignores such features, even when the road network, or other information, might be available as a GIS overlay.

Eric Wood of Princeton University was also a visitor to Wallingford in the late 1970s and a collaboration based around TOPMODEL ensued, resulting in some applications to US catchments (including Beven and Wood, 1983; Sivapalan *et al.*, 1990).

Many of the TOPMODEL runs were also used to explore how far different parameter sets could fit a set of calibration data. Random parameter sets were chosen from prior distributions (usually uniform distributions within some range, lacking better knowledge) and evaluated using different performance measures. It was often found that there were many different parameter sets that provided equally good fits to the calibration data, certainly using the Nash-Sutcliffe Efficiency measures (Nash and Sutcliffe, 1970) that has been very widely used then and since. This work had been started at the University of Virginia but was continued at Wallingford. It was the origin of the equifinality concept (see Beven, 2006a) and the generalised likelihood uncertainty estimation (GLUE) methodology (Beven and Binley, 1992, Beven, 2009) that was the result of trying to implement the suggestion by Beven (1989) referred to earlier. GLUE has since been very widely used (and widely criticised, see discussions in Beven, 2006c, 2008, Beven and Binley, 2013). Indeed, one of the very first applications of this approach to uncertainty estimation was in an application of IHDM4 to the Plynlimon catchments (Beven and Binley, 1992).

One of the uncertainty issues in the application of both distributed and semi-distributed models is how to define the *effective* parameter values for the model at the scale of discretisation. That issue was raised in Beven (1989) in respect of the classical Darcy-Richards-based models; more recently it has been suggested that we should be taking quite a different discrete, rather than continuum, approach to model structure development (Beven, 2006b). Even with a fully 3D fine resolution model (with stochastically generated random parameter fields to impose some realistic heterogeneity at the discretisation), it was not possible to define equivalent homogeneous effective parameter values of the Darcy-Richards equation unambiguously, especially when surface and subsurface flow processes were interacting (Binley *et al.*, 1989a,b).

There is clearly no simple solution to the problem of uncertainty in distributed and semi-distributed models in hydrology. This is in part because it is often the case that the uncertainties arise from lack of knowledge rather than random variability; lack of knowledge about model inputs, model structures (however physically based) and the observations with which we evaluate our models. Some of these issues were anticipated in the 'Changing ideas in hydrology' paper (Beven, 1989) and an understanding of the issues has continued to develop since (see Beven *et al.*, 2011; Beven, 2012a; Beven and Alcock, 2012; Beven and Smith, 2014) but there is still a need to learn more about how to handle knowledge or epistemic uncertainties that result in modelling errors that do not have simple statistical properties.

There is now also a very large literature on the use of 'ensembles' for showing uncertainties in forecasts of future flows, and for calculating rainfall fields from climate models which can be converted to forecasts of flow by rainfall–runoff models. These include Bayesian methods to combine forecasts from different models.

7.4   Discussion and Conclusions

The remarkable advances in hydrological modelling during the period since the formation of the original HRU at Wallingford in the 1960s cannot be seen in isolation. They have taken place hand-in-hand with tremendous technological advances. Complex numerical calculations that once could only be run over hours or days on a large mainframe or supercomputer can now be run in minutes or seconds on a PC or laptop. New types of instrumentation and data recording devices have enabled a far wider range of observations, to a much higher accuracy, and at a much greater frequency. Early attempts at distributed modelling, for example had to make use of digitised chart records from individual recording raingauges, each prone to timing errors. Now with the advent of weather radar, rainfall–runoff models can use continuous rainfall fields for flow forecasting in near real time. Intermittent manual measurements of soil moisture status by neutron probe or gravimetric sampling have been largely superseded with continuous records using capacitance or time domain reflectometry (TDR) probes,

or more recently naturally occurring cosmic rays (Chapter 2).

The advances in data capture and storage have been central for testing and developing our understanding of hydrological processes. The crucial role of interception in forest evaporation was only properly revealed through detailed micrometeorological measurements, and soil physics was liberated from studies of artificial 'soils' in the laboratory by the development of new classes of field instruments.

Ultimately, the performance of the PBDMs such as the IHDM and SHE will be constrained by the complexity of real hydrological systems, by the data available and by the approximations in both the process representation and the numerical solution. Their value is probably greatest where there is a need to predict a future change within the catchment, particularly relating to vegetation, or land management. The complexity of their model formulation led to early false expectations of their accuracy and of the availability of suitable data (Clarke, 2005). For simpler purposes, such as flow reconstruction or flood forecasting, simpler distributed models are easier to apply and still capable of giving performances that are just as good.

Recent serious flood events have encouraged the development of real-time flood forecasting systems using data from multi-parameter weather radars, with complex data processing algorithms and real-time updating of model parameters and the estimation of uncertainty in model forecasts.

Just as expectations of the accuracy of model predictions have become more realistic, there is a growing awareness that hydrological models need to be part of a wider interdisciplinary approach to flood management; this involves stakeholders in flood-affected areas with local knowledge, planners and asset managers responsible for engineered infrastructure such as flood defences and barriers, and software tools able to help visualise and analyse different options for flood risk assessment.

7.5   References

Abbott, M.B., Bathurst, J.C., Cunge, J.A., O'Connell, P.E. & Rasmussen, J. (1986a) An introduction to the European Hydrological System – Système Hydrologique Européen (SHE): 1. History and philosophy of a physically based, distributed modelling system. *Journal of Hydrology*, **87**, 45–59.

Abbott, M.B., Bathurst, J.C., Cunge, J.A., O'Connell, P.E. & Rasmussen, J. (1986b) An introduction to the European

Hydrological System – Système Hydrologique Européen, (SHE): 2. Structure of a physically-based, distributed modelling system. *Journal of Hydrology*, **87**, 61–77.

Bathurst, J.C. (1986a) Physically-based distributed modelling of an upland catchment using the Système Hydrologique Européen. *Journal of Hydrology*, **87**, 79–102.

Bathurst, J.C. (1986b) Sensitivity analysis of the Système Hydrologique Européen for an upland catchment. *Journal of Hydrology*, **87**, 103–123.

Bathurst, J.C. (2011) Predicting impacts of land use and climate change on erosion and sediment yield in river basins using SHETRAN. In: Morgan, R.P.C. & Nearing, M.A. (eds), *Handbook of Erosion Modelling*. Blackwell, pp. 263–288.

Bathurst, J.C., Amezaga, J., Cisneros, F. *et al.* (2010) Forests and floods in Latin America: science, management, policy and the EPIC FORCE project. *Water International*, **35**, 114–131.

Bathurst, J.C., Ewen, J., Parkin, G., O'Connell, P.E. & Cooper, J.D. (2004) Validation of catchment models for predicting land-use and climate change impacts. 3. Blind validation for internal and outlet responses. *Journal of Hydrology*, **287**, 74–94.

Bayliss, A. (1999) Catchment descriptors. In: *Flood Estimation Handbook*. Vol. **5**. Institute of Hydrology Wallingford.

Beck, M.B. (1991) Forecasting environmental change. *Journal of Forecasting*, **10**, 3–19.

Bell, V.A., Kay, A.L., Jones, R.G. & Moore, R.J. (2007) Use of a grid-based hydrological model and regional climate model outputs to assess changing flood risk. *International Journal of Climatology*, **27**, 1657–1671.

Bergstrom, S. (2014) From the early days of hydrological modelling to present-day operational hydrological tools – a Swedish perspective. In: O'Sullivan, J.J. & Bruen, M. (eds), *The Grand Challenges Facing Hydrology in the 21st-Century*. Dooge Nash International Symposium, Dublin Castle.

Beven, K.J. (1977) Hillslope hydrographs by the finite element method. *Earth Surface Processes and Landforms*, **2**, 13–28.

Beven, K.J. (1986a) Hillslope runoff processes and flood frequency characteristics. In: Abrahams, A.D. (ed), *Hillslope Processes*. Allen and Unwin, Boston, pp. 187–202.

Beven, K.J. (1986b) Runoff production and flood frequency in catchments of order *n*: an alternative approach. In: Gupta, V.K., Rodriguez-Iturbe, I. & Wood, E.F. (eds), *Scale Problems in Hydrology*. Reidel, Dordrecht, pp. 107–131.

Beven, K.J. (1987) Towards the use of catchment geomorphology in flood frequency predictions. *Earth Surface Processes and Landforms*, **12**, 69–82.

Beven, K.J. (1989) Changing ideas in hydrology: the case of physically-based models. *Journal of Hydrology*, **105**, 157–172.

Beven, K.J. (1997) TOPMODEL: a critique. *Hydrological Processes*, **11**, 1069–1085.

Beven, K.J. (2000) Uniqueness of place and process representations in hydrological modelling. *Hydrology and Earth System Sciences*, **4**, 203–213.

Beven, K.J. (2001) Dalton medal lecture: how far can we go in distributed hydrological modelling? *Hydrology and Earth System Sciences*, **5**, 1–12.

Beven, K.J. (2002a) Towards an alternative blueprint for a physically-based digitally simulated hydrologic response modelling system. *Hydrological Processes*, **16**, 189–206.

Beven, K.J. (2002b) Towards a coherent philosophy for environmental modelling. *Proceedings of the Royal Society London*, **A458**, 2465–2484.

Beven, K.J. (2006a) A manifesto for the equifinality thesis. *Journal of Hydrology*, **320**, 18–36.

Beven, K.J. (2006b) The holy grail of scientific hydrology: $Qt = H(S\ R)\ A$ as closure. *Hydrology and Earth System Sciences*, **10**, 609–618.

Beven, K.J. (2006c) On undermining the science? *Hydrological Processes*, **20**, 3141–3146.

Beven, K.J. (2008) On doing better hydrological science. *Hydrological Processes*, **22**, 3549–3553.

Beven, K.J. (2009) *Environmental Modelling: An Uncertain Future?* Routledge, London, UK.

Beven, K.J. (2012a) Causal models as multiple working hypotheses about environmental processes. *Comptes Rendus Geoscience, Académie de Sciences, Paris*, **344**, 77–88. doi:10.1016/j.crte.2012.01.005

Beven, K.J. (2012b) *Rainfall-Runoff Modelling – The Primer*, 2nd edn. Chichester, UK, Wiley-Blackwell.

Beven, K.J. & Alcock, R. (2012) Modelling everything everywhere: a new approach to decision making for water management under uncertainty. *Freshwater Biology*. **56**, 124–132 doi:10.1111/j.1365–2427.2011.02592.x

Beven, K.J. & Binley, A.M. (1992) The future of distributed models: model calibration and uncertainty prediction. *Hydrological Processes*, **6**, 279–298.

Beven, K.J. & Binley, A.M. (2013) GLUE, 20 years on. *Hydrological Processes*. **28(24)**, 5897–5918 doi:10.1002/hyp.10082

Beven, K.J. & Clarke, R.T. (1986) On the variation of infiltration into a homogeneous soil matrix containing a population of macropores. *Water Resources Research*, **22**, 383–388.

Beven, K.J., Calver, A. & Morris, E.M. (1987) *The Institute of Hydrology Distributed Model*. Institute of Hydrology, Wallingford, UK, IH Report No. 98.

Beven, K.J. & Cloke, H.L. (2012) Comment on 'Hyperresolution global land surface modeling: meeting a grand challenge for monitoring Earth's terrestrial water' by Eric Wood *et al.*, 2012. *Water Resources Research*, **48**, W01801. doi:10.1029/2011WR010982

Beven, K.J. & Germann, P.F. (1981) Water flow in soil macropores, II. A combined flow model. *Journal Soil Science*, **32**, 15–29.

Beven, K.J. & Germann, P.F. (1982) Macropores and water flow in soils. *Water Resources Research*, **18**, 1311–1325.

Beven K. & Germann P. (2013) *Macropores and water flow in soils revisited Water Resources Research*, **49**(6): 3071–3092 DOI:10.1002/wrcr.20156.

Beven, K.J. & Kirkby, M.J. (1979) A physically based, variable contributing area model of basin hydrology. *Hydrological Sciences Bulletin*, **24**, 43–69.

Beven, K.J., Kirkby, M.J., Schoffield, N. & Tagg, A. (1984) Testing a physically-based flood forecasting model (TOPMODEL) for three UK catchments. *Journal of Hydrology*, **69**, 119–143.

Beven, K.J. & O'Connell, P.E. (1982) *On the role of physically-based models in hydrology*. Institute of Hydrology, Wallingford, UK Report No. 81.

Beven, K.J. & Smith, P.J. (2014) Concepts of information content and likelihood in parameter calibration for hydrological simulation models. *ASCE Journal of Hydrologic Engineering*. doi:10.1061/(ASCE)HE.1943-5584.0000991

Beven, K., Warren, R. & Zaoui, J. (1980) SHE: towards a methodology for physically-based distributed forecasting in hydrology. In: *Hydrological Forecasting*. IAHS, pp. 133–137, Publication No. 129.

Beven, K.J. & Wood, E.F. (1983) Catchment geomorphology and the dynamics of runoff contributing area. *Journal of Hydrology*, **65**, 139–158.

Beven, K., Smith, P.J. & Wood, A. (2011) On the colour and spin of epistemic error (and what we might do about it). *Hydrology and Earth System Science*, **15**, 3123–3133.

Binley, A.M., Elgy, J. & Beven, K. (1989a) A physically based model of heterogeneous hillslopes. I. Runoff production. *Water Resources Research*, **25**, 1219–1226.

Binley, A.M., Beven, K.J. & Elgy, J. (1989b) A physically-based model of heterogeneous hillslopes. II. Effective hydraulic conductivities. *Water Resources Research*, **25**, 1227–1233.

Bird, S.C., Walsh, R.P.D. & Littlewood, I.G. (1990) Catchment characteristics and basin hydrology: their effects on stream acidity. In: Edwards, R.W., Gee, A.S. & Stoner, J.H. (eds), *Acid Waters in Wales*. Kluwer Academic Press, pp. 203–221.

Birkinshaw, S.J., Bathurst, J.C., Iroumé, A. & Palacios, H. (2011) The effect of forest cover on peak flow and sediment discharge – an integrated field and modelling study in central-southern Chile. *Hydrological Processes*, **25**, 1284–1297.

Blackie, J.R. (1979) The use of conceptual models in the catchment research. *East African Agricultural and Forestry Journal*, **43**, 36–42.

Blackie, J.R. & Eeles, C.W.O. (1985) Lumped catchment models. Chapter 11. In: Anderson, G. & Burt, T.P. (eds), *Hydrological Forecasting*. John Wiley and Sons Ltd, pp. 311–345.

Blazkova, S. & Beven, K.J. (2000) Flood frequency estimation by continuous simulation for an ungauged catchment with fuzzy possibility uncertainty estimation. *Water Resources Research*, **292**, 153–172.

Blazkova, S. & Beven, K.J. (2004) Flood frequency estimation by continuous simulation of subcatchment rainfalls and discharges with the aim of improving dam safety assessment in a large basin in the Czech Republic. *Journal of Hydrology*, **292**, 153–172.

Blazkova, S. & Beven, K.J. (2009) A limit of acceptability approach to model evaluation and uncertainty estimation in flood frequency estimation by continuous simulation: Skalka catchment, Czech Republic. *Water Resources Research*, **45**, W00B16. doi:10.1029/2007WR006726

Boorman, D.B. & Sefton, C.E.M. (1997) Recognising the uncertainty in the quantification of the effects of climate change on hydrological response. *Climate Change*, **35**, 415–434.

Calver, A. (1988) Calibration, sensitivity and validation of a physically-based rainfall–runoff model. *Journal of Hydrology*, **103**, 103–115.

Calver, A. & Cammeraat, L.H. (1993) Testing a physically-based runoff model against field observations on a Luxembourg hillslope. *Catena*, **20**, 273–288.

Calver, A. & Lamb, R. (1996) Flood frequency estimation using continuous rainfall–runoff modelling. *Physics and Chemistry of the Earth*, **20**, 479–483.

Cameron, D.S., Beven, K.J., Tawn, J., Blazkova, S. & Naden, P. (1999) Flood frequency estimation by continuous simulation for a gauged upland catchment with uncertainty. *Journal of Hydrology*, **219**, 169–187.

Cameron, D., Beven, K. & Naden, P. (2000) Flood frequency estimation under climate change with uncertainty. *Hydrology and Earth System Sciences*, **4**, 393–405.

Cameron, D., Beven, K.J. & Tawn, J. (2001) Modelling extreme rainfalls using a modified random pulse Bartlett-Lewis stochastic rainfall model with uncertainty. *Advances in Water Resources*, **24**, 203–211.

Clarke, R.T. (1973) A review of some mathematical models used in hydrology, with observations on their calibration and use. *Journal of Hydrology*, **19**, 1–20.

Clarke, R.T. (2005) The PUB decade: how should it evolve? Invited commentary. *Hydrological Processes*, **19**, 2865–9.

Crawford. N.H. & Linsley. R.K. 1966. *Digital simulation in hydrology: stanford watershed model IV*. Technical Report 39, Department of Civil Engineering, Stanford University.

Croke, B.F.W. & Jakeman, A.J. (2004) A catchment moisture deficit module for the IHACRES rainfall-runoff model. *Environmental Modelling and Software*, **19**, 1–5.

Croke, B.F.W. & Littlewood, I.G. (2005) Comparison of alternative loss modules in the IHACRES model: an application to 7 catchments in Wales. In: Zerger, A. & Argent, R.M. (eds), *MODSIM 2005 International Congress on Modelling and Simulation*. Modelling and Simulation Society of Australia and New Zealand, 2910, pp. 2904.

Croke, B.F.W., Andrews, F., Jakeman, A.J., Cuddy, S.M. & Luddy, A. (2006) IHACRES Classic Plus: a redesign of the IHACRES rainfall-runoff model. *Environmental Modelling and Software*, **21**, 426–427.

Daluz Vieira, J.H. (1983) Conditions governing the use of approximations for the St. Venant equations for shallow surface water flow. *Journal of Hydrology*, **60**, 43–58.

Davies, J., Beven, K.J., Nyberg, L. & Rodhe, A. (2011) A discrete particle representation of hillslope hydrology: hypothesis testing in reproducing a tracer experiment at Gårdsjön, Sweden. *Hydrological Processes*, **25**, 3602–3612. doi:10.1002/hyp.8085

Davies, J., Beven, K.J., Rodhe, A., Nyberg, L. & Bishop, K. (2013) Integrated modelling of flow and residence times at the catchment scale with multiple interacting pathways. *Water Resources Research*, **49**, 4738–4750. doi:10.1002/wrcr.20377

Dawdy, D.R. & O'Donnell, T. (1965) Mathematical models of catchment behaviour. *Journal of the Hydraulics Division: Proceedings of American Society of Civil Engineers*, **91** (**HY4**), 123–127.

Douglas, J.R., Clarke, R.T. & Newton, S.G. (1976) The use of likelihood functions to fit conceptual models with more than one dependent variable. *Journal of Hydrology*, **29**, 181–198.

Ewen, J. & Parkin, G. (1996) Validation of catchment models for predicting land-use and climate change impacts. 1. Method. *Journal of Hydrology*, **175**, 583–594.

Ewen, J., Parkin, G. & O'Connell, P.E. (2000) SHETRAN: distributed river basin flow and transport modelling system. *Proceedings of American Society of Civil Engineers: Journal of Hydrologic Engineering*, **5**, 250–258.

Ewen, J., O'Donnell, G., Bulygina, N., Ballard, C. & O'Connell, E. (2013) Towards understanding links between rural land management and the catchment flood hydrograph. *Quarterly Journal of the Royal Meteorological Society*, **139**, 350–357.

Faulkner, D. (1999) Rainfall frequency estimation. In: *Flood Estimation Handbook*. Vol. **2**. Institute of Hydrology Wallingford.

Freer, J. & Beven, K.J. (2001) Equifinality, data assimilation, and uncertainty estimation in mechanistic modelling of complex environmental systems using the GLUE methodology. *Journal of Hydrology*, **249**, 11–29.

Freeze, R.A. & Harlan, R.L. (1969) Blueprint for a physically-based, digitally-simulated hydrologic response model. *Journal of Hydrology*, **9**, 237–258.

Horton, R.E. (1933) The role of infiltration in the hydrologic cycle. *Transactions American Geophysical Union*, **14**, 446–460.

Houghton-Carr, H. (1999) Restatement and application of the Flood Studies Report rainfall-runoff method. In: *Flood Estimation Handbook*. Vol. **4**. Institute of Hydrology Wallingford.

IH (1999) *Flood Estimation Handbook*. Vol. **5**. Institute of Hydrology, Wallingford.

Jain, S.K., Storm, B., Bathurst, J.C., Refsgaard, J.C. & Singh, R.D. (1992) Application of the SHE to catchments in India. Part 2. Field experiments and simulation studies with the SHE on the Kolar subcatchment of the Narmada river. *Journal of Hydrology*, **140**, 25–47.

Jakeman, A.J. & Hornberger, G.M. (1993) How much complexity is warranted in a rainfall-runoff model? *Water Resources Research*, **29**, 2637–2649.

Jakeman, A.J., Littlewood, I.G. & Whitehead, P.G. (1990) Computation of the instantaneous unit hydrograph and identifiable component flows with application to two small upland catchments. *Journal of Hydrology*, **117**, 275–300.

Jakeman, A.J., Littlewood, I.G. & Symons, H.D. (1992a) IHACRES: a PC program for identification of unit hydrographs and component flows from rainfall, evapotranspiration and streamflow data. *Water and the Environment, Newsletter of the Water Research Foundation of Australia*, **307**, 1–4.

Jakeman, A.J., Hornberger, G.M., Littlewood, I.G., Whitehead, P.G., Harvey, J.W. & Bencala, K.E. (1992b) A systematic approach to modelling the dynamic linkage of climate, physical catchment descriptors and hydrologic response components. *Mathematics and Computers in Simulation*, **33**, 359–366.

Jakeman, A.J., Littlewood, I.G. & Symons, H.D. (1991) Features and applications of IHACRES: a PC program for identification of unit hydrographs and component flows from rainfall, evapotranspiration and streamflow data. In: Vichnevetsky, R. & Miller, J.J.H. (eds), *Proceedings of the 13th World IMACS World Congress on Computation and Applied Mathematics*. Trinity College, Dublin, pp. 1963–1967, 22–26, July 1991.

Lamb, R. (1999) Calibration of a conceptual rainfall–runoff model for flood frequency estimation by continuous simulation. *Water Resources Research*, **35**, 3103–3114.

Lamb, R. & Kay, A.L. (2004) Confidence intervals for a spatially generalised continuous simulation flood frequency model for Great Britain. *Water Resources Research*, **40**, W07501.

Legates, D.R. & McCabe, G.J. (1999) Evaluating the use of 'goodness-of-fit' measures in hydrologic and hydroclimatic model validation. *Water Resources Research*, **35**, 233–241.

Linsley, R.K. (1982) Rainfall-runoff models – an overview. In: Singh, V.P. (ed), *Proceedings of the International*

Symposium on Rainfall-Runoff Modelling. Mississippi State University, pp. 3–22.

Linsley, R.K. & Crawford, N.H. (1960) Computation of a synthetic streamflow records on a digital computer. *Hydrological Sciences Bulletin IASH Publication,* **51**, 526–38.

Littlewood, I.G. (2003) Improved unit hydrograph identification for seven Welsh rivers: implications for estimating continuous streamflow at ungauged sites. *Hydrological Sciences Journal,* **48**, 743–762.

Littlewood, I.G. (2007) Rainfall-streamflow models for ungauged basins: uncertainty due to modelling time-step. In: Pfister, L. & Hoffmann, L. (eds), *Uncertainties in the 'monitoring-conceptualisation-modelling' sequence of catchment research. Proceedings of the Eleventh Biennial Conference of the Euromediterranean Network of Experimental and Representative Basins,* UNESCO Technical Documents in Hydrology Series No. 81, UNESCO, Paris, pp. 149–155.

Littlewood, I.G. & Croke, B.F.W. (2008) Data time-step dependency of conceptual rainfall-streamflow model parameters: an empirical study with implications for regionalisation. *Hydrological Sciences Journal,* **53**, 685–695.

Littlewood, I.G. & Croke, B.F.W. (2013) Effects of data time-step on the accuracy of calibrated rainfall-streamflow model parameters: practical aspects of uncertainty reduction. *Hydrology Research,* **44**, 430–440.

Littlewood, I.G. & Jakeman, A.J. 1991. *Hydrograph separation into dominant quick and slow flow components.* Proceedings of the Third National Hydrology Symposium, British Hydrological Society, 1991, University of Southampton, pp. 3.9–3.16.

Littlewood, I.G. & Jakeman, A.J. (1993) Characterisation of quick and slow streamflow components by unit hydrographs for single- and multi-basin studies. In: Robinson, M. (ed), *Proceedings of the Fourth General Assembly of the European Network of Experimental and Representative Basins,* Oxford, September 29. October 2, Institute of Hydrology Report 120.

Littlewood, I.G. & Jakeman, A.J. (1994) A new method of rainfall-runoff modelling and its applications in catchment hydrology. In: Zannetti, P. (ed), *Environmental Modelling (Volume II).* Computational Mechanics Publications, Southampton, UK, pp. 143–171.

Littlewood, I.G., Down, K., Parker, J.R. & Post, D.A. 1997. *The PC version of IHACRES for Catchment-Scale Rainfall – Streamflow Modelling: User Guide. Institute of Hydrology Software Report,* p. 89. www.ceh.ac.uk /products/software/cehsoftware-pc-ihacres.htm

Littlewood, I.G., Croke, B.F.W., Jakeman, A.J. & Sivapalan, M. (2003) The role of 'top-down' modelling for Prediction in Ungauged Basins (PUB). *Hydrological Processes,* **17**, 1673–1679.

Littlewood, I.G., Clarke, R.T., Collischonn, W. & Croke, B.F.W. (2007a) Predicting daily streamflow using rainfall forecasts, a simple loss module and unit hydrographs: two Brazilian catchments. *Environmental Modelling & Software,* **22** (**9**), 1229–1239.

Littlewood, I.G., Jakeman, A.J., Croke, B.F.W., Kokkonen, T.S. & Post, D.A. (2007b). *Unit hydrograph characterisation of flow regimes leading to streamflow estimation in ungauged catchments (regionalisation). Predictions in Ungauged Basins: PUB Kick-off (Proceedings of the PUB Kick-Off Meeting, Brasilia, 20–22 November 2002).* IAHS Publ. 309. www.iahs.info/uploads/dms/309006.pdf.

Littlewood, I.G., Young, P.C. & Croke, B.F.W. (2010). *Preliminary comparison of two methods for identifying rainfall–streamflow model parameters insensitive to data time-step: the Wye at Cefn Brwyn, Plynlimon, Wales. Proceedings of the Third International Symposium, British Hydrological Society,* 19–23 July 2010, Newcastle University, UK. www.hydrology .org.uk/assets/2010%20papers/084Littlewood_etal.pdf

Loague, K. (2010) *Rainfall-Runoff Modelling.* IAHS Press, Wallingford, UK, IAHS Benchmark Papers in Hydrology No 4.

Mandeville, A.N., O'Connell, P.E., Sutcliffe, J.V. & Nash, J.E. (1970) River flow forecasting through conceptual models: III – The Ray catchment at Grendon Underwood. *Journal of Hydrology,* **11**, 109–128.

Moore, R.J. (1985) Probability-distributed principle and runoff prediction at point and basin scales. *Hydrological Sciences Journal,* **30**, 273–297.

Moore, R.J. & Clarke, R.T. (1981) A distribution function approach to rainfall runoff modeling. *Water Resources Research,* **17**, 1367–1382.

Moore, R.J. (2007) The PDM rainfall-runoff model. *Hydrology and Earth System Sciences,* **11**, 483–499.

Moore, R.J., Cole, S.J., Bell, V.A. & Jones, D.A. (2006) Issues in flood fore-casting: ungauged basins, extreme floods and uncertainty. In: Tchiguirinskaia, I., Thein, K.N.N. & Hubert, P. (eds), *Frontiers in Flood Research, 8th Kovacs Colloquium.* Vol. **305**. International Association of Hydrological Sciences Publishers, Paris, pp. 103–122.

Morris, E.M. (1980) *Forecasting flood flows in grassy and forested basins using a deterministic distributed mathematical model.* Vol. **129**. International Association of Hydrological Sciences Publishers, Paris, pp. 247–55.

Morris, E.M. & Woolhiser, D.A. (1980) Unsteady one-dimensional flow over a plane: partial equilibrium and recession hydrographs. *Water Resources Research,* **16** (**2**), 355–360.

Mulvany, T.J. (1850) On the use of self-registering rain and flood gauges. *Proceedings Institute of Civil Engineering (Dublin),* **4**, 1–8.

Nash, J.E. & Sutcliffe, J.V. (1970) River flow forecasting through conceptual models: I – a discussion of principles. *Journal of Hydrology,* **10**, 282–290.

Nelder, J.A. & Mead, R. (1965) A simplex method for function minimization. *Computer Journal*, **7**, 308–313. doi:10.1093/comjnl/7.4.308

NERC (1975) *Flood Studies Report (Five Volumes)*. UK Natural Environment Research Council, London.

O'Connell, P.E. (1991) A historical perspective. In: Bowles, D.S. & O'Connell, P.E. (eds), *Recent Advances in the Modelling of Hydrological Systems*. Kluwer Academic Publishers, Norwell, pp. 3–30.

O'Connell, P.E., Nash, J.E. & Farrell, J.P. (1970) River flow forecasting through conceptual models part II – the Brosna catchment at Ferbane. *Journal of Hydrology*, **10**, 317–29.

O'Sullivan, J.J. & Bruen, M. (eds) (2014) *The Grand Challenges Facing Hydrology in the 21st-Century*. Dooge Nash International Symposium, Dublin Castle, 23–26 April 2014, 515 pp.

Penman, H.L. (1948) Natural evaporation from open water, bare soil and grass. *Proceeding of the Royal Society Series A*, **193**, 120–145.

Post, D.A. & Jakeman, A.J. (1996) Relationships between catchment attributes and hydrological response characteristics in small Australian mountain ash catchments. *Hydrological Processes*, **10**, 877–892.

Post, D.A., Jones, J.A. & Grant, G.E. (1998) An improved methodology for predicting the daily hydrologic response of ungauged catchments. *Environmental Modelling and Software*, **13**, 395–403.

Refsgaard, J.C., Seth, S.M., Bathurst, J.C. *et al.* (1992) Application of the SHE to catchments in India – Part 1: general results. *Journal of Hydrology*, **140**, 1–23.

Refsgaard, J.C., Storm, B. & Clausen, T. (2010) Système Hydrologique Européen (SHE): review and perspectives after 30 years development in distributed physically-based hydrological modelling. *Hydrology Research*, **41**, 355–377. doi:10.2166/nh.2010.009

Robinson, M. & Stam, M.H. (1993) A study of soil moisture controls on streamflow for the Ock basin, United Kingdom. *Acta Geologica Hispánica*, **28** (**2–3**), 75–84.

Rosenbrock, H.H. (1960) An automatic method for finding the greatest or least value of a function. *The Computer Journal*, **3**, 175–184. doi:10.1093/comjnl/3.3.175

Schreider, S.Y., Whetton, P.H., Jakeman, A.J. & Pittock, A.B. (1997) Runoff modelling for snow-affected catchments in the Australian alpine region, eastern Victoria. *Journal of Hydrology*, **200**, 1–23.

Sefton, C.E.M. & Howarth, S.M. (1998) Relationships between dynamic response characteristics and physical descriptors of catchments in England and Wales. *Journal of Hydrology*, **211**, 1–16.

Sefton, C.E.M., Whitehead, P.G., Eatherall, A., Littlewood, I.G. & Jakeman, A.J. (1995) Dynamic response characteristics of the Plynlimon catchments and preliminary

relationships to physical descriptions. *Environmetrics*, **6**, 465–472.

Sherman, L.K. (1932) Streamflow from rainfall by the unit graph method. *Engineering News Record*, **108**, 501–505.

Sivapalan, M., Wood, E.F. & Beven, K.J. (1990) On Hydrologic Similarity, 3. A dimensionless flood frequency distribution. *Water Resources Research*, **26**, 43–58.

Smith, R.N.B., Blyth, E.M., Finch, J.W., Goodchild, S., Hall, R.L. & Madry, S. (2006) Soil state and surface hydrology diagnosis based on MOSES in the Met Office NIMROD nowcasting system. *Meteorological Applications*, **13**, 89–109.

Steel, M.E., Black, A.R., Werritty, A. & Littlewood, I.G. (1999) Reassessment of flood risk for Scottish rivers using synthetic runoff data. In: Gottschalk, L., Olivry, J.-C., Reed, D. & Rosbjerg, D. (eds), *Hydrological Extremes: Understanding, Predicting, Mitigating*. Vol. **255**. International Association of Hydrological Sciences, Publishers, Paris, pp. 209–215.

Sugawara, M. (1961) An analysis of runoff structure about several Japanese rivers. *Japanese Journal of Geophysics*, **2**, 10–17.

Thorne, M.C., Degnan, P., Ewen, J. & Parkin, G. (2000) Validation of a physically based catchment model for application in post-closure radiological safety assessments of deep geological repositories for solid radioactive wastes. *Journal of Radiological Protection*, **20**, 403–421.

Whitehead, P.G., Young, P.C. & Hornberger, G.M. (1979) A systems model of stream flow and water quality in the Bedford-Ouse River – 1. Stream flow modelling. *Water Research*, **13**, 1159–1169.

Wood, W.L. & Calver, A. (1990) Lumped versus distributed mass matrices in the finite element solution of subsurface flow. *Water Resources Research*, **26**, 819–825.

Wood, W.L. & Calver, A. (1992) Initial conditions for hillslope hydrology modelling. *Journal of Hydrology*, **130**, 379–397.

Wood, E.F., Roundy, J.K., Troy, T.J. *et al.* (2011) Hyperresolution global land surface modeling: meeting a grand challenge for monitoring Earth's terrestrial water. *Water Resources Research*, **47**, W05301. doi:10.1029/2010WR010090

Wood, E.F. *et al.* (2012) Reply to Comment by Keith J. Beven and Hannah L. Cloke on 'Hyperresolution global land surface modeling: meeting a grand challenge for monitoring Earth's terrestrial water'. *Water Resources Research*, **48** (**1**). doi:10.1029/2011WR011202

Woolhiser, D.A. (1996) Search for physically based runoff model-A hydrologic El Dorado? *Journal of Hydraulic Engineering*, **122**, 122–129.

Young, P.C. & Jakeman, A.J. (1979) Refined instrumental variable methods of recursive time-series analysis: Part I, single input, single output systems. *International Journal of Control*, **29**, 130.

8 Water Quality

RICHARD WILLIAMS[1], COLIN NEAL[1], HELEN JARVIE[1],
ANDREW JOHNSON[1], PAUL WHITEHEAD[2], MIKE BOWES[1],
AND ALAN JENKINS[1]

[1]Centre for Ecology and Hydrology, Wallingford, Oxfordshire, UK
[2]School of Geography and the Environment, University of Oxford, Oxford, UK

8.1 Background

Since the early days when CEH Wallingford was the Hydrological Research Unit and then the Institute of Hydrology, in the 1960s, water quality studies have been a strategically important component of the hydrological sciences and have extended both in terms of measurement and modelling into the arenas of the Earth and environmental sciences. These studies encompassed, over the past 40 years, the major developments within these and the hydrological sciences from the local to the national and international scales that relate to historic and contemporary issues, such as land use change, environmental pollution associated with acidic oxide deposition from industry and cars, nutrient and pesticide contamination from agriculture, nutrient contamination from wastewater and urban population, as well as metal, nutrient and organic pollution from industry (Likens *et al.*, 1979; Billen *et al.*, 2007). From this has evolved a range of priorities: from pure research through to environmental management, from individual to community-based research; from local to

Progress in Modern Hydrology: Past, Present and Future, First Edition. Edited by John C. Rodda and Mark Robinson.
© 2015 John Wiley & Sons, Ltd. Published 2015 by John Wiley & Sons, Ltd.

basin-wide focus and, in recent years, from science to policy and to ecosystem functioning. With these came dual recognition that the key environmental impacts related to aquatic biological health, but that the remediation efforts required could only be undertaken within affordable and sustainable social and economic costs (Bateman *et al.*, 2006). This work has impinged on key environmental legislation such as the wastewater and water framework directive of the European Union (CEC, 1991, 2000). Further, the period spanned from times when computing was very basic, information was largely contained on paper, and has evolved through to personal computers, spreadsheets, databases and the world wide web that nowadays is indispensable (Neal, 2001). Here, given the sheer breadth of the research and achievements gained, the focus is primarily on research at Wallingford, but fully acknowledging the breadth, vitality and contributions and collaborations with the wider research community in the UK and internationally. Also acknowledged is the extent to which IH/CEH research has been influenced over so many years by international researchers who have led major developments in key research areas such as environmental modelling (e.g. Jack Cosby from the USA, Nils Christopheresen, Norway and James Kirchner, Switzerland).

In the early days, a core feature was the development of a chemical laboratory, with associated analytical chemistry development (Truesdale and Smith, 1975a, 1975b). There were related hydrochemical studies that linked to issues such as cation exchange reactions (Neal and Cooper, 1983; Neal *et al.*, 1982) and the hydrochemistry of iodine (Neal and Truesdale, 1976; Jones and Truesdale, 1984). Going along with this was the recognition of the value of chemical tracers in hydrology, for example, in determining accurate flows using dilution gauging (Neal and Truesdale, 1976). Indeed, with the strong hydrological consultancy activities on site, water quality measurements were an important component in dealing with issues such as water potability. With these consultancy studies came the opportunity to undertake water quality research in remote parts of the world and contribute to initiatives of major research and environmental management importance, where access could be restricted or otherwise difficult. For example, in Northern Oman, there was the then rare opportunity to study extremely high pH waters emerging at the crust/mantle boundary where the ocean floor thrust over the edge of the Arabian Shield to form the Semail ophiolite nappe and the spine of the main mountain range of the Oman Peninsular. This nappe represents the largest and best-exposed fragment of oceanic lithosphere found on land in the world and it is a prime area for studying the hydrolysis of ultrabasic rocks with the formation of serpentine and the production of not only waters of pH up to 12, but also hydrogen gas (Neal and Stanger, 1983, 1985). Such consultancy activities lasted for many years in relation to areas such as the strategic water resources issues of seasonal wetted dambo areas of central, southern and eastern Africa (McCartney *et al.*, 1998; McCartney and Neal, 1999) and the Himalayan areas of Nepal. These studies focussed on important environmental issues. For example, the shallow extensive wetlands in higher rainfall flat plateau areas that constitute dambos are extremely important to local communities in many ways, such as rushes for thatching, soils for building materials and pottery, and critically agriculture, especially in drought years (the soils and shallow groundwater usually retain enough moisture to produce a harvest even when the seasonal rains fail). The work in the Himalayas was at least partly driven by the need to establish the chemical and biological quality of streams in 'pristine' areas as a reference for the acid rain impacts on soil and surface waters in Europe. IH scientists secured funding from the United States Environment Protection Agency to undertake a survey of streams draining un-glaciated catchments in the Nepal Himalaya (Jenkins *et al.*, 1995). The relevance of the work prompted further funding from the UK Department of the Environment (now Defra) Darwin Initiative and the survey areas were expanded into the Himalayan regions of Northern India (Jenkins 2001). At the same time, in 1992, the then Overseas Development Administration funded a 3-year study to quantify the impacts of agriculture on water quality in upland Himalayan environments. This led to the establishment of five monitored catchments in the Likhu Khola to the north of Kathmandu (Collins and Jenkins 1996).

Four major features helped with the shaping of water quality studies at Wallingford. Firstly, there was the development of UK upland research, particularly hydrological research initiatives in both Wales and Scotland (Neal *et al.*, 1986a; Roberts *et al.*, 1984, 1986; Morris and Thomas, 1987; Cooper *et al.*, 1987; Jenkins, 1989). The second was arrival of the British Geological Survey

Hydrogeology Group at Wallingford in 1977. They provided both a major impetus relating to the study of groundwater–surface water interactions and an extensive laboratory resource that, combined with the CEH resource, provided the opportunity for very wide-ranging and extensive analysis covering major elements, nutrients, minor, trace and ultra-trace elements. The third was the strong research link developed in the early 1980s between the Institute of Terrestrial Ecology at Bangor (now CEH Bangor) and Wallingford in the area of hydrochemical research, especially for the Plynlimon catchments in mid-Wales since the early 1980s (Reynolds *et al.*, 1986, 1988, 2001; Neal *et al.*, 1989). This formed the backbone of collaborative research in the uplands to the present day (Neal *et al.*, 2010d, 2011b, 2012, 2013).The fourth was the developing linkage between measurement and modelling, especially in the area of acid deposition impacts (Christophersen and Neal, 1990; Neal *et al.*, 1986b; Robson *et al.*, 1991; Whitehead *et al.*, 1986a,b; Hill *et al.*, 2002), water quality transport (Robson *et al.*, 1992, 1995; Page *et al.*, 2007) and later for lowland systems (Wade *et al.*, 2005, Johnson *et al.*, 2009a). Figure 8.1 shows the catchments on which water quality scientists at Wallingford have carried out the major part of their research.

8.2   The Chemistry of the Uplands

Uplands hydrochemical research in the UK flourished from the 1970s, especially within the context of acid deposition and land use change. In essence, much of the UK uplands are acidic and sensitive to atmospheric deposition and also to the influence of conifer plantations that were introduced for timber production after the first and second World Wars (Neal *et al.*, 2004). The issues lay with problems of thin acidic and organic-rich soils on low permeability and cation-depleted bedrock. For forestry, which was planted on such soils, there were issues of base cation uptake by the growing vegetation and scavenging of atmospheric pollution, both of which could lead to soil and then stream acidification. Further, with felling, there was the additional issue of nutrient releases (nitrogen in particular) and acidic flushing.

All the work on water quality cannot be covered here, so the focus is on research for the studies within the upper River Severn catchments of mid-Wales (Plynlimon), by way of illuminating

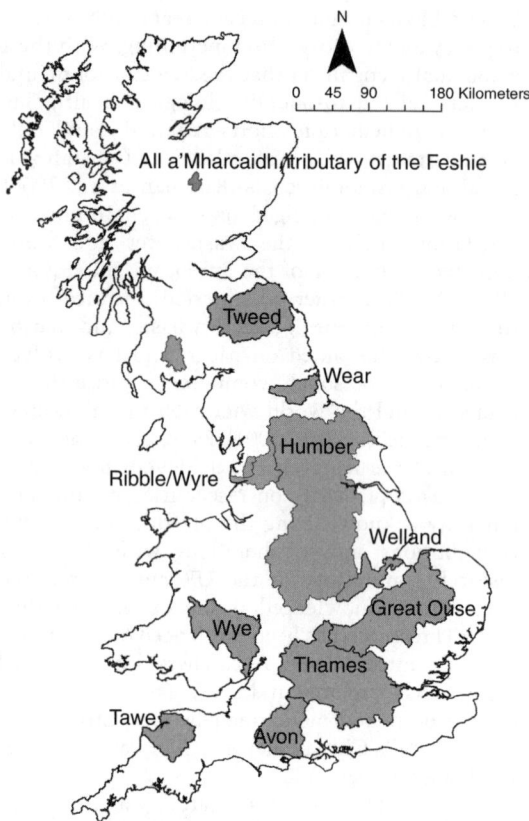

Fig. 8.1 Locations of catchments in UK that have been the focus of water quality research carried out at Wallingford. (Source: Reproduced with permission of CEH.)

case studies. This is done, because of the longevity and breadth of the research (more than 30 years), the wide implications to hydrochemical research and the major initiative to determine the impacts of conifer harvesting for the uplands (Neal and Reynolds, 1999). For this study, the research on catchment water quality functioning and acidic deposition developed with international partnerships, such as the Surface Water Acidification Programme (SWAP) and the European Network of Catchments (ENCORE). In particular, strong research links developed with Norway (Christophersen and Neal, 1990; Christophersen *et al.*, 1990), France (Durand *et al.*, 1991, 1992), Spain (Avila *et al.*, 1992), Germany (Neal *et al.* (1997f)) and Brazil (Cornu *et al.*, 2001; Forti *et al.*, 1999).

These in turn cross-linked with modelling studies (Durand *et al.*, 1992; Lundquist *et al.*, 1990; Neal *et al.*, 1992b). It also links to the United Kingdom Acid Waters Monitoring Network (UKAWMN), since its inception in 1988, when CEH Wallingford undertook extensive water quality analysis as part of a study to monitor the chemical and ecological impact of acid deposition in areas of the UK sensitive to acidification. The UKAWMN database now provides an extremely valuable long-term record of water chemistry and biology which is unique for upland freshwater systems in the UK (Monteith, 2005). This data set has been used extensively to assess and predict acidification improvement with the reduction in acidic emissions both at the national scale and within European settings (Evans *et al.*, 2001; Jenkins and Cullen, 2001).

8.3 Plynlimon

The Plynlimon study in the upper River Severn (see Chapter 2) relates to two main tributaries, the Afon Hafren and the Afon Hore, and to a smaller sub-catchment, the Nant Tanllwyth. The upper parts of the Hafren and the Hore comprise hill top plateau dominated by blanket peat whilst areas of seasonally saturated peat and gley soil occur in the valley bottoms. In the lower parts of the two catchments were first-generation conifer plantations of the Hafren forest (mainly Sitka with some Norway spruce, Lodgepole pine and Larch), but much of the area is now felled. Most of the soils are thin, acidic and acid sensitive and they overlie 'hard-rock' geology (Cambrian to Silurian slate and shale). Weekly monitoring of rainfall and stream water quality began in May 1983. Initially only the lower Hafren and the lower Hore were monitored together with rainfall. However, over the years, the monitoring was extended considerably to assess the following:

- Stemflow and throughfall.
- Drainage from small streams to assess felling impacts (these included control and fell response sites).
- Continuous monitoring of pH and electrical conductivity.
- Cloud water (mist/occult deposition) to assess the total and not just 'wet' deposition.
- The upper Hafren (to provide information on the chemistry of runoff from the moorland area).

- The Tanllwyth (to provide data for a completely forested catchment and baseline information for an experimental fell).
- Groundwater samples were collected across a network of sites.

The underlying approach was to study a wide range of chemical constituents to assess the hydrological flow pathways at the catchment level and the impacts of conifer harvesting and replanting. Almost 25 years after the instigation of the monitoring began, a major new initiative was undertaken to examine the short-term variations in water quality. This aspect is covered separately as are issues of chemical tracers and the hydrological implications to the work.

The publication list is extensive and it is not feasible to review the work fully. However, a detailed coverage is made elsewhere (Neal *et al.*, 2010d, 2011b, 2012, 2013) . Here the research on chemical tracers and high-resolution monitoring is dealt with.

8.3.1 Background water quality patterns (not influenced by felling)

Rainfall has a variable chemistry from event to yearly and longer time steps due to a variable input of sea salts from the North Atlantic and pollutants (transition metals, nitrate, ammonium and non-marine sulphur) from the UK and European mainland, as well as further afield. Of the total deposition to the catchments, cloud water can provide a highly significant input to the total wet deposition with values typically of the order of 40% for the forested areas and about a half of this value for moorland areas. Cloud water volumes are much lower than rainfall, but concentrations are generally much higher.

Within the streams, the rainfall signal is damped for components such as sodium and chloride which are of relatively low reactivity. Nonetheless, strong variations in the water quality of the streams can be linked to the inputs of water from two distinct hydrochemical provenances. Firstly, under baseflow conditions, the streams are non-acidic and of calcium and magnesium bicarbonate type characteristic of waters that have reacted with the minerals in the bedrock (weathering processes). Secondly, under high flows the waters are acidic and aluminium bearing and of low bicarbonate content due to greater inputs from the acidic soils in the area, the aluminium being supplied from aluminium oxides/hydroxides in the soil that are leached under acidic conditions.

For nutrient components (nitrate, bromine and iodine), annual cyclical patterns occur that reflect changes in decomposition rates in the soil and uptake by the vegetation throughout the year. There are also longer term patterns of water quality change within the rivers for arsenic, chromium, beryllium, dissolved organic carbon (DOC), iron, acidity (as measured as acid neutralisation capacity, ANC) and sulphate. Critically, the acidification reversal is significant in relation to declining atmospheric deposition of pollutants. This is shown especially by the sulphate (non-marine) and the ANC. This improvement is particularly noticed for the forested systems as it is these that capture the greatest proportion of such pollutants. In contrast, nitrate concentrations exhibited no systematic trends, but concentrations were highest for both moorland and forested areas during the earlier part of the record (1983–1986) and around 1996. This pattern of change includes surface acidification monitoring sites across the UK and it seems that the changes are associated with climatic variations associated with factors such as spring peaks that follow colder low North Atlantic Oscillation (NAO) winters. Further, sodium and chloride show year to year variations that also seem to be driven by the fluctuation of the NAO.

Many of the trends were unexpected. In the case of arsenic, there were declines observed within the Plynlimon rainfall and streams. Subsequently, this proved to link with long-term declines in global emissions. However, the most unexpected result came with regard to DOC. DOC is a critical component within the acidification debate as it is a component that influences the acid buffering of waters and there has been major controversy over how DOC would evolve with acidification reversal and the extent to which acidic waters link to humic and fulvic acids that are derived from organic matter within catchments (Krug and Frink, 1983). Major DOC increases have been observed over time with acidification reversal especially for forested systems where acidification reversal will be greatest, due to the increased scavenging of acidic oxides by trees. Indeed, such declines are observed for many upland organic-rich environments and it is often associated with long-term temperature increase that results in enhanced decomposition of peat, but other mechanisms may

be operating (Worrall *et al.*, 2004). As observed especially at Plynlimon, iron shows a similar pattern of behaviour to DOC and this probably links to colloid formation and its stabilisation by DOC in the water (Neal *et al.*, 2005). This is relevant to water resources issues in relation to colour and potability. It demonstrates that there are other features of acidification reversal that were unexpected when the acidification debate was at its height in the 1980s and 1990s.

The Plynlimon catchments now provide a major resource for examining long-term effects of both climate and atmospheric pollutants. This results not only from the longevity of the monitoring programme (as linked to the quality and consistency of the measurements), but that a multi-element approach has been taken and so a far wider range of determinands can be examined in terms of trend. One of the features of our results is that simple non-linear trends occur and the interplay of climate, pollutant change and within-catchment processes produces complexity and patterns that are often hard to resolve.

8.3.2 *Forest harvesting impacts on water quality*

There are two major catchment responses to tree felling. Firstly, there is an increase in nitrate, DOC, potassium, aluminium and acidity due to the disruption of the hydrobiogeochemical cycles in the soil. Secondly, there is a decrease in 'sea salt' components such as sodium and chloride, which derive from atmospheric sources, due to reduced scavenging as the surface area of the vegetation is greatly reduced. Both types of change recede after a few years: As the new trees become established, the nutrients are taken up by the developing vegetation and the surface area of the vegetation for atmospheric scavenging increases. Indeed, for many years post felling, there is better water quality than prior to felling.

One of the important findings in terms of forestry management is that phased felling is a particularly beneficial thing. Indeed, where there was phased felling over many years, the water quality response is hardly noticeable and is often within the noise of the longer term changes associated with climate variability. Further, it was very uncertain that there was any noticeable forestry effect on acidification change in relation to cation uptake by the growing vegetation. However, there was a clear impact from increased scavenging by the trees and this resulted in a greater rate of acidification recovery and non-marine sulphate decline.

8.3.3 Hydrological implications to the hydrochemical studies

The dynamics of water quality change for the Plynlimon streams, as elsewhere, is related to hydrology, with major issues relating to water pathways and water source areas with distinct chemistry and water storage. The observation of contrasting baseflow and stormflow chemistry led to the development of methodologies for chemical splitting of the hydrograph and the development of endmember mixing research (Christophersen *et al.*, 1990). From earlier studies at CEH Bangor, acidic soil waters had been directly measured at Plynlimon (Reynolds *et al.*, 1988) and so this end-member could be characterised. Groundwater had not been measured in the early 1990s as significant groundwater storage was considered by hydrologists to be largely insignificant for such 'hard-rock' areas due to low permeability. However, it was realised, based both on endmember mixing and modelling studies of chloride, that shallow groundwater storage could be significant (Neal and Kirchner, 2000; Neal *et al.*, 1988b). Directed by these observations, exploratory boreholes were drilled in the Hafren catchment and groundwater was confirmed to be widespread (Neal *et al.*, 1997a,b). The groundwater resided in fractures within the rock and its chemistry ranged between that observed for the streams under baseflow and stormflow, but extending to high alkalinities representative of high weathering rates. Thus, the groundwater fractures acted as conduits for the transport of soil waters under wet conditions. Overall, endmember mixing had proved its worth. However, for the soil, groundwater and small streams, water quality proved to be highly variable: Everywhere there seemed to be water mixing taking place and no endmember of constant chemistry could be observed and the endmembers were thus 'notional' (Reynolds *et al.*, 1988; Hill and Neal, 1997; Neal *et al.*, 1997c) and water transport through catchments is complex and highly variable. Further, based on strong collaborative research with the modelling community, simulations of water quality changes within the time span of hydrological events (sub-daily) using the Birkenes model (Christophersen *et al.*, 1982) revealed a critical mismatch. For components largely occurring from within-catchment sources

that varied markedly as a function of flow, the model required a high soil water storage component. In contrast, atmospheric components (e.g. chloride) that exhibited a damped response within the stream required a high groundwater component. This proved to be a very thorny problem and there were major issues over the degree to which models were over-parameterised and the modelling exercise started to become a fitting procedure rather than an interrogation of the internal functioning of catchments (Christophersen and Neal, 1990). This issue is only now coming to a resolution with the major input of James Kirchner, based in Switzerland, who has been key in the development of analytical procedures for time series analysis and the developing theme of fractal catchment responses.

Since chloride has a low chemical reactivity and the degree of damping of the rainfall by the catchment is high, it is clear that groundwater storage issues must be important (predicted by the Birkenes model). The chloride variations in the streams exhibit a '1/f' fractal structure even though the rainfall input shows only 'white noise' (Kirchner *et al.*, 2000). This fractal pattern for the streams links directly to the internal functioning of catchments and the observation of a very variable soil and groundwater chemistry is critical. Thus, the fractal response comes from localised inputs within the highly heterogeneous catchment system, a wide variation of water residence times, and advection and dispersion processes (Kirchner *et al.*, 2001). A key point here is that by allowing variable water residence times within the catchments and incomplete water mixing, the volumetric storage may actually be relatively small compared to the standard stirred tank models that unrealistically predict large storage volumes.

The modelling work, as well as continuous monitoring studies, indicated that the chemical signals within the stream can be highly complex and that standard monitoring exercises might well miss such dynamics (Kirchner *et al.*, 2004). Further, if the fractal dynamics applied to chloride, why should it not apply to other chemicals? The relevance of all of this is that with a fractal structure, recovery rates from pollutant inputs may take a lot longer than previous modelling work might imply (Kirchner *et al.*, 2000). At the time, such high-resolution monitoring across a wide range of chemicals had not been undertaken. However, this was to change and this aspect is covered in the next section.

8.3.4 High-resolution monitoring: The road to fractal functioning

In 2007, a major exploratory study was undertaken for the upper and lower Afon Hafren to examine high-frequency dynamics of rainfall and stream water quality: The study complemented over 20 years of weekly records for the same monitoring sites. The study lasted for 18 months and constituted an act of discovery with a central issue of 'What components of stream water quality dynamics do we miss with conventional weekly or monthly monitoring and ad hoc, event-based high-frequency measurement campaigns?'

The results of the study indicate that hydrochemical responses to major hydrological and biological drivers of short-term variability in rainfall and rivers are not captured by conventional lower frequency monitoring programmes. Rather, a wealth of flow-related, flow-independent, diurnal, seasonal and annual fluctuations have been revealed (Neal *et al.*, 2012; Halliday *et al.*, 2012). Despite the complexity of the chemical dynamics being visually obvious, there appears to be no clear way of translating this complexity into a simple algorithm: The findings indicate a cacophony of interactions within the catchment and stream. The work indicates a complex structure of catchment functioning and provides new insights into hydrogeochemical functioning and a novel resource for catchment modelling.

With regard to fractal functioning, analysis is now revealing that the fractal signal is universal across a very wide range of water quality determinands irrespective of their chemical reactivity within catchments (Neal *et al.*, 2012, 2013; Kirchner and Neal, 2013). This clearly poses new questions for describing and modelling complexity of catchment functioning that ultimately links to the hydrological functioning of catchments and water flow pathways (Neal, 2014). The work also shows that there is huge value in comparing the long- and short-term chemical signals as different inferences can be gauged in terms of contrasts between longer duration climate signals and individual storm events or series of storm events.

All of this material (weekly to high-resolution monitoring) is now available to researchers (https://gateway.ceh.ac.uk) and the data is augmented with information on sampling and analytical chemistry protocols, detection limits for each method and both 'raw' and edited data with comprehensive information on all the editing undertaken and the rationale (Neal *et al.*, 2013). It is a major and unique resource for new modelling initiatives and hydrochemical exploration.

8.4 Lowland River Water Quality

Increasing emphasis is being placed on characterising the water quality of agricultural and urban/industrially impacted river systems as part of the environmental management schemes of regulatory agencies such as the Environment Agency (EA) of England, Natural Resources Wales (NRW) and the Scottish Environmental Protection Agency (SEPA) (Robson and Neal, 1997; Jarvie *et al.*, 2003). Indeed, a central plank of the environmental policy drivers has been the European Union (EU) legislation under the Urban Waste Water and the Water Framework directives (CEC, 1991, 2000) that much of our lowland work has focussed on since the early 1990s. At national and international levels, increasing focus is placed upon basin-scale management particularly within policy frameworks (Neal and Jarvie, 2005; Billen *et al.*, 2007). Importantly, there are the strategic needs for considering social and economic factors that relate to cost-effective and viable environmental solutions (Bateman *et al.*, 2006). This is linked, for example to population increase and mobility, water resources, and the changes associated with fluctuating climate and economics (Johnson *et al.*, 2009b), as well as impinging on other issues and challenges such as scarcer resources (e.g. fertilisers), carbon footprints and food security. For the UK, a critical point is that the issues are multifaceted (Neal, 2001). This is due to a relatively high population density and a landscape that is impacted by many factors over many timescales (Neal, 2001).

Many of the key environmental pressures for UK rivers concern the lowlands which have the highest population density, industrialisation and agriculture. For CEH, work in this area stems from several research programmes beginning in the early 1990s. These programmes include

1 Land Ocean Interaction Study (LOIS; Leeks and Jarvie, 1998)

2 Phosphorus and Sediment Yield Characterisation In Catchments (Jarvie *et al.*, 2008a)

3 The Urban Regeneration and the Environment programme (URGENT, Leeks *et al.*, 2006)

4 The Lowland Catchment Research Programme (LOCAR, Wheater and Peach, 2006)

5 European Catchments (EUROCAT; Cave *et al.*, 2003)

6 Rural Economy and Land Use Programme (RELU, Bateman *et al.*, 2006)

7 Phosphorus from Agriculture: Riverine Impacts Study (PARIS, Jarvie *et al.*, 2010)

8 Core science budget activities for the River Thames (Bowes *et al.*, 2012a, 2012b) and the Ribble/Wyre systems (Neal *et al.*, 2011a)

9 Macronutrient Cycles Programme

These monitoring programmes were funded by NERC, Defra and the EU and include river basins which reach from the rural Tweed on the Scotland-England border to the Humber draining the industrial heartlands of north-eastern England and to the agricultural areas of the Thames Basin in south-eastern England. The monitoring data produced typically cover one to several years duration and are supplemented by data from the EA (Davies and Neal, 2007; Jarvie *et al.*, 2005a) and sewage treatment works (STWs) final effluent data across the Thames Basin. These large-scale river water quality monitoring activities have also provided a platform for detailed process studies and experiments, which have developed a deeper mechanistic understanding of the controls on water quality, chemical flux transformations through river systems and the ecological impacts of spatial and temporal variability in water quality.

A core feature of the LOIS study was that it was an early example of integrated basin-wide research dealing with major issues of water quality in relation to contamination from historic and contemporary drivers such as agriculture, industry and population growth. Such issues are considered below. During this time, advances in computing techniques facilitated collation and analysis of very large spatial water quality data sets, bringing together of a database combining EA data sets with research data and GIS tools for data interrogation. Here we do not go into details of this transition, but reference is given to key publications where such data is explored graphically and statistically that previously could not be achieved (Robson and Neal, 1997; Jarvie *et al.*, 1997). The LOIS as a community research programme was fundamental in the development of new collaborations and strands of research that became increasingly important through the subsequent years. One example of this was the introduction of methodologies for examining the transfer of phosphorus between sediments and overlying waters (House and Warwick, 1999; House and Denison, 2002), coupled with the need for examining the speciation of phosphorus in the sediment and the water column (House and Warwick, 1998; Jarvie *et al.*, 2005a,b; Whitton and Neal, 2011).

8.4.1 General overview

As all these studies show, water quality varies considerably in relation to catchment typologies (upland, rural, agricultural and urban/industrial). The waters become less acidic with increases in calcium and from the upland to the lowland and this links to a gradual transition from older hard-rock areas more depleted in weathering components to lowland calcareous sedimentary rocks such as limestone and chalk. Further, with much higher population densities in the lowlands, there is enrichment in pollutants such as nutrients from agriculture and sewage effluent discharges and historical contamination by trace metals from the industrial and urban areas. Nonetheless, for the upland areas, aluminium, manganese and iron concentrations are especially high due to the acidic and organic-rich conditions linked to a combination of organic and acidic soils, atmospheric deposition and acid mine drainage (Neal *et al.*, 2011a,b). Despite this, for the UK lowlands, surface waters are generally less acidic and high in bicarbonate due to the weathering of minerals within the soil/aquifer matrix bedrock by atmospheric and soil-generated CO_2 and the bicarbonate is balanced mainly by calcium and magnesium. The importance of dissolved CO_2 levels in the less acidic lowland rivers has a profound influence on the pH of the waters and pH can rise up to almost pH 11, during algal bloom events for the cleanest of rivers, due to the dominance of photosynthesis when dissolved CO_2 levels can be extremely low (Neal *et al.*, 1997d). Correspondingly, for the more polluted environments where respiration dominates, pH can be lowered by high CO_2 concentrations (Neal *et al.*, 1998).

Across the lowland region, the influence of humans is profound. For example, even for the rural agricultural areas, there is evidence of sewage contamination in the rivers (Neal *et al.*, 2005, 2011a; Jarvie *et al.*, 2006a) with enrichment of many elements such as nitrogen, phosphorus and boron (a reliable indicator of sewage effluent; Neal *et al.*, 2010b). As there is minimum dilution under low flows, pollutant concentrations from such effluents are particularly high at such times. However, sewage effluents can also be enriched at high flows in rural areas because of the flushing of septic tank effluents when the catchment wets up (Neal *et al.*, 2010a,b). Nonetheless, it is not simply the contaminants that dilute with increasing flow so to do the weathering components.

There is an issue of how long a pollutant may be stored in a catchment both in relation to reservoirs and groundwater (Neal *et al.*, 2010a,b; Jarvie *et al.*, 2011). Across the rivers monitored, a very useful effluent marker is boron as it is highly enriched in effluents and it is chemically conserved within the water column (Neal *et al.*, 2010b, 2011a). Boron concentrations have declined over the past 20 years due to much reduced use in detergents and soaps, and the relative changes within the river is less particularly at high flows. The results indicate that there is a strong sewage effluent component to the diffuse signal. Further, the results point to the effluent component in the diffuse signal being maintained at the decadal scale even when the effluent inputs reduce during that time. This in turn implies decadal-scale within-catchment storage (Neal *et al.*, 2010b).

Several important case studies of particular contaminants arise from the work carried out at Wallingford: nutrients, steroid oestrogens and pesticides.

8.5   Nutrients: Nitrogen and Phosphorus

For nitrogen (N), in the rivers nitrate (NO_3) is dominant and catchment sources largely come from agriculture (Jarvie *et al.*, 1997). For the more permeable agricultural catchments, concentration variations are relatively low within the rivers due to within-river uptake of NO_3 during the spring and the summer when biological activity is high and there is a seasonal fall in the water table. However, NO_3 concentrations have increased substantially over the last 50 years in response to increases in fertiliser inputs during the first half of the twentieth century. They result from the long-term release of contamination from the unsaturated, saturated and groundwater zones, because within-catchment attenuation and aquifer storage result in long water residence times. In terms of environmental strategies for reducing nitrate pollution, more recent reductions in fertiliser application will take a very

long time to translate into major reductions in NO_3 within the rivers (Wheater *et al.*, 2006; Smith *et al.*, 2010; Howden *et al.*, 2011). The diffuse component of NO_3 for the agricultural catchments is usually greater than 80% although effluents can be highly significant for low permeability cases under baseflow conditions when inputs are high and/or denitrification processes are limited (Neal *et al.*, 2006e, 2011a). Not all sources of nitrogen that enter catchments come from direct fertiliser usage. For example, atmospheric inputs of N to the catchments may be large, but they are dominated by ammonia/ammonium ions (NH_3/NH_4^+) that are strongly bound within the catchment (Neal *et al.*, 2004).

Phosphorus (P) occurs in rivers in both dissolved and particulate forms, and of these forms dissolved inorganic phosphorus (known as soluble reactive phosphorus, SRP) typically dominates in lowland rivers. For UK lowland river waters, SRP generally correlates with effluent inputs and population density indicating the importance of sewage sources (Davies and Neal, 2004, 2007; Neal *et al.*, 2005; Jarvie *et al.*, 2006a) and hence for the more contaminated rivers, SRP concentrations dilute with increasing flow. However, for the more rural areas, SRP concentrations can increase with increasing flow as diffuse inputs from septic tanks and agriculture are flushed from the catchment when they are wet. However, there are complications. For example, there are interactions between the water, sediment and biota, with uptake from the water column in the growing season. In addition, storage times can be especially long (Neal *et al.*, 2011a). For many UK river basins, there have been major removal of SRP in effluents for the main sewage treatment works since the late 1990s and this has led to corresponding reductions in SRP concentrations within the lowland rivers (Neal *et al.*, 2010a; Bowes *et al.*, 2010). These reductions occur particularly at low flows and the reductions can be less than expected due to contaminated groundwater storage, with interchange to/from the river and there may well be several types of store within the catchment that are released into the river during rainfall events (Neal *et al.*, 2010a, 2011a; Jarvie *et al.*, 2004, 2011). Our monitoring has shown the importance of within-river processes in retaining P along the river continuum (Jarvie *et al.*, 2005a, 2005b, 2012a). Further, immediately after point-source mitigation, there may also be a net release of SRP from in-stream sediments (Jarvie *et al.*, 2006b).

The need to explain the transformations in phosphorus along the river continuum resulted in new process studies on the cycling of phosphorus between river bed sediments and the overlying water column by means of laboratory-based studies of sediment–water exchange and the role of uptake by biota (Jarvie *et al.*, 2005b, 2006b). This then led to *in situ* experiments using probes to examine diffusion-driven fluxes of SRP across the sediment–water interface (Jarvie *et al.*, 2008b) and expanded the research to examine coupled cycling of N and P in rivers and wetlands (Palmer-Felgate *et al.*, 2010; Palmer-Felgate *et al.*, 2011a,b). New technologies in *in situ* high-resolution monitoring have vastly expanded our understanding of nutrient dynamics and fluxes, and particularly the reciprocal interactions between biological processes and nutrient uptake and release in rivers (Jarvie *et al.*, 2003; Palmer-Felgate *et al.*, 2008; Bowes *et al.*, 2009, 2012b). Furthermore, there is also increasing realisation that lags in response to remediation also apply to P and that chronic release of both P and N from 'legacy' stores in catchments and water bodies poses a major challenge in terms of meeting water quality targets. Indeed, nutrient legacies from past land management may continue to impair water quality over timescales of decades, perhaps longer (Jarvie *et al.*, 2013a; Sharpley *et al.*, 2013).

8.5.1 *Linking nutrients and ecological status of rivers*

Phosphorus in high concentrations within rivers has often been assumed to lead to ecological damage such as excessive plant and algal growth and low night-time dissolved oxygen concentrations within rivers. Agricultural and STW sources have both been implicated. Thus, it is not surprising that there have been great efforts to reduce P loss from agricultural land and reduce P concentrations within effluents discharged to rivers from STWs even though the financial costs have been high (Neal *et al.*, 2010c).

However, in order to assess biological impacts and recovery, all parts of the aquatic food chain need consideration. Further, there are physical effects such as temperature and flow that critically affect biological functioning and most UK lowlands rivers have been greatly affected by river and water management such as abstractions for water supply

and river straightening that have changed the flow regime of the river. Other changes include bank-side clearance of trees that increase light levels and temperature and these all affect the functioning of the river ecosystem and can destroy bank-side refuges for the zooplankton and invertebrates. There are other indirect influences of humans and these include high fish stocking levels designed to enhance the fishing amenity value of many of our rivers, and impoundments/sluices that reduce flow velocities, promoting algal growth. Many fish feed on the zooplankton and invertebrates that feed on the phytoplankton and invertebrates. The net effect has thus been to increase algal development, and it is therefore inappropriate to simply target/vilify SRP as the culprit of biological decline or the solution to the problem if SRP levels are reduced. The complexities in aquatic ecosystem response to changing nutrient concentrations may also explain the 'inconvenient truth' that, despite four decades or more of remediation measures to reduce P inputs, in many cases there has not yet been overwhelming success in reducing nuisance algal growth. In some cases, there has been decoupling of algal growth responses to river P loading in eutrophically impaired rivers, and recovery trajectories may be non-linear and characterised by thresholds and alternative stable states (Jarvie *et al.*, 2013b). With the Water Framework Directive requirement to achieve good ecological status, better understanding of the role of changing nutrient concentrations (in relation to other physiochemical and biological controls) is vital in designing appropriate P concentration standards to protect and improve river water quality. Experimental flumes that allow detailed control of nutrient concentrations, light levels and flow have been designed to explore impacts of both nutrient addition and nutrient removal on nuisance algal growth (Bowes *et al.*, 2007, 2012b).

8.6 Steroid Oestrogens

In the 1990s, much effort in monitoring and modelling focused on the challenge of pesticide contamination of rivers and the exposure of wildlife to persistent organic pollutants (POPs). In these cases, knowledge of the physico-chemical properties of the pollutant chemical was vital. For pesticides, predictive modelling required an appreciation of the complexities of soil hydrology and physics together with the processes of biodegradation, sorption and volatilisation (see Section 8.6).

For the POPs, the physico-chemical parameters would tell you into which environmental compartment the chemicals would partition. In both cases, the environment tended to be 'generic'.

Environmental scientists tended to think of sewage effluent simply as a source of BOD and nutrients to the aquatic environment, whilst the mysteries of sewage treatment technology were given scarcely a thought. However, in the early 1990s, the issue of endocrine disruption, where male fish in proximity to sewage effluents were discovered to be displaying female characteristics came as a great shock (Purdom *et al.*, 1994). In due course, the most probable causative agents were identified as the natural female oestrogen hormones and the active ingredient of the contraceptive pill ethinyloestradiol (Desbrow *et al.*, 1998). An important question was how widespread could this problem be in the UK? Finding out by a monitoring programme would mean killing thousands of fish and examining the gonads of the males was both expensive and awkward if trying to preserve fish was the aim. Measuring ng/L concentrations in sewage effluent and sub-ng/L concentrations in rivers was and remains extremely challenging (and expensive) for analytical chemists to achieve. On the other hand, as the source was the human population, perhaps oestrogen concentrations were predictable? For Wallingford scientists to make a contribution they needed to come to terms and understand the medical literature on the human excretion of hormones in order to estimate likely loading at sewage treatment works. Then entering into the murky field of STWs efficacy in removing these compounds, to understand how they worked and influenced what came out in the effluent. The final step was to appreciate the importance of dilutions and the need to connect with GIS-based water resource models to map out concentrations and hence risk.

Probably one of the most important developments that assisted this endeavour was the timely availability of almost all peer-reviewed scientific literature on the Internet via such sites as the 'Web of Science'. This allows scientists to explore a new field to discover and research the literature through the use of keywords without the need to trawl through libraries and struggle to photocopiers with heavy journal volumes. Through this electronic-assisted review of the literature, models could be developed to predict human excretion. Wallingford involvement in UK, European projects and reviews of the literature meant that sewage treatment removal rates could then be incorporated into the model (Johnson *et al.*, 2005; Johnson and Sumpter, 2001; Johnson *et al.*, 2007b). The approach of studying consumption of a pharmaceutical (or personal care product), excretion and sewage removal was an approach that would be followed to study many similar contaminants emanating from the home in due course. Scientists could then predict the concentration of estrogens in sewage effluent, which was seen as a function of the human population served, the flow through the STW and the removal based on the treatment type (Johnson and Williams, 2004). Thus, the oestrogen discharge model only required the local population, flow and treatment type of the STWs to predict effluent values (Figure 8.2). On its own it could predict the degree of endocrine disruption in fish living immediately downstream of sewage effluent (Jobling *et al.*, 2006). This model was possibly the first of its kind to predict successfully the concentration of a micro-organic chemical excreted by humans from an STW. However, fish live in the river and not the sewage effluent, so how could predictions be made for estrogens along the length and breadth of Britain's river network?

Fortunately, Wallingford hydrologists had been working on and perfecting for many years geographic-based water resource models. Thirty years of rainfall data for river basins could be converted to runoff to give a probabilistic distribution of natural flows and the Low Flows2000 modelling system was created (Young *et al.*, 2003). To get the water balance correct also needed was the location and quantity of water discharged by STWs into the river network. Thus, the vital facility of identifying where chemicals from the human population would be discharged was available. Through collaboration with the UK water industry, it was possible to identify all the vital statistics of Britain's major STWs and the treatment type each used. The LF2000 model was further developed to include a water quality module (WQX) so that it could predict concentrations of a chemical in a river, not only based on input and dilution, but also through a natural loss rate, generally assumed to be degradation. Thus, given the volume and length of a river reach together with flow velocity, a residence time could be calculated. When combined with a loss rate, the amount of parent molecule surviving to the end of each river reach could be calculated.

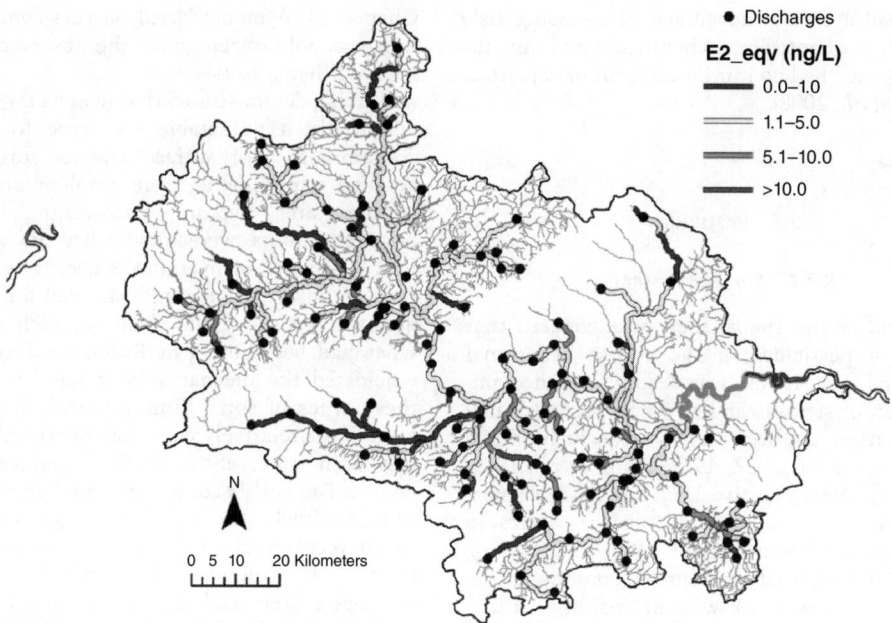

Fig. 8.2 Prediction of combined steroid oestrogen concentrations (estradiol equivalents, E2_eqv) in the Thames catchment originating from the human population and their sewage treatment plants for the month of August. Note concentrations above 1 ng/L suggest some level of endocrine disruption with more significant disruption occurring over 5 ng/L. Thin blue lines denote the river network not impacted by significant discharges. (Source: Reproduced with permission of CEH.) *(See insert for colour representation of this figure.)*

Most organic chemicals are not conservative and estrogens were no exception. Wallingford scientists took water samples from several rivers at several times of the year and spiked these with estrogens to assess possible degradation rates and from these rate constants could be derived (Jurgens *et al.*, 2002).

The excretion/discharge model could now be combined with a geographic-based water quality model to predict concentrations of estrogens throughout Britain's rivers (Williams *et al.*, 2009). These predictions were typically within an order of magnitude of measured concentrations (Huo and Hickey, 2007; Williams *et al.*, 2012). However, the vital question was 'how widespread could endocrine disruption be in British fish?' To this end, the hydrologists and environmental chemists at Wallingford developed a close collaboration with ecotoxicologists at Brunel University (Sumpter and Johnson, 2005). Thus, a concentration of a mixture of chemicals in any one point on a river could be converted into an expected degree of intersex in the

fish resident in that part of a river (Sumpter *et al.*, 2006).

Finally the question 'how widespread could endocrine disruption be in British fish' could be addressed without killing millions of fish or spending millions of pounds on laborious chemical analysis. The answer appeared to be startling – roughly 1/3 of river locations below major STWs in England and Wales had some degree of endocrine disruption predicted for the resident fish (Williams *et al.*, 2009). It was clear that a densely populated small island with small rivers would be highly exposed to chemicals from the human population.

Wallingford scientists, chemists and hydrologists working together and with the assistance of university ecotoxicologists made a significant contribution to an environmental issue of worldwide importance. The dominance of local hydrology in determining the exposure of wildlife to chemicals from sewage effluent was demonstrated (Johnson, 2010; Johnson *et al.*, 2011b). It was observed that models had a crucial advantage in assessing risk from minute quantities of chemical contaminants in rivers – they had no minimum limit of detection (Johnson *et al.*, 2008).

8.7   Pesticides

8.7.1   *Surface waters*

At the end of the 1980s, there was concern that the use of pesticides in the course of normal agricultural production was leading to the contamination of streams and rivers in the UK. Such contamination had been shown to occur in North America (Johnston *et al.*, 1967, Frank *et al.*, 1982, Wauchope, 1978). In the UK, only a few measurements had been made in surface waters in agricultural catchments (Croll, 1986, Clark *et al.*, 1991). These showed widespread contamination, but made no causal link with normal agricultural applications, rather residues were attributed to spills and amenity use (Croll, 1986). Wallingford scientists proposed research to monitor a headwater agricultural catchment at the Agricultural Development Advisory Service (ADAS) research farm at Rosemaund, Herefordshire, specifically to study pesticide runoff resulting from normal agricultural applications and the role played by hydrology in these runoff processes (Williams *et al.*, 1991, Matthiessen, *et al.*, 1992, Williams *et al.*, 1995, Turnbull *et al.*, 1997). This work was funded by the EA. Subsequently, two other groups from the Ministry of Agriculture Fisheries and Food (MAFF, now Defra) and the Building Research Establishment collaborated in research at the same location. The studies showed the direct causal link between the use of pesticides in agriculture and contamination of a headwater stream. Moreover, the pollution was shown to be episodic, that is, driven by rainfall events. The amounts leaving the fields during these events were a very small fraction of that applied (a few per cent) but transient concentrations were high and could have an ecological impact (Matthiessen *et al.* 1995). It was established that the likely mechanism of pesticides transport was through the high permeable layers in and around the permanent drainage channels below the fields. These were activated under heavy rainfall, with water flowing laterally through the surface layers to be intercepted by the drains (see

Chapter 4). A model based on this conceptualisation was able to reproduce the observed pesticide loss (Williams, 1998).

Having demonstrated that heavy clay soils were potentially a vulnerable soil type for pesticide contamination of surface waters, further more detailed experiments were implemented at the another site, Wytham, Oxfordshire. Here, two large soil plots were established in which the flow pathways of overland, subsurface and drain flow could be isolated and sampled for pesticides (Johnson *et al.*, 1995). The research confirmed what had been found at Rosemaund and further elucidated the mechanisms of rapid transport in these types of soils. Simply, during high rainfall events, the relatively porous and permeable topsoil fills with water above the less permeable lower layers. The soil becomes saturated and the water can flow freely over and through the soils directly to intercept cracks and other macro-pore features in the clay structure, which link directly to the drainage system and thus move rapidly to surface water bodies (Haria *et al.*, 1994; Johnson *et al.*, 1998). The significance of this is that a direct route for water draining pesticide-rich soils to drains 1 m below the surface had been identified. There was little interaction between the soil and the pesticide during runoff events, thus the normal means of ameliorating the pesticides concentrations were not present. Only if there was a sufficient time between the pesticide application and the first large rainfall event, there was sufficient time for soil degradation to reduce the amount of pesticide available for leaching (Heppell *et al.*, 1999).

8.7.2   *Groundwater*

The idea that (shrink-swell) cracking in clay soils gave rise to rapid transport routes for pesticides encouraged Wallingford researchers to collaborate with their BGS colleagues to consider if such mechanisms might exist for groundwater contamination. The Chalk with its well-known fractured nature was an obvious candidate and so a field site was established on the Chalk near Wonston in southern England, which had shallow soils and a relatively shallow water table. The thinking was that if rapid flows of pesticides to groundwater would happen anywhere, they would happen in such locations. Two herbicides were studied: isoproturon (IPU) and chlortoluron (CTU) in both groundwater and shallow unsaturated zone sediments. Concentrations of IPU in groundwater samples varied from < 0.05 to 0.23 µg/L over a 5-year period of monitoring, with variations rising and falling in response to applications and timing of rainfall events. Concentrations of pesticides in groundwater samples collected during periods of rising water table were significantly higher than pumped samples and suggested that rapidly infiltrating recharge water contains higher herbicide concentrations than the native groundwater, confirming the rapid flow mechanisms hypothesised in advance of the study. Significant variations in herbicide concentrations were observed over a 3-month period in groundwater samples collected by an automated system, with concentrations of IPU ranging from 0.1 to 0.5 µg/L, and concentrations of a recent application of CTU ranging from 0.2 to 0.8 µg/L. The results of this study indicated that rapid transport of IPU and CTU through the unsaturated zone to shallow groundwater occurs and that this transport increases immediately following herbicide application (Gooddy *et al.*, 2001; Haria *et al.*, 2001, 2003; Johnson *et al.*, 2001; Chilton *et al.*, 2002, 2005)

The concentrations observed in the rapidly infiltrating water, although higher than the native grounds water, were still generally low and although they sometimes exceeded the 0.1 µg/L drinking water limit, this would be a local phenomenon. By the time water had travelled to water supply boreholes, natural attenuation processes would most likely have reduced these concentrations considerably. The studies had shown previously that there was a small but significant potential for IPU to sorb to the upper chalk (Johnson *et al.*, 1998). Although little (but highly variable) degradation of IPU was observed in laboratory tests systems containing unsaturated chalk, there was significant degradation measured in the groundwater itself (Johnson *et al.*, 1998). Further laboratory studies using chalk cores extracted from the site and irrigated with sterile groundwater showed retardation of IPU in the breakthrough from the bottom of the column, compared to that of bromide (an inert tracer), thus supporting the case for some sorption to the chalk. When non-sterile groundwater was used, the breakthrough was similar, but the amount of IPU recovered was significantly reduced suggesting degradation was indeed a possible loss route (Besien *et al.*, 2000).

The research undertaken at Wallingford highlighted the huge importance of the water flow mechanisms in pesticide transport. This applied equally to translocation to surface and groundwater with similar mechanisms occurring in both. Soil degradation rates of pesticides were well known at this point as an important amelioration mechanism for transport to surface and groundwaters, but the occurrence of significant degradation in groundwater was new. Strikingly, this degradation rate could be faster than that in the unsaturated zone, something that was not expected. The Wallingford contribution to this research area spanned a relatively short period of time but was significant for all that.

8.8 Water quality modelling

Modelling of water quality has not happened in isolation from other research at IH and CEH. It will have been apparent from the previous sections that modelling has arisen in monitoring and process studies, to help explain observations and to understand the effects of processes on catchment-scale behaviour of a wide range chemicals (natural and anthropogenic in source). This section therefore gives the background to how water quality modelling emerged as a research and management tool in its own right and highlights particular important studies that have been undertaken in following that road.

Water quality modelling research was very limited at IH prior to the 1980s. The IH mandate was focused on the water quantity aspects of research within the Plynlimon, Coalburn and Balquidder experiments (see Chapter 2). The main issue in these experiments was to determine how changes in land use such as forestation and deforestation would affect water balances in upland catchments. However, in 1980, a new water quality modelling research group was formed at IH, following two successful bids for funding from Anglian Water and Thames Water.

The project for Anglian Water involved applying the Bedford Ouse dynamic water quality model (Whitehead *et al.*, 1981) to the Ouse and to set this up as part of a real-time forecasting system to protect the water supplies for Bedford (Whitehead *et al.*, 1984; Tester *et al.*, 1992). This application used for the first time a mini computer to provide pollution control officers with an online system, which they could use to predict pulses of pollution moving down the river system. It was brought into play for real when a major pollution event from upstream Milton Keynes threatened public water supplies at Bedford. Fortunately, the system worked and Anglian Water was able to switch in alternative supplies.

The second project for Thames Water involved modelling the whole of the Thames River system from Cricklade to Teddington in order to assess nitrogen issues on the river and to help with management strategies to reduce nitrogen. At that time, nitrate-N was above the EU limit of 11.3 mg N/L on occasions, which caused serious problems with public water supply. A water quality model was developed that for the first time related agricultural runoff loads to in-stream water quality (Whitehead and Williams, 1984, Whitehead, 1990). This led to a focus on the ecological impacts of nutrients in the Thames, and Whitehead and Hornberger (1984) developed a new dynamic model of algal growth in the Thames. Stochastic techniques such as the extended Kalman filter and Monte Carlo analysis were used to quantify parametric uncertainty and identify the key parameters controlling algal behaviour.

This developing modelling expertise led to the development of the QUASAR model at Wallingford, which simulated the dynamics and processes of dissolved oxygen, biochemical oxygen demand, nitrate, ammonia and phytoplankton (Whitehead *et al.*, 1997a). Many applications of QUASAR were undertaken including modelling pollution in catchments as part of the LOIS research programme, assessing ecological flows for the Teddington Weir (for Thames Water) and assessing the impacts of the Maidenhead Flood Relief Channel on the River Thames (for the EA). In addition, an environmental impact assessment was undertaken on the upgrade of Didcot power station, investigating potential impacts on water quality of the Thames. There were also EU projects on water quality modelling in rivers systems to study process dynamics and pollution control. The model QUASAR then became the QUESTOR model following a major overhaul of the code.

QUESTOR was developed directly from the LOIS research programme as a means of linking water quality concentrations and loads arising from diffuse and point sources across the whole of North East England to models of the estuary and near coast. This application was further developed in the EU-funded project CHESS (Climate, Hydrochemistry and Economics of Surface-water Systems) which introduced the biological sub-model

(Boorman, 2003a–c). A full sensitivity analysis of the QUESTOR model applications to the Rivers Aire and Ouse was made as part of the EU project BMW (Benchmarking Models for the Water Framework Directive). This showed that in the urban River Aire discharges from point sources were more influential than in-stream processes, whereas for the rural River Ouse, tributaries and process model parameters had equal weight on the model predictions (Deflandre *et al.*, 2006). The QUASAR model also led to the development of another model INCA (Whitehead *et al.*, 1998a,b) as part of an NERC project and 3 EU projects, namely EU INCA, EUROLIMPACS and REFRESH. The focus of these projects was to assess the impacts of land use and climate change on hydrology, water quality and ecology across Europe. INCA utilises the basic in-stream modelling of QUASAR but has a processes-based dynamic model for the terrestrial system, so that soil chemistry and water movements through the landscape and groundwater could be modelled.

More recent water quality modelling developments have concentrated on systems designed for national- and continental-scale assessments. The LF2000-WQX model was developed as a means of predicting environmental concentrations of chemicals released down the drain, so that risk assessment could be made for chemicals that are used routinely in daily lives. The simulation steps involve estimating point-source load data for a chemical, simulating in-stream removal and transport processes using a first-order decay coupled with a Monte Carlo dynamic mass balance of inputs, which are described probabilistically, followed by post-processing of the simulation results to provide both tabulated values of predicted environmental concentrations (PECs) and a graphical interpretation of the PECs (CEH, 2003). LF2000-WQX can also model conservative and non-conservative determinands based on measured data and model dissolved oxygen concentrations using a modified Streeter–Phelps model. This functionality was added so that any conceptual representation of the river system and the point-source discharges to that system can be evaluated by modelling general quality assessment (GQA)-measured water quality determinands collected routinely by the EA. The application of this model to steroid oestrogens (chemicals that have been shown to have a feminisation effect on male fish) is described in Section 8.6.

The need to investigate the effects of policies, socio-economic forces and combinations of these two, which act at the scale of the European Union prompted the development of continental-scale water quality models. Within the EU SCENES project, a WQX was developed for the already established water resources model, GWAVA (Meigh *et al.*, 1999). The WQX calculates the water concentrations throughout Europe representing the land areas as a series of 177,470 grid squares of approximately 6×9 km (5 by 5 arc minutes) dimensions connected by a river network. The model uses geographic data on the location and size of the human EU population and their association with STWs. The flows through these STWs are incorporated with other flows and abstractions into the hydrological model. The hydrology is driven by monthly climate over the period 1970–2000. The discharge of the chemicals from the sewage treatment works and from diffuse sources is calculated for each grid cell and is diluted in each grid square of the model by the 31 years of flow (1970–2000) calculated in monthly time steps (Dumont *et al.*, 2012). The initial application mapped water scarcity across the EU combining 'traditional' water scarcity measures with measures of water quality which influenced whether the (untreated) water was of potable quality (Dumont *et al.*, 2012). The model has shown itself to be particularly useful for examining the distributions of the concentrations of 'down-the-drain' chemicals across Europe and to make risk assessment by making comparisons with known effect levels (Johnson *et al.*, 2013). Further applications are expected to investigate the emerging contaminants that are referred to in the last section of this chapter.

In parallel with the development of the lowland river models, there was increasing attention on the uplands and the impacts of acidification across the UK and Northern Europe (UKAWRG, 1988). Following the announcement of the Royal Society Surface Waters Acidification Project, Sir John Mason, the programme Director, decided to accept that British researchers should be involved in the programme and asked for catchment studies in the UK to assess impacts of changing air pollution causing acid rain on water quality and ecology in lakes and rivers and also to develop models for assessing the science and management of acidification. One of the catchments set up was Allt a'Mharcaidh in the Cairngorms

in Scotland, which provided the hydrological and hydrochemical data to underpin advances in process understanding to enable further development of hydrochemical models (Jenkins *et al.*, 1994).

In the early 1980s, a key model describing the onset and recovery of acidification of soil and surface water (MAGIC) was developed in the USA by Jack Cosby and colleagues at the University of Virginia (Cosby *et al.*, 1985). Collaboration with this group proved extensive and highly rewarding in the field of acidification and its reversal. Joint work began with an application of the model to the Loch Dee system in South West Scotland (Cosby *et al.*, 1986). This reconstructed the historical behaviour of the acidification processes in the catchments and was used to predict future change under alternative sulphur reduction strategies. Several applications to catchments and regions followed, and Figure 8.3 shows a typical application to Scoat Tarn in the Lake District (Whitehead *et al.*, 1997b). It is remarkable that the model has been shown to reproduce the main trends in the recovery process as indicated in Figure 8.3. Further research was undertaken with Ormerod *et al.* (1990) to link the chemistry outputs from MAGIC to the biology of streams, so that the impact on Trout populations and invertebrate diversity could be evaluated, and the opportunities for recovery assessed. At the same time, a vigorous debate with the Forestry Commission was in progress over the impacts of upland forestation

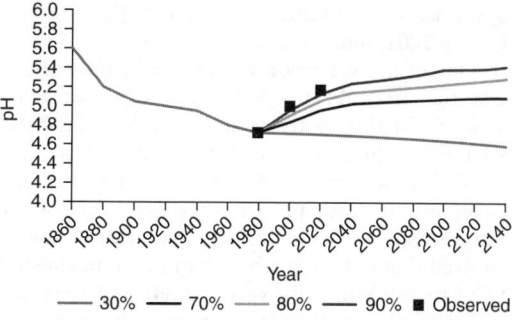

Fig. 8.3 Reconstruction of past pH at Scoat Tarn in the Lake District together with predictions for various sulphur reductions – plus (in red) the observed pH from the Acid Waters Monitoring Network at CEH. (Source: Reproduced with permission of CEH.) *(See insert for colour representation of this figure.)*

on acidification, and the MAGIC model proved exemplary in showing that forests had a significant effect by capturing increased loads of anions of sulphur, chloride and nitrate with conifer forestation doubling the pollution load. This caused much controversy at the time, but has now largely been accepted and supported by modelling studies (Jenkins *et al.*, 1990; Cosby *et al.*, 1990) and by experimental evidence. This has resulted in forest management practices being updated to remove large 'even-aged' and single species plantations in upland areas (as discussed previously in this chapter).

Since the early 1980s, the growing understanding of the link between atmospheric emission, transport, deposition and impacts of sulphur and nitrogen compounds prompted negotiations on an international scale to reduce the acidification problem. Negotiations were undertaken under the auspices of the UNECE Convention on Long-Range Trans-boundary Air Pollution and the subsequent protocols were negotiated on the basis of reducing ecosystem effects and utilised the concept of critical loads. The critical load is defined as a quantitative estimate of the exposure to one or more pollutants below which significant harmful effects on specified sensitive elements of the environment do not occur. Under this definition, the critical load does not incorporate any measure of time for the deposition reduction to be implemented or for the time involved for the ecosystem response. To supplement the negotiations, several scenarios were undertaken using MAGIC on a range of catchments in the UK and Europe in order to demonstrate the advantages of pollution reduction (Jenkins and Cullen 2001; Jenkins *et al.*, 2003a).

One of the great beauties of the MAGIC model was that it could produce predictions that otherwise could not easily be visualised or described within the time frames of decision making. The MAGIC model provided a great learning tool and gave insights into process that otherwise would have been missed. For example, the MAGIC model showed that even though calcium and magnesium were hydrochemically very similar and took part in both cation exchange and weathering reactions, their concentrations could vary in different ways with long-term changes in acidic deposition and forestry plantation/harvesting cycles (Neal *et al.*, 1995).

With the original MAGIC modelling, a one-box model was used that integrated the processes that occurred within the soil and groundwater regions even though they could have very distinct chemistries. Also, as a consequence of this lumping, the model predicted average annual chemistries rather than extremes and thus could then miss the acidic extremes that determine biological deterioration within the stream. Spurred by the work on endmember mixing as discussed earlier and an extremely dynamic modelling research environment, the MAGIC model was extended to include the surface and groundwater regimes and thereby give estimates of the chemistries at both baseflow and, more crucially, the acidic stormflow endmember (Neal *et al.*, 1992a). The MAGIC model was applied across Europe and further afield with a core modelling initiative being here (and Virginia of course). One unusual but telling application of the modelling was to examine strategically important areas of the world such as Amazonia where the acidification and climate change research were just developing (Neal *et al.*, 1992b).

As a final step, the MAGIC model was restructured (Cosby *et al.*, 2001) to incorporate detailed nitrogen dynamics and to enable the calculation of target loads (Jenkins *et al.*, 2003b). Target loads enable determination of the level of emission reduction needed to restore an acidified soil or water by a certain time in the future. At the time of writing this chapter, sulphur emissions have declined to around 10% of their peak levels in the 1970s, surface water acidity is declining across Europe and North America and biological systems are recovering in many areas. This is perhaps one of the greatest success stories for environmental science.

Magic was not the only international model used in relation to surface water acidification. Another, which considered in much more detail dynamic, short-term changes in water quality was the Birkenes model of Nils Christophersen and co-workers in Norway (Christophersen *et al.*, 1982). This led to many lines of new research including a deeper examination of how catchments actually functioned (Christophersen and Neal, 1990) and ultimately to issues of fractal functioning that was covered earlier in this chapter. Another line of modelling followed from collaboration with Keith Beven of the University of Lancaster, which linked streamwater dynamics to both time series analysis and linkages with a major hillslope hydrological framework, TOPMODEL (Robson *et al.*, 1992, 1994).

8.9 Future Challenges and Opportunities

It has already been noted for the uplands that high-frequency monitoring gives greater insights into the way chemicals move through catchments and the way that short-term variability looks different from that caught at weekly or monthly time steps (Kirchner *et al.*, 2000, 2004). Instrumentation that automatically measures and records simple water quality indicators such as pH, dissolved oxygen, temperature and conductivity have been around for quite a while (e.g. Whitehead *et al.*, 1984, Tester *et al.*, 1992; Williams *et al.*, 2000); however, recently automatic nutrient analysers and nitrogen probes have become available, and CEH were at the forefront of adopting this new technology, applying the hourly concentration data it produced on the River Kennet to investigate diurnal chemical and biological cycles (Palmer-Felgate *et al.*, 2008) and to determine changing nutrient sources through storm events (Bowes *et al.*, 2012c). This work has advanced in recent years, in collaboration with the University of Reading and the EA, through EPSRC funding, resulting in the design of specialised monitoring kiosks, integrating nutrient auto-analysers with the latest probe technology for delivering continuous high-quality pH, temperature, dissolved oxygen, turbidity, chlorophyll, nitrate and ammonium concentrations and transmitting real-time data via telemetry (Wade *et al.*, 2012). This monitoring kiosk design has now been adopted at multiple sites across the country, in monitoring programmes such as the Defra Demonstration Test Catchments project. Such systems are likely to bring great advances in our understanding of nutrient dynamics and how these affect such things as algal growth and species succession.

Predicting the effect of water chemistry and other factors on the ecological response of a river system is a challenge and one that the EU Water Framework Directive (WFD) forces us to address. This is a current focus of research at Wallingford by incorporating algal growth models into the QUESTOR model (Hutchins *et al.*, 2010, Waylett *et al.*, 2013). These models have shown promise in modelling algal growth in the Thames on a yearly basis, but finding parameters that apply over all years is problematic. This suggests that there are important processes that are missing from the model algorithms: Current thinking has attributed this to the effect of benthic grazers. Although more research is clearly needed, the model is still able to produce useful information on responses of algal growth to external pressures. A QUESTOR model application for the Thames was made for climate periods of 1961 to 1991 and 2040 to 2069 to study the effects of climate change on algal growth. The main drivers of change were river flows and temperature, which were taken from the EA future flows project (Prudhomme *et al.*, 2013). The QUESTOR model was then used to simulate phytoplankton levels and other affected variables, under both flow and temperature regimes for 11 different, but equally likely, present and future climates. The results showed that the number of days when temperature, dissolved oxygen, biochemical oxygen demand and phytoplankton exceeded undesirable values set in the WFD is likely to increase, even assuming no change in nutrient loads (Hutchins *et al.*, 2013).

Crucially, the identification of fractal stream water quality and the observed complexity of both water quality and biological functioning represent a major challenge for the modelling communities. In this regard, one journey ends and an exciting new one starts. Modelling of fractal systems will require new approaches and different ways of thinking. The new lines of measurement and the vast volume of data collected within CEH water quality studies are in the vanguard for change. A fundamental question is to how this challenge will be met within the modelling community.

Trace metal contamination is observed for many lowland UK rivers associated with urban and industrial conurbations and there is enrichment in a number of trace metals. Many of these contaminants come from historic sources rather than from present day inputs, but not all. For example, lead, zinc and copper contamination can occur in regions such as the upper parts of the Humber basin because of flood-plain contamination from mining activities from Roman times to the Industrial Revolution of the eighteenth and nineteenth centuries (Hudson-Edwards *et al.*, 1997; Macklin *et al.*, 1997; Neal *et al.*, 1997e). The realisation that a significant proportion of what is measured as 'dissolved' ($<0.45\,\mu m$) trace metals (and also interestingly P) is colloidal has provided a springboard for new studies on colloid and nanoparticle transport in rivers, using cross flow ultrafiltration (Jarvie *et al.*, 2012b) and small-angle neutron scattering (Jarvie and King, 2007). These techniques, used to probe the behaviour of natural aquatic colloids and nanoparticles, have been applied to novel research on the fate and behaviour of manufactured/engineering nanoparticles from consumer products (Jarvie *et al.*, 2009; Jarvie and King, 2010). This growing area of research in Environmental Nanoscience has widened to catchment modelling studies and environmental risk assessment studies (Johnson *et al.*, 2011a), linked to ecotoxicology studies as is being done currently within the EU NanoFATE project.

New chemicals of concern are always coming onto the radar of scientists and regulators. Indeed, much of the aforementioned research has been targeted at such examples (acidification, steroid oestrogens, pesticides). The current crop of emerging contaminants includes pharmaceuticals, personal care products and nanoparticles. CEH is developing modelling techniques to predict likely exposures in rivers to these chemicals. The main driver for this is the difficulty in analysing these compounds at low levels (almost impossible for nanoparticles), which mean actual measured values are few and far between and insufficient to draw wide conclusions. The methods developed for steroid oestrogens are being developed for other compounds. For example, the techniques learned were put to good use to predict exposure of wildlife in British rivers to pharmaceuticals (Johnson *et al.*, 2007a; Singer *et al.*, 2007) including anti-cancer drugs (Johnson *et al.*, 2008; Rowney *et al.*, 2009), then through the GWAVA model to predict exposure to these same chemicals throughout the continent of Europe (Johnson *et al.*, 2013). Ultimately, by refining spatial modelling, these exposure levels can be compared with known effect levels to produce a prediction of where aquatic species are expected to be at risk from these chemicals. The aim, of course, is to improve the confidence in these model predictions for new chemicals by using emerging measured data to check predictions (Williams *et al.*, 2012). The use of both models and observations in these assessments is a practical and powerful approach (Johnson *et al.*, 2008).

8.10 References

Avila, A., Pinol, J., Roda, F. & Neal, C. (1992) Storm solute behaviour in a montane Mediterranean forested catchment. *Journal of Hydrology*, **140**, 143–161.

Bateman, I.J., Brouwer, R., Davies, H. *et al.* (2006) Analysing the agricultural costs and non-market benefits of implementing the water framework directive. *Journal of Agricultural Economics*, **57**, 221–237.

Besien, T.J., Williams, R.J. & Johnson, A.C. (2000) The transport and behaviour of isoproturon in unsaturated chalk cores. *Journal of Contaminant Hydrology*, **43**, 91–110.

Billen, G., Garnier, J., Mouchel, J.-M. & Silvestre, M. (2007) The seine system: Introduction to a multidisciplinary approach of the functioning of a regional river system. *Science of the Total Environment*, **375**, 1–12.

Boorman, D.B. (2003a) LOIS in-stream water quality modelling. Part 1. Catchments and methods. *Science of the Total Environment*, **314–316**, 379–395.

Boorman, D.B. (2003b) LOIS in-stream water quality modelling. Part 2. Results and scenarios. *Science of the Total Environment*, **314–316**, 397–409.

Boorman, D.B. (2003c) Climate, Hydrochemistry and Economics of Surface-water Systems (CHESS): adding a European dimension to the catchment modelling experience developed under LOIS. *Science of the Total Environment*, **314–316**, 411–437.

Bowes, M.J., Smith, J.T., Hilton, J., Sturt, M.M. & Armitage, P.D. (2007) Periphyton biomass response to changing phosphorus concentrations in a nutrient-impacted river: a new methodology for phosphorus target setting. *Canadian Journal of Fisheries and Aquatic Sciences*, **64** (2), 227–238.

Bowes, M.J., Smith, J.T. & Neal, C. (2009) The value of high-resolution nutrient monitoring: A case study of the River Frome, Dorset, UK. *Journal of Hydrology*, **378** (1–2), 82–96.

Bowes, M.J., Neal, C., Jarvie, H.P., Smith, J.T. & Davies, H.N. (2010) Predicting phosphorus concentrations in British rivers resulting from the introduction of improved phosphorus removal from sewage effluent. *Science of the Total Environment*, **408**, 4239–4250.

Bowes, M.J., Gozzard, E., Johnson, A.E. *et al.* (2012a) Spatial and temporal changes in chlorophyll a concentrations in the River Thames Basin: Are phosphorus concentrations beginning to limit phytoplankton biomass? *Science of the Total Environment*, **426**, 45–55.

Bowes, M.J., Ings, N.L., McCall, S.J. *et al.* (2012b) Nutrient and light limitation of periphyton in the River Thames: Implications for catchment management. *Science of the Total Environment*, **434**, 201–212.

Bowes, M.J., Palmer-Felgate, E.J., Jarvie, H.P. *et al.* (2012c) High-frequency phosphorus monitoring of the River Kennet, UK: are ecological problems due to intermittent sewage treatment works failures? *Journal of Environmental Monitoring*, **14**, 3137–3145.

Cave, R.R., Ledoux, L., Turner, K., Jickells, T., Andrews, J.E. & Davies, H. (2003) The Humber catchment and its coastal area: from UK to European perspectives. *Science of the Total Environment*, **314**, 31–52.

CEC. Council of European Communities (1991) *Directive concerning urban waste water treatment (91/271/EEC)*. Official J, L135/40.

CEC. Council of European Communities (2000). *Water Framework Directive (2000/60/EC)*. Official J, L327/72

Centre for Ecology and Hydrology (2003) *GREAT-ER II: Development of the integrated water resources and water quality data modelling system – phase II. Unpublished draft final report*. Centre for Ecology and Hydrology, Wallingford, UK.

Chilton, P.J., Stuart, M., Gooddy, D., Williams, R.J., Johnson, A.C. & Haria, A.H. (2002) *Transport and fate of pesticides in the Chalk aquifer of southern England*. British Geological Survey, Keyworth, Nottingham.

Chilton, P.J., Stuart, M.E., Gooddy, D.C., Williams, R.J. & Johnson, A.C. (2005) Pesticide fate and behaviour in the UK Chalk aquifer, and implications for groundwater quality. *Quarterly Journal of Engineering Geology and Hydrogeology*, **38**, 65–81.

Christophersen, N., Seip, H.M. & Wright, R.F. (1982) A model for streamwater chemistry at Birkenes, Norway. *Water Resources Research*, **18**, 977–996.

Christophersen, N. & Neal, C. (1990) Linking hydrochemical, geochemical and soil chemical processes on the catchment scale: an interplay between modelling and field work. *Water Resources Research*, **26**, 3077–3086.

Christophersen, N., Neal, C., Hooper, R.P., Vogt, R.D. & Andersen, S. (1990) Modelling streamwater chemistry as a mixture of soilwater end-members - a step towards second-generation acidification models. *Journal of Hydrology*, **116**, 307–320.

Collins, R. & Jenkins, A. (1996) The impact of agricultural land use on stream chemistry in the Middle Hills of the Himalayas, Nepal. *Journal of Hydrology*, **185**, 71–86.

Cooper, D.M., Morris, E.M. & Smith, C.J. (1987) Precipitation and streamwater chemistry in a sub-arctic Scottish catchment. *Journal of Hydrology*, **93**, 221–240.

Cornu, S., Neal, C., Ambrosi, J.P. *et al.* (2001) The environmental impact of heavy metals from sewage sludge in ferralsols (Sao Paulo, Brazil). *Science of the Total Environment*, **271**, 27–48.

Cosby, B.J., Hornberger, G.M., Galloway, J.N. & Wright, R.F. (1985) Modelling the effects of acid deposition: assessment of a lumped-parameter model of soil water and streamwater chemistry. *Water Resources Research*, **21**, 51–63.

Cosby, B.J., Whitehead, P.G. & Neale, R. (1986) A preliminary model of long term changes in stream acidity in South Western Scotland. *Journal of Hydrology*, **84**, 381–401.

Cosby, B.J., Jenkins, A., Ferrier, R.C., Miller, J.D. & Walker, T.A.B. (1990) Modelling stream acidification in afforested catchments: long-term reconstructions at two sites in central Scotland. *Journal of Hydrology*, **120**, 143–162.

Cosby, B.J., Ferrier, R.C., Jenkins, A. & Wright, R.F. (2001) Modelling the effects of acid deposition: refinements, adjustments and inclusion of nitrogen dynamics in the MAGIC model. *Hydrology and Earth System Sciences*, **5**, 499–517.

Clark, L., Gomme, J. & Hennings, S. (1991) Study of pesticides in waters from a chalk catchment, Cambridgeshire. *Pesticide Science*, **32**, 15–33.

Croll, B.T. (1986) The effects of the agricultural use of pesticides on freshwater. In: Solbe, J.F.D.L.G. (ed), *Effects of Land use on Freshwaters: Agriculture, Forestry, Mineral Exploitation*. Urbanisation. Ellis Horwood, Chichester, pp. 201–206.

Davies, H.N. & Neal, C. (2007) Estimating nutrient concentrations from catchment characteristics. *Hydrology and Earth System Sciences*, **11**, 550–558.

Deflandre, A., Williams, R.J., Elorza, F.J., Mira, J. & Boorman, D.B. (2006) Analysis of the QUESTOR water quality model using a Fourier amplitude sensitivity test (FAST) for two UK rivers. *Science of the Total Environment*, **360**, 290–304.

Desbrow, C., Routledge, E.J., Brighty, G.C., Sumpter, J.P. & Waldock, M. (1998) Identification of estrogenic chemicals in STW effluent. 1. Chemical fractionation and in vitro biological screening. *Environmental Science & Technology*, **32**, 1549–1558.

Dumont, E., Williams, R., Keller, V., Voß, A. & Tattari, S. (2012) Modelling indicators of water security, water pollution and aquatic biodiversity in Europe. *Hydrological Sciences Journal*, **57**, 1378–1403.

Durand, P., Neal, C., Lelong, F. & Didon-Lescot, J.F. (1991) Hydrochemical variations in spruce, beech and grassland areas, Mont-Lozere, Southern France. *Journal of Hydrology*, **129**, 57–70.

Durand, P., Lelong, F. & Neal, C. (1992) Comparison and significance of annual hydrochemical budgets in three small granitic catchments with contrasting vegetation (Mont Lozere, Southern France). *Environment and Pollution*, **75**, 223–228.

Evans, C.D., Cullen, J.M., Alewell, C. *et al.* (2001) Recovery from acidification in European surface waters. *Hydrology and Earth System Sciences*, **5**, 283–298.

Forti, M.C., Boulet, R., Melfi, A.J. & Neal, C. (1999) Hydrogeochemistry of a small catchment in northeastern Amazonia: a comparison between natural with deforested parts of the catchment (Serra do Navio, Amapa State, Brazil). *Water, Air, and Soil Pollution*, **118**, 263–279.

Frank, R., Braun, H.E., Holdrinet, M.V.H., Sirons, G.J. & Ripley, B.D. (1982) Agriculture and water quality in the Canadian Great Lakes Basin: V. Pesticide use in 11 agricultural watersheds and presence in stream water, 1975–1977. *Journal of Environmental Quality*, **11**, 497–505.

Gooddy, D.C., Bloomfield, J.P., Chilton, P.J., Johnson, A.C. & Williams, R.J. (2001) Assessing herbicide concentrations in the saturated and unsaturated zone of a Chalk Aquifer in southern England. *Ground Water*, **39**, 262–71.

Halliday, S.J., Wade, A.J., Skeffington, R.A. *et al.* (2012) An analysis of long-term trends, seasonality and short-term dynamics in water quality data from Plynlimon, Wales. *Science of the Total Environment*, **434**, 186–200.

Haria, A.H., Johnson, A.C., Bell, J.P. & Batchelor, C.H. (1994) Water-movement and isoproturon behavior in a drained heavy clay soil. 1. Preferential flow processes. *Journal of Hydrology*, **163**, 203–216.

Haria, A.H., Johnson, A.C. & Hodnett, M.G. (2001) Pesticide penetration and groundwater recharge on a chalk hillslope in southern England. In: Walker, A. (ed), *Pesticide Behaviour in Soils and Water*. Proceedings No. 78 of British Crop Protection Council, pp. 189–194.

Haria, A.H., Hodnett, M.G. & Johnson, A.C. (2003) Mechanisms of groundwater recharge and pesticide penetration to a chalk aquifer in southern England. *Journal of Hydrology*, **275**, 122–137.

Heppell, C.M., Burt, T.P., Williams, R.J. & Haria, A.H. (1999) The influence of hydrological pathways on the transport of the herbicide, isoproturon, through an underdrained clay soil. *Water Science and Technology*, **39**, 77–84.

Hill, T. & Neal, C. (1997) Spatial and temporal variation in pH, alkalinity and conductivity in surface runoff and groundwater for the Upper River Severn catchment. *Hydrology and Earth System Sciences*, **1**, 697–716.

Hill, T., Whitehead, P. & Neal, C. (2002) Modelling long-term stream acidification in the chemically heterogeneous Upper Severn catchment, Mid-Wales. *Science of the Total Environment*, **286**, 215–232.

House, W.A. & Warwick, M.S. (1998) Intensive measurements of nutrient dynamics in the River Swale. *Science of the Total Environment*, **210**, 111–137.

House, W.A. & Warwick, M.S. (1999) Interactions of phosphorus with sediments in the River Swale, Yorkshire, UK. *Hydrological Processes*, **13**, 1103–1115.

House, W.A. & Denison, F.H. (2002) Exchange of inorganic phosphate between river waters and bed-sediments. *Environmental Science and Technology*, **36**, 4295–4301.

Howden, N.J.K., Burt, T.P., Worrall, F., Whelan, M.J. & Bieroza, M. (2011) Nitrate concentrations and fluxes in the River Thames over 140 years (1868–2008): are increases irreversible? *Hydrological Processes*, **23**, 2657–2662. doi:10.1002/hyp.7835.

Hudson-Edwards, K.A., Macklin, M.G. & Taylor, M. (1997) Tracing historic metal mining related sediment inputs to the Tees river basin from source to fluvial sink. *Science of the Total Environment*, **194/195**, 437–445.

Huo, C.X. & Hickey, P. (2007) EDC demonstration programme in the UK Anglian Water's approach. *Environmental Technology*, **28**, 731–741.

Hutchins, M.G., Johnson, A.C., Deflandre-Vlandas, A., Comber, S., Posen, P. & Boorman, D. (2010) Which offers more scope to suppress river phytoplankton blooms: Reducing nutrient pollution or riparian shading? *Science of the Total Environment*, **408**, 5065–5077.

Hutchins, M., Elliot, A., Caillout, L. & Williams, R.J. (2013) *Understanding the effects of climate change on water quality: A case-study assessment on rivers and lakes in England, Defra, Water Availability and Quality R&D programme*: p 29.

Jarvie, H.P., Neal, C., Leach, D.V., Ryland, G.P., House, W.A. & Robson, A.J. (1997) Major ion concentrations and the inorganic carbon chemistry of the Humber rivers. *Science of the Total Environment*, **194/195**, 285–302.

Jarvie, H.P., Neal, C., Withers, P.J.A., Robinson, A. & Salter, N. (2003) Nutrient water quality of the Wye catchment, UK: exploring patterns and fluxes using the Environment Agency data archives. *Hydrology and Earth System Sciences*, **7**, 722–743.

Jarvie, H.P., Neal, C. & Williams, R.J. (2004) Assessing changes in phosphorus concentrations in relation to in-stream plant ecology in lowland permeable catchments: Bringing ecosystem functioning into water quality monitoring. *Water, Air, and Soil Pollution*, **4**, 641–655.

Jarvie, H.P., Neal, C., Withers, P.J.A., Wescott, C. & Acornley, R.M. (2005a) Nutrient hydrochemistry for a groundwater-dominated catchment: The Hampshire Avon, UK. *Science of the Total Environment*, **344**, 143–158.

Jarvie, H.P., Jurgens, M.D., Williams, R.J. *et al.* (2005b) Role of river bed sediments as sources and sinks of phosphorus across two major eutrophic UK river basins: the Hampshire Avon and Herefordshire Wye. *Journal of Hydrology*, **304 (1–4)**, 51–74.

Jarvie, H.P., Neal, C. & Withers, P.A. (2006a) Sewage-effluent phosphorus: A greater risk to river eutrophication than agricultural phosphorus? *Science of the Total Environment*, **306**, 243–253.

Jarvie, H.P., Neal, C., Jurgens, M.D. *et al.* (2006b) Within-river nutrient processing in Chalk streams: The Pang and Lambourn, UK. *Journal of Hydrology*, **330 (1–2)**, 101–125.

Jarvie, H.P. & King, S.M. (2007) Small-angle neutron scattering study of natural aquatic nanocolloids. *Environmental Science & Technology*, **41 (8)**, 2868–2873.

Jarvie, H.P., Withers, P.J.A., Hodgkinson, R. *et al.* (2008a) Influence of rural land use on streamwater nutrients and their ecological significance. *Journal of Hydrology*, **350 (3–4)**, 166–186.

Jarvie, H.P., Mortimer, R.J.G., Palmer-Felgate, E.J., Quinton, K.S., Harman, S.A. & Carbo, P. (2008b) Measurement of soluble reactive phosphorus concentration profiles and fluxes in river-bed sediments using DET get probes. *Journal of Hydrology*, **350 (3–4)**, 261–273.

Jarvie, H.P., Al-Obaidi, H., King, S.M. *et al.* (2009) Fate of silica nanoparticles in simulated primary wastewater treatment. *Environmental Science & Technology*, **43 (22)**, 8622–8628.

Jarvie, H.P., Withers, P.J.A., Bowes, M.J. *et al.* (2010) Streamwater nutrients and their sources across a gradient in rural-agricultural land use intensity. *Agriculture, Ecosystems and Environment*, **135**, 238–252.

Jarvie, H.P. & King, S.M. (2010) Just scratching the surface? New techniques show how surface functionality of nanoparticles influences their environmental fate. *Nano Today*, **5** (4), 248–250.

Jarvie, H.P., Neal, C., Withers, P.J.A., Baker, D.B., Richards, R.P. & Sharpley, A.N. (2011) Quantifying phosphorus retention and release in rivers and watersheds using extended end-member mixing analysis (E-EMMA). *Journal of Environmental Quality*, **40**, 492–504.

Jarvie, H.P., Sharpley, A.N., Scott, J.T., Haggard, B.E., Bowes, M.J. & Massey, L.B. (2012a) Within-river phosphorus retention: accounting for a missing piece in the watershed phosphorus puzzle. *Environmental Science & Technology*, **46** (24), 13284–13292.

Jarvie, H.P., Neal, C., Rowland, A.P. et al. (2012b) Role of riverine colloids in macronutrient and metal partitioning and transport, along an upland-lowland land-use continuum, under low-flow conditions. *Science of the Total Environment*, **434**, 171–185.

Jarvie, H.P., Sharpley, A.N., Spears, B., Buda, A.R., May, L. & Kleinman, P.J.A. (2013a) Water quality remediation faces unprecedented challenges from "Legacy Phosphorus". *Environmental Science & Technology*, **47** (16), 8997–8998.

Jarvie, H.P., Sharpley, A.N., Withers, P.J.A., Scott, J.T., Haggard, B.E. & Neal, C. (2013b) Phosphorus Mitigation to Control River Eutrophication: Murky Waters, Inconvenient Truths, and "Postnormal" Science. *Journal of Environmental Quality*, **42** (2), 295–304.

Jenkins, A. (1989) Storm period hydrochemical response in an unforested Scottish catchment. *Hydrological Sciences Journal*, **34**, 393–404.

Jenkins, A. (2001) The sensitivity of headwater streams in the Hindu Kush Himalayas to acidification. *Water, Air, and Soil Pollution*, **2**, 181–189.

Jenkins, A., Cosby, B.J., Ferrier, R.C., Walker, T.A.B. & Miller, J.D. (1990) Modelling stream acidification in afforested catchments: an assessment of the relative effects of acid deposition and afforestation. *Journal of Hydrology*, **120**, 163–181.

Jenkins, A., Ferrier, R.C., Harriman, R. & Ogunkoya, Y.O. (1994) A case study in catchment hydrochemistry: conflicting interpretations from hydrological and chemical observations. *Hydrological Processes*, **8**, 335–349.

Jenkins, A., Sloane, W.T. & Cosby, B.J. (1995) Stream chemistry in the middle hills and high mountains of the Himalayas, Nepal. *Journal of Hydrology*, **166**, 61–79.

Jenkins, A. & Cullen, J.M. (2001) An Assessment of the potential impact of the Gothenburg Protocol on surface water chemistry using the dynamic MAGIC model at acid sensitive sites in the UK. *Hydrology and Earth System Sciences*, **5**, 529–542.

Jenkins, A., Camarero, L., Cosby, B.J. et al. (2003a) A modelling assessment of acidification and recovery of European surface waters. *Hydrology and Earth System Sciences*, **7**, 447–455.

Jenkins, A., Cosby, B.J., Ferrier, R.C., Larssen, T. & Posch, M. (2003b) Assessing emission reduction targets with dynamic models: deriving target load functions for use in integrated assessment. *Hydrology and Earth System Sciences*, **7**, 609–617.

Jobling, S., Williams, R., Johnson, A. et al. (2006) Predicted exposures to steroid estrogens in UK rivers correlate with widespread sexual disruption in wild fish populations. *Environmental Health Perspectives*, **114**, 32–39.

Johnson, A.C., Haria, A.H., Cruxton, V.L. et al. (1995) Isoproturon and Anion Transport by Preferential Flow through a Drained Clay Soil. In: Walker, A., Allen, R., Bailey, S.W., Blair, A.M., Brown, C.D., Gunther, P., et al. (eds), *Pesticide Movement to Water*. British Crop Protection Council Monograph No 62, pp. 105–10.

Johnson, A.C., Hughes, C.D., Williams, R.J. & Chilton, P.J. (1998) Potential for aerobic isoproturon biodegradation and sorption in the unsaturated and saturated zones of a chalk aquifer. *Journal of Contaminant Hydrology*, **30**, 281–97.

Johnson, A.C. & Sumpter, J.P. (2001) Removal of endocrine-disrupting chemicals in activated sludge treatment works. *Environmental Science & Technology*, **35**, 4697–4703.

Johnson, A.C., Besien, T.J., Bhardwaj, C.L. et al. (2001) Penetration of herbicides to groundwater in an unconfined chalk aquifer following normal soil applications. *Journal of Contaminant Hydrology*, **53** (1–2), 101–117.

Johnson, A.C. & Williams, R.J. (2004) A model to estimate influent and effluent concentrations of estradiol, estrone, and ethinylestradiol at sewage treatment works. *Environmental Science & Technology*, **38**, 3649–3658.

Johnson, A.C., Aerni, H.R., Gerritsen, A. et al. (2005) Comparing steroid estrogen, and nonylphenol content across a range of European sewage plants with different treatment and management practices. *Water Research*, **39**, 47–58.

Johnson, A.C., Keller, V., Williams, R.J. & Young, A. (2007a) A practical demonstration in modelling diclofenac and propranolol river water concentrations using a GIS hydrology model in a rural UK catchment. *Environmental Pollution*, **146**, 155–165.

Johnson, A.C., Williams, R.J., Simpson, P. & Kanda, R. (2007b) What difference might sewage treatment performance make to endocrine disruption in rivers? *Environmental Pollution*, **147**, 194–202.

Johnson, A.C., Jürgens, M.D., Williams, R.J., Kümmerer, K., Kortenkamp, A. & Sumpter, J.P. (2008) Do cytotoxic chemotherapy drugs discharged into rivers pose a risk to the environment and human health? An overview and UK case study. *Journal of Hydrology*, **348**, 167–175.

Johnson, A.C., Ternes, T., Williams, R.J. & Sumpter, J.P. (2009a) Assessing the concentrations of polar organic microcontaminants from point sources in the aquatic environment: Measure or model? *Environmental Science & Technology*, **42**, 5390–5399.

Johnson, A.C., Acreman, M.C., Dunbar, M.J. *et al.* (2009b) The British river of the future: How climate change and human activity might affect two contrasting river ecosystems in England. *Science of the Total Environment*, **407**, 4787–4798.

Johnson, A.C. (2010) Natural variations in flow are critical in determining concentrations of point source contaminants in rivers: An estrogen example. *Environmental Science & Technology*, **44**, 7865–7870.

Johnson, A.C., Bowes, M.J., Crossley, A. *et al.* (2011a) An assessment of the fate, behaviour and environmental risk associated with sunscreen TiO2 nanoparticles in UK field scenarios. *Science of the Total Environment*, **409**, 2503–2510.

Johnson, A.C., Yoshitani, J., Tanaka, H. & Suzuki, Y. (2011b) Predicting national exposure to a point source chemical: Japan and endocrine disruption as an example. *Environmental Science & Technology*, **45**, 1028–1033.

Johnson, A.C., Oldenkamp, R., Dumont, E. & Sumpter, J.P. (2013) Predicting concentrations of the cytostatic drugs cyclophosphamide, carboplatin, 5-fluorouracil, and capecitabine throughout the sewage effluents and surface waters of Europe. *Environmental Toxicology and Chemistry*, **32**, 1954–1961.

Johnston, W.R., Ittihadieh, F.T., Craig, K.R. & Pillsbury, A.F. (1967) Insecticides in tile drainage effluent. *Water Resources Research*, **3**, 525–537.

Jones, S.D. & Truesdale, V.W. (1984) Dissolved iodine species in British freshwater systems. *Limnology and Oceanography*, **29**, 1016–1024.

Jurgens, M.D., Holthaus, K.I.E., Johnson, A.C., Smith, J.J.L., Hetheridge, M. & Williams, R.J. (2002) The potential for estradiol and ethinylestradiol degradation in English rivers. *Environmental Toxicology and Chemistry*, **21**, 480–488.

Kirchner, J.W., Feng, X. & Neal, C. (2000) Fractal stream chemistry and its implications for contaminant transport in catchments. *Nature*, **403**, 524–527.

Kirchner, J.W., Feng, X. & Neal, C. (2001) Catchment-scale advection and dispersion as a mechanism for fractal scaling in stream tracer concentrations. *Journal of Hydrology*, **254**, 81–100.

Kirchner, J.W., Feng, X., Neal, C. & Robson, A.J. (2004) The fine structure of water-quality dynamics: the (high-frequency) wave of the future. *Hydrological Processes*, **18**, 1353–1359.

Kirchner, J.W. & Neal, C. (2013) Universal fractal scaling in stream chemistry and its implications for solute transport and water quality trend detection. *Proceedings of the National Academy of Sciences*, **110**(30), 12213–12218. doi:10.1073/pnas.1304328110.

Krug, E.C. & Frink, C.R. (1983) Acid rain on acid soil: a new perspective. *Science*, **221**, 520–525.

Leeks, G.J.L. & Jarvie, H.P. (1998) Introduction to the Land-Ocean Interaction Study (LOIS). Rationale and international context. *Science of the Total Environment*, **210/211**, 5–20.

Leeks, G.J.L., Jones, T.P. & Hollingworth, T.N. (2006) Forward: an introduction to UK research on the urban environment. *Science of the Total Environment*, **360**, 1–4.

Likens, G.E., Wright, R.F., Galloway, J.N. & Butler, T.J. (1979) Acid Rain. *Scientific American*, **241**, 43–51.

Lundquist, D., Christophersen, N. & Neal, C. (1990) Towards developing a new short-term model for the Birkenes catchment lessons learned. *Journal of Hydrology*, **116**, 391–401.

Macklin, M.G., Hudson-Edwards, K.A. & Dawson, E.J. (1997) The significance of pollution from historic metal mining in the Pennine orefields on river sediment contaminant fluxes to the North Sea. *Science of the Total Environment*, **194/195**, 391–397.

Matthiessen, P., Allchin, C., Williams, R.J., Bird, S.C., Brooke, D. & Glendinning, P.J. (1992) The Translocation of Some Herbicides between Soil and Water in a Small Catchment. *Journal of the Institution of Water and Environmental Management*, **6**, 496–504.

Matthiessen, P., Sheahan, D., Harrison, R. *et al.* (1995) Use of a Gammarus-pulex bioassay to measure the effects of transient carbofuran runoff from farmland. *Ecotoxicology and Environmental Safety*, **30**, 111–119.

McCartney, M.P. & Neal, C. (1999) Water flow pathways and the water balance within a headwater catchment containing a dambo: inferences drawn from hydrochemical investigations. *Hydrology and Earth System Sciences*, **3**, 581–591.

McCartney, M.P., Neal, C. & Neal, M. (1998) Use of deuterium to understand runoff generation in a headwater catchment containing a dambo. *Hydrology and Earth System Sciences*, **2**, 65–76.

Meigh, J.R., McKenzie, A.A. & Sene, K.J. (1999) A grid-based approach to water scarcity estimates for eastern and southern Africa. *Water Resources Management*, **13**, 85–115.

Monteith, D.T. (2005) The United Kingdom Acid Waters Monitoring Network: a review of the first 15 years and introduction to the special issue. *Environmental Pollution*, **137**, 3–13.

Morris, E.M. & Thomas, A.G. (1987) Transient acid surges in an upland stream. *Water, Air, and Soil Pollution*, **34**, 429–438.

Neal, C. (2001) The water quality of eastern UK rivers: the study of a highly heterogeneous environment. In: Huntley, D.A., Leeks, G.J.L. & Walling, D.E. (eds), *Land Ocean Interaction: Measuring and Modelling Fluxes from River Basins to Coastal Seas*. IWA Publishing, London, pp. 69–104.

Neal, C. (2014) Catchment water quality: the inconvenient but necessary truth of fractal functioning. *Hydrological Processes*, **27**, 3516–3520.

Neal, C. & Truesdale, V.W. (1976) The sorption of iodate and iodide by riverine sediments. *Journal of Hydrology*, **3**, 281–291.

Neal, C. & Cooper, D.M. (1983) Extended version of Gouy-Chapman electrostatic theory as applied to the exchange behaviour of clay in natural waters. *Clays and Clay Minerals*, **31**, 367–376.

Neal, C. & Stanger, C. (1983) Hydrogen generation from mantle source rocks in Oman. *Earth and Planetary Science Letters*, **66**, 315–320.

Neal, C. & Stanger, G. (1985) Past and present serpentinisation of ultramafic rocks: an example from the Semail Ophiolite Nappe of Northern Oman. In: Drever, J.I. (ed), *The Chemistry of Weathering*. Springer, Netherlands, pp. 249–275.

Neal, C. & Reynolds, B. (1999) The Impact of Conifer Harvesting and Replanting on Upland Water Quality. R&D Technical report to the Environment Agency: Report P211, 137pp

Neal, C. & Kirchner, J.W. (2000) Sodium and chloride levels in rainfall, mist, streamwater and groundwater at the Plynlimon catchments, mid-Wales: inferences on hydrological and chemical controls. *Hydrology and Earth System Sciences*, **4**, 295–310.

Neal, C., Thomas, A.G. & Truesdale, V.W. (1982) Thermodynamic characterisation of clay/electrolyte solutions. *Clays and Clay Minerals*, **30**, 291–296.

Neal, C., Smith, C.J., Walls, J. & Dunn, C.S. (1986a) Major, minor and trace element mobility in the acidic upland forested catchment of the upper River Severn, Mid-Wales. *Quarterly Journal of the Geological Society of London*, **143**, 635–648.

Neal, C., Whitehead, P.G., Neale, R. & Cosby, B.J. (1986b) Modelling the effects of atmospheric deposition and conifer afforestation in the British uplands. *Journal of Hydrology*, **86**, 15–26.

Neal, C., Whitehead, P.G. & Jenkins, A. (1988a) Are present UK SO emission declines sufficient to reverse the long-term stream acidification in the British uplands. *Nature*, **334**, 109–110.

Neal, C., Christophersen, N., Neale, R., Smith, C.J., Whitehead, P.G. & Reynolds, R. (1988b) Chloride in precipitation and streamwater for the upland catchment of the River Severn, Mid-Wales, some consequences for hydrochemical models. *Hydrological Processes*, **2**, 156–165.

Neal, C., Reynolds, B., Stevens, P. & Hornung, M. (1989) Hydrogeochemical controls for inorganic aluminium in acidic stream and soil waters at two upland catchments in Wales. *Journal of Hydrology*, **106**, 155–175.

Neal, C., Robson, A., Reynolds, B. & Jenkins, A. (1992a) Prediction of future short term stream chemistry - a modelling approach. *Journal of Hydrology*, **130**, 87–103.

Neal, C., Forti, M.C. & Jenkins, A. (1992b) Towards modelling the impact of climate change and deforestation on stream water quality in Amazonia: a perspective based on the MAGIC model. *Science of the Total Environment*, **127**, 225–241.

Neal, C., Avila, A. & Roda, F. (1995) Modelling the long term impacts of atmospheric pollution deposition and repeated forestry cycles on stream water chemistry for a Holm Oak forest in northeastern Spain. *Journal of Hydrology*, **168**, 51–71.

Neal, C., Robson, A.J., Shand, P. et al. (1997a) The occurrence of groundwater in the Lower Palaeozoic rocks of upland Central Wales. *Hydrology and Earth System Sciences*, **1**, 3–18.

Neal, C., Hill, T., Alexander, S. et al. (1997b) Stream water quality in acid sensitive UK upland areas, an example of potential water quality remediation based on groundwater manipulation. *Hydrology and Earth System Sciences*, **1**, 185–196.

Neal, C., Hill, T., Hill, S. & Reynolds, B. (1997c) Acid neutralization capacity measurements in surface and ground waters in the Upper River Severn, Plynlimon: from hydrograph splitting to water flow pathways. *Hydrology and Earth System Sciences*, **1**, 687–696.

Neal, C., Robson, A.J., Harrow, M.L. et al. (1997d) Major, minor, trace element and suspended sediment variations in the River Tweed: results from the LOIS core monitoring programme. *Science of the Total Environment*, **194/195**, 193–205.

Neal, C., Robson, A.J., Jeffery, H.A. et al. (1997e) Trace element inter-relationships for the Humber rivers: inferences for hydrological and chemical controls. *Science of the Total Environment*, **194/195**, 321–343.

Neal, C., House, W.A., Jarvie, H.P. & Eatherall, A. (1998) The significance of dissolved carbon dioxide in major lowland rivers entering the North Sea. *Science of the Total Environment*, **210/211**, 187–203.

Neal, C., Ormerod, S.J., Langan, S.J., Nisbet, T.R. & Roberts, J. (2004) Sustainability of UK forestry: contemporary issues for the protection of freshwaters, a conclusion. *Hydrology and Earth System Sciences*, **8**, 589–595.

Neal, C. & Jarvie, H.P. (2005) Agriculture, community, river eutrophication and the Water Framework Directive. *Hydrological Processes*, **19**, 1895–1901.

Neal, C., Robson, A.J., Neal, M. & Reynolds, B. (2005) Dissolved organic carbon for upland acidic and acid sensitive catchments in mid-Wales. *Journal of Hydrology*, **304**, 203–220.

Neal, C., Lofts, S., Evans, C.D., Reynolds, B., Tipping, E. & Neal, M. (2008) Increasing iron concentrations in UK upland waters. *Aquatic Geochemistry*, **14**, 263–288.

Neal, C., Jarvie, H.P., Williams, R.J. et al. (2010a) Declines in phosphorus concentration in the upper River Thames (UK): Links to sewage effluent cleanup and extended end member mixing analysis. *Science of the Total Environment*, **408**, 1315–1330.

Neal, C., Williams, R.J., Bowes, M.J. *et al.* (2010b) Decreasing boron concentrations in UK rivers: Insights into reductions in detergent formulations since the 1990s and within-catchment storage issues. *Science of the Total Environment*, **408**, 1315–1330.

Neal, C., Jarvie, H.P., Withers, P.J.A., Whitton, B.A. & Neal, M. (2010c) The strategic significance of wastewater sources to pollutant phosphorus levels in English rivers and to environmental management for rural, agricultural and urban catchments. *Science of the Total Environment*, **408**, 1485–1500.

Neal, C., Robinson, M., Reynolds, B. *et al.* (2010d) Hydrology and water quality of the headwaters of the River Severn: Stream acidity recovery and interactions with plantation forestry under an improving pollution climate. *Science of the Total Environment*, **408**, 5035–5051.

Neal, C., Rowland, P., Scholefield, P., Vincent, C., Woods, C. & Sleep, D. (2011a) The Ribble/Wyre observatory: Major, minor and trace elements in rivers draining from rural headwaters to the heartlands of the NW England historic industrial base. *Science of the Total Environment*, **409**, 1516–1529.

Neal, C., Reynolds, B., Norris, D. *et al.* (2011b) Three decades of water quality measurements from the Upper Severn experimental catchments at Plynlimon, Wales: an openly accessible data resource for research, modelling, environmental management and education. *Hydrological Processes*, **25**(14), 3818–3820. doi:10.1002/hyp.8191.

Neal, C., Reynolds, B., Rowland, P. *et al.* (2012) High-frequency water quality time series in precipitation and streamflow: from fragmentary signals to scientific challenge. *Science of the Total Environment*, **434**, 3–12.

Neal, C., Reynolds, B., Kirchner, J.W. *et al.* (2013) High-frequency precipitation and stream water quality time series from Plynlimon, Wales: an openly accessible data resource spanning the periodic table. *Hydrological Processes*, **27**, 2531–2539.

Neal, M., Neal, C. & Brahmer, G. (1997f) Stable oxygen isotope variations in rain, snow and streamwaters at the Schluchsee and Villingen sites in the Black Forest, SW Germany. *Journal of Hydrology*, **190**, 102–110.

Ormerod, S.J., Weatherly, N.S., Merrett, W.J., Gee, A.S. & Whitehead, P.G. (1990) Restoring streams in upland Wales modelling comparison of the chemical biological effects of liming and reduced sulphate deposition. *Environmental Pollution*, **64**, 67–85.

Page, T., Beven, K.J., Freer, J. & Neal, C. (2007) Modelling the chloride signal at Plynlimon, Wales, using a modified dynamic TOPMODEL incorporating conservative chemical mixing (with uncertainty). *Hydrological Processes*, **21**, 292–307.

Palmer-Felgate, E.J., Bowes, M.J., Stratford, C., Neal, C. & MacKenzie, S. (2011a) Phosphorus release from sediments in a treatment wetland: Contrast between DET and EPC0 methodologies. *Ecological Engineering*, **37**, 826–832.

Palmer-Felgate, E.J., Jarvie, H.P., Williams, R.J., Mortimer, R.J.G., Loewenthal, M. & Neal, C. (2008) Phosphorus dynamics and productivity in a sewage-impacted lowland chalk stream. *Journal of Hydrology*, **351**, 87–97.

Palmer-Felgate, E.J., Mortimer, R.J.G., Krom, M.D. & Jarvie, H.P. (2010) Impact of Point-Source Pollution on Phosphorus and Nitrogen Cycling in Stream-Bed Sediments. *Environmental Science & Technology*, **44**, 908–914.

Palmer-Felgate, E.J., Mortimer, R.J.G., Krom, M.D. *et al.* (2011b) Internal loading of phosphorus in a sedimentation pond of a treatment wetland: Effect of a phytoplankton crash. *Science of the Total Environment*, **409** (11), 2222–2232.

Prudhomme, C., Giuntoli, I., Robinson, E.L. *et al.* (2013) Hydrological droughts in the 21st century, hotspots and uncertainties from a global multimodel ensemble experiment. *Proceedings of the National Academy of Sciences*, **111**(9), 3262–3267.

Purdom, C.E., Hardiman, P.A., Bye, V.J., Eno, N.C., Tyler, C.R. & Sumpter, J.P. (1994) Estrogenic effects of effluents from sewage treatment works. *Chemistry and Ecology*, **8**, 275–285.

Reynolds, B., Neal, C., Hornung, M. & Stevens, P.A. (1986) Baseflow buffering of streamwater acidity in five mid-Wales catchments. *Journal of Hydrology*, **87**, 167–185.

Reynolds, B., Neal, C., Hornung, M., Hughes, S. & Stevens, P.A. (1988) Impact of afforestation on the soil solution chemistry of stagnopodzols in Mid-Wales. *Water, Air, and Soil Pollution*, **38**, 55–70.

Reynolds, B., Neal, C. & Norris, D.A. (2001) Evaluation of regional acid sensitivity predictions using field data: issues of scale and heterogeneity. *Hydrology and Earth System Sciences*, **5**, 75–82.

Roberts, G., Hudson, J.A. & Blackie, J.R. (1984) Nutrient inputs and outputs in a forested grassland catchment at Plynlimon, mid-Wales. *Agricultural Water Management*, **9**, 177–191.

Roberts, G., Hudson, J.A. & Blackie, J.R. (1986) Effect of upland pasture improvement on nutrient release in flows from a natural lysimeter and field drain. *Agricultural Water Management*, **11**, 231–245.

Robson, A.J., Jenkins, A. & Neal, C. (1991) Towards predicting future episodic changes in stream chemistry. *Journal of Hydrology*, **125**, 161–174.

Robson, A., Beven, K. & Neal, C. (1992) Towards identifying sources of subsurface flow: a comparison of components identified by a physically-based runoff model and those determined by chemical mixing techniques. *Hydrological Processes*, **6**, 199–214.

Robson, A.J., Neal, C. & Beven, K.J. (1995) Linking mixing techniques to a hydrological framework-an upland application. In: S, T. (ed), *Solute Processes in Catchment Systems*. J. Wiley and Sons Ltd., Chapter 13, pp. 347–369.

Robson, A.J. & Neal, C. (1997) A summary of regional water quality for Eastern UK rivers. *Science of the Total Environment*, **194/195**, 15–37.

Rowney, N.C., Johnson, A.C. & Williams, R.J. (2009) Cytotoxic drugs in drinking water: A prediction and risk assessment exercise for the Thames catchment in the United Kingdom. *Environmental Toxicology and Chemistry*, **28**, 2733–2743.

Sharpley, A., Jarvie, H.P., Buda, A., May, L., Spears, B. & Kleinman, P. (2013) Phosphorus legacy: Overcoming the effects of past management practices to mitigate future water quality impairment. *Journal of Environmental Quality*, **42** (5), 1308–1326.

Singer, A.C., Nunn, M.A., Gould, E.A. & Johnson, A.C. (2007) Potential risks associated with the proposed widespread use of Tamiflu. *Environmental Health Perspectives*, **115**, 102–106.

Sumpter, J.P. & Johnson, A.C. (2005) Lessons from endocrine disruption and their application to other issues concerning trace organics in the aquatic environment. *Environmental Science & Technology*, **39**, 4321–4332.

Sumpter, J.P., Johnson, A.C., Williams, R.J., Kortenkamp, A. & Scholze, M. (2006) Modeling effects of mixtures of endocrine disrupting chemicals at the river catchment scale. *Environmental Science & Technology*, **40** (17), 5478–5489.

Smith, J.T., Clarke, R.T. & Bowes, M.J. (2010) Are groundwater nitrate concentrations reaching a turning point in some chalk aquifers? *Science of the Total Environment*, **408** (20), 4722–4732.

Tester, D.J., Waldron, P.A., Williams, R. *et al.* (1992) The great ouse automatic quality monitoring network. *Journal of the Institution of Water and Environmental Management*, **6** (2), 165–171.

Truesdale, V.W. & Smith, P.J. (1975a) The automatic determination of iodide or iodate in solution by catalytic spectrophotometry, with particular reference to river water. *Analyst*, **100**, 111–112.

Truesdale, V.W. & Smith, C.J. (1975b) The formation of molybdosilicic acids from mixed solutions of molybdate and silicate. *Analyst*, **100**, 203–212.

Turnbull, A.B., Harrison, R.M., Williams, R.J. *et al.* (1997) Assessment of the fate of selected adsorptive pesticides at ADAS Rosemaund. *Water and Environment Journal*, **11**, 24–30.

UKAWRG. United Kingdom Acid Waters Review Group (1988) second report, 'Acidity in United Kingdom fresh waters', Her Majesty's Stationary Office London:1–61.

Wade, A.J., Neal, C., Whitehead, P.G. & Flynn, N.J. (2005) Modelling nitrogen fluxes from the land to the coastal zone in European systems: a perspective from the INCA project. *Journal of Hydrology*, **304**, 413–429.

Wade, A.J., Palmer-Felgate, E.J., Halliday, S.J. *et al.* (2012) Hydrochemical processes in lowland rivers: insights from in situ, high-resolution monitoring. *Hydrology and Earth System Sciences*, **16** (11), 4323–4342.

Wauchope, R.D. (1978) The Pesticide Content of Surface Water Draining from Agricultural Fields—A Review. *Journal of Environmental Quality*, **1978** (7), 459–472.

Waylett, A.J., Hutchins, M.G., Johnson, A.C., Bowes, M.J. & Loewenthal, M. (2013) Physico-chemical factors alone cannot simulate phytoplankton behaviour in a lowland river. *Journal of Hydrology*, **497**, 223–233.

Wheater, H.S., Neal, C. & Peach, D. (2006) Hydro-ecological functioning of the Pang and Lambourn catchments, UK, An introduction to the special issue. *Journal of Hydrology*, **330** (1–2), 1–9.

Whitehead, P.G. (1990) Modelling nitrate from agriculture into public water supplies. *Philosophical Transactions of the Royal Society of London, Series B: Biological Sciences*, **329**, 403–410.

Whitehead, P.G., Beck, M.B. & O'connell, P.E. (1981) A system model of flow and water quality in the Bedford Ouse River system: Part II, Water Quality Modelling. *Water Research*, **15**, 1157–1171.

Whitehead, P.G., Caddy, D.E. & Templeman, R.F. (1984) An on-line monitoring data management and forecasting system for the Bedford Ouse River Basin. *Water Science and Technology*, **16**, 295–314.

Whitehead, P.G. & Hornberger, G.E. (1984) Modelling algal behaviour in the River Thames. *Water Research*, **18**, 945–953.

Whitehead, P.G. & Williams, R. (1984) Modelling nitrate and algal behaviour in the River Thames. *Water Science and Technology*, **16**, 621–633.

Whitehead, P.G., Neal, C. & Neale, R. (1986a) Modelling the effects of hydrological changes on stream water acidity. *Journal of Hydrology*, **84**, 353–364.

Whitehead, P.G., Neal, C., Seden-Perriton, S., Christophersen, N. & Langan, S. (1986b) A time-series approach to modelling stream acidity. *Journal of Hydrology*, **85**, 281–303.

Whitehead, P.G., Williams, R.J. & Lewis, D.R. (1997a) Quality simulation along rivers (QUASAR) Part 1: Model theory and development. *Science of the Total Environment*, **194/195**, 447–456.

Whitehead, P.G., Barlow, J., Howarth, E.Y. & Adamson, I.K. (1997b) Acidification in Three Lake District Tarns: Historical long term trends and modelled future behaviour under changing sulphate and nitrate deposition. *Hydrology and Earth System Sciences*, **1**, 197–204.

Whitehead, P.G., Wilson, E.J. & Butterfield, D. (1998a) A semi-distributed Integrated Nitrogen model for multiple source assessment in Catchments (INCA): Part I - Model structure and process equations. *Science of the Total Environment*, **210/211**, 547–558.

Whitehead, P.G., Wilson, E.J., Butterfield, D. & Seed, K. (1998b) A Semi-distributed Integrated Nitrogen Model for Multiple source assessment in Catchments (INCA): Part II Application to large River Basins in South Wales and Eastern England. *Science of the Total Environment*, **210/211**, 559–583.

Whitton, B.A. & Neal, C. (2011) Organic phosphate in UK rivers and its relevance to algal and bryophyte surveys. *International Journal of Limnology*, **47**, 3–10.

Worrall, F., Harriman, R., Evans, C.D. *et al.* (2004) Trends in dissolved organic carbon in UK rivers and lakes. *Biogeochemistry*, **70**, 369–402.

Williams, R.J., Bird, S.C. & Clare, R.W. (1991) Simazine concentrations in a stream draining an agricultural catchment. *Journal of the Institution of Water and Environmental Management*, **5**, 80–84.

Williams, R.J., Brooke, D.N., Matthiessen, P., Mills, M., Turnbull, A. & Harrison, R.M. (1995) Pesticide Transport to Surface Waters within an Agricultural Catchment. *Journal of the Institution of Water and Environmental Management*, **9**, 72–81.

Williams, R.J. (1998) Modelling pesticide run-off to surface waters. Part I: Model theory and development. *Pesticide Science*, **54**, 113–130.

Williams, R.J., White, C., Harrow, M.L. & Neal, C. (2000) Temporal and small-scale spatial variations of dissolved oxygen in the Rivers Thames, Pang and Kennet, UK. *Science of the Total Environment*, **251**, 497–510.

Williams, R.J., Keller, V.D.J., Johnson, A.C. *et al.* (2009) A national risk assessment for intersex in fish arising from steroid estrogens. *Environmental Toxicology and Chemistry*, **28**, 220–230.

Williams, R.J., Churchley, J.H., Kanda, R. & Johnson, A.C. (2012) Comparing predicted against measured steroid estrogen concentrations and the associated risk in two United Kingdom river catchments. *Environmental Toxicology and Chemistry*, **31**, 892–898.

Young, A.R., Grew, R. & Holmes, M.G.R. (2003) Low flows 2000: a national water resources assessment and decision support tool. *Water Science and Technology*, **48**, 119–126.

9 **Ecohydrology**

MIKE C. ACREMAN[1], JAMES R. BLAKE[1],
LAURENCE R. CARVALHO[2], MIKE J. DUNBAR[1],
IAIN D. M. GUNN[2], ALAN GUSTARD[3], IAN D. JONES[4],
CEDRIC LAIZÉ[1], STEPHEN C. MABERLY[4], ELEANOR B.
MACKAY[4], LINDA MAY[2], J. OWEN MOUNTFORD[1], BRYAN
M. SPEARS[2], CHARLIE J. STRATFORD[1], STEPHEN J.
THACKERAY[1], AND IAN J. WINFIELD[4]

[1]Centre for Ecology and Hydrology, Wallingford, Oxfordshire, UK
[2]Centre for Ecology and Hydrology, Edinburgh, Midlothian, UK
[3]Water Resource Associates, Henley-on-Thames, UK
[4]Centre for Ecology and Hydrology, Bailrigg, Lancaster, UK

9.1 Introduction

It is often said that water is essential to all life on Earth (e.g. Gleick, 1993; Young et al., 2004). It is not surprising then that hydrologists are long used to cross-disciplinary links with other fields that link water to life such as aquatic ecology. The term 'hydroecology' has been coined to describe the science defining the water needs for aquatic ecosystems (Acreman, 2001b), whereas another term, 'ecohydrology', was initially focused on sustainable water resource management and the ability of ecosystems to improve water quality (Zalewski *et al.*, 1997; Zalewski and Harper, 2001). However, ecohydrology has developed a much broader scope, covering all the interactions between water and ecosystems (Hannah *et al.*, 2004) and the term now encompasses water-based ecosystem services and environmental flows.

Much of our current thinking about ecosystems was distilled at the United Nations Conference on Environment and Development in Rio de Janeiro in 1992, where it was recognised that the lives of people and the environment are inherently interrelated (UN, 1992). Ecological processes maintain the planet's capacity to deliver goods and services, such as water, food and medicines and much of what we call 'quality of life' (Acreman, 2001a). The Millennium Development Goals (UN, 2014) included the need for environmental sustainability, such as reducing the rate of loss of species threatened with extinction. The concept of ecosystem services (Barbier, 2009; Fisher *et al.*, 2009) brought to prominence in the Millennium Ecosystem Assessment (2005), demonstrates that healthy freshwater ecosystems provide economic security, for example fish, medicines and timber (Emerton and Bos, 2004; Cowx and Portocarrero, 2011); social security, for example, protection from natural hazards, such as floods and ethical security, for example, upholding the rights of people and other species to water (Acreman, 2001a). Thus, understanding the links between the water cycle and the environment also supports people by maintaining the ecosystem services on which we depend (Acreman, 1998; MEA, 2005). The Rio+20 meeting (http://www.uncsd2012.org/) called for action to protect and sustainably manage ecosystems (including maintaining water quantity and quality) and recognised that the global loss of biodiversity and the degradation of ecosystems undermines global development (Costanza and Daly, 1992), affecting food security and nutrition, the provision of and access to water and the health of the rural poor. Rio+20 also launched a process to develop a set of Sustainable Development Goals (SDGs) that will build upon the Millennium Development Goals and converge with the post 2015 development agenda.

These International initiatives have highlighted the need for a rapid expansion of research at the interface of ecology and hydrology. This has led to the launch of new dedicated journals, for example, *Ecohydrology & Hydrobiology* (Elsevier) launched in 2001 and *Ecohydrology* (Wiley) launched in 2008, in addition to a plethora of articles in hydrological and ecological journals, such as *Hydrological Science Journal, Journal of Hydrology and Freshwater Biology*. CEH have made major contributions to this research in ecohydrology particularly in river, wetland and lake ecosystems. Key achievements in these fields are summarised below.

9.2   Water Requirements of River Ecosystems

9.2.1   *Introduction*

The first river flow management for ecosystems focused on the concept of a minimum flow for diluting polluted discharges, based on the notion that as long as the flow is maintained at or above a critical minimum, the river ecosystem will be conserved (Acreman *et al.*, 2014a). The UK Water Resources Act 1963 required minimum acceptable flows to maintain natural beauty and fisheries. The US Clean Water Act in 1972 set the objective of restoring and maintaining the chemical, physical and biological integrity of the nation's waters. Water allocations for ecosystem maintenance have been incorporated into Integrated Water Resources Management (IWRM; Falkenmark, 2003), environmental impact assessment (Wathern, 1998) and the Ecosystem Approach (Maltby *et al.*, 1999). Recognising the importance of water for the environment is now part of the policies and laws of many countries (Le Quesne *et al.*, 2010).

Flow releases from UK reservoirs have been made since the 1800s under Acts of Parliament. They were made to provide water for downstream riparian users, including mills and navigation on the expanding canal system and were termed compensation flows, with minimal attention to impacts on river ecosystems. The term 'compensation flow' was defined in the 1976 Drought Act as 'water which any water authority or statutory water company are under an obligation to discharge into a river, stream or brook or other running water or into a canal as a condition of carrying out their undertaking'. In 1985, the Department of the Environment commissioned a major review of compensation flow policy in the

United Kingdom, to assess the pattern of releases below 261 United Kingdom reservoirs and if appropriate to provide guidelines of how releases should be set (Gustard, 1989). The staff of the Freshwater Biological Association (subsequently joining the Institute of Freshwater Ecology and then CEH) undertook surveys of macroinvertebrate fauna below 29 reservoirs in the United Kingdom. The field sampling appeared to demonstrate that invertebrate fauna were remarkably resilient to impacts arising from flow regulation (Armitage et al., 1987). There were no noticeable detrimental effects of regulation at the family level of identification and no obvious relationship between fauna and the level of compensation flow. However, it was notable that the regulated sites tended to be similar to natural sites on different rivers which had lower slope, at a lower altitude and further downstream. This was a clear indication of the stabilising effect of the regulated regime on macroinvertebrate fauna (Bass and Armitage, 1987).

9.2.2 *Physical habitat modelling*

A key recommendation of the Compensation Flow Study (Gustard, 1989) was to explore the use of the Instream Flow Incremental Methodology (IFIM) and in particular its component tool the Physical Habitat Simulation (PHABSIM) system developed by the US Fish and Wildlife Service (Waters, 1976; Bovee, 1982). This recognised that river habitat is in part defined by hydraulics, including water depth and velocity rather than flow *per se*, that is, discharge in m/s (Dunbar et al., 2012). PHABSIM produces a relationship between river flow and useable physical habitat (Fig. 9.1) for target species such as trout and salmon (Bovee, 1982). This approach is a legal requirement in some US states for setting instream flows. However, it had not been applied in the United Kingdom or in mainland Europe and was considered to be an innovative approach to determining environmental flows by combining hydrology, hydraulics and ecology. PHABSIM is based on simplified channel hydraulics and does include detailed biological process; nevertheless, no better approach was available at the time. Initial development of PHABSIM for UK rivers was carried out by IH for the Department of the Environment (Bullock et al., 1991), with the first application for an operational water resource problem on the River Allen, Dorset, England (Johnson et al., 1995). The River Allen was

regarded as a classic example of a chalk stream with a high-quality trout and salmon fishery. Since 1946 the river had been subjected to reduced flow caused by increased groundwater abstraction. Significant reductions in salmonid populations prompted the National Rivers Authority (NRA) to initiate detailed hydrological, biological and fisheries studies with the objective of assessing an 'ecologically acceptable flow regime'. Time series of physical habitat produced by PHABSIM were used in negotiations with Bournemouth Water plc to reduce groundwater abstraction rates by up to 50%. This investigation was typical of many studies carried out on a list of 'Top 40 Rivers' identified by the NRA to have acute low flow problems (National Rivers Authority, 1993). The NRA gave considerable support to IH for the implementation of PHABSIM in England and Wales. This initiated cooperation, which extended over more than three decades, between the then Institute of Freshwater Ecology and Institute of Hydrology in applying and developing the methodology. By 1996 PHABSIM had been applied at over 35 sites in England and Wales (Gustard and Elliott, 1997) for assessment of the implications of flow alteration for salmonid fish. In addition, PHABSIM was used to assess the habitat implications of channel modification in restoration of the rural River Wey (Acreman and Elliott, 1996) and effects of channelisation in urban areas of the River Tame (Booker et al., 2003). This habitat modelling approach was also developed to consider flow management for whole catchments, such as the River Itchen (Booker et al., 2004a), rather than river reaches for which PHABSIM had been used previously.

Physical habitat suitability is measured empirically by recording the depth and velocity of the river at which fish are located (Dunbar et al., 2001). This raised a question of why fish prefer certain physical conditions. CEH started collaboration with Dr Thom Hardy of Utah State University (now at Texas State University) in 1996, who suggested that feeding juvenile salmonid fish preferentially select areas of river where the velocity is high as this provides an energy gain when feeding on invertebrates drifting downstream with the current, but move to lower velocity areas when resting because swimming against a strong current expends considerable energy (Guensch et al., 2001). Bioenergetic modelling at CEH demonstrated that fish locations coincided with areas where they acquire a net energy gain (Booker et al, 2004b).

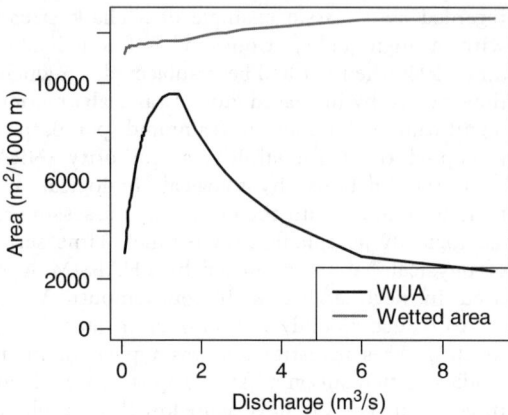

Fig. 9.1 River discharge *v* Weighted Usable Area of habitat (lower curve) compared with the total wetted area of river bed (upper curve) that defines a maximum. (Source: Reproduced with permission of CEH.)

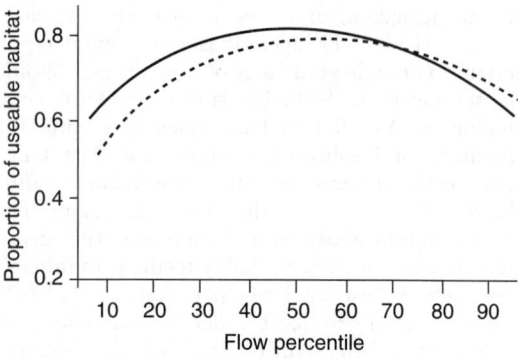

Fig. 9.2 Rapid assessment of flow *v* habitat curves from PHABSIM (black-solid) and RAPHSA (green-dashed). (Source: Reproduced with permission of CEH.) *(See insert for colour representation of this figure.)*

The late 1990s saw the beginnings of a shift from relatively detailed assessment of specific river reaches, typified by the application of PHABSIM, towards a requirement for more generic environmental flow methods which could be undertaken over much larger areas including at the catchment scale, for which PHABSIM would be prohibitively expensive to apply. CEH undertook a review of environmental flow methods from other countries (Dunbar *et al.*, 1998) which became highly cited itself, as well as forming the basis of other publications (Acreman and Dunbar, 2004; Dyson *et al.*, 2003). A limitation of PHABSIM is that it requires collection of hydraulic data (river depth and water velocity) at multiple levels of river flow (ideally covering high, medium and low flows) at a set of cross sections that represent the hydraulic variations along the impacted river reach. However, the Environment Agency needed to assess the sensitivity of river ecosystems to abstraction for many hundreds of sites as part of the definition of Catchment Abstraction Management Strategies (CAMS; Environment Agency, 2001). CEH analysis of results from 65 UK PHABSIM studies showed some consistency in the derived flow–habitat relationships for different types of river reach. This led to the development of a tool for Rapid Assessment of Physical Habitat Sensitivity to Abstraction (RAPHSA – Booker *et al.*, 2005) based on measurements of river width and depth that could be collected relatively quickly and easily

during one short site visit (Booker and Acreman, 2007). Figure 9.2 shows the RAPHSA estimate of the flow versus habitat curve for juvenile salmon compared with that from PHABSIM.

The Water Framework Directive of the European Union came into force on 22 December 2000 (CEC, 2000) and requires member states to achieve at least Good Ecological Status (GES) in all waterbodies. The Ecological Status of any waterbody is defined by the degree of deviation of the waterbody from biological reference conditions, measured primarily through assessment of taxonomic composition of fish, invertebrates, macrophytes and algae. It does not use the term environmental flows explicitly, but it is accepted that ecologically appropriate hydrological regimes are necessary to achieve GES and that implementing environmental flows will be a key measure for restoring and managing river ecosystems (Acreman & Ferguson, 2010). Where waterbodies are designated as Heavily Modified (HMWB) or artificial, an alternate objective of at least Good Ecological Potential (GEP) can be applied. This objective takes account of the constraints imposed by physical modifications to the waterbody such as dams. GEP is the best example of biological conditions in a similar waterbody with the same modifications, that is where reasonable mitigation measures have been implemented and best practice management is applied. CEH undertook a major study for the Environment Agency to develop an approach to assess HMWBs (Dunbar *et al.*, 2002).

9.2.3 *Environmental Flows*

The term environmental flows is now widely used to describe the quantity, quality and timing of water flows required to sustain freshwater and estuarine ecosystems and the human livelihoods and well-being that depend on these ecosystems (Brisbane Declaration http://www.eflownet.org/). The concept highlights the indirect benefits to people of providing water to ecosystems (such as food, recreation and cultural identity) in addition to the direct benefits of water used for drinking, growing food and supporting industry (Acreman, 1998).

Between 2005 and 2007, CEH led a series of projects to define environmental flow requirements for UK rivers to meet the WFD. Since the concept of the WFD is based on minor deviation from a natural baseline, it seemed appropriate to address environmental flows by taking the natural flow paradigm (Poff *et al.*, 1997). This is based on the argument that organisms evolved and adapted, and communities were assembled and are maintained by a natural flow regime (Lytle and Poff, 2004) and that modification of the natural flow regime will alter riverine, riparian and floodplain species (including plants, fish, invertebrates and algae) and processes. It also assumes that there may be limits to hydrological change beyond which significant (or unacceptable) ecological alteration takes place (Richter *et al.*, 1996; Bunn and Arthington, 2002; Arthington *et al.*, 2006). To set these limits of allowable hydrological alteration, a panel of UK river scientists specialising in fisheries, macroinvertebrate ecology, macrophyte plant ecology, hydrology and water resource management was created from universities, research institutes including CEH, consultancies and environment protection agencies. The main forum of interaction for the panel was a series of workshops. The panel also provided information and references for a literature review and commented on drafts of the various documents produced, including workshop and project reports. The project produced two main outputs; a means of classifying rivers into types based on the characteristics of the catchments draining to them and a set of look-up tables for each river type specifying the maximum abstraction allowable at different flows. The analysis resulted in four broad types with different permitted abstraction levels. The maximum levels of abstraction ranged from 7.5% to 35% of the natural flow depending on river type and flow rate (Acreman *et al.*, 2008). These figures were incorporated into the Environment Agency's Resource Assessment and Management (RAM) framework (Environment Agency, 2002), the technical assessment component of the CAMS used to manage abstraction licences. It is also the basis of hydrological status classification for the WFD.

It was recognised that releasing water from a reservoir to achieve environmental flows involves active management (Acreman and Dunbar, 2004) and is a very different issue than limiting abstraction. Consequently, a second project was established to provide best practice guidance for setting flow releases from impoundments, where the waterbody would be considered 'heavily modified' under WFD and the target would be GEP. This study concluded that no simple generic rules could be set that would apply to all UK rivers. Instead, it was recommended that each waterbody below a dam required individual analysis of its ecosystem and flow regime and identification of key elements of the flow regime required to conserve different components of the river ecosystem including birds, fish, invertebrates, plants, amphibians, reptiles, sediments and nutrient dynamics (Acreman *et al.*, 2009b). It develops the Building Block Methodology initiated in South Africa (Tharme and King, 1998). The magnitude of a given block may be defined using methods such as PHABSIM. A flow regime for GEP can, thus, be constructed by combining these building blocks (Fig. 9.3). Reporting of the work to experts in Scandinavia led to testing of this approach in Norway (Alfredson *et al.*, 2012) and to assessment of environmental flow releases from dams on the Zhangxi River, China (Acreman *et al.*, 2010). The approach has been recommended for implementation by the UK Technical Advisory Group on WFD (UKTAG, 2013).

Tharme (2003) identified 207 methods for assessing environmental flows, and Acreman and Dunbar (2004) classified these into four groups: (1) look-up tables, (2) desktop hydrological analysis, (3) functional analysis and (4) habitat modelling. These are all applied in the United Kingdom. A look-up table (Group 1) was produced for abstraction management to achieve GES, ecologically relevant hydrological indicators; Group 2 was used to assess the risk of significant change in rivers ecosystems across the United Kingdom, Europe as part of the SCENES EU project (Fig. 9.4; Laizée *et al.*, 2014) and the Mekong in collaboration with UCL (Thompson *et al.*, 2014), the Building

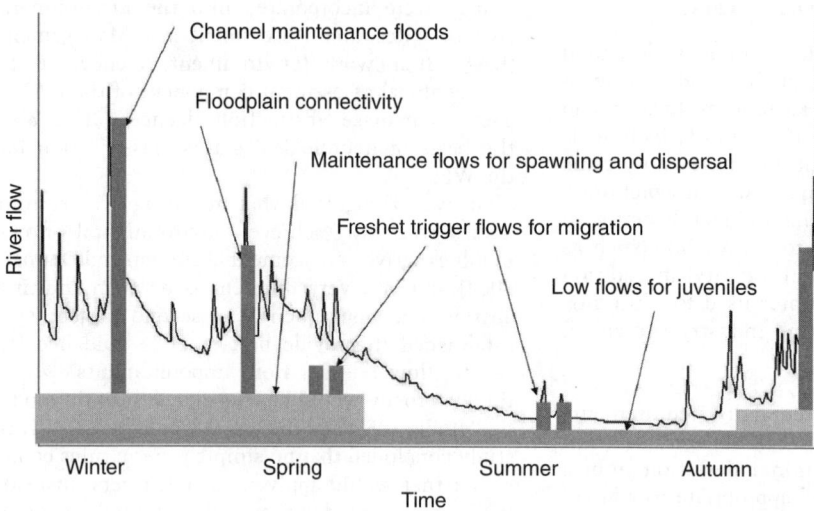

Fig. 9.3 Basic elements of flow regime required to deliver components of the river ecosystem (after Acreman *et al.*, 2009a, 2009b). (Source: Adapted from Acreman *et al.* 2009.)

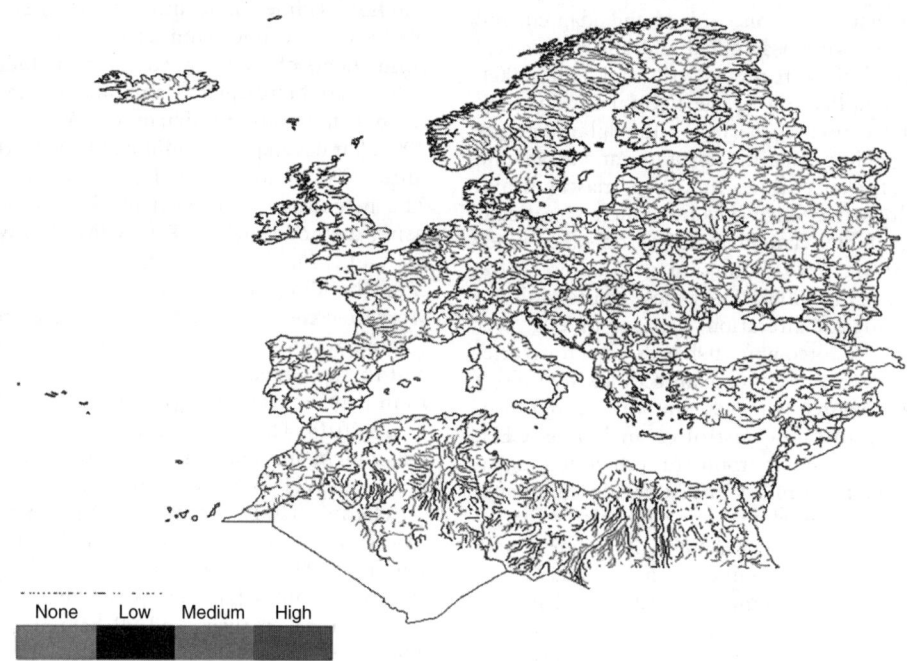

Fig. 9.4 Risk of river ecosystem change in European rivers by 2050. (Source: Adapted from Laizé et al. 2014. Reproduced with permission of John Wiley & Sons.) *(See insert for colour representation of this figure.)*

Block Methodology is an example of functional analysis (Group 3) and PHABSIM is a habitat model (Group 4).

In 2009, a team of 18 international environmental flow experts, including from CEH, came together to define a framework entitled the 'Ecological Limits of Hydrologic Alteration' (ELOHA). This brought together best practice for synthesising hydrological and ecological databases from many rivers within a user-defined region to develop scientifically defensible and empirically testable relationships between flow alteration and the response of fish, macroinvertebrates and plants (Poff *et al*, 2010). ELOHA is based on the deviation of the altered flow regime from a static baseline of natural conditions and thus needs to adapted for rivers that have been heavily managed for centuries and natural is no longer an appropriate baseline. Acreman *et al*. (2014b) identified two basic environmental flow approaches based on either constraining alteration from a natural flow baseline to maintain biodiversity and ecological integrity or designing flow regimes to achieve specific ecological and ecosystem service outcomes. They argue that the former is more applicable to natural and semi-natural rivers where the primary objective and opportunity is ecological conservation. The latter is appropriate for modified and managed rivers where return to natural conditions is no longer feasible and the objective is to maximise ecosystem services, such as water resources, pollutant removal, birds and aesthetic beauty, that support economic growth, recreation or cultural history, permitting elements of ecosystem design and adaptation to environmental change. Under a future with altered climate and heavy regulation, where hybrid and novel aquatic ecosystems predominate, the designer approach may become the only feasible option. This results from insufficient natural ecosystems from which to draw analogues and the need to support broader socio-economic benefits and valuable configurations of natural and social capital.

9.2.4 Linking river ecosystems and the flow regime

Recent research has recognised that changes in external forces on ecosystems tend to occur in synchrony rather than as individual pressures (Ormerod *et al*., 2010). There is a need to improve our knowledge of the links between changes in flow, channel morphology and water quality and to assess whether impacts are additive, synergistic or antagonistic (Acreman *et al*., 2014a). Beginning in 2001, CEH contributed to the development of ecological metrics and assessment tools which directly link historical river flow to ecological response.

In the mid-1990s, the Environment Agency began to recognise that although it undertook considerable river ecological monitoring across a range of anthropogenic pressures, its understanding of the impacts of flow alteration lagged significantly behind its understanding of the impacts of organic pollution. This led to the development of an ecological metric, Lotic Invertebrate Indicator for Flow Evaluation (LIFE; Extence *et al*., 1999). CEH became involved in the testing of LIFE in 2000. In 2006, further research by CEH for the Environment Agency highlighted how an existing statistical modelling procedure, linear regression with mixed effects, was able to elegantly describe the generic response of LIFE to antecedent flow and variation around that generic response. Starting from a small dataset from the East Midlands, the approach was broadened across unregulated rivers in England and Wales. The analysis also showed how some of the aforementioned variation in ecological response to low flows appeared to be linked to the extent of channel modifications (re-sectioning, widening, straightening and deepening) of the river channel (Fig. 9.5a,b). This is consistent with the theory that more natural river channels are ecologically more resilient to low flow stress because they have a greater range of physical habitat niches across a range of flows (Dunbar *et al*., 2010a, 2010b).

Dunbar *et al*., (2010b) analysed macroinvertebrate data for numerous sites across England and found a strong interaction between flow and channel morphology in determining ecological response. Figure 9.6 shows that the flow–ecosystem relationship for natural channels (0% modification to morphology) has a shallow slope showing that the river ecosystem is robust, or insensitive, to flow change. Conversely, the relationship for modified channels has a steeper slope demonstrating that the river ecosystem is more sensitive to flow change. This is because natural rivers tend to have diverse morphology, so that whatever the flow, there will be a diversity of physical habitats (deep, shallow, fast and slow). In contrast modified river channels, such as those straightened and deepened for flood defence, tend to have homogenous, often

(a)

(b)

Fig. 9.5   (a) Natural river channel with complex channel geometry and dense bank vegetation. (Source: Photograph by Mike Acreman.) (b) Highly modified river channel with straight, reinforced banks and sparse vegetation. (Source: Photograph by Mike Acreman.)

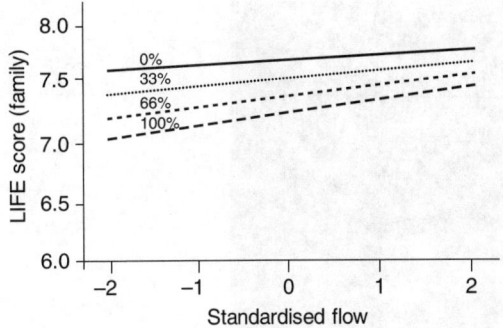

Fig. 9.6 Variations in flow–river ecosystem relationships with channel alteration (annotated as % modification from natural). (Source: Reproduced with permission of CEH.) *(See insert for colour representation of this figure.)*

trapezoidal, morphology in which good habitat only occurs at certain flows.

Much of this multiple pressure research has focused primarily on external drivers (such as flow and channel geometry). In addition, ecosystems have internal processes, such as biotic interaction and trophic relationships that govern flows of energy and carbon and thus also control ecosystem type, health and status. Freshwater ecologists at the River Laboratory, Dorset, including CEH staff and other collaborators, studied the food web dynamics of rivers. Ledger *et al.* (2011, 2013a,b) used channel mesocosm experiments (Ledger *et al.*, 2006, Brown *et al.*, 2011) to show how the intensification of drought may alter the underlying structure and functioning (biomass flux dynamics) of river food webs through species traits in benthic algae (Ledger *et al.*, 2008) and invertebrates (Woodward *et al.*, 2012). The role of some key river invertebrate species in governing ecosystem function and food web stability was demonstrated in a field manipulation experiment in the Bere Stream, Dorset (Woodward *et al.*, 2008). In another manipulation, Atlantic salmon were reintroduced to the Tadnoll Brook, Dorset (Edwards *et al.*, 2009a), with further studies identifying complex stoichiometric relationships between nutrients, detritus, biofilms and consumers (Lauridsen *et al.*, 2014), while identifying fundamental issues with food web quantification (Edwards *et al.*, 2008, 2009b, Lauridsen *et al.*, 2012). The datasets collected as part of these studies contributed to a number of meta-analyses and reviews of food web

ecology (Gilljam *et al*, 2011, Ings *et al.*, 2009, Brose *et al.*, 2006, Woodward *et al.*, 2010). As part of Integrated Assessment of the 2007 Countryside Survey (Smart *et al.*, 2010), flow time series were generated for river monitoring sites and these showed a negative relationship with ecological quality. This work demonstrates that there is a strong need for biologists and ecologists to work more closely with hydrologists to address the challenges of combining flow effects with internal ecosystem dynamics that underpins ecohydrology.

In 2007, CEH purchased 600m of the River Lambourn and 24 acres of associated water meadows at Boxford, Berkshire to establish the Lambourn Observatory (Fig. 9.7). The River Lambourn is a groundwater-fed chalk stream draining one of the least modified catchments of southern England. The river and its floodplain are designated Sites of Special Scientific Interest (SSSI) and Special Areas of Conservation (SAC). Research at Boxford has focused on fundamental understanding of processes in groundwater-fed rivers and wetlands. The Observatory provides a platform for research studies and model testing. The first assessments are being undertaken on the suitability of the MIKE-SHE/MIKEII suit to model the complex relationships between the aquifers, river channels and floodplain wetlands in collaboration with UCL. Notable research outputs have been the assessment of hydraulic and ecohydrological implications for weed cuts, which increased velocity and lowered water depth, reducing physical habitat for young trout and grayling young (Old *et al.*, 2014). Complementary work at Boxford by the University of Southampton in collaboration with CEH has focused on exploring the use of habitats of different temperature by trout.

9.3 Wetlands

9.3.1 Introduction

CEH, with others (particularly Open University, UCL and Royal Holloway), have undertaken significant work on enhancing understanding of the ecohydrological functioning of wetlands. This work developed new scientific thinking, for example, on hydrologically defined niches for plant communities (e.g. Silvertown *et al.*, 1999) and on hydrological functions of wetlands (Bullock

Fig. 9.7 River Lambourn Observatory, Boxford, UK. (Source: Photograph by Mike Acreman.)

and Acreman, 2003), supported major wetland restoration projects, such as the Great Fen (Mountford *et al.*, 2002) and Baston and Thurlby Fens (Mountford *et al.*, 2004) and input to the Ramsar, International Convention on Wetlands including provision of guidance on groundwater-fed wetlands (Acreman, 2005).

Wetlands occupy the transitional zones between permanently wet and generally drier areas; they share characteristics of both environments yet cannot be classified unambiguously as either fully aquatic or terrestrial (Acreman and José, 2000). It is the presence of water for some significant period of time which creates the soil, its microorganisms and the plant and animal communities, such that the land functions in a different way from either aquatic or dry habitats. Mitsch and Gosselink (1993) recognised that *'hydrology is probably the single most important determinant for the establishment and maintenance of specific types of wetlands and wetland processes ... when hydrologic conditions in wetlands change even slightly, the biota may respond with massive changes in species richness and ecosystem productivity'*.

The international Convention on Wetlands, the intergovernmental treaty established in Ramsar, Iran in 1971 (Davis, 1993), provides the framework for national action and international cooperation for the conservation and wise use of wetlands and their resources. The 'Ramsar' Convention adopts an extremely broad approach and defines wetlands as: *'areas of marsh, fen, peatland or water, whether natural or artificial, permanent or temporary, with water that is static or flowing, fresh, brackish or salt, including areas of marine water, the depth of which at low tide does not exceed six metres'*. Such a definition thus includes many ecosystems from coral reefs to lakes in underground caves.

In general, wetlands include a range of soils (e.g. peat in fens, alluvium in floodplains and marine clays in estuaries), vegetation communities (e.g. grasslands, forests, mangroves and reed-beds), animals (e.g. fish, reptiles and amphibians) and microbes (e.g. methane-producing bacteria). Many local terms are applied to wetlands (Thompson and Finlayson, 2001), including general anglicised terms such as 'marsh', 'swamp', 'bog' and regionally specific terms such as aapa mires (rheotrophic mires of the boreal zone), fadamas (floodplain farmland in Nigeria) and dambos (headwater wetlands in southern Africa). There is no known means of providing a direct association between local terms and hydrological type in a fully inclusive manner.

9.3.2 Hydrological services of wetlands

One of the most commonly quoted regulating services of wetlands is flood reduction; some wetlands are said to 'act like a sponge', storing water during wet periods and releasing it during dry periods (e.g. Bucher *et al.*, 1993). Maltby (1991) reports ' ... the case for wetland conservation is made in terms of ecosystem functioning, which result in a wide range of values including groundwater recharge and discharge, flood flow alteration, sediment stabilization and water quality'. Wetland conservation has often been promoted as a potential means of flood management by organisations such as IUCN (Dugan, 1990), Wetlands International (Davies and Claridge, 1993) and the Ramsar Convention on Wetlands of International Importance (Davis, 1993). They have influenced international wetland policy (OECD, 1996) and its uptake at the national (e.g. Zimbabwe – Mazvimavi 1994, and Uganda – Republic of Uganda 1995), and continental levels (e.g. Europe – CEC, 1995; Blackwell and Maltby 2006; and Asia – Howe *et al.*, 1992).

Many wetlands exist because they overlie impermeable soils or rocks, and there is little interaction with groundwater. In a review of several hundred papers by CEH, Bullock and Acreman (2003) found 69 statements referring to groundwater recharge, 32 concluded merely that recharge takes place and 18 concluded there is no recharge. There are similar numbers of studies that report wetlands either to recharge more (6) or less than (9) other land types. Some wetlands, such as floodplains in India and West Africa on sandy soils, recharge groundwater when flooded (Thompson and Hollis, 1995).

Many studies show that floodplain wetlands reduce or delay floods, with examples from all regions of the world (Novitski, 1978, Bedinger, 1981, Bruland and Richardson, 2005, Acreman *et al.*, 2003). CEH research found less consistent evidence for wetlands in the headwaters of river systems, for example, bogs and river margins (Acreman and Holden, 2013). Indeed some (27 of 66) of headwater wetlands increase flood peaks. These studies were mostly from Europe, but included work from West Africa and Southern Africa. Around half of the statements (11 of 20 for flood event volumes and 8 of 13 for wet period flows) show that headwater wetlands increase the immediate response of rivers to rainfall, generating higher volumes of flood flow, even if the peak flow is not increased.

The lateral extent of floodplain wetlands and their often rough vegetation has an important hydraulic effect, slowing down water movement and absorbing energy during river floods. Modelling of the River Cherwell, UK, by CEH for the European Commission project 'The Wise Use of Floodplains' (Acreman *et al.*, 2003) showed that removal of embankments, separating the river from its floodplain, resulted in a reduction in downstream flood magnitude of 132% (Fig. 9.8).

9.3.3 Water requirements of wetlands

CEH and other UK academic organisations (Open University and University of Sheffield) joined forces to bring together a considerable body of theory and practice in wetland ecohydrology into objective and quantitative guidelines for the maintenance and restoration management of wetlands (Mountford *et al.*, 2005; Wheeler *et al.*, 2004). Within the United Kingdom, the development of such tools has

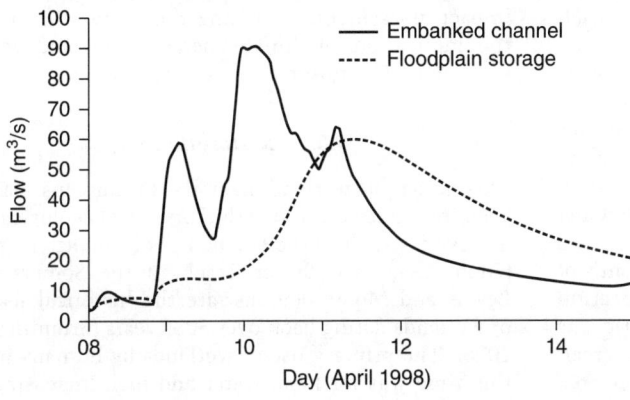

Fig. 9.8 Flood hydrographs at Oxford for April 1998 with and without (embanked) floodplain storage upstream (after Acreman *et al.*, 2003.) (Source: Reproduced with permission of CEH.) *(See insert for colour representation of this figure.)*

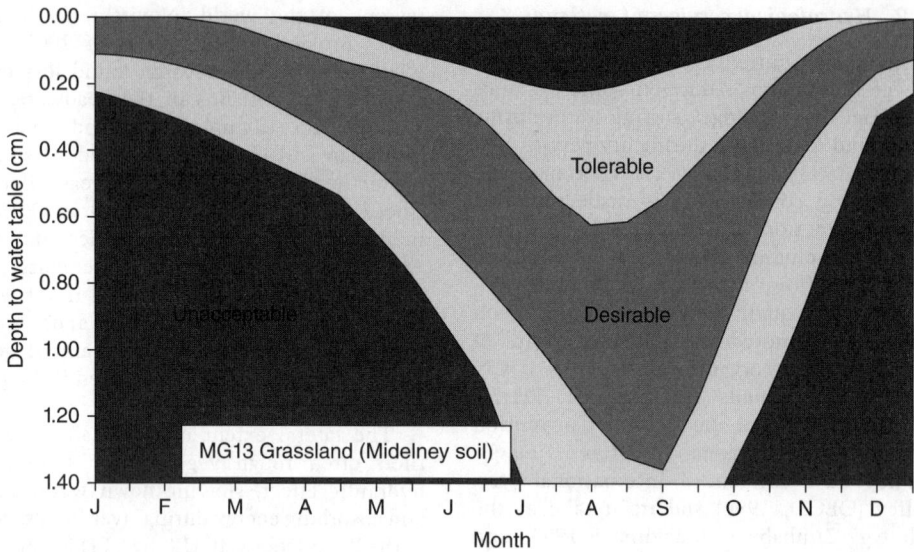

Fig. 9.9 Water-regime requirements of MG13 wetland plant community (Wheeler *et al.*, 2004). (Source: Wheeler *et al.* 2004. Reproduced with permission of CEH.) *(See insert for colour representation of this figure.)*

been stimulated by agrienvironment schemes and nature protection designations, notably in flood-plain grasslands (Gowing *et al.*, 2002) and grazing marsh, as at Tadham Moor (Mountford *et al.*, 1996). These studies began with investigations of altered nutrient regimes and then proceeded to experimentally manipulate water levels, integrating the hydrological and nutrient aspects to produce the guidelines. Working on the basis of vegetation or habitat types, these ecohydrological guidelines provide information on (a) water supply mechanisms and a conceptual model of hydrological functioning, (b) water regime, (c) nutrient regime, (d) management regime, (e) vulnerability, (f) the restorability of the type and (g) gaps in the knowledge. For each season, the water-regime parameters indicate the water depths (maximum, minimum, duration of exposure, *etc.*) that are favourable for the habitat, the depth ranges that the habitat can tolerate for short periods, and those that are damaging for the habitat. Figure 9.6 shows a 'traffic light-based' water level regime zones diagram (known in CEH circles as 'underpants' diagrams) that depicts the mean water table requirements for each month of the year for the National Vegetation Classification (NVC) MG13 (*Agrostis stolonifera – Alopecurus geniculatus grassland*) community. The green area shows preferred conditions. The amber area

denotes conditions that are not ideal, but which the plants can withstand for short periods, whereas the red area marks conditions which the vegetation cannot tolerate. CEH and UCL (Duranel *et al.*, 2007) developed a method for assessing the suitability of floodplains for species-rich meadow restoration. Figure 9.9 shows a traffic light, the method was tested on the floodplain of the River Thames floodplain in central southern England and it was found that both the maximum duration of flood events in autumn and winter and the depth of the groundwater table during the summer exceeded the requirements of the target species. It has subsequently been applied within hydroecological impact assessments including those investigating the implications of climate change on lowland wet grasslands (Thompson *et al.*, 2009).

9.3.4 Restoration

The value of wetland habitats to humans has been recognised through history. Archaeological discoveries such as the bog bodies of iron age man (Glob, 1969) and Sweet Track in the Somerset Levels and Moors demonstrate the historical use of wetlands dating back over 5000 years (Brunning, 2006). The primary use of wetlands by humans in this time was for food, water and fuel. Increasing awareness of the often high fertility associated with wetland areas, brought with it a desire to exploit these areas of the landscape. Now, as a result of years of drainage, development and pollution, only about 10% of the original wetland area in England remains (Wetland Vision, 2008). The industrial revolution and increasing use of machines to do manual tasks greatly sped up the activities that led to loss of wetlands, converting inaccessible wet and boggy areas into highly fertile agricultural land. Elsewhere in the world, the story is much the same with wetland areas shrinking as a result of multiple human pressures (Thompson and Finlayson, 2001).

It is only now that the impact of this loss of wetlands is being fully recognised. We now understand that wetlands provide a range of other useful services such as storage and release of water and chemicals, providing aesthetically pleasing and restful places, cycling of carbon and nutrients and providing areas of high biodiversity (Millennium Ecosystem Assessment, 2005). The efficiency with which wetlands provide these services is in some cases dramatically reduced once the natural hydroecological functioning of the wetland has been damaged (Peh *et al*, 2013). Reduced capacity for storing water can increase downstream flood risk (Acreman *et al.*, 1999), exposure of carbon-rich soils increases carbon dioxide fluxes to the atmosphere (Lloyd, 2006) and the presence hydroecological niches is important in supporting biodiversity (Araya *et al.*, 2011).

With this in mind, measures have been taken in recent years to conserve remaining areas of wetland and where possible restore previously degraded wetlands or expand wetland areas through their creation on previously dry sites. CEH has played a key role in understanding ecohydrological relationships on which management actions have been based. At Tadham Moor in Somerset, water levels in ditches have been manipulated to reflect a 'natural' cycle, with high water tables in the winter and low water tables in the summer. In East Anglia, the Great Fen Project (Mountford *et al.*, 2002) seeks to restore around 3000 ha area of fenland. At Chimney Meadows on the River Thames, an area of floodplain wetland is being rejuvenated through removal of river embankments and creation of new 'wet' features (Stratford, 2014). At Braunton Burrows, on the north Devon coast activities are underway to improve rare dune slack wetland habitats through a programme of land cover management (Stratford, 2014).

Ongoing studies at each of these sites aim to investigate the effectiveness of different restoration techniques, through observing the hydroecological regime of restored and unrestored areas. The difference in depth to the water table between restored and unrestored (drained) areas at Tadham Moor (Fig. 9.7) was found to be up to 50 cm in winter months (Fig. 9.10). Acreman *et al.* (2002) found that within the span of the experiment (1994–2002), the implementation of raised water

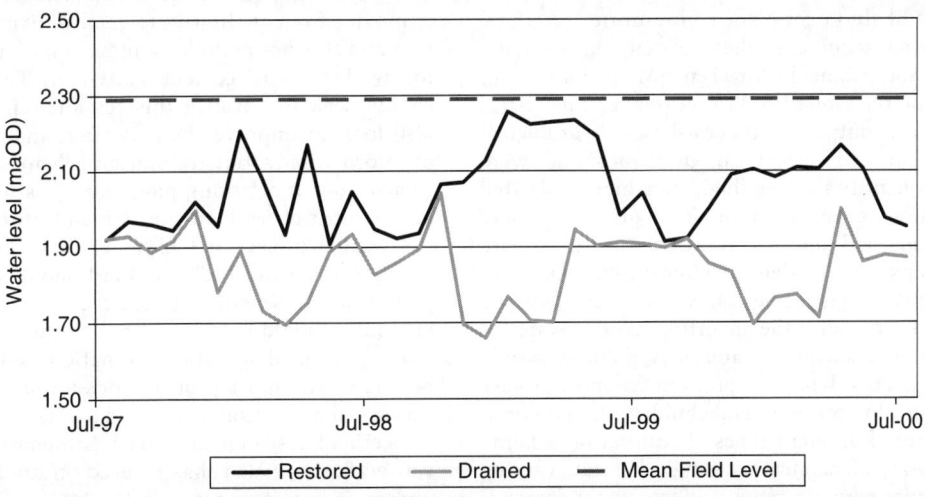

Fig. 9.10 Time series of water table elevations in restored and drained plots at Tadham Moor, from July 1997 to July 2000. (Source: Reproduced with permission of CEH.) *(See insert for colour representation of this figure.)*

Fig. 9.11 Tadham Moor wet grassland Somerset. (Source: Photograph by Mike Acreman.)

levels led to the partial replacement of an old meadow vegetation (NVC MG5 and MG8) by a ruderal community (NVC OV28), swamp (*NVC S6, S22*) or inundation grassland (NVC MG13) (Fig. 9.11).

This work and others has helped to inform the development of the previously discussed hydroecological guidelines for various plant communities (Wheeler *et al.*, 2004; Davy *et al.*, 2010). These guidelines have since been used to assess monitored and modelled water level data sets in order to identify likely target plant communities. Wetland restoration scoping studies carried out by CEH for the South Lincolnshire Fens (Mountford *et at.*, 2004) and the Great Fen Project (Blake *et al.*, 2012) linked the output of process-based hydrological models to hydroecological guidelines and were able to quantify the likelihood of achieving desired restoration goals both under current climate conditions and under a range of future climate conditions. The models developed by Blake *et al.* (2012) give a high degree of flexibility and allow the user to set rules for the apportionment of water in years when inadequate water is available to satisfy all areas. This dynamic approach has proved very useful in illustrating to stakeholders the potential feasibility of different types of restoration scheme (i.e. mosaic of habitats vs. single habitat, varying the balance between more- and less-water demanding vegetation communities, the requirement for water storage facilities) and has been applied in a

recent integrated study in the South Lincolnshire Fens (Stratford *et al.*, 2014)

The pressures on wetland habitats are only likely to increase in the future as the gap between demand and availability of water continues to increase. The importance of conserving and restoring what we can will similarly increase. There will be a greater need to show good evidence for how and why a programme of measures will be successful in providing benefits. Research into wetland hydrocology, has up to now, made considerable progress in moving from qualitative to quantitative information and this has provided a more solid foundation for wetland management activities. The future should seek to advance this research and should also look to improve the collection and collation of information about restoration schemes. A well thought-out monitoring programme, as developed at a number of wetland sites which feature within the research projects of CEH and their collaborators (e.g. Great Fen and Tadham Moor), should be a core part of any restoration scheme and presentation of results should be an expected output. Increased sharing of good quality information is likely to be one of the most positive forward steps in the successful protection of wetland habitats.

Wetland research at CEH's Lambourn Observatory (Section 9.2.4) has focused on fundamental understanding of processes in the 24-acre wet grassland site adjacent to the river. A new conceptual model has been produced of water and chemical fluxes between the chalk, gravel and peat deposits using a combination, ground-penetrating radar, Electrical Resistivity Tomography and temperature anomalies. Temperature data were collected in winter at multiple depths down to 0.9 m and transformed into a three-dimensional model via ordinary kriging. Anomalous warm zones indicated distinct areas of groundwater upwelling which were concurrent with relic channel structures and coincided with vegetation associated with groundwater sources (House *et al.*, in review). All data are being drawn together within a 3D hydrological model to simulated implications of groundwater abstraction and climate change.

9.3.5 *Climate Change Impacts*

It is widely accepted that the Earth's climate is changing more rapidly than it has in the past and that over the next 100 years temperatures will rise and patterns of precipitation will be altered (Kundzewicz *et al.*, 2007). These predictions for the future have important implications for all ecosystems, particularly those, such as wetlands, whose ecological character is very dependent on hydrological regime (Baker *et al.*, 2009). The potential impacts of climate change on wetland hydrology are of interest to a wide range of stakeholders from wetland managers to international policy makers. Ecohydrological models that combine climate changes, hydrological processes and ecological responses provide a means of estimating what might happen to some wetland functions and species in the future. CEH developed a framework that can be used for combining models and available data at a regional scale and is appropriate for different wetlands, in different countries and for different levels of data availability (Acreman *et al.*, 2009a). The simple models are based on broad conceptual understanding of wetland hydrology and are intended to describe basic ecohydrological processes within the constraints of data availability; they are thus fit for the purpose of general assessment of different wetland types, indexed by their plant communities and do not pretend to provide precise results for specific wetland sites. Data from Great Britain (GB) have been used to demonstrate each step in the framework for two temperate wetland types: rain-fed wetlands (wet heaths or degraded raised mires) and floodplain margins with projections for 2080 (Acreman *et al.*, 2009a). Whilst GB may be considered to

be relatively data rich, we believe that sufficient information is available in many countries to apply this framework for the regional assessment of climate change impacts on wetlands. Although the models successfully represent the baseline conditions, it is not possible to test whether they accurately predict the future vulnerability of the selected areas. Results for GB suggest that predictions of reduced summer rainfall and increased summer evaporation will put stress on wetland plant communities in late summer and autumn with greater impacts in the south and east of GB (Fig. 9.12). In addition, impacts on rain-fed wetlands will be greater than on those dominated by river inflows.

CEH undertook further development of the wetland models with university (UCL, Exeter and Open University), agencies (Environment Agency, Natural England and English Heritage) and NGOs (RSPB and WWT) to produce a web-based tool (Acreman *et al.*, 2013) that projects the potential impacts of climate change on a range of wetlands across the river basin regions of England and Wales. Because of the uncertainty in climate change projections and the different results obtained from using different climate models, the UKCP09 output of the UK Climate Projections programme (Murphy *et al.*, 2009) provides 10,000 different realisations of future climate for each future time slice, emissions scenario and location, each of which is equally likely (or unlikely). The model outputs therefore take the form of histograms of possible future wetland conditions. Thresholds of significant alteration in a range of impact metrics were developed, including hydrological metrics and those indexing vegetation, birds and archaeology. Figure 9.10 shows the projected impact on gross water balance as measured by changes in mean annual rainfall minus evaporation for an NCV M16 wet heath in East Anglia. It can be seen that under baseline conditions, mean annual rainfall exceeds actual evaporation by 116.4 mm. There is a 33% chance that under climate change by the 2050s, the water balance will only experience a minor change (green). However, there is a 42% chance (35% + 7%) that the water balance will experience an intermediate impact (amber). This could be the result of slightly drier or slightly wetter conditions. Additionally, there is a 25% (23% + 2%) chance of a major impact, consisting of a 23% chance of significantly drier conditions and a 2% chance of significantly wetter conditions (Fig. 9.13).

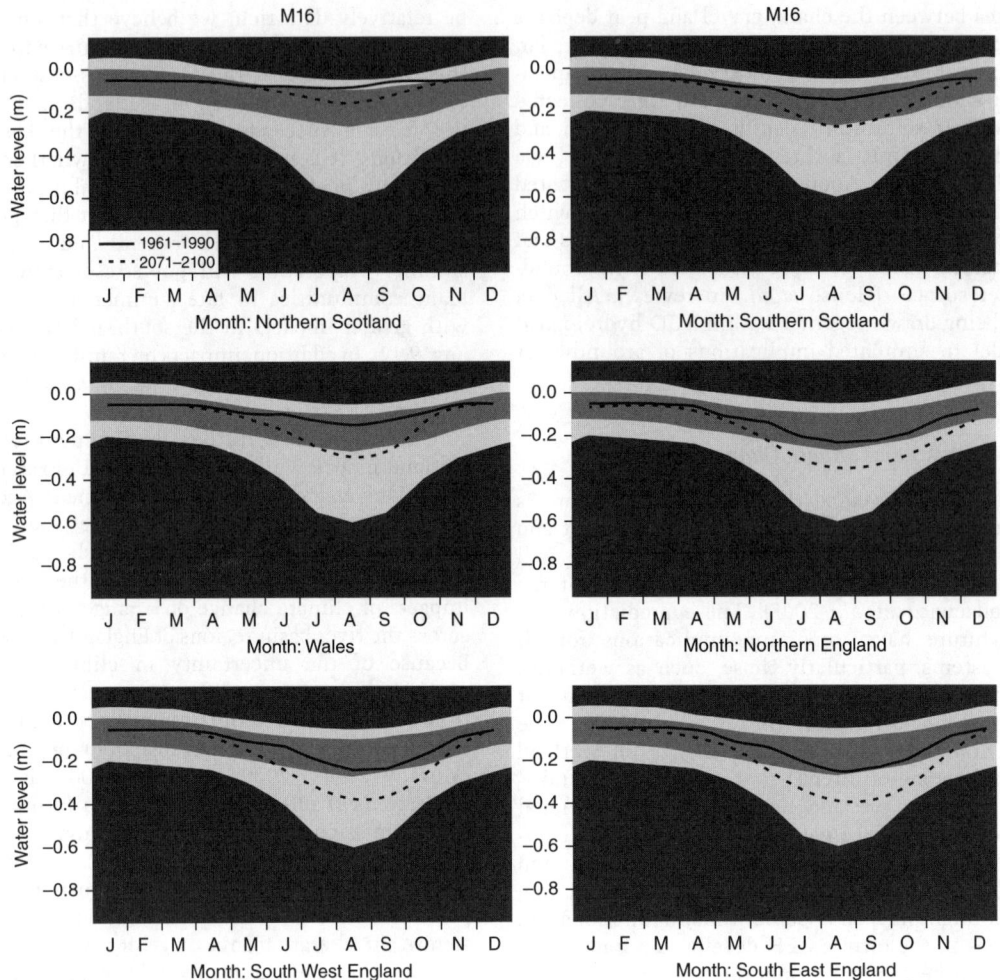

Fig. 9.12 Monthly mean water table levels (*m*) for the rain-fed wetland type, superimposed on the M16 wet heath water level requirements zone graphs, at the six sample sites across GB. The solid lines represent baseline (1961–1990) climate, the dashed lines show the projected values for the 2080s (2071–2100) using UKCIP02 (Hulme *et al.*, 2002) climate change data for the 'Medium–High' emissions scenario. (Source: Reproduced with permission of CEH.) *(See insert for colour representation of this figure.)*

9.4 Lakes

A significant body of research by CEH and its predecessors (namely, the Freshwater Biological Association [FBA], the Institute of Terrestrial ecology [ITE] and the Institute of Freshwater Ecology [IFE]) has focused on lake ecosystems and their response to changes in hydrological drivers. Water can enter and leave lakes by various pathways:

rivers, groundwater, subsurface drainage, overland flow, direct precipitation and evaporation (Winter, 2003). Each of these varies spatially and temporally with respect to climate, catchment soils, geology, land use and topography, and the characteristics of the lake itself.

Functionally, lakes can be separated into two main hydrological classes: (1) closed-basin lakes, which have no significant outflow and in which

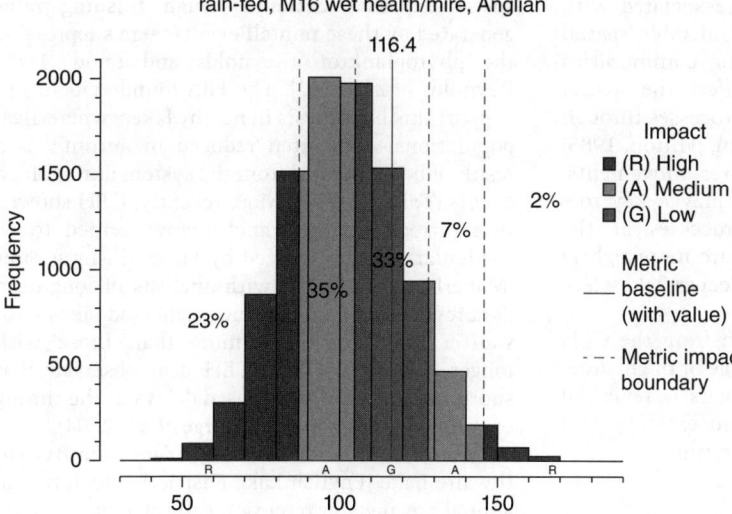

Percentage probability of impact
rain-fed, M16 wet health/mire, Anglian

Fig. 9.13 'Wetland Tool for Climate Change' example output showing the potential impact of climate change on wetlands for the 2050s under the 'Medium' emissions scenario – Gross mean annual water balance for a rain-fed M16 wet heath in East Anglia. (Source: Reproduced with permission of the Environment Agency.) *(See insert for colour representation of this figure.)*

water loss is dominated by evapotranspiration and (2) open-basin lakes, where water loss is largely through surface drainage or seepage to groundwater (Hutchinson, 1957). Most studies within the United Kingdom have been focused on open-basin lakes, because this is the most common type of lake in this area.

Although often referred to as 'standing waters', the hydrology of lakes is actually very dynamic (Hutchinson, 1957). The flow of water through a lake may be much slower than along a river, but it still has an important impact on ecology. It delivers sediment, nutrients and pollutants to the lake from its catchment, and determines in-lake responses to those inputs. The speed at which influent water passes through a lake is referred to as its flushing rate (calculated as lake volume/inflow volume, per year) or its inverse, water retention time (Winter, 2003).

The overall water balance of a lake, that is, inflow volume minus outflow, abstraction and evaporative losses, affects the water level of the lake. Where losses are greater than inflow volumes, lake shores can become exposed; the reverse situation can result in high water levels and flooding of marginal areas. Both can have important repercussions for the plant and animal communities that live there (see Section 9.4.2).

Within the lake itself, water movement is mainly driven by inputs of thermal and mechanical (wind) energy. Thermal energy is the main driver of stratification, a process that results in a region of rapid temperature change (the thermocline) within the water column, which partially restricts the transfer of material between the upper, warmer, layer (epilimnion) and the lower, cooler, layer (hypolimnion) (Hutchinson, 1957). Stratification occurs when the net heat input to a lake, which causes surface layers to become less dense, is greater than the input of turbulent kinetic (mixing) energy from wind stress at the lake surface. At the local scale, this is affected by the size and shape of the lake and weather conditions. However, CEH were instrumental in demonstrating that stratification was also affected by larger-scale processes, such as teleconnexions driven by the Gulf Stream (George and Taylor, 1995), the North Atlantic Oscillation (George *et al.*, 2004) and, potentially, the jet-stream (Strong and Maberly, 2011).

In contrast to rivers, water movement in lakes is essentially driven by wind rather than gravity. In addition to degrading stratification, wind energy generates other types of water movements such as surface waves, currents and seiches (internal waves). These types of wind-induced disturbances can have a large impact, either directly or indirectly, on the ecological structure and functioning of a lake (see Section 9.4.4). CEH showed, for example, that water movements associated with internal seiches can generate considerable spatial heterogeneity within lake plankton communities (Colebrook, 1960a, 1960b) and affect the spatial heterogeneity of biogeochemical processes through sediment resuspension and focusing (Hilton, 1985, Mackay *et al.*, 2012). Other water movements, such as currents and waves, can play a key role in regulating biogeochemical processes at the sediment–water interface, which are more tightly coupled in shallow lakes than deeper lakes (see Section 9.4.5).

The contribution made by staff from the CEH and its forerunners to our knowledge of the hydroecology of lakes over the last 50 years is reviewed below, focusing on the following processes:

- Flushing rate and water retention time
- Water level
- Stratification
- Wind-induced disturbance
- Currents, waves and sediment biogeochemical processes

9.4.1 Flushing rate and water retention time

Lake flushing rate or its inverse, water retention time, critically affects a lake's sensitivity to environmental change, especially in relation to eutrophication and acidification. It is a key component of lake models that are still widely used to predict lake nutrient concentrations and algal standing crop from nutrient inputs (Vollenweider, 1975; Dillon & Rigler, 1974; OECD, 1979). Data collected by ITE from Loch Leven, Scotland, contributed to the development of the Vollenweider and OECD models.

Further work by ITE on Loch Leven (1968–1985) showed that variation in flushing rate also affected in-lake temperature regimes and nutrient dynamics (Bailey-Watts *et al.*, 1990). It was found that, through such changes, flushing rate indirectly controlled the major features of phytoplankton succession, including the development and collapse of particular algal species and the zooplankton that graze on them. In particular, small phytoplankton with relatively high growth rates tended to dominate during seasons when retention times were lowest (Bailey-Watts *et al.*, 1990).

In the 1980s, work by the FBA showed that the phytoplankton population in Grasmere, English Lake District, changed in years of heavy rainfall in comparison to more 'normal' years. It was speculated that extremely high flushing rates generated by these rainfall events were suppressing the phytoplankton (Reynolds and Lund, 1988; Reynolds *et al.*, 2012). The FBA found evidence to support this hypothesis in nearby lakes, where algal populations were often reduced in autumn as a result of being flushed from the system during flood events (Talling, 1993). More recently, CEH showed that nitrogen-fixing cyanobacteria seemed to be particularly disadvantaged by faster flushing rates (Maberly *et al.*, 2002), with analysis of long-term data revealing that wet winters affected lakes with shorter retention times more than those with longer retention times. CEH also observed that shorter retention times caused delays in the timing of spring diatom blooms (George *et al.*, 2004).

More recent analyses of CEH data from Bassenthwaite Lake (English Lake District), which has an annual mean water retention time of about 20 days, showed that algal populations fell dramatically when seasonal (especially autumn and winter) retention times fell below 20 days (Jones *et al.*, 2011). This was due to wash out. In contrast, CEH found that, in lakes with longer retention times of, say, 20 to 100 days, effects of flushing on algal biomass and composition were mainly mediated through impacts on nutrient concentration and stratification. This is illustrated in CEH studies from Loch Leven (retention time about 180 days), where the lowest algal concentrations were recorded in the wettest summers (1985, 2007, 2008) and the highest in the driest summers (1994, 1995, 2006) (Carvalho *et al.*, 2012).

The underlying mechanisms responsible for the observed link between algal biomass and flushing rate were investigated using PROTECH, a numerical model developed by staff at CEH that predicts the accumulation of algal biomass in lakes in response to inputs of nutrients, light and thermal energy (Reynolds *et al.*, 2001; Elliott *et al.*, 2010). This study showed that, not only were there direct impacts of flushing on the algae, but also there were also indirect influences acting through changes in nutrient concentrations and thermal stratification (Jones and Elliott, 2007). An upper effect threshold of about 100 days was postulated; changes in flushing rate seemed to have little direct effect on the ecology of lakes with longer residence times.

A number of CEH studies have highlighted that flushing rate is especially important in the development of harmful blooms of cyanobacteria in lakes, modifying the relationship between nutrient availability and biomass accumulation (e.g. Carvalho *et al.*, 2013). Carvalho *et al.* (2011) demonstrated that summer concentrations of cyanobacteria were generally low in 134 UK lakes that had retention times of less than 30 days, but increased as retention time increased. CEH added a further dimension to this by showing that retention time was more important than water temperature in increasing the frequency of algal blooms to levels above World Health Organisation thresholds (Elliott, 2009), and intensive long-term monitoring at Grasmere revealed that the filamentous cyanobacterium, *Anabaena*, only occurred during long, dry summers (Reynolds and Lund, 1988). Globally, the effects of changing climate on flow are predicted to be highly variable, geographically (Milly *et al.*, 2005), and seasonally dependent (Nijssen *et al.*, 2001). However, within the United Kingdom, a depletion of river flow in future summers is the most likely scenario (Fowler and Kilsby, 2007; UKCP09, 2013). This would result in a decrease in lake flushing rate, suggesting that a further reduction in nutrient inputs would be needed to avoid consequent increases in algal bloom formation.

CEH also explored the impact of flushing rate on in-lake nutrient concentrations in relation to the relative importance of nutrient inputs from diffuse and point sources. It was found that if nutrients from external sources were supplied diffusely, the level of input (load) tended to reflect the variation in inflow (and thus flushing) rates, especially during high rainfall events (Defew *et al.*, 2013). However,

in contrast, if nutrients were from point or internal sources, their load would remain more or less constant as inflow rates changed, with nutrient concentrations within the lake tending to increase as flushing rates decrease (Elliott *et al.*, 2009). Overall, inflow rates and flushing rates can have a marked effect on in-lake nutrient concentrations and this, in turn, affects the ecology of the system.

9.4.2 Water level

Changes in lake water levels are caused by a variety of factors, including floods and droughts, water abstraction and changes in evaporation rates. These may occur seasonally, weekly, or even daily under some circumstances (Smith *et al.*, 1987). As lakes are relatively closed systems, changes in water level affect biological communities across the lake, but this is especially true around the shoreline. Here, previously inundated areas can become dry and exposed (e.g. Fig. 9.14), or previously dry and exposed areas can become inundated.

The effect of a significant, and permanent, reduction in lake water level was documented by May and Spears (2012), of CEH, who reviewed historical documents summarising the impacts of a 1.4 m reduction in water depth at Loch Leven, Scotland, in the mid-1800s when sluice gates were installed on the outflow. The effect on the ecology of the lake was dramatic. Macrophyte beds were destroyed through desiccation and this reduced habitat and food availability for macroinvertebrates, fish and aquatic birds. The local fishery was particularly badly affected, its economic value

Fig. 9.14 Haweswater, Cumbria, UK, August 1995, showing exposed shoreline after exceptionally dry conditions. (Source: Photograph by Ian J. Winfield, CEH.)

being reduced by about one-third. CEH used these outcomes to highlight the fact that unintended, and mostly unwelcome, consequences may result from lake management activities that focus on only one issue at a time (May and Spears, 2012).

The more general impact of variation in water level on littoral benthic communities was studied by ITE. Smith *et al.* (1987) recorded macrophyte and macroinvertebrate communities in 27 Scottish lakes and compared them to long-term records of water level. They found that the flora and fauna of the littoral zones were strongly affected by the type of water level fluctuations that had occurred in previous years. In particular, they found that lakes with relatively minor water level fluctuations (<5 m per year) had variable and abundant macrophyte and zoobenthos communities, whereas lakes with major fluctuations in water level (≥5 m per year) had relatively impoverished macrophyte and zoobenthos communities. In extreme cases, for example, at Sloy reservoir where annual water level fluctuations of more than 29 m were recorded between 1975 and 1980, CEH reported that there were no flora or fauna at all along the shoreline.

The littoral zone is often a critical habitat for lake fish populations, even for species that use it as spawning habitat for only a few days or weeks each year (Winfield, 2004). Study of the effects of changing water levels on such transitory events is extremely difficult, but Winfield *et al.* (2004) reported CEH studies that successfully incorporated long-term water level data into fish population models to demonstrate that more frequent falls in water level during the egg incubation period would result in a long-term decrease in the abundance of whitefish (*Coregonus lavaretus*) in Haweswater, English Lake District. The modelling was also able to predict the likely consequences of alternative water level regimes on the recovery of this protected fish species, thus informing the future management of this strategically important reservoir. As a result, an artificial mobile spawning substrate system has been developed for incorporation into the system when needed and, as a precaution, a refuge population for the threatened whitefish of this lake has been established (Winfield *et al.*, 2013).

9.4.3 *Stratification*

Stratification drastically alters the structure and function of lakes by affecting biogeochemical

processes such as 'benthic–pelagic coupling' (Spears *et al.*, 2007b), and the biology of phytoplankton, zooplankton and fish. When rates of oxygen transport across the thermocline are low compared to rates of decomposition and respiration, oxygen consumption can lead to anoxia at depth. CEH found that this may be stimulated by nutrient enrichment and climate change (Foley *et al.*, 2012). The FBA were amongst the first to demonstrate that deep water anoxia affected lake biota and promoted the release of nutrients, especially phosphorus and nitrogen, from the sediment surface into the overlying water (Mortimer, 1941). More recently, CEH has shown that this internal load of nutrients, which may have entered the lake decades earlier, is an important factor in controlling the rate of recovery of lakes when external inputs have been reduced (Spears *et al.*, 2012).

When a lake is unstratified, algae are mixed throughout the water column. However, early work by the FBA showed that, during stratification (Fig. 9.15), the mixing depth reduces and this tends to restrict algae to the warmer, and better illuminated, epilimnion (Talling, 1971). This can alter the underwater light climate experienced by the algae, favouring different functional groups. CEH research then showed that, as such, the onset of stratification is a major cause of the spring phytoplankton bloom in many lakes (Neale *et al.*, 1991) and that changes in the timing of this event can cause phenological changes within lake ecosystems (Thackeray *et al.*, 2008).

The separation of productivity in the epilimnion from nutrient regeneration in the hypolimnion and sediment can restrict the nutrient supply to phytoplankton in lakes, affecting species composition and abundance. For example, research by CEH and its collaborators suggested that strong stratification may reduce species richness in Esthwaite Water by restricting nutrient supply or encouraging competition (Madgwick *et al.*, 2006). Further work showed that interannual variation in summer phytoplankton biomass in this lake may be driven by changes in the position of the Gulf Stream by affecting summer wind-induced mixing, which controls the supply of nutrients from deeper to surface waters (George, 2002). CEH also showed that, in lakes where stratification is stable, for example, Priest Pot, a small lake at the head of Esthwaite Water in the English Lake District, motile algae can aggregate at preferred depths in response to gradients in light, oxygen,

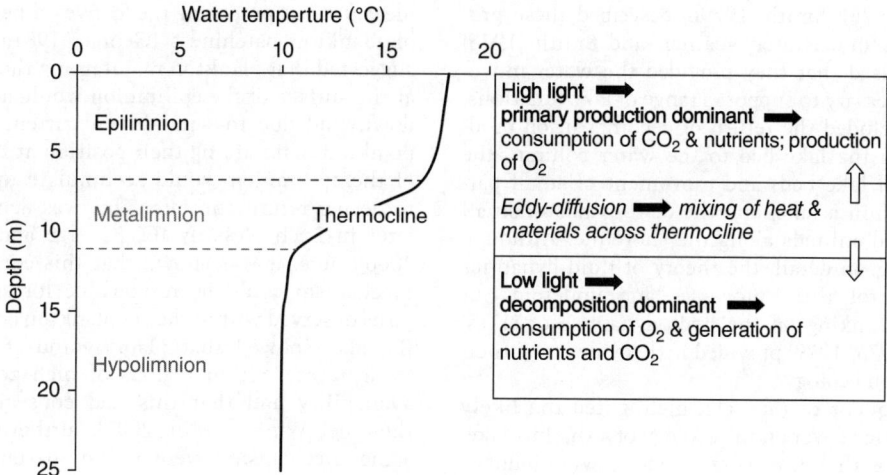

Fig. 9.15 Hypothetical change in water temperature with depth in a lake during stratification and lake zones associated with this process. (Source: Reproduced with permission of CEH.)

carbon dioxide and temperature, improving their competitive advantage (Clegg *et al.*, 2007).

As well as affecting phytoplankton, stratification can also influence the vertical distribution of zooplankton. This is because density stratification induces vertical heterogeneity in many of the environmental factors that zooplankton respond to, including water temperature, phytoplankton concentration/composition and oxygen concentration. Early work by the FBA showed that the zooplankton species *Ceriodaphnia quadrangula* tended to aggregate near vertical gradients in temperature and oxygen in Esthwaite Water, which has relatively weak stratification (Smyly, 1974). Later work by CEH and FBA showed that, in Windermere, which has much stronger thermal stratification, the vertical distribution of the dominant crustacean zooplankton species tends to be more heterogeneous (Thackeray *et al.*, 2006). As a result of this, and species differences in depth selection, there is a marked spatial turnover in zooplankton community composition with depth (Thackeray *et al.*, 2005).

CEH studies have also shown that stratification can influence the depth zones occupied by fish populations through its effects on the provision of appropriate temperature and oxygen conditions. In temperate lakes, these effects are most marked on species such as salmonids and their relatives, which have relatively strict environmental requirements.

Jones *et al.* (2008), using oxygen and hydroacoustic data, showed that the vertical distribution of Arctic charr (*Salvelinus alpinus*) in Windermere has been affected by a long-term decrease in oxygen availability at depth. In fact, in recent years, these fish have begun to migrate vertically to avoid oxygen concentrations of less than 2.3–3.1 mg/l, which now occur more widely at greater depths. CEH explored habitat constrictions on fish distributions in relation to vendace (*Coregonus albula*) in Bassenthwaite Lake, English Lake District (Elliott and Bell, 2011). It was concluded that a marked loss of appropriate habitat for these fish was likely to occur under projected conditions of climate change, which included a mean increase of more than two degrees centigrade in water temperature and a 10% decline in oxygen concentration. Of particular concern was the conclusion that the volume of habitat available to these fish will decline greatly under the future climate change scenarios tested. All of the 20 years simulated had periods of more than seven consecutive days when there was no habitat available for these fish, mainly as a result of high temperature.

9.4.4 *Wind-induced disturbance*

ITE was amongst the first to recognise that efficient mixing processes provided by turbulent motion play an important part in lake ecology

(Smith, 1975). Smith (1979b) described these processes as 'underwater weather' and Smith (1975) hypothesised that they provided the water movement necessary to support a range of lake functions. These included the return of decomposition products from the lake bed to the water column, the erosion of lake beds and movement of small particles within a lake, and the disturbance of small plants and animals along the shoreline. Although describing, in detail, the theory of fluid dynamics that control these processes and underpin our current thinking on the hydrodynamics of lakes, Smith (1975, 1979) provided little evidence of their impacts on ecology.

Subsequent to this, ITE highlighted the likely effects of ice cover on the ecology of a shallow lake, suggesting that this would reduce wind-induced disturbance and encourage deoxygenation (Lyle, 1982). Lyle (1982) suggested that this in turn would increase nutrient release from the sediments and promote the rapid development of algal blooms when the ice melted. It was suggested that the presence or absence of ice may also affect overwintering wildfowl. Lyle (1982) also postulated that the large numbers of greylag geese (*Anser anser*), pink-footed geese (*Anser brachyrhynchus*) and whooper swans (*Cygnus cygnus*) that overwinter on Loch Leven (as described by Allison and Newton, 1974) may be sufficient to maintain small areas of open water in all but the coldest of weather, thus implying a feedback mechanism between lake ecology and hydrology.

George (1981a, 1993) showed that phytoplankton and zooplankton were distributed heterogeneously throughout lakes and reservoirs as a result of the action of, and responses to, scale-dependent physical, chemical and biological processes. Further CEH, IFE and FBA studies of UK lakes led to important insights into the ecohydrological drivers of heterogeneity. These revealed that an important driver of 'plankton patchiness' is continuous, wind-induced water movement.

At the FBA, Colebrook (1960a,b) reasoned that wind-induced circulation and internal waves (seiches) should passively transport plankton throughout lakes, with considerable spatial heterogeneity being generated by the interactions between these processes and the vertical migration behaviour of the plankton themselves. These expectations were confirmed in a series of surveys of the Windermere zooplankton community (Colebrook, 1960a). Subsequent studies developed these

ideas further, proposing the 'conveyor belt' concept of plankton patchiness (George, 1981a,b), which suggested that plankton maintaining their position at the surface of the epilimnion would accumulate downwind due to wind-driven currents, whereas plankton maintaining their position at the bottom of the epilimnion would accumulate upwind due to deeper return currents. This was demonstrated later in Loch Ness by IFE (George and Winfield, 2000); here, it was shown that this conveyor belt mechanism could be responsible for spatial patterns observed within the zooplankton community. IFE also showed that planktivorous fish tended to aggregate within regions of high zooplankton availability and that this had consequences for their diet (Winfield *et al.*, 2002). Furthermore, these same mechanisms were shown to operate over the more restricted spatial scales of small (5 m) and large (45 m) in-lake experimental enclosures, which has implications for the design of sampling protocols for these systems (George, 1989).

More recently, attempts have been made to quantify the effect of wind-driven water circulation on observed patterns of plankton distribution, and to determine how much this might vary temporally with changes in the intensity of wind forcing. CEH studied the plankton community in Windermere and, using subsurface water temperature as a tracer of displaced water (Thackeray *et al.*, 2004; Fig. 9.16), showed that the direct effects

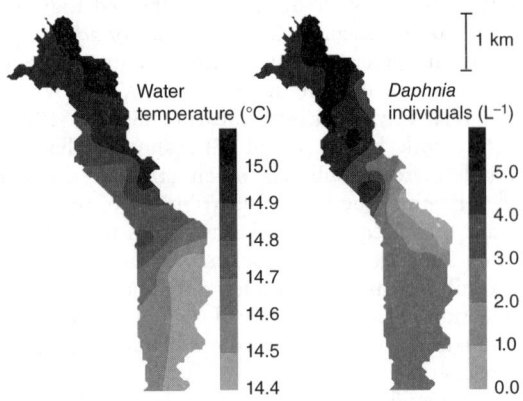

Fig. 9.16 Spatial variations in subsurface water temperature and *Daphnia* population densities in the north basin of Windermere, English Lake District, on 20th June 2001. (Source: Adapted from Thackeray *et al.* 2004. Reproduced with permission of John Wiley & Sons.)

of wind-induced water circulation patterns upon zooplankton distribution probably varied over time. Using Esthwaite Water as a case study, a follow on study by Mackay *et al.* (2011a) found that downwind accumulation of phytoplankton could be detected even under modest wind forcing. However, near a lake inflow, any such wind-driven pattern could be obscured by the 'dilution effect' of inflowing water, which could cause local reductions in plankton concentrations (Mackay *et al.*, 2011b). Hydrodynamic modelling by CEH of some of these ecohydrological processes is now showing promise as a means of understanding, and simulating, constraints on spatial heterogeneity within the pelagic environment (Hedger *et al.*, 2004).

Wind-induced water movements can also affect the distribution of bed sediments and other substrates within a lake, especially along the shoreline (Fig. 9.17), and this has implications for the distribution of biological communities (Jupp and Spense, 1977). In Loch Leven, ITE showed that this resulted in two types of macroinvertebrate communities, one associated with sandy sediments and the other with mud (Maitland, 1979). In larger lakes, such as Lochs Lomond, Awe, Ness, Morar and

Shiel, ITE found that macrophytes and macroinvertebrates were more commonly associated with sheltered silty bays than stony, wave washed shores (Bailey-Watts and Duncan, 1981; Smith *et al.*, 1981). These studies were some of the first to show this type of relationship, clearly illustrating the indirect relationship between water movement and macroinvertebrate distribution.

Heterogeneity in the horizontal distributions of macrophytes and types of substrate often results in patchy distributions of littoral fish species; this was demonstrated by CEH for the adult stages of widespread and abundant species such as roach (*Rutilus rutilus*) (Winfield, 2004). Such spatial focusing has its most extreme impacts on species that require hard, coarse substrates for spawning. These areas are usually very limited in extent and relatively poorly described for many species because of sampling difficulties. However, with contributions from CEH scientists, Miller *et al.* (2015) have recently used novel hydroacoustic and underwater video techniques to revisit Arctic charr (*Salvelinus alpinus*) spawning grounds in Windermere that were originally described in the 1960s. Spawning ground characteristics were found to be

Fig. 9.17 Wind-induced disturbance along the shoreline of Loch Girlsta, Scotland. (Source: Photograph by Ian J. Winfield, CEH.)

more complex than originally thought and affected by unsuitable soft, fine sediments that have been transported from other areas of the lake by internal water movements. In the more eutrophicated south basin of Windermere, suitable habitat for Arctic charr spawning was found to be extremely limited in availability.

9.4.5 *Currents, waves and sediment biogeochemical processes*

The importance of phosphorus release from bed sediments into the water column as a driver of lake ecosystem structure and function was identified in Loch Leven, Scotland, by Holden and Caines (1974). It was concluded that imbalances in the phosphorus budget when comparing inflow and outflow discharges were probably the result of fluxes between the water column and bed sediments under hydrological conditions that favour sediment deposition.

Water temperature and oxygen concentrations near the sediments are strongly affected by in-lake hydrological processes and are, in turn, important drivers of the biological and chemical processes that are responsible for liberating nutrients from lake sediments. During stratification, surface and deep waters are largely isolated from each other, although some vertical mixing continues to occur through vertical diffusion across the thermocline and/or the entrainment of hypolimnetic waters as the epilimnion deepens. More recently, CEH has shown that the rate of vertical diffusion and the number and size of entrainment events are affected by atmospheric forcing, causing changes to deep water anoxia, phosphorus accumulation in the hypolimnion and subsequent fluxes of phosphorus into the epilimnion (Mackay et al., 2014).

CEH has also been instrumental in demonstrating that energy, surface heating and lake transparency determine the patterns of vertical mixing that occur within lakes and the extent of stratification (Persson and Jones, 2008; Jones et al., 2005), mediating diffusive exchanges between sediments and overlying water, and impacting upon many biogeochemical processes that are known to regulate key lake functions. More recently, CEH has shown that these coupled processes can drive water quality and the ecological structure and function of lakes (Spears et al., 2012), regulate the delivery of nutrients and other pollutants to downstream systems (Spears et al., 2007a) and contribute to global scale climate regulating processes

(Maberly et al., 2013). Drivers of these processes are related to a mosaic of interacting physicochemical and biogeochemical drivers that vary in space and time (Spears et al., 2007b), most of which are sensitive to various combinations of the external and in-lake hydrological processes outlined above.

Fluxes of heat and mechanical energy across the surface of a lake can affect sediment resuspension and nutrient release, both directly and indirectly (Smith, 1974; George, 1981b). Early work in Windermere by the FBA demonstrated that wind stress, in particular, resulted in the downwind tilting of isotherms and currents, followed by internal waves (or seiches) along the thermocline once the wind had dropped (Mortimer, 1952). Later, researchers at ITE examined changes in current speed with depth, and the effect of wind speed and fetch on wave height, length, steepness and period (Smith, 1979; Smith and Sinclair, 1972). They found that, for wind-driven currents, an exponential decay in current speed occurred with increasing water depth, with the conservation of mass within a lake resulting in a return current in the opposite direction (Smith, 1979).

The wave mixed depth of a lake is generally about half of the wave length in areas where waves reach the lake bed, that is, generally around the shore (Smith and Sinclair, 1972) (Fig. 9.18), with 'Ekman' spirals and eddy patches forming across lakes even within relatively slow moving water (George, 1981b; George and Allen, 1994). CEH showed that, in shallow lakes such as Loch Leven, Scotland, a large area of the lake bed can be disturbed by this process (Spears and Jones, 2010; Fig. 9.19). A number of CEH studies have considered the effects of wave action and currents on sediment resuspension and redistribution in lakes (Hilton, 1985; Hilton and Gibbs, 1985; Hilton et al., 1986; Mackay et al., 2012). These showed that sediment accumulation rates and the distribution of heavy metals and nutrients in sediments vary spatially and are often linked to sediment focusing, that is, where small and/or low-density particles are moved preferentially into the deepest parts of a lake by water currents (Hilton and Gibbs, 1985; Mackay et al., 2012).

The exchange of water and nutrients between littoral and pelagic areas in lakes is another example of an ecohydrological process that contributes to nutrient cycling and the linkage between different habitats within lakes. George (2000) of CEH, using remote sensing of surface water temperature in

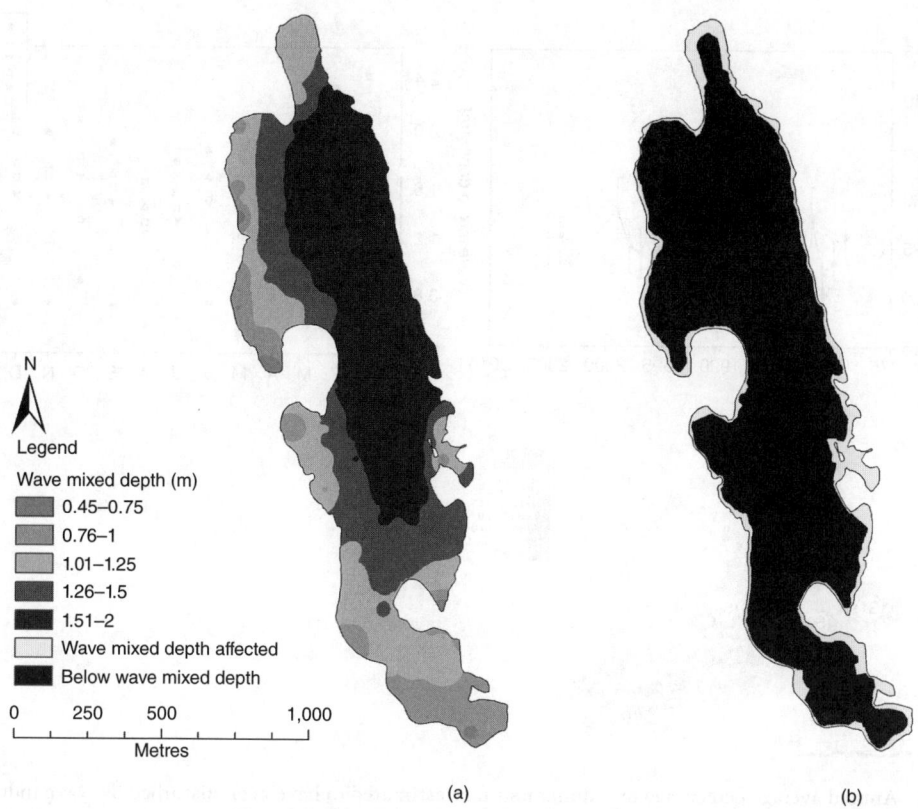

Fig. 9.18 Esthwaite Water (*a*) average maximum wave mixed depth (*m*) for 2005–2009 and (b) wave affected area based on this wave mixed depth. (Source: Adapted from Mackay *et al.*, 2012. Reproduced with permission of John Wiley & Sons.) *(See insert for colour representation of this figure.)*

Esthwaite Water, traced water movements from reed canopies in the littoral zone towards the pelagic zone and suggested that this exchange of water, which continued to occur under relatively low wind conditions, may contribute to large and episodic pulses of phosphorus reaching the pelagic zone.

Our understanding of the key chemical pathways of phosphorus cycling in lakes is based on evidence from long-term monitoring studies and targeted experimental manipulations set within a comprehensive hydrological framework. In addition, CEH has undertaken similar work in rivers (House, 2003; Jarvie *et al.*, 2005), which has helped improve our understanding of these processes in lakes, especially in relation to sediment phosphorus dynamics. Studies of sediment phosphorus in Loch Leven over the past 35 years, for example,

have provided a unique insight into the drivers of phosphorus fluxes between the lake bed and the water column; this is the main driver of prolonged ecological problems in lakes that are recovering from eutrophication (Sharpley *et al.*, 2013; Jarvie *et al.*, 2013).

Farmer *et al.* (1994) described the key ecosystem scale processes that were responsible for phosphorus cycling in shallow lakes, with ITE (Bailey-Watts and Kirika, 1987, 1993, 1999) highlighting the combined role of sediment phosphorus release and low summer flushing rates in regulating recovery processes (Spears *et al.*, 2007a). Further work by CEH found that only a very small proportion of stored sediment phosphorus at the whole lake scale was likely to be responsible for the significant ecological effects observed at the ecosystem scale (Spears *et al.*, 2006, 2012). This illustrates the

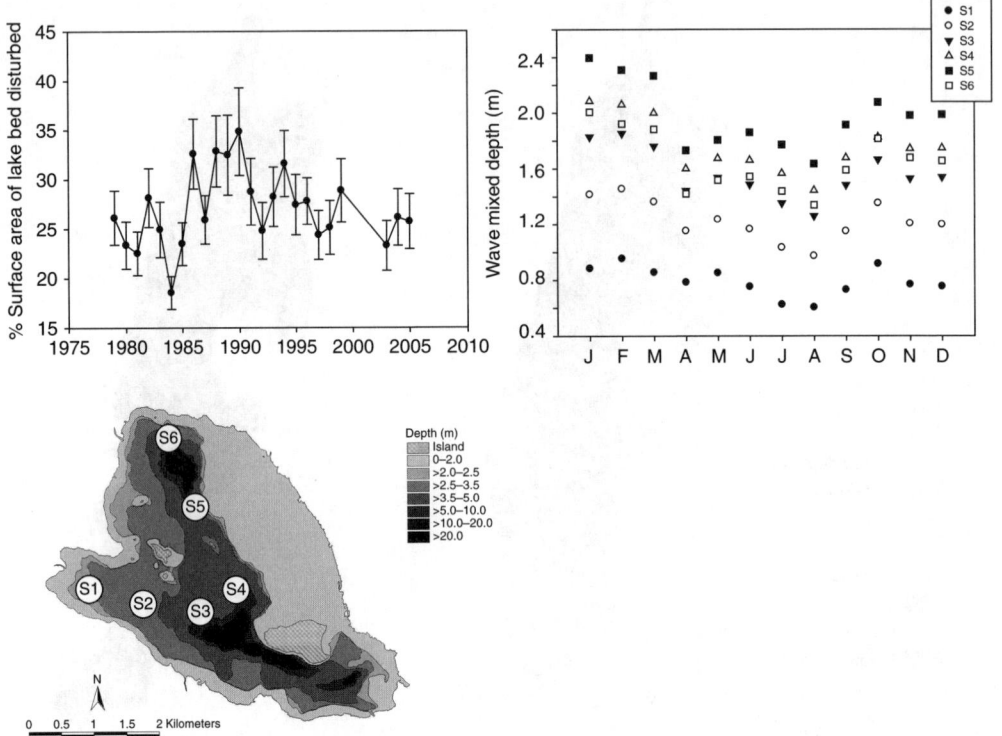

Fig. 9.19 Annual average percentage of sediment surface estimated to have been disturbed by wave induced mixing in Loch Leven, Scotland, 1979–2005 (upper left); seasonal variation in 30-year average, modelled, wave mixed depth across six sites in the lake (upper right and lower left) between 1979 and 2005. (Source: Spears & Jones 2010. Reproduced with permission of Springer Science and Business Media.)

importance of processes that buffer sediment phosphorus release, including stratification, aeration of the sediment surface and the presence of biological communities, all of which are, at least in part, regulated by hydrological processes (Spears *et al.*, 2006, 2007a, 2007b).

CEH has identified a significant knowledge gap in relation to the role of benthic biological communities in regulating the physicochemical and hydrological conditions that control phosphorus cycling between bed sediments and the overlying water column (Poulickova *et al.*, 2008; Spears *et al.*, 2008; Letovsky *et al.*, 2012). These include the hydrological regulation and the maintenance of habitats that are resistant to wind-induced mixing and other erosive forces. The capacity for benthic algal communities to 'biostabilise' bed sediments is likely to have less of a role in lakes than in estuarine communities, where low levels of salinity tend to

reduce the cohesivity of bed sediments and make them more sensitive to erosive hydrological forces (Spears *et al.*, 2008).

In contrast to microalgae, macrophytes can regulate the hydrodynamics of bottom waters very effectively and can also play an important role in the regulation of phosphorus cycling in lakes. CEH and its collaborators have shown that this depends upon their spatial coverage (Spears *et al.*, 2009; O'Connell *et al.*, 2007). Evidence of spatial variation in phosphorus fractions, and of concentration gradients across the sediment–water interface, highlights the functional role of these communities as they vary with lake depth. In shallow water of Loch Leven, CEH demonstrated that dissolved phosphorus concentration gradients were lowest where macrophyte biomass was highest (i.e. 3.5 m depth) and highest in deeper sediments where macrophytes were absent (i.e. 25 m), during summer stratification (Spears *et al.*, 2007b). This work, which is still being developed, provides a biological supplement to the maps of physical and hydrological processes that were produced by ITE (Smith, 1974).

9.5 References

Acreman, M.C. (1998) Principles of water management for people and the environment. In: de Shirbinin, A. & Dompka, V. (eds), *Water and Population Dynamics*. American Association for the Advancement of Science, pp. 321.

Acreman, M.C. & José, P. (2000a) Wetlands. In: Acreman, M.C. (ed), *The Hydrology of the UK – a study of change*. Routledge, London.

Acreman, M.C. (2001a) Ethical aspects of water and ecosystems. *Water Policy Journal*, 3 (3), 257–265.

Acreman, M.C. (ed) (2001b) *Hydro-ecology: linking hydrology and aquatic ecology*. IAHS Pub. No. 266. . IAHS, pp. 213.

Acreman, M.C. & Elliott, C.R.N. (1996). Evaluation of the river Wey restoration project using the Physical HABitat SIMulation (PHABSIM) model. *Proceedings of the MAFF Conference of River and Coastal Engineers, Keele, 3–5 July 1996*.

Acreman, M.C., Riddington, R. & Booker, D.J. (1999) Hydrological impacts of floodplain restoration: a case study of the River Cherwell, UK. *Hydrology and Earth System Sciences*, 7 (1), 75–85.

Acreman, M.C., Harding, R.J., Lloyd, C.R. & McNeil, D.D. (2003) Evaporation characteristics of wetlands; experience from a wet grassland and a reedbed using eddy correlation measurements. *Hydrology and Earth System Science*, 7 (1), 11–22.

Acreman, M.C., Mountford, J.O., McCartney, M.P., Wadsworth, R.A., Swetnam, R.D., McNeil, D.D., Manchester, S.J., Myhill, D.G. & Broughton, R.K. (2002) *Integrating wetland management, catchment hydrology and ecosystem functions*. Final report of the CEH Integrating Fund Round 5. CEH, Wallingford.

Acreman, M.C. (2005) *Guidelines for the management of groundwater to maintain wetland ecological character*. Report to the International Convention on Wetlands (Ramsar). Centre for Ecology and Hydrology, Wallingford.

Acreman, M.C., Blake, J.R., Booker, D.J. *et al.* (2009a) A simple framework for evaluating regional wetland ecohydrological response to climate change with case studies from Great Britain. *Ecohydrology*, 2, 1–17.

Acreman, M.C. & Dunbar, M.J. (2004) Methods for defining environmental river flow requirements – a review. *Hydrology and Earth System Sciences.*, 8 (5), 861–876.

Acreman, M.C., Dunbar, M.J., Hannaford, J. *et al.* (2008) Developing environmental standards for abstractions from UK rivers to implement the water framework directive. *Hydrological Sciences Journal*, 53 (6), 1105–1120.

Acreman, M.C., Aldrick, J., Binnie, C. *et al.* (2009b) Environmental flows from dams; the water framework directive. *Engineering Sustainability*, **162**, 13–22.

Acreman, M.C. & José, P. (2000b) Wetlands. In: Acreman, M.C. (ed), *The Hydrology of the UK – a study of change*. Routledge, London.

Acreman, M.C. & Ferguson, A. (2010) Environmental flows and European water framework directive. *Freshwater Biology*, **55**, 32–48.

Acreman, M.C., Liu, Z., Peng, R., Luo, Y., Gong, F.J, Chen, M.R., Lin, X. & Rameshwaran, P. (2010) Use of hydraulic rating to set environmental flows in the Zhangxi River, China. *International Symposium on The role of hydrology in managing consequences of a changing global environment*. British Hydrological Society, Newcastle, UK.

Acreman, M.C., Blake, J.R., Mountford, O., Stratford, C., Prudhomme, C., Kay, A., Bell, V., Gowing, D., Rothero, E., Thompson, J., Hughes, A., Barkwith, A. & van de Noort, R. (2013) *Guidance on using the Wetland Toolkit for Climate Change. A contribution to the Wetland Vision Partnership*. Centre for Ecology and Hydrology, Wallingford. http ://www.ceh.ac.uk/sci_programmes/water/wetlands /climatechangeassessmenttoolforwetlands.html

Acreman, M.C. & Holden, J. (2013) Do wetlands reduce floods? *Wetlands*, **33**, 773–786.

Acreman, M.C., Overton, I., King, J. *et al.* (2014a) The changing role of science in environmental flows. *Hydrological Sciences Journal.*, **59** (3–4), 433–450. doi:10.1080/02626667.2014.886019

Acreman, M.C., Arthington, A.H., Colloff, M.J. *et al.* (2014b) Environmental flows – natural, hybrid and novel riverine ecosystems. *Frontiers in Ecology and Environment*, **12** (8), 466–473.

Alfredson, K., Harby, A., Linnansaari, T. & Ugedal, O. (2012) Development of an inflow-controlled environmental flow regime for a Norwegian river. *Rivers Research and Applications*, **28**, 731–739.

Allison, G.A. & Newton, I. (1974) Waterfowl at Loch Leven, Kinross. *Proceedings of the Royal Society of Edinburgh. Section B. Biology*, **74**, 365–381.

Armitage, P.D., Gunn, R.J.M., Furse, M.T., Wright, J.F. & Moss, D. (1987) The use of prediction to assess macroinvertebrate response to river regulation. *Hydrobiologia*, **144**, 25–32.

Arthington, A.H., Bunn, S.E., Poff, N.L. & Naiman, R.J. (2006) The challenge of providing environmental flow rules to sustain river ecosystems. *Ecological Applications*, **16**, 1311–1318.

Araya, Y.N., Silvertown, J., Gowing, D.J., McConway, K.J., Peter Linder, H. & Midgley, G. (2011) A fundamental, eco-hydrological basis for niche segregation in plant communities. *New Phytologist*, **189** (1), 253–258.

Bailey-Watts, A.E. & Duncan, P. (1981) The ecology of Scotland's largest lochs: Lomond, Awe, Ness, Morar and Shiel. 5. A review of macrophyte studies. *Monographiae Biologicae*, **44**, 119–34.

Bailey-Watts, A.E. & Kirika, A. (1987) A Re-Assessment of Phosphorus Inputs to Loch Leven (Kinross, Scotland): Rationale and an Overview of Results on Instantaneous Loadings with Special Reference to Run-Off. Trans. Royal Soc. *Edinburgh: Earth Science*, **78**, 351–367.

Bailey-Watts, A.E. & Kirika, A. (1993). *Loch Leven NNR:Water Quality 1992–1994 with Special Reference to Nutrients and Phytoplankton, and an Assessment of Phosphorus Levels in the Loch Sediments.* Interim Report to Scottish Natural Heritage. Institute of Freshwater Ecology, Edinburgh.

Bailey-Watts, A.E. & Kirika, A. (1999) Poor water quality in Loch Leven (Scotland) in 1995, in spite of reduced phosphorus loading since 1985: the influences of catchment management and inter-annual weather variation. *Hydrobiologia*, **403**, 135–151.

Bailey-Watts, A.E., Kirika, A., May, L. & Jones, D.H. (1990) Changes in phytoplankton over various time scales in a shallow, eutrophic loch; the Loch Leven experience with special reference to the influence of flushing rate. *Freshwater Biology*, **23**, 85–111.

Baker, C., Thompson, J.R. & Simpson, M. (2009) Hydrological dynamics I: Surface waters, flood and sediment dynamics. In: Maltby, E.B. & Barker, T. (eds), *The Wetlands Handbook*. Chichester, Wiley-Blackwell, pp. 120–168.

Barbier, E.B. (2009) Ecosystems as Natural Assets. *Foundations and Trends in Microeconomics*, **4** (8), 611–681.

Bass, J.A.B. & Armitage, P.D. (1987) Observed and predicted occurrence of blackflies (*Diptera: Simuliidae*) at fifty reservoir outlets in Britain. *Regul. Rivers: Res. & Mgmt*, **1**, 247–255.

Bedinger, M.S. (1981) Hydrology of bottomland hardwood forests of the Mississippi Embayment. In: Clark, J.R. & Benforado, J. (eds), *Wetlands of bottomland hardwood forests*. Elsevier, pp. 161–176.

Blackwell, M.S.A. & Maltby, E. (2006) Ecoflood Guidelines: How to use floodplains for flood risk reduction. *European Commission D.G. Research, Brussels p144.* http://levis.sggw.waw.pl/ecoflood/

Blake, J.R., Blyth, K., Mountford, J.O., Stratford, C.J., Roberts, C. & Acreman, M.C. (2012). *Great Fen Project: Hydro-ecological model development, sustainability assessment and refined vision (Stage 2).* Report to BCNP Wildlife Trust. CEH Project Number NEC03768.

Booker, D.J., Dunbar, M.J., Shamseldin, A., Durr, C.S. & Acreman, M.C. (2003) Physical habitat assessment in urban rivers under future flow scenarios. *Journal of the Chartered Institution of Water and Environmental Management.*, **17** (4), 251–256.

Booker, D.J., Dunbar, M.J., Acreman, M.C., Akande, K & Declerck, C. (2004a). Habitat assessment at the catchment scale; application to the River Itchen, UK. In: Webb, B., Acreman, M., Maksimovic, C., Smithers, H. & Kirby, C. (eds) *Hydrology: Science and practice for the 21st Century, vol. II, Proceedings of the British Hydrological Society International Conference July 2004.*

Booker, D.J., Dunbar, M.J. & Ibbotson, A. (2004b) Predicting juvenile salmonid drift-feeding habitat quality using a three-dimensional hydraulic-bioenergetic model. *Ecological modelling*, **177**, 157–177.

Booker, D.J., Acreman, M.C., Dunbar, M.J., Hardy, T., Goodwin, T., Young, A.R. and Gowing, I. (2005) *Rapid Assessment of the Physical Habitat Sensitivity to Abstraction. Interim Technical Report 5.* Report to the Environment Agency and the Centre for Ecology and Hydrology. CEH, Wallingford, p. 21.

Booker, D.J. & Acreman, M.C. (2007) Generalisation of physical habitat-discharge relationships. *Hydrology and Earth System Sciences*, **11** (1), 141–157.

Bovee, K.D., (1982). *A guide to stream habitat analysis using the IFIM.* – US Fish and Wildlife Service Report FWS/OBS-82/26. Fort Collins, USA

Brose, U., Jonsson, T., Berlow, E.L. et al. (2006) Consumer-resource body-size relationships in natural food webs. *Ecology*, **87**, 2411–2417.

Bucher, E.H., Bonetto, A., Boyle, T. et al. (1993) *Hidrovia- an initial environmental examination of the Paraguay – Parana waterway.* Wetlands for the Americas Publication No. 10, Manomet, MA, USA, pp. 72.

Brown, L.E., Edwards, F.K., Milner, A.M., Woodward, G. & Ledger, M.E. (2011) Food web complexity and allometric scaling relationships in stream mesocosms: implications for experimentation. *Journal of Animal Ecology*, **80**, 884–895.

Bruland, G.L. & Richardson, C.J. (2005) Hydrologic, edaphic, and vegetative responses to microtopographic reestablishment in a restored wetland. *Restoration Ecology*, **13**, 515–523.

Bullock, A. & Acreman, M.C. (2003) The role of wetlands in the hydrological cycle. *Hydrology and Earth System Sciences*, **7** (3), 75–86.

Bullock, A., Gustard, A. & Grainger, E.S. (1991) *Instream flow requirements of aquatic ecology in two British rivers: application and assessment of the instream flow incremental methodology using the PHABSIM system.* IH Report no.115, Institute of Hydrology, Wallingford, 150pp.

Bunn, S.E. & Arthington, A.H. (2002) Basic principles and ecological consequences of altered flow regimes for aquatic biodiversity. *Environmental Management*, **30** (4), 492–507.

Brunning, R. (2006) A window on the Past – The prehistoric archaeology of the Somerset Moors. In: Hill-Cottingham, P., Briggs, D., Brunning, R., King, A. & Rix, G. (eds), *The Somerset Wetlands*. Somerset Books, Tiverton, Devon, UK, pp. 39–46.

Carvalho, L., Ferguson, C.A., Gunn, I.D.M., Bennion, H., Spears, B. & May, L. (2012) Water quality of Loch Leven: responses to enrichment, restoration and climate change. *Hydrobiologia*, **681**, 35–47.

Carvalho, L., McDonald, C., de Hoyos, C. *et al.* (2013) Sustaining recreational quality of European lakes: minimising the health risks from algal blooms through phosphorus control. *Journal of Applied Ecology*, **50**, 315–323.

Carvalho, L., Miller, C.A., Scott, E.M., Codd, G.A., Davies, P.S. & Tyler, A.N. (2011) Cyanobacterial blooms: Statistical models describing risk factors for national-scale lake assessment and lake management. *Science of the Total Environment*, **409**, 5353–5358.

Clegg, M.R., Maberly, S.C. & Jones, R.I. (2007) Behavioural response as a predictor of seasonal depth distribution and vertical niche separation in freshwater phytoplanktonic flagellates. *Limnology & Oceanography*, **52**, 441–455.

Colebrook, J.M. (1960a) Plankton and water movements in Windermere. *Journal of Animal Ecology*, **29**, 217–240.

Colebrook, J.M. (1960b) Some observations of zooplankton swarms in Windermere. *Journal of Animal Ecology*, **29**, 241–242.

CEC, 1995. *Wise use and conservation of wetlands*. Communication from the Commission to the Council and the European Parliament, Commission of the European Communities COM(95) 189 (final), 54p.

European Commission and Parliament (2000) *Directive 2000/60/EC of 20 December 2000 establishing a framework for community action in the field of water policy*. European Parliament and Council, Luxembourg.

Costanza, R. & Daly, H.E. (1992) Natural capital and sustainable development. *Conservation Biology*, **6**, 37–46.

Cowx, I.G. & Portocarrero Aya, M. (2011) Paradigm shifts in fish conservation: moving to the ecosystem services concept. *Journal of Fish Biology*, **79**, 1663–1680.

Davis, T.J. (ed) (1993) *The Ramsar Convention Manual: a guide to the Convention on wetlands of international importance especially as waterfowl habitat*. Ramsar Convention Bureau, Gland, Switzerland.

Davies, J. & Claridge, C.F. (1993) *Wetland benefits: the potential for wetlands to support and maintain development*. Asian Wetland Bureau Publication No. 87: IWRB Spec. Publ. 27: Wetlands for the America Publication No. 11. Asian Wetland Bureau, Kuala Lumpur, Malaysia, pp. 45.

Davy, A.J., Hiscock, K.M., Jones, M.L.M., Low, R., Robins, N.S. & Stratford, C. (2010) *Protecting the Plant Communities and Rare Species of Dune Wetland Systems: Ecohydrological Guidelines for Wet Dune GHabitats. Phase 2*. Environment Agency, Bristol, UK, pp. 113.

Defew, L.H., May, L. & Heal, K. (2013) Uncertainties in estimated phosphorus loads as a function of different sampling frequencies and common calculation methods. *Marine and Freshwater Science*, **64**, 373–386.

Dillon, P.J. & Rigler, F.H. (1974) The phosphorus-chlorophyll relationship in lakes. *Limnology and Oceanography*, **19**, 767–773.

Dugan, P.J. (1990) *Wetland conservation – a review of current issues and required action*. IUCN – The World Conservation Union, Gland, Switzerland, pp. 96.

Dunbar, M.J., Acreman, M.C., Gustard, A. & Elliott, C.R.N. (1998) *Overseas Approaches to Setting River Flow Objectives. Phase I Report to the Environment Agency*. Environment Agency R&D Technical Report W6-161, 82pp.

Dunbar, M.J., Ibbotson, A.T., Gowing, I.M., McDonnell, N., Acreman, M. & Pinder, A. (2001) *Ecologically Acceptable Flows Phase III: Further validation of PHABSIM for the habitat requirements of salmonid fish*. Final R&D Technical Report (project W6-036) to the Environment Agency and CEH (project C00962). 137pp + appendices.

Dunbar, M.J. (ed), Ash, J., Bass, J., Booker, D., Dawson, H. Fenn, T., Gozlan, R., Latimer, P., Postle, M., Rogerson, H., & Welton, S. (2002). *Guidelines for the Identification and Designation of Heavily Modified Water Bodies in England and Wales*. Environment Agency Project Record P2-260/3. 166pp + appendices.

Dunbar, M.J., Pedersen, M.L., Cadman, D. *et al.* (2010a) River discharge and local scale physical habitat influence macroinvertebrate LIFE scores. *Freshwater Biology*, **55**, 226–242.

Dunbar, M.J., Warren, M., Extence, C. *et al.* (2010b) Interaction between macroinvertebrates, discharge and physical habitat in upland rivers. *Aquatic Conserv: Mar. Freshw. Ecosyst.*, **20**, S31–S44. doi:10.1002/aqc.1089

Dunbar, M.J., Alfredsen, K. & Harby, A. (2012) Hydraulic-habitat modelling for setting river flow needs for salmonids. *Fisheries Management and Ecology*, **19** (6), 500–517. doi:10.1111/j.1365-2400.2011.00825.x

Duranel, A., Acreman, M.C., Stratford, C., Thompson, J.R. & Mould, D. (2007) Assessing hydrological suitability of the Thames floodplain for species-rich meadow restoration. *Hydrology and Earth System Sciences*, **11** (1), 170–179.

Dyson, M., Bergkamp, G. & Scanlon, J. (eds) (2003) *Flow. The essentials of environmental flows*. IUCN, Gland, Switzerland.

Edwards, F.K., Lauridsen, R.B., Armand, L., Vincent, H.M. & Jones, J.I. (2009a) The relationship between length, mass and preservation time for three species of freshwater leeches (Hirudinea). *Fundamental and Applied Limnology*, **173**, 321–327.

Edwards, F.K., Lauridsen, R.B., Duerdoth, C.R. & Jones, J.I. (2008) Dry and ash-free dry mass to length relationships of bullhead (Cottus gobio L.). *Fundamental and Applied Limnology*, **172**, 315–316.

Edwards, F.K., Lauridsen, R.B., Fernandes, W.P.A. *et al.* (2009b) Re-introduction of Atlantic salmon, Salmo salar L., to the Tadnoll Brook, Dorset. *Proceedings of the Dorset Natural History and Archeological Society*, **130**, 1–8.

Elliott, J.A. (2009) The seasonal sensitivity of Cyanobacteria and other phytoplankton to changes in flushing rate and water temperature. *Global Change Biology*, **16**, 864–876.

Elliott, J.A. & Bell, V.A. (2011) Predicting the potential long-term influence of climate change on vendace (*Coregonus albula*) habitat in Bassenthwaite Lake, U.K. *Freshwater Biology*, **56**, 395–405.

Elliott, A.E., Irish, A.E. & Reynolds, C.S. (2010) Modelling Phytoplankton Dynamics in Fresh Waters: Affirmation of the PROTECH Approach to Simulation. *Freshwater Reviews*, **3**, 75–96.

Elliott, J.A., Jones, I.D. & Page, T. (2009) The importance of nutrient source in determining the influence of retention time on phytoplankton: an explorative modelling study of a naturally well-flushed lake. *Hydrobiologia*, **627**, 129–142.

Emerton, L. & Bos, E. (2004) *Value. Counting Ecosystems as an Economic Part of Water Infrastructure.* IUCN, Gland, Switzerland and Cambridge, UK, pp. 88.

Environment Agency (2001) *Managing Water Abstraction: the Catchment Abstraction Management Strategy Process.* Environment Agency, Bristol, UK.

Environment Agency (2002) *Resource Assessment and Management Framework.* Report and User Manual Version 3. Environment Agency, Bristol.

Extence, C., Balbi, D.M. & Chadd, R.P. (1999) River flow indexing using British benthic macro-invertebrates: a framework for setting hydro-ecological objectives. *Regulated River Research and Management.*, **15**, 543–574.

Falkenmark, M. (2003) *Water Management and Ecosystems: Living with Change.* Global Water Partnership Technical Committee Paper No 9.. GWP, Stockholm.

Farmer, J.G., Bailey-Watts, A.E., Kirika, A. & Scott, C. (1994) Phosphorus fractionation and mobility in Loch Leven sediments. *Aquatic Conservation: Marine Freshwater Ecosystems*, **4**, 45–56.

Fisher, B., Turner, R.K. & Morling, P. (2009) Defining and classifying ecosystem services for decision making. *Ecological Economics*, **68**, 643–653.

Foley, B., Jones, I.D., Maberly, S.C. & Rippey, B. (2012) Long-term changes in oxygen depletion within a small temperate lake: Effects of climate change and eutrophication. *Freshwater Biology*, **57**, 278–289.

Fowler, H.J. & Kilsby, C.G. (2007) Using regional climate model data to simulate historical and future river flows in northwest England. *Climatic Change*, **80**, 337–367.

George, D.G. (1981a) Zooplankton patchiness. *Reports of the Freshwater Biological Association*, **49**, 32–44.

George, D.G. (1981b) *Wind-induced water movements in the South Basin of Windermere.* Freshwater Biological Association, Windermere, Vol. **11**, pp. 37–60.

George, D.G. (1989) Zooplankton patchiness in enclosed and unenclosed areas of water. *Journal of Plankton Research*, **11**, 173–184.

George, D.G. (1993) Physical and chemical scales of pattern in freshwater lakes and reservoirs. *The Science of the Total Environment*, **135**, 1–15.

George, D.G. (2000) Remote sensing evidence for the episodic transport of phosphorus from the littoral zone of a thermally stratified lake. *Freshwater Biology*, **43**, 571–578.

George, D.G. (2002) Regional-scale influences on the long-term dynamics of lake plankton. In: Williams, P.J.L.B., Thomas, D.N. & Reynolds, C.S. (eds), *Phytoplankton Production.* Blackwell Scientific, Oxford, pp. 265–290.

George, D.G. & Allen, C.M. (1994) Turbulent mixing in a small thermally stratified lake. In: Beven, K.J., Chatwin, P.C. & Millbank, J.H. (eds), *Mixing and Transport in the Environment.* John Wiley & Sons Ltd.

George, D.G., Maberly, S.C. & Hewitt, D.P. (2004) The influence of the North Atlantic Oscillation on the physics, chemistry and biology of four lakes in the English Lake District. *Freshwater Biology*, **49**, 760–774.

George, D.G. & Taylor, A.H. (1995) U.K. lake plankton and the Gulf Stream. *Nature*, **378**, 139.

George, D.G. & Winfield, I.J. (2000) Factors influencing the spatial distribution of zooplankton and fish in Loch Ness, UK. *Freshwater Biology*, **43**, 557–570.

Gilljam, D., Thierry, A., Edwards, F.K. *et al.* (2011) Seeing Double: Size-Based and Taxonomic Views of Food Web Structure. In: Belgrano, A. & Reiss, J. (eds), *Advances in Ecological Research, Vol 45: The Role of Body Size in Multispecies Systems*, Elsevier, pp. 67–133.

Gleick, P.H. (1993) *Water in crisis. A guide to the World's Fresh Water Resources.* Oxford University Press, pp. 473.

Glob, O.P.V. (1969) *The Bog People: Iron Age Man Preserved.* Faber and Faber.

Gowing, D.J.G., Lawson, C.S., Youngs, E.G., Barber, K.R., Rodwell, J.S., Prosser, M.V., Wallace, H.L., Mountford, J.O. & Spoor, G. (2002) *The Water Regime Requirements and the Response to Hydrological Change of Grassland Plant Communities DEFRA-commissioned project BD1310 Final report to the Department for Environment, Food and Rural Affairs.*

Guensch, G.R., Hardy, T.B. & Addley, R.C. (2001) Examining feeding strategies and position choice of drift-feeding salmonids using an individual-based, mechanistic foraging model. *Can. J. Fish. Aquat. Sci.*, **58**, 446–457.

Gustard, A. (1989) Compensation flows in the UK: A hydrological review. *Regulated Rivers: Research and Management*, **3**, 49–59.

Gustard, A. & Elliott, C.R.N. (1997) *The role of hydro-ecological models in the development of sustainable water resources*, IAHS Publ. No 240. IAHS, pp. 407–417.

Hannah, D.M., Sadler, J.P. & Wood, P.J. (2004) Hydroecology and ecohydrology: a potential route forward? *Hydrological Processes*, **21** (24), 3385–3390.

Harris, G.R., Sexton, D.M.H., Booth, B.B., Collins, M. & Murphy, J.M. (2013) Probabilistic projections of transient climate change. *Climate Dynamics*, **40**, 2937–2972. doi:10.1007/s00382-012-1647-y

Hedger, R.D., Olsen, N.R.B., George, D.G., Malthus, T.J. & Atkinson, P.M. (2004) Modelling spatial distributions of *Ceratium hirundinella* and *Microcystis* spp. in a small productive British lake. *Hydrobiologia*, **528**, 217–227.

Hilton, J. (1985) A conceptual framework for predicting the occurrence of sediment focusing and sediment redistribution in small lakes. *Limnology & Oceanography*, **30**, 1131–1143.

Hilton, J. & Gibbs, M.M. (1985) The horizontal distribution of major elements and organic matter in the sediment of Esthwaite Water, England. *Chemical Geology*, **47**, 57–83.

Hilton, J., Lishman, P. & Allen, V. (1986) The dominant processes of sediment distribution and focussing in a small, eutrophic, monomictic lake. *Limnology & Oceanography*, **31**, 125–133.

Holden, A.V. & Caines, L.A. (1974) Nutrient Chemistry of Loch Leven, Kinross. Proceedings of the Royal Society of Edinburgh. Series B. *Biology*, **74**, 101–121. doi:10.1017/S0080455X00012340

House, W.A. (2003) Geochemical cycling of phosphorus in rivers. *Applied Geochemistry*, **18**, 739–748.

Howe, C.P., Claridge, G.F., Hughes, R. & Zuwendra (1992) *Manual of guidelines for scoping EIA in tropical wetlands*, 2nd Edition edn. Asian Wetland Bureau-Indonesia, Bogor, Indonesia, pp. 261.

Hulme, M., Jenkins, G.J., Lu, X., Turnpenny, J.R., Mitchell, T.D., Jones, R.G., Lowe, J., Murphy, J.M., Hassell, D., Boorman, P., McDonald, R. & Hill, S. (2002). *Climate Change Scenarios of the United Kingdom: The UKCIP02 Scientific Report*. Tyndall Centre for Climate Change Research, School of Environmental Sciences, University of East Anglia: Norwich; 120.

Hutchinson, G.E. (1957) *A Treatise on Limnology*. Vol. **1**. John Wiley & Sons, NY, pp. 1015.

Ings, T.C., Montoya, J.M., Bascompte, J. *et al.* (2009) Ecological networks – beyond food webs. *Journal of Animal Ecology*, **78**, 253–269.

Jarvie, H.P., Jürgens, M.D., Williams, R.J. *et al.* (2005) Role of river bed sediments as sources and sinks of phosphorus across two major eutrophic UK river basins: the Hampshire Avon and Herefordshire Wye. *Journal of Hydrology*, **304**, 51–74.

Jarvie, H.P., Sharpley, A.N., Spears, B.M., Buda, A.R., May, L. & Kleinman, P.J.A. (2013) Water quality remediation faces unprecedented challenges from "Legacy Phosphorus". *Environmental Science & Technology*, **47**, 8997–8998. doi:10.1021/es403160a

Johnson, I.W., Elliot, C.R.N. & Gustard, A. (1995) Modelling the effect of groundwater abstraction on salmonid habitat availability in the River Allen, Dorset, England. *Regulated Rivers: Research and Management*, **10**, 229–238.

Jones, I.D. & Elliott, J.A. (2007) Modelling the effects of changing retention time on abundance and composition of phytoplankton species in a small lake. *Freshwater Biology*, **52**, 988–997. doi:10.1111/j.1365-2427.2007.01746.x

Jones, I., George, G. & Reynolds, C. (2005) Quantifying the Effects of Phytoplankton on the Summer Heat Budget of Large Limnetic Enclosures. *Freshwater Biology*, **50**, 1239–1247.

Jones, I.D., Page, T., Elliott, J.A., Thackeray, S.J. & Heathwaite, A.L. (2011) Increases in lake phytoplankton biomass caused by future climate-driven changes to seasonal river flow. *Global Change Biology*, **17**, 1809–1820. doi:10.1111/j.1365-2486.2010.02332.x

Jones, I.D., Winfield, I.J. & Carse, F. (2008) Assessment of long-term changes in habitat availability for Arctic charr (*Salvelinus alpinus*) in a temperate lake using oxygen profiles and hydroacoustic surveys. *Freshwater Biology*, **53**, 393–402.

Jupp, B.P. & Spense, D.H.N. (1977) Limitations of macrophytes in a eutrophic lake, Loch Leven. II. Wave action, sediments and waterfowl grazing. *Journal of Ecology*, **65**, 431–446.

Kundzewicz, Z.W., Mata, L.J., Arnell, N.W. *et al.* (2007) Freshwater resources and their management. In: Parry, M.L., Canziani, O.F., Palutikof, J.P., van der Linden, P.J. & Hanson, C.E. (eds), *Climate Change 2007: Impacts, Adaptation and Vulnerability. Contribution of Working Group II to the Fourth*. Cambridge University Press, Cambridge.

Laizée, C.L.R., Acreman, M.C., Schneider, C. *et al.* (2014) Projected flow alteration and ecological risk for pan-European rivers. *Rivers Research and Applications*, **30**, 299–314.

Lauridsen, R.B., Edwards, F.K., Bowes, M.J. *et al.* (2012) Consumer-resource elemental imbalances in a nutrient-rich stream. *Freshwater Science*, **31**, 408–422.

Lauridsen, R.B., Edwards, F.K., Cross, W.F., Woodward, G., Hildrew, A.G. & Jones, J.I. (2014) Consequences of inferring diet from feeding guilds when estimating and interpreting consumer-resource stoichiometry. *Freshwater Biology*, **59**, 1497–1508.

Ledger, M.E., Brown, L.E., Edwards, F.K., Hudson, L.N., Milner, A.M. & Woodward, G. (2013a) Chapter Six – Extreme Climatic Events Alter Aquatic Food Webs: A Synthesis of Evidence from a Mesocosm Drought Experiment. In: Guy, W. & Eoin, J.O.G. (eds), *Advances in Ecological Research*. Academic Press, pp. 343–395.

Ledger, M.E., Brown, L.E., Edwards, F.K., Milner, A.M. & Woodward, G. (2013b) Drought alters the structure and functioning of complex food webs. *Nature Clim. Change*, **3**, 223–227.

Ledger, M.E., Edwards, F.K., Brown, L.E., Milner, A.M. & Woodward, G.U.Y. (2011) Impact of simulated drought on ecosystem biomass production: an experimental test in stream mesocosms. *Global Change Biology*, **17**, 2288–2297.

Ledger, M.E., Harris, R.M.L., Armitage, P.D. & Milner, A.M. (2008) Disturbance frequency influences patch dynamics in stream benthic algal communities. *Oecologia*, **155**, 809–819.

Ledger, M.E., Harris, R.M.L., Milner, A.M. & Armitage, P.D. (2006) Disturbance, biological legacies and community development in stream mesocosms. *Oecologia*, **148**, 682–691.

Letovsky, E., Heal, K., Carvalho, L. & Spears, B. (2012) Intracellular Versus Extracellular Iron Accumulation in Freshwater Periphytic Mats Across a Mine Water Treatment Lagoon. *Water, Air, & Soil Pollution*, **223**, 1519–1530.

Le Quesne, T., Kendy, E. & Weston, D. (2010). *The implementation challenge: Taking stock of government policies to protect and restore environmental flows*. WWF and The Nature Conservancy. http://conserveonline.org/workspaces/eloha/documents/wwf-tnc-e-flow-policies-report.

Lloyd, C.R. (2006) Annual carbon balance of a managed wetland meadow in the Somerset Levels, UK. *Agricultural and Forest Meteorology*, **138** (1), 168–179.

Lyle, A. (1982) Ten years of ice records for Loch Leven, Kinross. *Weather*, **36**, 116–125.

Lytle, D.A. & Poff, N.L. (2004) Adaptation to natural flow regimes. *Trends in Ecology and Evolution*, **19** (2), 94–100.

Maberly, S.C., Barker, P.A., Stott, A.W. & De Ville, M.M. (2013) Catchment productivity controls CO_2 emissions from lakes. *Nature Climate Change*, **3**, 391–394. doi:10.1038/nclimate1748

Maberly, S.C., King, L., Dent, M.M., Jones, R.I. & Gibson, C.E. (2002) Nutrient limitation of phytoplankton and periphyton growth in upland lakes. *Freshwater Biology*, **47**, 2136–2152.

Mackay, E., Folkhard, A.M. & Jones, I.D. (2014) Interannual variations in atmospheric forcing determine trajectories of hypolimnetic soluble reactive phosphorus supply in a eutrophic lake. *Freshwater Biology*, **59**, 1646–1658. doi:10.1111/fwb.12371

Mackay, E.B., Jones, I.D., Folkard, A.M. & Barker, P.A. (2012) Contribution of sediment focusing to heterogeneity of organic carbon and phosphorus burial in small lakes. *Freshwater Biology*, **57**, 290–304.

Mackay, E.B., Jones, I.D., Folkard, A.M. & Thackeray, S.J. (2011a) Transition zones in small lakes: the importance of dilution and biological uptake on lake-wide heterogeneity. *Hydrobiologia*, **678**, 85–97.

Mackay, E.B., Jones, I.D., Thackeray, S.J. & Folkard, A.M. (2011b) Spatial heterogeneity in a small, temperate lake during archetypal weak forcing conditions. *Fundamental and Applied Limnology*, **179**, 27–40.

Madgwick, G., Jones, I.D., Thackeray, S.J., Elliott, J.A. & Miller, H. (2006) Phytoplankton communities and antecedent conditions: high resolution sampling in Esthwaite Water. *Freshwater Biology*, **51**, 1978–1810.

Maitland, P.S. (1979) The distribution of zoobenthos and sediments in Loch Leven, Kinross, Scotland. *Archive Hydrobiology*, **85**, 98–125.

Maltby, E. (1991) Wetland management goals: wise use and conservation. *Landscape Urban Planning*, **20**, 9–18.

Maltby, E., Holdgate, M., Acreman, M.C. & Weir, A. (eds) (1999) *Ecosystem management; questions for science and society*. Sibthorp Trust.

May, L. & Spears, B.M. (2012) Managing ecosystem services at Loch Leven, Scotland, UK: actions, impacts and unintended consequences. *Hydrobiologia*, **681**, 117–130.

Mazvimavi, D. (1994). A review of the hydrology of dambos in Zimbabwe. In: Matiza, T. and Crafter, S.A. (eds) *Wetlands ecology and priorities for conservation in Zimbabwe*. Proceedings of a Seminar on Wetlands ecology and priorities for conservation in Zimbabwe, Harare, 13–15 January, 1992. International Union for Conservation of Nature and Natural Resources, Geneva, pp. 47–53.

Millennium Ecosystem Assessment (2005) *Ecosystems and human well-being*. Island Press, Washington DC, USA.

Miller, H., Winfield, I.J., Fletcher, J.M. *et al.* (2015) Distribution, characteristics and condition of Arctic charr (*Salvelinus alpinus*) spawning grounds in a differentially eutrophicated twin-basin lake. *Ecology of Freshwater Fish*, **24**, 32–43. doi:10.1111/eff.12122

Milly, P.C.D., Dunne, K.A. & Vecchia, A.V. (2005) Global pattern of trends in streamflow and water availability in a changing climate. *Nature*, **438**, 347–350.

Mitsch, W.J. & Gosslink, J.G. (1993) *Wetlands*, 2nd edn. Van Nostrand Reinhold, New York.

Mortimer, C.H. (1941) The exchange of dissolved substances between mud and water in lakes. *Journal of Ecology*, **29**, 280–329.

Mortimer, C.H. (1952) Water movements in lakes during summer stratification; evidence from the distribution of temperature in Windermere. *Philosophical Transactions of the Royal Society of London B*, **236** (635), 355–398.

Mountford, J.O., McCartney, M.P., Manchester, S.J. & Wadsworth, R.A. (2002) *Wildlife habitats and their requirements within the Great Fen Project*. Report to the Great Fen Project Steering Group, Centre for Ecology & Hydrology, Monks Wood.

Mountford, J.O., Folwell, S.S., Manchester, S.J., Meigh, J.R., Wadsworth, R.A. & McCartney, M.P. (2004) *Feasibility study for wetland restoration at Baston and Thurlby Fens*. Final Report to the Baston and Thurlby Fens Project Steering Group.

Mountford, J.O., Rose, R.J. & Bromley, J. (2005) *Development of eco-hydrological guidelines for wet heaths – Phase 1*. English Nature Research Report no. 620. Peterborough: English Nature.

Mountford, J.O., Lakhani, K.H. & Holland, R.J. (1996) Reversion of vegetation following the cessation of fertiliser application. *Journal of Vegetation Science*, **7**, 219–228.

Murphy, J.M., Sexton, D.M.H., Jenkins, G.J. *et al.* (2009) *UK Climate Projections Science Report: Climate change projections*. Met Office Hadley Centre, Exeter.

National Rivers Authority (1993) *Facts on the top 40 low flow rivers in England and Wales*. National Rivers Authority, Bristol.

Neale, P.J., Talling, J.F., Heaney, S.I., Reynolds, C.S. & Lund, J.W.G. (1991) Long time series from the English Lake District: irradiance-dependent phytoplankton dynamics during the spring maximum. *Limnology & Oceanography*, **36**, 751–760.

Nijssen, B., O'Donnell, G.M., Hamlet, A.F. & Lettenmaier, D.P. (2001) Hydrologic sensitivity of global rivers to climate change. *Climatic Change*, **50**, 143–175.

Novitski, R.P. (1978). Hydrologic characteristics of Wisconsin's wetlands and their influence on floods, streamflow and sediment. In: *Wetland functions and values: the state of our understanding*, Amer. Water. Resour. Assoc., Minneapolis, MN, USA, 377–388.

OECD (1979) *Shallow Lakes and Reservoirs*. Final Report Vol. 1 & 2 to OECD Cooperative Programme for Monitoring of Inland waters (Eutrophication Control)

OECD (1996) *Guidelines for aid agencies for improved conservation and sustainable use of tropical and sub-tropical wetland*. Organisation for Economic Cooperation and Development Development Assistance Committee: Guidelines on Aid and Environment No.9, 69pp.

O'Connell, M.J., Ward, R.M., Onoufriou, C. *et al.* (2007) Integrating multi-scale data to model the relationship between food resources, waterbird distribution and human activities in freshwater systems: preliminary findings and potential uses. *Ibis*, **149**, 65–72.

Old, G., Naden, P.S., Rameshwaran, P., Acreman, M.C., Baker, S., Edwards, F.K., Sorensen, J.P., Mountford, J.O., Gooddy, D.C., Stratford, C.J., Scarlett, P.M., Newman, J.R. & Neal, M., in press. *Instream and riparian implications of weed cutting in a chalk river* Ecological Engineering.

Ormerod, S.J., Dobson, M., Hildrew, A.G. & Townsend, C.R. (2010) Multiple stressors in freshwater ecosystems. *Freshwater Biology*, **55** (1), 1–4.

Peh, K.S.H., Balmford, A., Bradbury, R.B. *et al.* (2013) TESSA: A toolkit for rapid assessment of ecosystem services at sites of biodiversity conservation importance. *Ecosystem Services*, **5**, 51–57.

Persson, I. & Jones, I.D. (2008) The effect of lake colour on lake hydrodynamics: a modelling study. *Freshwater Biology*, **53**, 2345–2355.

Poff, N.L., Allan, J.D., Bain, M.B. *et al.* (1997) The natural flow regime: a paradigm for river conservation and restoration. *Bioscience*, **47**, 769–784.

Poff, N.L., Richter, B., Arthington, A.H. *et al.* (2010) The ecological limits of hydrologic alteration (ELOHA): A new framework for developing regional environmental flow standards. *Freshwater Biology*, **55**, 147–170.

Poulickova, A., Hasler, P., Lysakova, M. & Spears, B. (2008) The ecology of freshwater epipelic algae: an update. *Phycologia*, **47**, 437–450.

Reynolds, C.S. & Lund, J.W.G. (1988) The phytoplankton of an enriched, soft-water lake subject to intermittent hydraulic flushing (Grasmere, English Lake District). *Freshwater Biology*, **19**, 379–404.

Reynolds, C.S., Irish, A.E. & Elliott, J.A. (2001) The ecological basis for simulating phytoplankton responses to environmental change (PROTECH). *Ecological Modelling*, **140**, 271–291.

Reynolds, C.S., Maberly, S.C., Parker, J.E. & De Ville, M.M. (2012) Forty years of monitoring water quality in Grasmere (English Lake District): separating the effects of enrichment by treated sewage and hydraulic flushing on phytoplankton ecology. *Freshwater Biology*, **57**, 384–399.

Richter, B.D., Baumgartner, J.V., Powell, J. & Braun, D.P. (1996) A Method for Assessing Hydrological Alteration within Ecosystems. *Conserv. Biol.*, **10**, 1163–1174.

Sharpley, A., Jarvie, H.P., Buda, A., May, L., Spears, B.M. & Kleinman, P. (2013) Phosphorus Legacy: Overcoming the Effects of Past Management Practices to Mitigate Future Water Quality Impairment. *Journal of Environmental Quality*, **42**, 1308–1326.

Silvertown, J., Dodd, M.E., Gowing, D. & Mountford, O. (1999) Hydrologically-defined niches reveal a basis for species-richness in plant communities. *Nature*, **400**, 61–63.

Smart, S., Dunbar, M.J., Emmett, B.A., Marks, S., Maskell, L.C., Norton, L.R., Rose, P. & Simpson, I.C. (2010). *An Integrated Assessment of Countryside Survey data to investigate Ecosystem Services in Great Britain.* Technical Report No. 10/07 NERC/Centre for Ecology & Hydrology 230pp. (CEH Project Number: C03259).

Smith, B.D., Maitland, P.S. & Pennock, S.M. (1987) A comparative study of water level regimes and littoral benthic communities in Scottish lochs. *Biological Conservation*, **39**, 291–316.

Smith, B.D., Maitland, P.S., Young, M.R. & Carr, M.J. (1981) The ecology of Scotland's largest lochs: Lomond, Awe, Ness, Morar and Shiel, 7. The littoral zoobenthos. *Monographiae Biologicae*, **44**, 155–204.

Smith, I.R. (1974) The structure and physical environment of Loch Leven, Scotland. *Proceedings of the Royal Society of Edinburgh B*, **74**, 81–100.

Smith, I.R. (1975) *Turbulence in lakes and rivers*. FBA Scientific Publication no. 29. Freshwater Biological Association, Ambleside, pp. 79.

Smith, I.R. (1979) Hydraulic conditions in isothermal lakes. *Freshwater Biology*, **9**, 119–145.

Smith, I.R. & Sinclair, I.J. (1972) Deep water waves in lakes. *Freshwater Biology*, **2**, 387–399.

Smyly, W.J.P. (1974) Vertical distribution and abundance of *Ceriodaphnia quadrangula* (O. F. Müller) (Crustacea, Cladocera). *Freshwater Biology*, **4**, 257–266.

Spears, B.M., Carvalho, L. & Paterson, D.M. (2007a) Phosphorus partitioning in a shallow lake: implications for water quality management. *Water Environment Journal*, **21**, 47–53.

Spears, B.M., Carvalho, L., Perkins, R., Kirika, A. & Paterson, D.M. (2006) Spatial and historical variation in sediment phosphorus fractions and mobility in a large shallow lake. *Water Research*, **40**, 383–391.

Spears, B.M., Carvalho, L., Perkins, R., Kirika, A. & Paterson, D.M. (2007b) Sediment phosphorus cycling in a large shallow lake: spatio-temporal variation in phosphorus pools and release. *Hydrobiologia*, **584**, 37–48.

Spears, B.M., Carvalho, L., Perkins, R., Kirika, A. & Paterson, D.M. (2012) Long-term variation and regulation of internal phosphorus loading in Loch Leven. *Hydrobiologia*, **681**, 23–33. doi:10.1007/s10750-011-0921-z

Spears, B. & Jones, I. (2010) The long-term (1979–2005) effects of the North Atlantic Oscillation on wind-induced wave mixing in Loch Leven (Scotland). *Hydrobiologia*, **646**(1), 49–59.

Spears, B.M., Gunn, I.D.M., Carvalho, L. *et al.* (2009) An evaluation of methods for sampling macrophyte maximum colonisation depth in Loch Leven, Scotland. *Aquatic Botany*, **91**, 75–81.

Spears, B.M., Saunders, J.E., Davidson, I. & Paterson, D.M. (2008) Microalgal sediment biostabilisation along a salinity gradient in the Eden Estuary, Scotland: unravelling a paradox. *Marine and Freshwater Research*, **59**, 313–321.

Stratford, C.J., Mountford, J.O., Robins, N.S., Redhead, J., Blake, J., Bowes, M.J., Edwards, F. & Vincent, H. (2014) *River Glen Integrated Catchment Management Study – Phase 1: Project Appraisal*. CEH final report to South Lincolnshire Fens Partnership. 123pp. CEH Project Number NEC05053.

Stratford, C. (2014) NEC04635 – Wetland Core Monitoring – Overview of current monitoring sites. Centre for Ecology and Hydrology. Internal Report

Strong, C. & Maberly, S.C. (2011) The influence of atmospheric wave dynamics on the surface temperature of lakes in the English Lake District. *Global Change Biology*, **17**, 2013–2022.

Talling, J.F. (1971) The underwater light climate as a controlling factor in the production ecology of freshwater phytoplankton. *Mitteilungen der Internationalen Vereinigung für Theoretische und Angewandte Limnologie*, **19**, 214–243.

Talling, J.F. (1993) Comparative seasonal changes, and inter-annual variability and stability, in a 26-year record of total phytoplankton biomass in four English lake basins. *Hydrobiologia*, **268**, 65–98.

Thackeray, S.J., George, D.G., Jones, R.I. & Winfield, I.J. (2005) Vertical heterogeneity in zooplankton community structure: a variance partitioning approach. *Archiv für Hydrobiologie*, **164**, 257–275.

Thackeray, S.J., George, D.G., Jones, R.I. & Winfield, I.J. (2006) Statistical quantification of the effect of thermal stratification on patterns of dispersion in a freshwater zooplankton community. *Aquatic Ecology*, **40**, 23–32.

Thackeray, S.J., George, D.G., Jones, R.I. & Winfield, I.J. (2004) Quantitative analysis of the importance of wind-induced circulation for the spatial structuring of planktonic populations. *Freshwater Biology*, **49**, 1091–1102.

Thackeray, S.J., Jones, I.D. & Maberly, S.C. (2008) Long-term change in the phenology of spring phytoplankton: species-specific responses to nutrient enrichment and climate change. *Journal of Ecology*, **96**, 523–535.

Tharme, R.E. (2003) A global perspective on environmental flow assessment : emerging trends in the development and application of environmental flow methodologies for rivers. *River Res. Appl.*, **19**, 397–441.

Tharme R. E. & King J.M. (1998) *Development of the building block methodology for instream flow assessments and supporting research on the effects of different magnitude flows on riverine ecosystems*. Report to Water Research Commission, 576/1/98. Cape Town, South Africa.

Thompson, J.R. & Hollis, G.E. (1995) Hydrological Modeling and the Sustainable Development of the Hadejia-Nguru Wetlands, Nigeria. *Hydrological Sciences Journal*, **40** (1), 97–116.

Thompson, J.R. & Finlayson, C.M. (2001) Freshwater Wetlands. In: Warren, A. & French, J.R. (eds), *Habitat Conservation: Managing the Physical Environment*. Wiley, Chichester, pp. 147–178.

Thompson, J.R., Gavin, H., Refsgaard, A., Refstrup Sørenson, H. & Gowing, D.J. (2009) Modelling the hydrological impacts of climate change on UK lowland wet grassland. *Wetlands Ecology and Management*, **17**, 503–523.

Thompson, J.R., Laize, C. & Acreman, M.C. (2014) Climate change uncertainty in environmental flows for the Mekong River. *Hydrological Sciences Journal*, **59** (3–4), 935–954. doi:10.1080/02626667.2013.842074

UKTAG (2013) River flow for good ecological potential Final recommendations UK Technical Advisory Group to WFD.

United Nations (1992) United Nations Conference on Environment and Development (UNCED), Rio de Janeiro, 3–14 June 1992 http://www.un.org/geninfo/bp/enviro.html

United Nations (2014) The Millennium Development Goals Report United Nations, New York. 59 pp.

Vollenweider, R.A. (1975) Input–output models; with special reference to the phosphate loading concept in limnology. *Schweizerische Zeitschrift für Hydrologie*, **37**, 53–84.

Wathern, P. (1998) *Environmental impact assessment: theory and practice*. Routledge, New York, pp. 402.

Wetland Vision Project (2008) *Wetland Vision: A 50 year vision for wetlands*. RSPB, Sandy, Bedfordshire, UK.

Wheeler, B.D., Gowing, D.J.G., Shaw, S.C., Mountford, J.O. & Money, R.P. (2004) Ecohydrological Guidelines for Lowland Wetland Plant Communities (A.W. Brooks, P.V. Jose, and M.I. Whiteman (eds)). Environment Agency (Anglian Region)

Winfield, I.J. (2004) Fish in the littoral zone: ecology, threats and management. *Limnologica*, **34**, 124–131.

Winfield, I.J., Bean, C.W. & Hewitt, D.P. (2002) The relationship between spatial distribution and diet of Arctic charr (*Salvelinus alpinus*) in Loch Ness, U.K. *Environmental Biology of Fishes*, **64**, 63–73.

Winfield, I.J., Fletcher, J.M. & James, J.B. (2004) Modelling the impacts of water level fluctuations on the population dynamics of whitefish (*Coregonus lavaretus* (L.))

in Haweswater, U.K. *Ecohydrology & Hydrobiology*, **4**, 409–416.

Winfield, I.J., Bean, C.W., Gorst, J., Gowans, A.R.D., Robinson, M. & Thomas, R. (2013) Assessment and conservation of whitefish (*Coregonus lavaretus*) in the U.K. *Advances in Limnology*, **64**, 305–321.

Winter, T.C. (2003) The hydrology of lakes. In: O'Sullivan, P.E. & Reynolds, C.S. (eds), *The Lakes Handbook*. Vol. 1. Wiley-Blackwell, pp. 61–78.

Waters, B.F. (1976) A methodology for evaluating the effects of different stream flows on salmonid habitat. In: Orsborn, J.F. & Allman, C.H. (eds), *Instream Flow Needs*, Proceedings of the Symposium and Speciality Conference on Instream Flow Needs. American Fisheries Society, Bethesda, USA, pp. 254–266.

Woodward, G., Blanchard, J., Lauridsen, R.B. *et al.* (2010) Individual-Based Food Webs: Species Identity, Body Size and Sampling Effects. In: Guy, W. (ed), *Advances in Ecological Research*. Academic Press, Proceedings of the Symposium and Speciality Conference on Instream Flow Needs. American Fisheries Society, Bethesda, USA, pp. 211–266.

Woodward, G., Brown, L.E., Edwards, F.K. *et al.* (2012) Climate change impacts in multispecies systems: drought alters food web size structure in a field experiment. *Philosophical Transactions of the Royal Society B-Biological Sciences*, **367**, 2990–2997.

Woodward, G., Papantoniou, G., Edwards, F.K. & Lauridsen, R.B. (2008) Trophic trickles and cascades in a complex food web: impacts of a keystone predator on stream community structure and ecosystem processes. *Oikos*, **117**, 683–692.

Young, G.J., Dooge, J.C.I. & Rodda, J.C. (2004) *Global water resources issues*. Cambridge University Press, pp. 194.

Zalewski, M., Janauer, G. & Jolankai, G. (1997) Ecohydrology, a new paradigm for the sustainable use of aquatic resources. In: *UNESCO IHP Tech. Doc. In Hydrology no 7, IHP-V projects 2.3/2.4*. UNESCO, Paris, France.

Zalewski, M. & Harper, D. (2001) Rationale Ecohydrology – the use of ecosystem properties as a management tool for enhancement of absorbing capacity of ecosystems against human impact. *Ecohydrol. Hydrobiol.*, **1**, 1–2.

10 Climate Change and Hydrology

RICHARD HARDING[1], NIGEL ARNELL[2], NICK REYNARD[1],
CHRISTEL PRUDHOMME[1], ELEANOR BLYTH[1], AND
CHRIS TAYLOR[1]

[1]Centre for Ecology and Hydrology, Wallingford, Oxfordshire, UK
[2]Walker Institute for Climate Research, University of Reading, Reading, UK

10.1 Introduction to Climate Change: Changes in Twentieth Century and Expected Impact of Increasing GHG on Global Water Cycle

The link between increases in atmospheric carbon dioxide and the warming of the Earth's climate was first proposed in the nineteenth century (Arrhenius, 1896); indeed, the Earth would be approximately 30°C cooler without its greenhouse gases (GHGs). A GHG is one which is transparent to radiation from the Sun but absorbs radiation at long-wave (or infrared) wavelengths (i.e. terrestrial radiation re-emitted by the surface). Carbon dioxide (CO_2) is the most well known of the GHGs, but there are others of global significance such as water vapour, methane, ozone and nitrous oxide (Fig. 10.1).

The Earth's climate system is powered by solar radiation, approximately half of the energy from the Sun is in the visible part of the electromagnetic spectrum. Of the incoming solar radiation about one half is absorbed by the Earth's surface (the rest being absorbed by the atmosphere or directly reflected the surface or clouds). In a stable climate, the net incoming solar energy must balance with outgoing radiation. The majority of the outgoing energy flux from the Earth is in the long-wave (or infrared) part of the spectrum. The long-wave radiation emitted from the Earth's surface is largely absorbed by the GHGs and clouds; all these emit long-wave radiation into all directions. The downward directed component adds heat to Earth's

Progress in Modern Hydrology: Past, Present and Future, First Edition. Edited by John C. Rodda and Mark Robinson.
© 2015 John Wiley & Sons, Ltd. Published 2015 by John Wiley & Sons, Ltd.

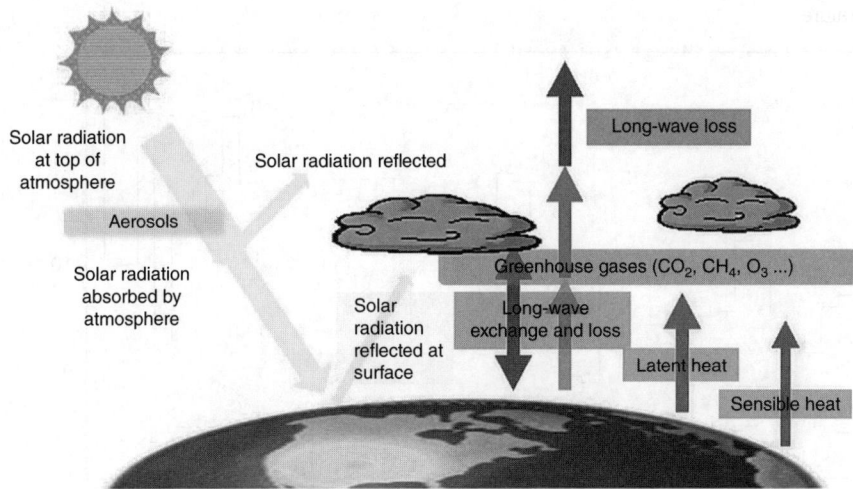

Fig. 10.1 Schematic of main energy flows in the atmosphere and greenhouse effect. (Source: Reproduced with permission of CEH.) *(See insert for colour representation of this figure.)*

surface and the lower layers of the atmosphere (the greenhouse effect). Ultimately, the dominant energy loss from the Earth is from long-wave radiation back to space from the upper layers of the troposphere. Energy moves upwards through the atmosphere by a cascade of long-wave exchanges and the evaporation of water at the surface and subsequent condensation (releasing latent heat) at higher levels. There are many feedbacks (positive and negative) within the climate system. Many of these feedbacks (and uncertainties) associated with climate change are linked with water. For example, changing snow and ice cover will change the amount of the Sun's energy reflected back to space. In addition, water vapour is itself a powerful GHG and clouds play a crucial part in the Earth's energy budget. Thus, water plays a fundamental role in the climate system and changes in the hydrological cycle are the most important impacts of climate change.

Carbon dioxide levels (and other GHGs) have increased substantially over the last century. Since pre-industrial times, the CO_2 concentration in the atmosphere has increased from 280 to 392 ppm (2012 average) and is growing by approximately 2 ppm/year (see http://www.esrl.noaa. gov/gmd/ccgg/trends/global.html). Other potent GHGs, such as methane and nitrous oxide, are also increasing. It is well established that this growth is due anthropogenic emissions. Parallel to these changes has been an increase in global temperature of 0.7 °C through the twentieth century (with 0.5 °C since 1950) and accompanying

reductions in glacier and sea ice extent and rises in sea level and atmospheric humidity. Since Institute of Hydrology (IH) was founded in 1968, the UK mean temperature has increased by 1.0 °C. Figure 10.2a shows that the mean air temperature at Wallingford also increased by 1 °C during the 1990s and 2000s, the equivalent time series in rainfall (Fig. 10.2b) shows no obvious trend but, as is typical with rainfall series, a considerable variability which may mask underlying changes in the rainfall regime.

Water-related issues are the most prominent amongst the commonly identified impacts of climate change. In particular, water-related extremes, storms, floods and droughts, are the likely most immediate and damaging manifestations of a changing climate. Given the critical role of water in the Earth's energy and radiation budgets, increasing GHGs are very likely to influence the hydrological cycle. The direct impact of increasing GHGs is increasing temperatures. The amount of water vapour the air can hold at saturation increases exponentially with temperature (7% per degree centigrade). This is very likely to lead to increasing evaporation from ocean regions and increasing evaporative demand over land. The global near surface air specific humidity has increased since the 1970s; however, because the air temperature has also increased a fairly widespread decrease in relative humidity near the surface is observed over the land (see e.g. Willett *et al.*, 2008). (For definitions of the various measures of humidity please see Box 10.1.) Increasing temperatures and

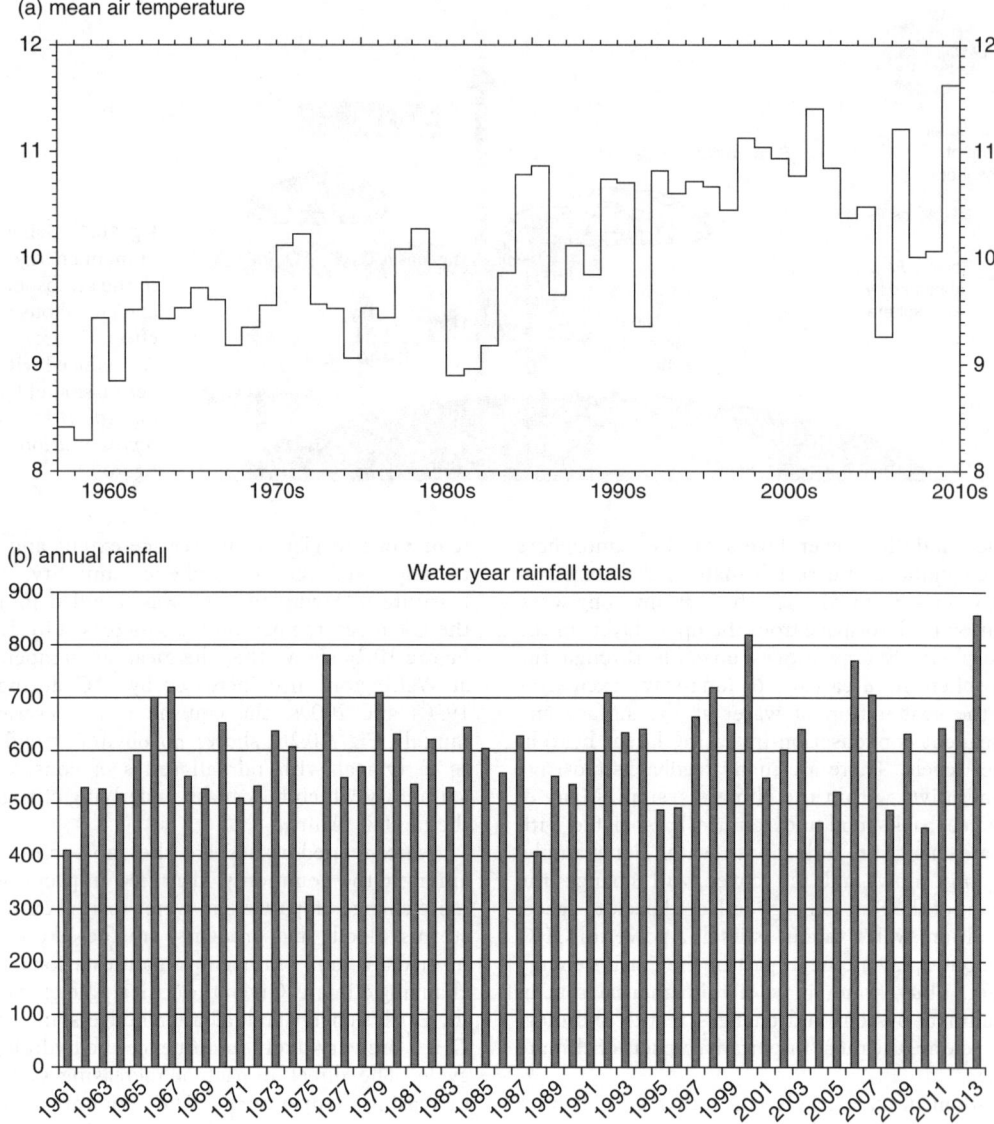

Fig. 10.2 Time series from the Wallingford Meteorological site of (a) mean air temperature and (b) annual rainfall. (Source: Reproduced with permission of CEH.)

decreasing relative humidity would be expected to increase evaporation; however, this trend is compounded by changes in wind speed over the ocean and radiation and soil water over the land. There are few long-term observations of evaporation, but the overall consensus is that actual evaporation has increased and this is contributing to the observed

increased atmospheric humidity. Increasing evaporation and atmospheric humidity should lead to increasing precipitation – over the long-term evaporation and precipitation must balance. The signal of increasing rainfall has, however, not yet been observed in the global mean rainfall datasets, most of which show only a small, but statistically

Box 10.1   Definition of atmospheric humidity

The amount of water held in gaseous phase in the atmosphere is important quantity as it influences atmospheric processes, such as evaporation, cloud amount, rainfall and energy and also human comfort. Atmospheric humidity can be expressed in a number of ways, the commonly used are:

Specific humidity is the ratio of mass of water vapour in the air relative to the mass of air in which it is contained. It is therefore a measure of the absolute amount of water in an air parcel.

Vapour pressure is the partial pressure of the water vapour in the air – is dependent on the absolute humidity and the atmospheric pressure.

Saturation vapour pressure is the vapour pressure of the humidity in equilibrium with a plain water surface.

The relative humidity (in percent) is the ratio of the actual humidity with that of the saturated humidity. (Thus, air at 100% relative humidity is completely saturated and that at 0% is completely dry). Relative humidity is thus a relative measure which depends on the temperature of the air as well as its water vapour content but is important because it directly influences important processes such as evaporation, cloud formation and human comfort.

insignificant, increase in the last two decades. Generally, studies indicate that precipitation over wet areas had increased but decreased over drier areas (Zhang *et al.*, 2007 and Allan *et al.*, 2010).

Increasing evaporation and atmospheric humidity will increase the energy exchanges within the atmosphere. The potential for increased latent heat release associated with cloud condensation is likely to increase the intensity of storms and enhance convective rainfall. This has in fact been observed in the northern hemisphere, where data is available, two-thirds or more areas show significant increases in daily precipitation intensity (defined typically by the 95th percentile of daily rainfall, Min *et al.*, 2011). Regionally, there are significant upward trends in northern and Mid Europe, north and Central America and India. Unfortunately

in many regions of the world, the observational record of hydrological extremes is not long enough to identify trends or the statistical tools are not sufficient to identify unambiguously trends in incomplete and noisy time series.

Droughts have profound impact not only on water supply but also on food production, biodiversity and human well-being in general. It is generally accepted that climate change will bring increasing occurrence of drought (see, e.g. the 2007 report of the Intergovernmental Panel on Climate Change, IPCC, 2007, see Box 10.2). There has been a suggestion that drought occurrence has already increased through the twentieth century (Dai *et al.*, 2004), but the magnitude of this change has been shown to depend on the methodology used to estimate evaporative losses (Sheffield *et al.*, 2012); hence, suggesting that prediction of future drought depends critically on its definition and method of calculation (see also Prudhomme *et al.*, 2014).

Many if not most large rivers are strongly affected by human influences such as dam construction, abstraction of water such as for irrigation, so apparent trends in river flow must be interpreted with caution. Dai *et al.* (2009) found that only about one-third of the top 200 rivers of the world show statistically significant trends during 1948–2004, with the rivers having downward trends (45) outnumbering those with upward trends (19). The changes in river flow are qualitatively consistent with regional changes in precipitation and temperature.

Intergovernmental Panel on Climate Change (IPCC) AR4 WGII concluded that there was no general global trend in the incidence of floods (IPCC, 2007; see also Box 10.2). An IPCC report in 2012 on extremes (IPCC, 2012) went further to suggest that there was low agreement and thus low confidence at the global scale regarding changes in the magnitude or frequency of floods (or even the sign of changes). Trends in floods are, of course, strongly influenced by changes in river management. Most evident flood trends appear to be in northern high latitudes, where observed warming trends have been largest (IPCC, 2013), even here no evidence of a trend in extreme flooding has been found in some regions, for example over Russia based on daily river discharge (Shiklomanov *et al.*, 2007). Other studies for Europe (such as those by Hannaford and Marsh, 2008) show evidence for upward, downward or no trend in the magnitude

---

Box 10.2   The IPCC process

The prediction that increasing CO_2 would lead to increasing air temperatures was first postulated in the nineteenth century (refs). Since the 1970s, climate models of ever-increasing sophistication have been used to explore this link and make estimates of possible impacts. In 1988, the IPCC was established by the World Meteorological Organization and the United Nations Environment Programme to provide governments with a clear view of the current state of knowledge about the science of climate change, potential impacts and options for adaptation and mitigation using the most recent information published in the scientific, technical and socioeconomic literature worldwide. The IPCC makes regular assessments of the science of climate change and assessments of possible changes in the future. The first assessment report was in 1990 (ref) and the fifth report has just been published (ref). The assessments use scenarios of possible increases of GHGs fed into a range of climate models to provide plausible predictions out in 2010. There is always a wide spread caused equally by the range of possible scenarios (based on possible socioeconomic storylines) and the spread of the climate model predictions. Broadly all five assessments have concluded a likely increase in temperature of between 1 and 6 °C by 2100 and the observed increase in the last two decades, since the first assessment, is in line with the early predictions. The assessments go on to explore other aspects of climate change (that of the hydrological cycle will be dealt with below), possible impacts and mitigation strategies.

The IPCC process has been important both scientifically and politically. It is probably unique in science in producing such a wide ranging and authoritative consensus about a scientific issue. It has its faults, it is time-consuming and conservative and makes occasional errors, but the widespread agreement of scientists (over 97% of scientists agree with its findings) give it considerable political impact.

---

and frequency of floods, so that there is currently no clear and widespread evidence for observed changes in flooding except for the earlier spring flow in snow dominated regions (IPCC, 2012). There is a perception amongst scientists, politicians and the public that extremes of weather (including floods and droughts) have increased in the past decade (Huntingford *et al.*, 2013). It remains to be seen whether this is a decadal variability or a long-term trend, but it also provides a challenge to hydrological scientists to marshal the data and understanding to make unambiguous statements about the impact of climate change on hydrological events.

10.2   Introduction to GCMs: From 1970s Onwards

To assess the sensitivity of the atmosphere to increasing GHG concentrations requires a numerical model. Simple models of the energy and water budgets of the Earth can predict the impact on the global temperature; however, to include feedbacks and the impacts on atmospheric dynamics, and hence regional changes, requires a complex spatially distributed model. Climate models were developed from weather forecasting models in the 1960s. All climate models have at their core an atmospheric model which solves the basic equations of motion and conservation of energy and water at high frequency (the routine UK Met Office GCM calculates every 15 minutes) on a three-dimensional grid. The resolution of the grid is constrained by computing resources, originally these grids were many degrees of latitude and longitude and 10–20 vertical levels, the most recent models have grids of less than 1 degree and many tens of atmospheric levels. In addition to the equations of motion, the models need to include a description of radiation absorption at different wavelengths and a representation of clouds and rainfall generation.

As climate models have developed, it has also proved necessary to include many additional components, including the land surface, ocean dynamics, sea and land ice and atmospheric chemistry. These more complex models are now often referred to as Earth system models.

In the UK pioneering work on climate models was done in the 1970s and 1980s at the UK Met Office. This was given a major boast in 1990 with the formation of the Hadley Centre for Climate Prediction and Research, an autonomous department within the UK Met Office.

10.3 The Development of Land Surface Models Within Climate Models: Short History

Scientists at Wallingford (and now Centre for Ecology and Hydrology (CEH)) have been central to the development of land surface component of climate models. A major function of the land surface model is to take the absorbed solar radiation and partition this into the fluxes of latent heat (evaporation), sensible heat and long-wave radiation. During the 1970s, IH scientists had been using the micrometeorological measurements from Thetford and Plynlimon (see Chapters 2 and 5) to develop improved evaporation equations. These were primarily designed for water resource estimates, but in the 1980s, as the land surface component in climate models became more complex, it became clear that climate science could learn from these hydrologically based models and Wallingford scientists held considerable knowledge and data which might be used to develop and test these models further. At this time, Wallingford scientists became involved in a series of field experiments in the Amazon, North America, Spain and the Arctic to generate observations which would help the development of land surface models. Also, an alliance was formed between the scientists from Wallingford and the newly formed Hadley Centre and regular meetings and collaborations ensured the best use of existing data and shaped some of the planned measurement campaigns. These collaborations were formalised further in December 2000 with the formation of the Joint Centre for Hydro-Meteorological Research (JCHMR) when a number of Met Office climate and flood scientists were based at Wallingford. The JCHMR brought together the complementary research skills of the Met Office and the CEH at Wallingford, at the interface of weather and flow forecasting and climate prediction, to support the better understanding and management of floods and droughts and their impacts and to facilitate the joint development of JULES (Joint UK Land Environment Simulator) land surface model (and its predecessor MOSES (Met Office Surface Exchange Scheme)), see Box 10.3.

During the 1990s, the focus of climate research shifted somewhat to carbon as scientists began to include the carbon cycle into the models. This enabled the feedbacks between the atmospheric CO_2 and the land surface (and oceans) to be included thus providing a better projections of the growth of CO_2 levels. At the same time, it became increasingly possible to make measurements of fluxes of water and energy, and ultimately carbon dioxide, in remote locations (see Chapter 5). The measurements of CO_2 fluxes, and submodels of photosynthesis and decomposition, have proved very valuable in the development, calibration and testing of land surface and climate models.

It was increasingly recognised that the disparity between the spatial scale of most measurements and models was a big issue. The early land surface models treated each grid cell, which could be many hundreds of kilometres across, as a single uniform surface and described the exchanges using

---

Box 10.3 The JULES land surface model (https://jules.jchmr.org/)

JULES is a community land surface model based on the MOSES. It is used both as a standalone model and as the land surface component in the Met Office Unified Model. JULES is a core component of both CEHs and the Met Office's modelling infrastructure and is an essential component of NERC's (Natural Environment Research Council) earth system modelling strategy, and it is continually improving and at the cutting edge of land surface modelling.

By allowing different land surface processes (surface energy balance, hydrological cycle, carbon cycle, dynamic vegetation, etc.) to interact, JULES provides a framework to assess the impact of modifying a particular process on the ecosystem as a whole, for example the impact of climate change on hydrology, and to study potential feedbacks.

JULES is available to any researcher, free of charge. It has a large and diverse community of users from across the globe, who study land surface processes on a wide variety of temporal and spatial scales. The JULES community has regular meetings where researchers using JULES can present results and discuss land surface science. The development of JULES is governed by a community process and is presided over by a management committee comprised of organisations providing significant resources for the ongoing development of JULES.

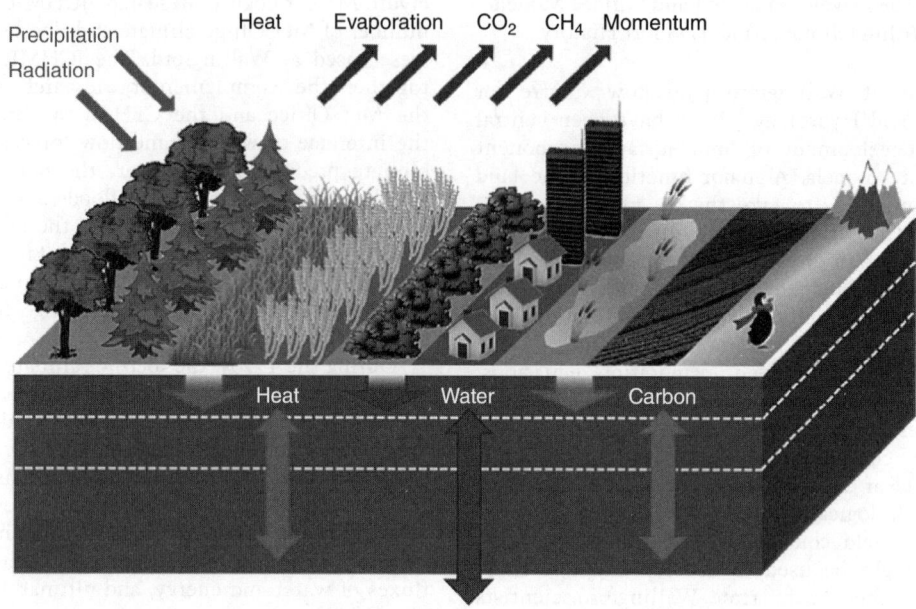

Overview of the science of JULES *(See insert for colour representation of this figure.)*

JULES has a tiled model of sub-grid heterogeneity with separate surface temperatures, short-wave and long-wave radiative fluxes, sensible and latent heat fluxes, ground heat fluxes, canopy moisture contents, snow masses and snow melt rates computed for each surface type in a grid-box.

Nine surface types are normally used:

- Five plant functional types (PFTs)
  - Broadleaf trees
  - Needleleaf trees
  - C3 (temperate) grass
  - C4 (tropical) grass
  - Shrubs
- Four non-vegetation types
  - Urban
  - Inland water
  - Bare soil

Air temperature, humidity and wind speed above the surface and soil temperatures and moisture contents below the surface are treated as homogeneous across a grid-box.

essentially point scale models. This was widely recognised as a weakness, but it is extremely difficult to include submodels which represent the sub-grid processes which can be applied across all terrestrial grid cells and within the computing constraints. There are a number of issues which need to be addressed: firstly, within a grid cell there are a range of vegetation, soil and topographic types, secondly, there are likely to be horizontal interactions between the sub-grid patches, both above and below ground, and, finally, there will be spatial rainfall variability within a grid, leading to spatial variability of soil moisture and evaporation. Through the 1980s and 1990s, there were a series of large-scale experiments whose overall purpose was to understand large-scale land surface/atmosphere interactions which had a major focus on these scaling issues (see Box 10.4). Major insights from these experiments were gained, for example into the rainfall variability (Taylor and Lebel, 1998) and small-scale interactions between bare soil and vegetation (Blyth, 1997). It has, however, proved very difficult to translate these insights into parameterisations within the land surface models. This is primarily to do with problems of complexity and scaling, coupled with the need for climate models to have globally applicable schemes.

The development of the hydrology within JULES has been a major concern at Wallingford. An early development was the partitioning of precipitation into interception, throughfall, runoff and infiltration (Dolman and Gregory 1992). The rainfall rate is assumed to be distributed exponentially across the area. In addition, if the rainfall is 'convective', then it is assumed to cover only 0.3 of the area. This part of the JULES model aims to overcome the problem of GCM-drizzle, whereby every time it rains, a small amount of water covers the entire area of the grid cell and all the water is held on the leaves of the vegetation and re-evaporates, without entering the soil matrix. In practice, the rainfall can be intense over small areas and this means the rainfall fall through the vegetation canopy and into the ground. A further development was the four-layer

soil water model in which soil moisture is distributed between four vertical layers with gradient (or Darcy) flow between them. This new model was calibrated with soil moisture measurement from the Wallingford site (Harding *et al.*, 2000).

Through the 1990s, however, major improvements were made. A 'tiling' scheme was introduced which allows a number of different patches of vegetation to exist within each grid cell, but without any interactions between the patches (Essery *et al.*, 2003). New soil water formulations were also introduced, appropriate to the semi-arid regions, for example a description of bare soil stores and evaporation was introduced coupled to the four-layer soil water model. Considerable work also went into the defining global parameters for these models, in particular parameters describing hydraulic conductivity and storage and aerodynamic exchange parameters appropriate to sparse vegetation.

The one-dimensional structure of the bucket (and four-layer) soil water models, while making some sense for the energy exchanges (and evaporation), makes no sense at all for runoff generation and routing. In fact, the one-dimensional structure leads to a substantial underestimate of runoff. To address this problem, two options have been developed for JULES to introduce sub-gridscale heterogeneity into the soil moisture and hence runoff generation.

Box 10.4 International land surface experiments and land surface models

During the 1980s the development of land surface and biosphere models increasingly required consistent observations at both the point and climate model grid cell scales. This required a major international effort. The International Satellite Land Surface Climatology Project (ISLSCP) began in 1983 driven by a group of interested scientists from the remote sensing, climate, and ecological communities. A field experiment programme was developed for numerous small-scale studies (at a scale of less than 1 km) embedded in experiments at a scale of about 10 km, and eventually to the scale of a GCM grid cell (approximately 100 km on a side). The first major experiment, the First ISLSCP Field Experiment (FIFE), took place in the Kansas prairies in the summer of 1987. Scientists from the IH took the newly developed HYDRA flux system and obtained an excellent data set which has been extensively used to test the grassland component of the MOSES/JULES model. Overall the experiment provided considerable insight into how the land surface interacts with the lower boundary layer.

The baton then passed to the European groups who developed the EFEDA experiment (1989) in central Spain and ultimately the HAPEX-Sahel experiment (1992) in West Africa. The UK involvement was supported by the NERC TIGER programme with substantial inputs from Wallingford. In particular the HAPEX-Sahel was an experiment designed to provide the field data needed to model the climate of the Sahel and its dependence on the land surface. The focus of these experiments was on the semi-arid environment and the substantial environmental problems associated with drought. A secondary focus was on how to represent the substantial variability found in semiarid areas across the 1 degree HAPEX-Sahel square.

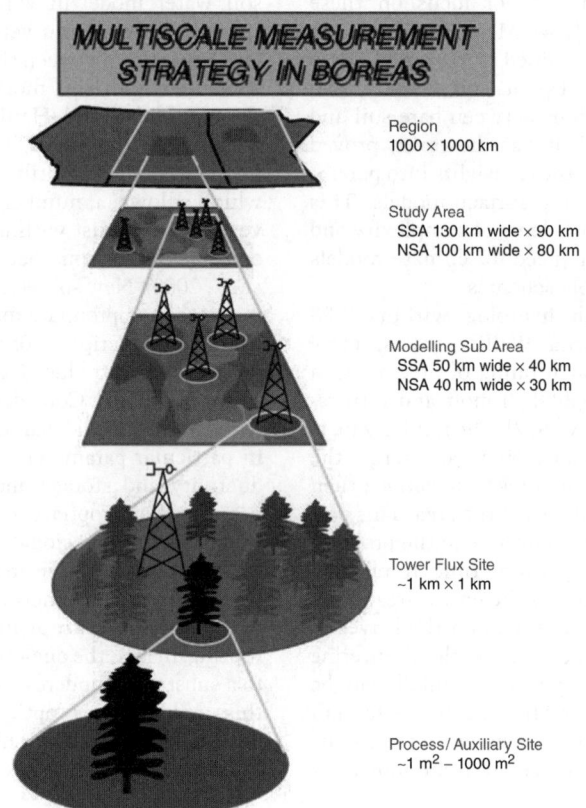

Region
1000 × 1000 km

Study Area
SSA 130 km wide × 90 km
NSA 100 km wide × 80 km

Modelling Sub Area
SSA 50 km wide × 40 km
NSA 40 km wide × 30 km

Tower Flux Site
~1 km × 1 km

Process/Auxiliary Site
~1 m² – 1000 m²

In the 1990s the North Americans developed the Boreal Ecosystem-Atmosphere Study (BOREAS). During 1995 and 1996 extensive measurements were taken in two study regions in northern Canada with a focus on boreal forest water and carbon fluxes and cold region processes. CEH sent a small team to measure fluxes over snow covered forest and lakes. The contrasts in the fluxes of energy and water over these surfaces were quite dramatic and the measurements led to a reassessment of how snow and vegetation were represented in land surface models.

After BOREAS the focus of International studies has shifted to continental-scale experiments (such as LBA in Amazonia and AMMA in W Africa) and networks of flux measurements (such as FLUXNET http://fluxnet.ornl.gov/). The emphasis has also continued to move from heat and water fluxes to carbon and more exotic biogeochemical fluxes and longer-term measurements. CEH now runs over a dozen flux sites routinely across the UK. Four of these sites are embedded in long-term 'Carbon Catchments', where both the vertical and lateral fluxes of carbon and water are measured. CEH also runs one of two UK EMEP super-sites where biogeochemical fluxes of interest to both greenhouse gas and pollution studies are made.

One, TOPMODEL (Beven and Kirkby, 1979), is the more complex scheme that represents this heterogeneity throughout the soil column, including aspects such as a shallow water table and the

capability to estimate wetland fractions (Gedney *et al.*, 2004). However, this scheme requires additional topographic ancillary information. The other option, a probability-distributed moisture (PDM) model, developed at Wallingford, Moore 1985 and Chapter 7) is a much simpler scheme that does not require as much information. This scheme is a statistical representation of the heterogeneity and is not explicitly linked to the water table depth so cannot affect the baseflow component. However, it can still be used to increase surface runoff and has been shown to improve subsequent river discharge when fed into a river routing scheme. The runoff generated at each grid cell can be routed using the TRIP (total runoff integrating pathways) model, Oki *et al.* (1999), or the CEH Grid2Grid model.

Other hydrology components have also been developed for JULES, such as an inundation model using an over-bank flow parameterisation. This submodel comprises the PDM for soil moisture and runoff production coupled with a discrete approximation to the 1D kinematic wave equation to route river water downslope (river flow module), Dadson *et al.* 2010. There are also parameterisations for deep groundwater, irrigation and reservoir storage and extractions. It is clear that a representation of these processes are necessary to model the dynamics of river flows and water storage; however, none of these hydrology submodels have yet been incorporated into the main version of JULES but remain as options, an illustration that no definitive solution has been found to the representation of hydrology at these large scales.

Increasingly, it has been recognised that robust assessments of effects of increasing GHGs on climate change require the incorporation of a whole range of physical, chemical and biological processes into climate models, the resulting integrated systems are often called 'Earth System Models'. CEH has been well placed to assist with the development of routines to describe the exchanges of a whole range of chemical species between the surface and the atmosphere. A good example is methane. Methane release from the land surface only occurs when the soil is saturated. It therefore becomes important to simulate the area and dynamics of wetlands globally. CEH and Met Office scientists have worked together to develop such schemes (e.g. Gedney *et al.*, 2004) and to develop datasets for their validation (see e.g. Bartsch *et al.*, 2012). Work is continuing at CEH to further develop more realistic schemes to describe methane release and to include the exchanges of nitrogen species and ozone, important GHGs which also have significant interactions with the land surface.

Improvements in land models is key to further progress towards understanding and predicting the terrestrial role in, and response to, weather, climate variability and climate change. There is broad agreement that there is a need for improved mechanisms for assessment of the quality and suitability of land surface models for a broad range of studies and uses, ranging from seasonal forecasting to water systems and ecosystem vulnerability to climate change to carbon cycle feedbacks. Several international projects, coming under the general banner of 'Benchmarking', have recently been initiated to try to improve the situation Fundamentally, the goal of a land model benchmarking process is a more substantive, detailed and systematic evaluation of land models and land model processes which will enable modellers to track progress, intercompare models and identify avenues for improvement and will provide information about model strengths and weaknesses to scientists who utilise land models in their research.

Over the past few years, CEH scientists have taken a lead in land model benchmarking engaging the International community and designing prototype benchmarking systems (Blyth *et al.*, 2011). The JULES modelling group have proposed that the water and the carbon cycle should be tested together and presented a simple suite of data to do that. This cross-over between the carbon and water cycles has been accepted by the International community (e.g. in the International programmes of iLEAPS, http://www.ileaps.org/ and GEWEX, http://www.gewex.org/).

10.4   Targets for Benchmarking

It is possible to simplify the problem. The energy balance is important (and should be benchmarked) at the hourly timescale, the water balance at the monthly timescale and the carbon balance at the annual timescale. However, in practice, it is not possible to get the hourly energy balances right, without getting also the hourly evaporation (water flux) right. Similarly, it is not possible to get the annual carbon balance right, without getting the daily carbon response to sunlight and soil moisture stress right. So although the final requirement looks reasonably straightforward, the processes that deliver it are not only complex but interrelated.

A comprehensive list of possible sources of data for testing the models includes fluxes of carbon dioxide and water from micrometeorological towers, river flows, satellite products and experiments.

Choosing between these data is a task in itself as each has inherent errors, may not be the right scale or requires some intermediate model to translate between model output and observation. But one thing is certain: despite the improved availability of data, the time required to gather and quality control these data for analysis against model output remains high and is one of the principal reasons that accessible and repeatable benchmarking systems are desirable.

The increase in relevant data availability is a boon to land modellers and needs to be better exploited. The number and length of datasets are, however, limited and it is often the case that the same dataset that is used to develop a parameterisation or to calibrate a model is used again in the model evaluation process. In some cases, a dataset can even be used a third time to weight or eliminate models within a multimodel database based on their skill at replicating some aspect of the climate system prior, for example to using the models to examine climate projections. Multiple uses of the same data during the model development and evaluation process are clearly a problem, but due to the scarcity of available data, it is often unavoidable.

It is often challenging to design a metric that tests a specific land model process. Often, real insight into these models is gained through comparison of the model against experimental data or case studies of particular extreme climatic events. As an example, Bonan *et al.* (2012) utilised a set of litter bag decomposition studies to evaluate simulated versus observed carbon loss over time through a controlled set of model experiments. They conclude that 'long-term litter decomposition experiments provide a real-world process-oriented benchmark to discriminate ecological fact from model fantasy'. Useful information about model behaviour can also be gleaned through analysis of models against available and future manipulation experiments including, for example rainfall exclusion, free-air CO_2 enrichment (FACE), nitrogen fertilisation and snow fence experiments.

10.5   Studies in Land/Atmosphere Interaction

It has long been recognised that exchanges of heat and moisture between the surface and the atmosphere have an important influence on weather and climate. At the large scale, the contrast between the land and the sea drive the monsoonal circulations and, similarly, at regional scales these contrasts generate sea and lake breezes. Water/land contrasts, however, provide probably the largest possible temperature gradients and it is not obvious whether gradients in vegetation and soil moisture can influence the climate. Indeed, it is often said 'forest attracts rainfall', but the hard observational evidence for this is difficult to find. Enhanced evaporation over forests has to lead to more rainfall somewhere but what size of forest in needed? and whether the enhanced rain is over the forest (or downwind), are difficult questions to answer. Studies with numerical models do suggest that large-scale deforestation (such as significant parts of the Amazon) will reduce rainfall considerably. Thus, it is not that the effect does not exist, but it is difficult to discern amongst the very large spatial and temporal variations of rainfall.

Some persuasive evidence was found during the HAPEX-Mobilhy experiment in 1986 when on satellite images distinctive cloud patterns were observed over the Landes forest in SW France – with increased cloud over the forest area (Fig. 10.3). Subsequent modelling (Blyth *et al.* 1994) suggested that rainfall might also be enhanced, but firm observational evidence of the effect on rainfall was still missing.

More evidence for land–atmosphere interactions was found in the ABRACOS (Anglo-Brazilian Amazonian Climate Observation Study) experiment in Amazonia (Chapter 5). Here, the cleared forest pastures in Rondonia dried out during dry periods, increasing sensible heat fluxes and leading to deeper planetary boundary layers. There was some evidence these deeper boundary layers had an influence on cloud amounts, but, again, it was difficult to see the impact on rainfall patterns (Culf *et al.*, 1996; Shuttleworth *et al.*, 1991).

It is, however, in semi-arid regions where soil moisture variability is most likely to affect surface fluxes of heat and water and rainfall. Hence, the HAPEX-Sahel experiment in 1992, to which CEH was a major contributor, was a good opportunity to look for evidence. One of the outcomes from this multinational experiment was a network of soil moisture and flux measurements – which showed that evaporation, sensible heat flux and surface temperature do depend on both the soil moisture and the vegetation cover (Taylor *et al.*, 1997). A dense network of rain gauges showed the expected large spatial variability from individual storms but

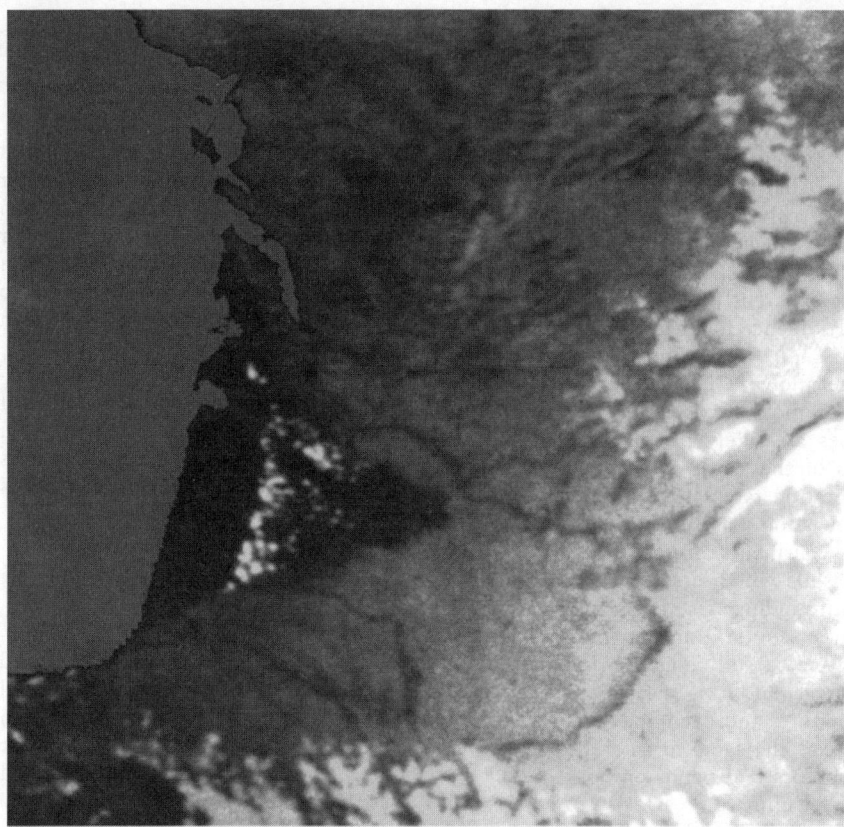

Fig. 10.3 Cloud developing over the Le Landes forest. (Source: NOAA – National Oceanic and Atmospheric Administration.) *(See insert for colour representation of this figure.)*

intriguingly also a persistence of rainfall – if an area received rain early in the wet season it continued to be favoured with rain (Taylor and Lebel, 1998). Again, subsequent modelling showed this was possible and that the Sahelian region was indeed a hotspot for land/atmosphere interactions.

More recent measurements by the African monsoon multidisciplinary analysis (AMMA) indicate that low evaporation due to soil water stress provides a potentially powerful feedback. Before the AMMA observational campaign analysis of satellite data suggested, perhaps surprisingly, that afternoon storm clouds were about 40% more prevalent over dry soils when compared with wetter areas. However, the mechanism underlying this feedback was unclear. New aircraft data showed increased temperature and reduced humidity in the lower atmosphere occurred over areas where it had not rained in the last few days. This was evident even when the dry areas were only 5 km

across, a surprising result given that turbulent eddies tend to mix the lower atmosphere rather effectively during the daytime. It was also found that where dry soils met wet soils, the differences in temperature were large enough to affect the winds, in a manner analogous to sea breezes, with the potential for triggering rain storms. This triggering of afternoon rain in the Sahel plays a critical role for the hydrology of the region. Some of these storms grow very rapidly, moving from a single cloud to a major storm covering an area the size of Wales in a matter of 2 or 3 hours.

Another area of potential interaction is around wetlands. Again CEH returned to study the Sahel and the Mali Wetlands. Satellite and modelling studies in this region show evaporation is enhanced over the wetland area, with a suppression of the sensible heat flux and evidence of the suppression of storm initiation over the wetland but with an enhancement at its boundaries, with a 'wetland breeze' effect (Taylor, 2010). The interaction with wetlands opens the possibility that there will also be an interaction between irrigated areas and climate. The highest density of irrigation is across the Ganges delta in Northern India. There is some modelling and trajectory evidence that there is an interaction, but it is clear that its magnitude and effect depends on spatial scale and the complex interactions between the land surface and monsoon circulations (Harding *et al.*, 2013). A major Anglo-Indian experiment is currently being planned to provide observational evidence for these drivers to the Indian monsoon.

Conditions in the Sahel allowed CEH researchers to isolate the impact of soil moisture on convective storms. Similar processes are likely to influence rain in the many parts of the world where soil moisture restricts evaporation rates. Taylor *et al.* (2012) were able to show, from satellite observations, that across all six continents afternoon rain falls preferentially over soils which are relatively dry compared to their surrounding area. The signal emerges most clearly over semi-arid areas, where surface fluxes are sensitive to soil moisture and convective rainfall is prevalent. It is clear from these studies that the nature, and even the sign, of the interaction between the land surface and rainfall is complex. It appears it depends on both the scale and the underlying climate.

All these studies demonstrate that climate and weather models, and water resource assessments, need to include some recognition of these land/atmosphere interactions (Dadson *et al.* 2013). But it is not easy to use our understanding to improve the representation of such feedbacks in climate and weather models, not least because of the coarse nature of models, which cannot represent individual clouds growing over locally realistic land conditions. An important ongoing development is the use of increasingly accurate land surface data, acquired by satellites, in weather models. In particular, the launch of the soil moisture and ocean salinity (SMOS) mission in 2009 is providing information globally about soil water that occurs several centimetres beneath vegetation and this is being followed by new missions, such as soil moisture active/passive (SMAP) and Sentinel 1. It is hoped that these will lead to real-time estimates of soil moisture leading to improved weather predictions, though the effective translation of new measurements into operational forecast systems can take many years.

10.6 Impacts of Climate Change on River Flows and Water Resources

10.6.1 Introduction

Research into the potential consequences of climate change for hydrological regimes and water management began at the IH in the mid-1980s. Over time, this research has developed methods for estimating impacts that are both robust and can produce results and information that are useful to water managers. This section outlines how research into climate change impacts developed at CEH, introduces the key issues that have been encountered over the years and presents some examples of impact assessments. Most of the research has been stimulated by the demand from water managers and regulators for information on potential changes in river flows.

10.6.2 The development of impacts research

Research at CEH into the potential consequences of climate change for water resources can be traced back to an early report produced for the World Meteorological Organisation (WMO) by Max Beran (Beran, 1984) and subsequently expanded in book chapters (Beran, 1986) and an IAHS (International Association of Hydrological Sciences from 1971) volume (Solomon *et al.*, 1987). These together drew on the emerging literature on the potential effects of increasing GHG concentrations on global climate and suggested that these effects could plausibly impact upon hydrological regimes. However, at this time, there had been very few attempts to quantify these potential impacts.

10.6.3 Key issues in estimating impacts on hydrological behaviour

In the late 1980s, the Department of the Environment (as was) commissioned CEH to review potential impacts on water resources in the UK. The first public report (Arnell *et al.*, 1990) explored past variability in river flows in the UK as well as looking at how river flows might change in the future. It used a small set of climate scenarios derived by expert judgement from the rather high-level output available from climate models at the time and considered a range of approaches to estimating impacts. Both temporal and spatial analogues were considered (using information either from the past or another place to infer future changes in a catchment) and found to have significant limitations. Simple empirical equations relating measures such as annual runoff to annual precipitation and evaporation were evaluated, and again found to have major limitations; different equations gave very different results. Finally, the report used a simple monthly water balance model calibrated in 15 representative catchments. The report demonstrated that climate change had the potential to substantially alter river flow regimes across the UK, and results were summarised in Arnell (1992). Six of the study catchments have subsequently been used in a large number of studies (most recently in Arnell *et al.* 2014).

This report was followed by a larger Department of the Environment project, completed in 1994, which was undertaken by a consortium and considered not only river flows but also river water quality, river water temperature, estuarine water quality and the demand for water. The river flow report concentrated on the application of catchment models with climate scenarios, this time using a daily water balance model – based on the PDM model (Moore, 1985 and Chapter 7) – and scenarios based more explicitly on the output from climate models. Again, the simulations demonstrated that plausible climate scenarios could lead to substantial changes in river flows, and the study also showed how different catchments responded differently to the same climate scenario (Arnell and Reynard, 1996). Upland and lowland catchments respond differently because the baseline water balance differs. Because rainfall and evaporation are of similar magnitudes in summer in upland British catchments, summer streamflow changes are sensitive to small changes in summer rainfall or evaporation. In lowland catchments, summer evaporation is typically considerably greater than summer rainfall, so changes in summer flows are more dependent on changes in rainfall in spring. Runoff changes in groundwater-dominated catchments are dominated by changes in recharge during the winter recharge season, which depends not only on winter rainfall but also on when soil moisture deficits start to decline in autumn and increase in spring.

Following this research, and after the experience of significant droughts in the early 1990s, the UK water industry and regulators (originally the National Rivers Authority, and subsequently the Environment Agency) initiated the development of methods to allow water supply companies to estimate the effects of climate change on the reliability of their supply systems. These methods were intended to be used in the Asset Management Plans submitted by supply companies to the 5-yearly Periodic Reviews conducted by the regulator Ofwat; the aim was to produce a consistent methodology used by all companies. In essence, the approach developed involved constructing regional 'flow factors' which could be used to perturb historical river flow records to represent the effects of a suite of climate scenarios. The regional flow factors were based on the average changes simulated in a number of modelled catchments in each region. The basic methodology, initially proposed in 1997, was subsequently revised for later periodic reviews to incorporate more recent climate scenarios, but the principles remain the same (Arnell, 2011a). Changes to river flow regimes in six study catchments (first introduced in 1990) under the UKCIP02 scenarios and later climate model-based scenarios have been published Arnell (2004, 2011b), using the same hydrological model as applied in the mid-1990s.

Since the 2000s, research into potential changes in river flows in the UK at CEH has followed two strands. The first involves further investigation into potential changes in river flow regimes, looking at the specific effects of different ways of constructing scenarios (e.g. Prudhomme and Davies 2009a,b) and at changes over the UK as a whole using large-scale hydrological models. This has culminated in the FutureFlows project (Prudhomme *et al.*, in press), which produced changes in river flows and groundwater levels for nearly 300 rivers catchments and 24 boreholes across Great Britain for a range of plausible climate change scenarios. Pioneering the use of Digital Object Identifier for environment change data, the associated time series are publically available to the research community free of charge (Prudhomme *et al.*, 2012a, 2013a), so that the cascading impact of climate change on water-related sectors can be explored by other research groups. FutureFlows Climate was the most downloaded NERC dataset from its DOI during September 2012 and August 2013. The results were used in the Environment Agency's evidence supporting the Government's 2012 Water White Paper (Environment Agency, 2012).

The second strand focuses on the potential impacts of climate change on flood flows in the UK. Early work included the development of the semi-distributed conceptual rainfall–runoff model CLASSIC to explore combined impact of climate and land use change in large catchments (Crooks and Davies, 2001; Crooks and Naden, 2007; Crooks *et al.*, 1996; Reynard *et al.*, 2001) and methodological work on sources of uncertainty (Prudhomme *et al.*, 2003). Bell *et al.* (2012) simulated the changes in flood frequency curves in catchments within the Thames basin, using – such as Future-Flows – scenarios derived from the regional climate model simulations underpinning the UKCP09 climate projections.

Since the late 1980s, the research into the implications of climate change for hydrological behaviour undertaken at CEH has led to significant developments in two areas. The first is in the development of hydrological models which can be used with confidence to simulate the effects of climate change. Hydrological models have, of course, been developing over many years, and this development is summarised in Chapter 7. Of particular, concern in the context of climate change is the ability of the model to represent reliably processes and flows which may be outside the range of calibration and the ability to simulate flows in ungauged locations; this is necessary because information on potential changes in flows is often needed at places where there are no observed hydrological data. Over the years, a suite of models has been developed for use in the UK, depending on local context. Models based on the PDM are used in small to medium-sized catchments, whereas network models such as Grid2Grid, or semi-distributed models such as CLASSIC (climate and land-use scenario simulation in catchments) (Crooks and Naden, 2007) or CERF (continuous simulation of river flow) (Young, 2006) are used for the larger British catchments or ungauged sites. These models have been shown in numerous studies to simulate well flows during dry or wet periods, and their parameterisation is based on physical principles.

The second key area is in the development of approaches to construct and apply climate scenarios. In principle, river flows can be taken directly from the climate models. This is, however, not practical primarily because the rainfall generated by climate models is biased in its total quantity (sometimes by a factor of 2) and with a wrong distribution in time (climate models tend to produce too much drizzle). The early impact assessments used the so-called delta method, which basically applies changes in mean monthly climate to observed daily weather series. Whilst this has the advantage that at least the baseline series is realistic, it is difficult

with this approach to incorporate potential changes in year-to-year and day-to-day variability. Much work has therefore been undertaken on the development of ways of appropriately using directly output from regional climate models; this involves applying empirical 'bias correction' procedures to correct for bias in model simulations of current climate (Lafon *et al.*, 2013). The key source of uncertainty in projected hydrological changes is the range in change in rainfall across different climate models. One way of addressing this is to use many climate models, but another is to use a 'scenario-neutral' way of defining changes in hydrological behaviour in terms of changes in generic measures of climate change. Prudhomme *et al.* (2010, 2013b,c) developed an approach to define change in flood characteristics in terms of changes in the amount and seasonal distribution of rainfall change – based on multiple applications of a hydrological model in many catchments – and used this to identify a set of characteristic flood responses to climate change. The 'response surfaces' can also be used to estimate rapidly the effects on flood properties of a given climate scenario, without the use of a hydrological model. Since then, similar techniques have been applied in other countries (Bastola *et al.*, 2011; Wetterhall *et al.*, 2011) or environment variables (Fronzek *et al.*, 2011). This radically different approach demonstrated a regional variation in the likelihood of changes in flood peak of a fixed magnitude across Great Britain, a resulted in an update of Defra's Supplementary Note to Operating authorities to be published (Environment Agency, 2011).

Figure 10.5 shows some example estimates of the potential impact of climate change on river flows in the UK. Figure 10.5a shows change in mean monthly runoff (by 2050) in six example catchments under the UKCIP02 climate scenarios, together with a 'cool-wet' and 'hot-dry' range (Arnell, 2004); the difference between catchments and range between scenarios is clear. Figure 10.5b shows change across the UK in seasonal runoff (again by the 2050s) under the 11 regional models used in UKCP09 (Prudhomme *et al.*, 2012b), again illustrating regional variability and both the similarities and differences between climate models. Figure 10.5c shows flood frequency curves for the River Thames at Kingston under current climate and, in the 2080s, with the 11 UKCP09 regional models (Bell *et al.*, 2012). The figure shows not only the uncertainty in projected changes, but also

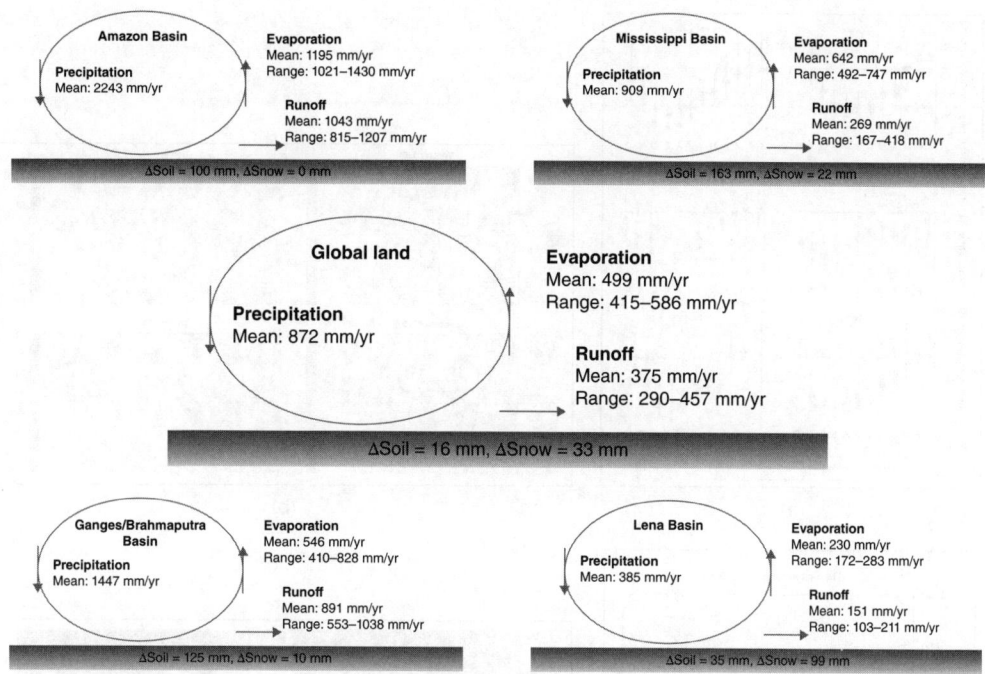

Fig. 10.4 Components of water fluxes and storages for global terrestrial land surface and four major basins representing different climate regimes (1985–1999). The numbers represent simulation results of 11 models participating in WaterMIP (Haddeland *et al.*, 2011). (Source: Reproduced with permission of CEH.)

that both the magnitudes of a given frequency flood are likely to increase, and the return period of a given magnitude flood is likely to decrease. Finally, Figure 10.5d shows the scenario-neutral response surfaces showing change in the 20-year flood peak in two catchments, with change in the amount and seasonal distribution of rainfall (Prudhomme *et al.*, 2010).

10.6.4 Climate change and global water resources

In the last decade, the climate community has been very active in producing new datasets of climate (rainfall and meteorology) and hydrology (river networks and dams). Wallingford scientists have been in the forefront of using these new global datasets with land surface and hydrology models to assess, in a consistent way, past and future global water resources. Between 2007 and 2011, CEH coordinated a major European Union funded project WATCH (Water and Global Change), http://www.eu-watch.tv/ and Harding *et al.* (2011).

The project brought together climate and water scientists from Europe, North America, Australia and Japan. The project developed a methodology to assess past and future water resources across the world in consistent manner. Central to this methodology was the set of climate data for use as input to hydrological models, the WATCH forcing data. This dataset was used in an international intercomparison of global hydrological models (GHMs) and land surface models (WaterMIP) and in the development of a bias correction methodology to enable hydrological simulations from climate model output. The WaterMIP experiment compared 11 global land surface and hydrology models and highlighted the very large range of outputs. The simulated range in global runoff was large: 45% of the mean value (290–457 mm/annum). Figure 10.4 shows with an even larger variability in individual basins, in particular in semi-arid regions (Haddeland *et al.*, 2011). Moreover, the 11 global models were shown to reproduce relatively well the regional development of flood and drought episodes (Prudhomme *et al.*, 2011), which means they are

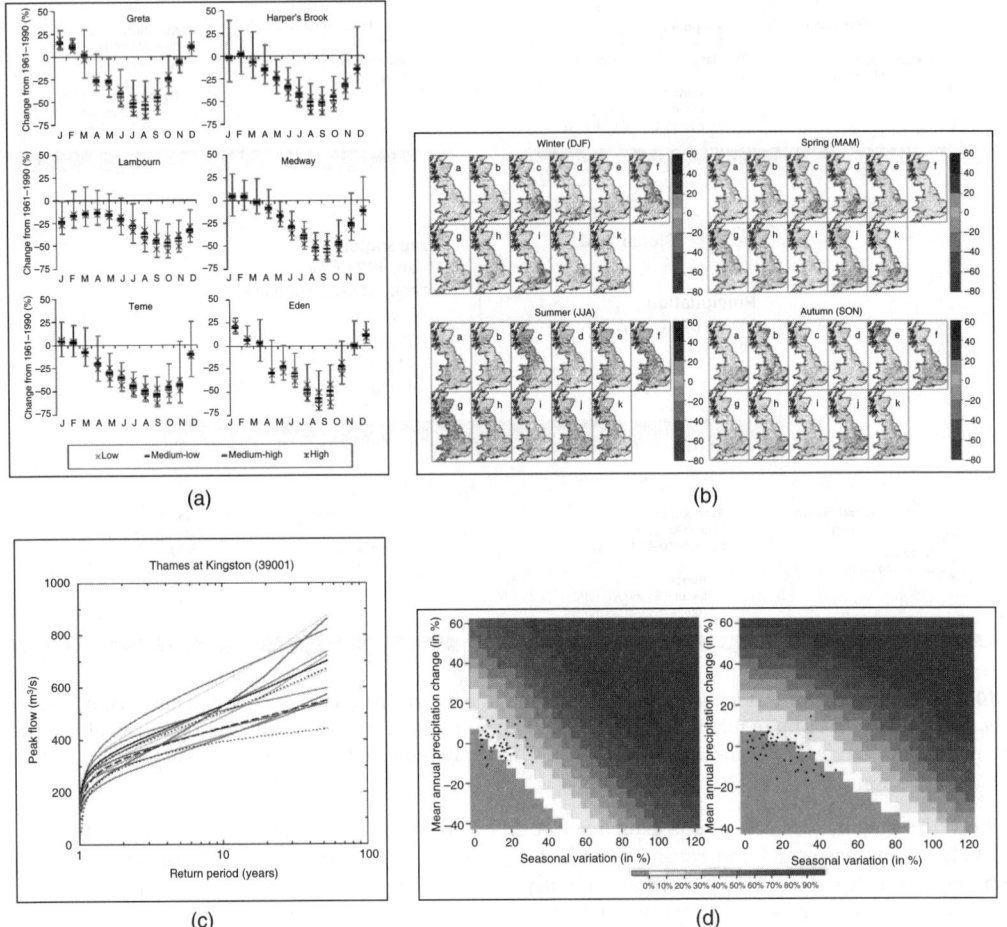

Fig. 10.5 Example impacts on river flows in the UK. (a) Change in mean monthly river flows in the 2050s, with UKCIP02 and other scenarios. (Source: Arnell 2004. Reproduced with permission of John Wiley & Sons.) (b) Change in seasonal runoff in the 2050s, under the 11 UKCP09 regional climate model scenarios (Prudhomme *et al.*, 2012b. Reproduced with permission of John Wiley & Sons.). (c) Flood frequency curves under the current climate (dashed line) and in the 2080s, under the 11 UKCP09 regional climate model scenarios. (Source: Bell *et al.* 2012. Reproduced with permission of Elsevier.) The dotted lines show the uncertainty range for the current frequency distribution. (d) Response surfaces for percentage change in the 20-year return period flood, against magnitude of annual rainfall change (vertical axis) and change in seasonal variation in rainfall (horizontal axis). (Source: Prudhomme *et al.* 2010. Reproduced with permission of Elsevier.) Left: Endrick catchment, North East Scotland and right: Roding catchment, South East England. *(See insert for colour representation of this figure.)*

useful tools to explore the impact of climate change on the hydrology at the regional to continental scale, not possible from catchment-scale models.

Subsequently, and using the WaterMIP template, the Inter-Sectoral Impact Model Intercomparison Project (ISI-MIP) has produced a comprehensive analysis of the uncertainties arising from the combination of climate models and impact models (Warszawski *et al.*, 2013). This major international effort has looked at the predictions and uncertainties in six sectors (water availability, river flooding, coastal flooding, agriculture, ecosystems and energy demands) using a coherent set of climatic and socioeconomic scenarios. In total, the study used over 30 impact models (CEH contributed to simulation for the water sector with JULES), five GCMs and four Representative Concentration Pathways (RCPs). The results of the ISI-MIP provide a unique and systematic overview of the state of the art of climate impact research across sectors.

To assess future water scarcity, ISI-MIP used 11 GHMs driven offline by the five climate models used for the fifth assessment report of the IPCC. Schewe *et al.* (2013) show that climate change is likely to exacerbate regional and global water scarcity. The ensemble average of the models suggest that a global warming of 2 °C above present (approximately 2.7 °C above pre-industrial) will severely deplete the water resources for an additional 15% of the global population and will increase the number of people living under absolute water scarcity ($<500 \, m^3$ per capita per year) by another 40% (according to some models, more than 100%), compared with the effect of population growth alone. For some of these impacts, the steepest increase is seen between the present day and 2 °C temperature rise, whereas indicators of very severe impacts increase unabated beyond 2 °C. At the same time, the study highlights large uncertainties associated with these estimates, with both global climate models and GHMs contributing to the spread. GHM uncertainty is particularly dominant in many regions affected by declining water resources.

The water scarcity study considered only the long-term average condition and did not consider potential changes in interannual and seasonal variability. But, it is often the extreme events, floods and droughts, which have the greatest impact on water, food production and livelihoods. Prudhomme *et al.* (2014) used the ISI-MIP framework to assess future droughts. Drought severity – essentially the proportion of land under drought conditions – is a measure of the time-integrated effect of several interlinked processes and stores including precipitation, evaporation and soil moisture storage. They concluded that an increase in global severity of hydrological drought is likely by the end of the twenty-first century, with systematically greater increases if no climate change mitigation policy is implemented. Drought frequency may increase by more than 20% in some regions of the globe by the end of the twenty-first century. Under the 'business as usual' scenario (an energy-intensive world due to high population growth and slower rate of technological

development), droughts exceeding 40% of analysed land area were projected by nearly half of the simulations carried out. This increase in drought severity has a strong signal–to-noise ratio at the global scale: while we are relatively confident that an increase in drought will happen, its exact magnitude is uncertain. Regionally, Southern Europe, Middle East, South East United States, Chile and South West Australia are identified as possible regions where water scarcity will become an increasing problem.

The study also showed that the different representations of terrestrial water cycle processes in GHMs are responsible for a much larger uncertainty in the response of hydrological drought to climate change than previously thought. One important source of uncertainty was identified based on a model sensitivity experiment done by CEH using JULES, which depends on whether the models allow plants to adapt to enriched carbon dioxide atmosphere; this is only incorporated currently in land surface models accounting for dynamic response of plants to CO_2 and climate. If this is accounted for, the increase in droughts due to warmer climate and changes in precipitation is mitigated by reduced evaporation from plants, because they are more efficient at capturing carbon during photosynthesis. The process of plant adaptation under an enriched carbon dioxide atmosphere is currently absent from the majority of conceptual hydrological models and only considered on a few land surface and ecology models. This finding highlights the importance of considering a diverse range of hydrological models to better capture uncertainty with these models when assessing the impact of climate change on hydrology. These studies also illustrate there is considerable potential for improved water resources simulations through hydrological model development.

10.7 Conclusion

Climate change is happening, very likely driven by man's emissions of carbon dioxide and other GHGs. Climate models are an essential tool to explore options for mitigation and adaption. Since the 1970s, it has been increasingly recognised that the interactions between the land surface (hydrology and vegetation) and the atmosphere play a fundamental role in the development of the climate. The deep understanding of hydrological processes at Wallingford, and the early recognition of the role of vegetation, enabled Wallingford scientists to play a leading role in the early development of land surface models for climate studies. In addition, their expertise in flux measurements, at Thetford, Brazil, Africa, the Arctic and elsewhere, meant it was possible to bring together modellers and experimentalists in one place. Visionary scientists, such as John Stewart, Jim Shuttleworth and John Gash, played their part nationally and internationally in a series of ground breaking campaigns across the world (Box 10.4).

The incorporation of the IH into CEH brought access to wider expertise in the modelling and measurement of the carbon, nitrogen and other biogeochemical cycles. This has again allowed Wallingford to be at the forefront of the development of climate (and now Earth System) models.

At the same time, the strong links between Wallingford hydrological science and practical flood estimation enabled Wallingford scientists to develop the science of coupling climate and hydrological models for practical risk and adaptation studies for the UK. Overall, these various studies demonstrate the value of bringing together climate scientists, with their understanding of climate models and access to global datasets and hydrological and ecological scientists, with their understanding of processes, uncertainty and data. This fusion is not trivial, these communities work at different scales and use different languages, but first the IH and now the CEH have played a valuable role in bringing these communities together.

The challenges of the future are many; climate change poses huge risks. Scientists have the responsibility to provide the best possible assessment of the science and associated risks to a, sometime, sceptical public and media. CEH and IH before it, have been able to provide a reasoned, unbiased and respected advice over the last five decades. The new challenge is to integrate the hydrological sciences with those of agriculture and social science to explore how 9 billion people can live on the planet while maintaining the ecosystem services vital to our lives.

10.8   References

Allan, R.P., Soden, B.J., John, V.O. *et al.* (2010) Current changes in tropical precipitation. *Environmental Research Letters*, **5**, 025205. doi:10.1088/1748-9326/5/2/025205

Arnell, N.W. (1992) Impacts of climate change on river flow regimes in the UK. *Journal of the Institution of Water and Environmental Management*, **6**, 432–442.

Arnell, N.W. (2004) Climate change impacts on river flows in Britain: the UKCIP02 scenarios. *Journal of the Chartered Institute of Water and Environmental Management*, **18**, 112–117.

Arnell, N.W. (2011a) Incorporating climate change into water resources planning in England and Wales. *Journal of the American Water Resources Association (JAWRA)*, **47** (3), 541–549. doi:10.1111/j.1752-1688.2011.00548.x

Arnell, N.W. (2011b) Uncertainty in the relationship between climate forcing and hydrological response in UK catchments. *Hydrology and Earth System Sciences*, **15**, 1–16. doi:10.5194/hess-15-1-2011

Arnell, N.W. & Reynard, N.S. (1996) The effects of climate change due to global warming on river flows in Great Britain. *Journal of Hydrology*, **183**, 397–424.

Arnell, N.W., Brown, R.P.C. & Reynard, N.S. (1990) *Impact of Climatic Variability and Change on River Flow Regimes in the UK*. Institute of Hydrology, Wallingford. Report 107, p. 150.

Arnell, N.W., Charlton, M.B. & Lowe, J.A. (2014) The effect of climate policy on the impacts of climate change on river flows in the UK. *Journal of Hydrology.* **510**, 424–435. doi:10.1016/j.jhydrol.2013.12.046

Arrhenius, S. (1896) On the influence of carbonic acid in the air upon the temperature of the ground. *Philosophical Magazine and Journal of Science* Series 5, **41**, 237–276.

Bartsch, A., Trofaier, A.M., Hayman, G. *et al.* (2012) Detection of open water dynamics with ENVISAT ASAR in support of land surface modelling at high latitudes. *Biogeosciences*, **9** (2), 703–714. doi:10.5194/bg-9-703-2012

Bell, V.A., Kay, A.L., Cole, S.J., Jones, R.G., Moore, R.J. & Reynard, N.S. (2012) How might climate change affect river flows across the Thames Basin? An area-wide analysis using the UKCP09 regional climate Model ensemble. *Journal of Hydrology*, **442**, 89–104.

Beran, M.A. (1984) *Climate Change and Variability: New Problems for Water Resources and Hydrology*. Paris, World Meteorological Organization, pp. 22pp.

Beran, M.A. (1986) The water resources impact of future climatic change and variability. In: Titus, J.G. (ed), *Effects of Changes in Stratospheric Ozone and Global Climate.* Vol. **1**. US EPA/UNEP, Washington DC, pp. 299–328.

Bastola, S., Murphy, C. & Sweeney, J. (2011) The sensitivity of fluvial flood risk in Irish catchments to the range of IPCC AR4 climate change scenarios. *The Science of the Total Environment*, **409** (24), 5403–5415.

Beven, K.J. & Kirkby, M.J. (1979) A physically based, variable contributing area model of basin hydrology/Un modèle à base physique de zone d'appel variable de l'hydrologie du bassin versant. *Hydrological Sciences Bulletin*, **24** (1), 43–69.

Blyth, E.M., Dolman, A.J. & Noilhan, J. (1994) The effect of forest on mesoscale rainfall - an example from HAPEX-MOBILHY. *Journal of Applied Meteorology*, **33** (**4**), 445–454.

Blyth, E.M. (1997) Representing heterogeneity at the Southern Super Site with average surface parameters. *Journal of Hydrology*, **189**, 869–877.

Blyth, E.M., Clark, D.B., Ellis, R. *et al.* (2011) A comprehensive set of benchmark tests for a land surface model of simultaneous fluxes of water and carbon at both the global and seasonal scale. *Geoscientific Model Development*, **4**, 255–269. doi:10.5194/gmd-4-255-2011

Bonan, G.B. *et al.* (2012) Evaluating litter decomposition in earth system models with long-term litterbag experiments: an example using the Community Land Model version 4. (CLM4). *Global Change Biology*, **3**, 957–974.

Crooks, S.M. & Davies, H.N. (2001) Assessment of land use change in the Thames catchment and its effect on the flood regime of the river. *Physics and Chemistry of the Earth (B)*, **26** (**7–8**), 583–591.

Crooks, S.M. & Naden, P.S. (2007) CLASSIC: a semi-distributed rainfall run-off modelling system. *HESS*, **11** (**1**), 516–531.

Crooks, S.M., Naden, P.S., Broadhurst, P. & Gannon, B. (1996) *Modelling the flood response of large catchments: initial estimates of the impacts of climate and land use change*. FD0412, Institute of Hydrology, Wallingford.

Culf, A.D., Esteves, J.l., Marques Filho, A.D. & Rocha, H.R. (1996) Radiation, temperature and humdity over forest and pasture in Amazonia. In: Gash, J.H.C., Nobre, C.A., Roberts, J.M. & Vitoria, R.L. (eds), *Amazonian Deforestation and Climate*. Wiley, pp. 175–191.

Dadson, S., Acreman, M. & Harding, R.J. (2013) Water security, global change and land-atmosphere feedbacks. Philosophical transactions. *Series A, Mathematical, Physical, and Engineering Sciences*, **371** (**2002**), 20120412.

Dadson, S.J., Ashpole, I., Harris, P. *et al.* (2010) Wetland inundation dynamics in a model of land surface climate: Evaluation in the Niger inland delta region. *Journal of Geophysical Research: Atmospheres*, **115**, D23114.

Dai, A., Qian, T.T., Trenberth, K.E. & Milliman, J.D. (2009) Changes in continental freshwater discharge from 1948 to 2004. *Journal of Climate*, **22**, 2773–2792.

Dai, A.G., Trenberth, K.E. & Qian, T.T. (2004) A global dataset of Palmer Drought Severity Index for 1870–2002: Relationship with soil moisture and effects of surface warming. *Journal of Hydrometeorology*, **5** (**6**), 1117–1130.

Dolman, A.J. & Gregory, D. (1992) The parameterization of rainfall interception in GCMs. *Quarterly Journal of the Royal Meteorological Society*, **118**, 455–467.

Environment Agency (2011) *Adapting to climate change: advice for flood and coastal erosion risk management authorities*. Environment Agency, Bristol.

Environment Agency (2012) *The case for change – current and future water availability*. Environment Agency Report GEHO1111BVEP-E-E

Essery, R.L.H., Best, M.J., Betts, R.A., Cox, P.M. & Taylor, C.M. (2003) Explicit representation of subgrid heterogeneity in a GCM land surface scheme. *Journal of Hydrometeorology*, **4**, 530–543.

Fronzek, S., Carter, T.R. & Luoto, M. (2011) Evaluating sources of uncertainty in modelling the impact of probabilistic climate change on sub-arctic palsa mires. *Natural Hazards and Earth System Sciences*, **11**, 2981–2995.

Gedney, N., Cox, P.M. & Huntingford, C. (2004) Climate feedback from wetland methane emissions. *Geophysical Research Letters*, **31** (**L20503**). doi:10.1029/2004GL020919

Haddeland, I., Clark, D.B., Franssen, W. *et al.* (2011) Multimodel estimate of the terrestrial global water balance: setup and first results. *Journal of Hydrometeorology*, **12**, 869–884. doi:10.1175/2011JHM1324.1

Hannaford, J. & Marsh, T. (2008) High-flow and flood trends in a network of undisturbed catchments in the UK. *International Journal of Climatology*, **28**, 1325–1338.

Harding, R.J., Best, M., Blyth, E. *et al.* (2011) Preface to the 'Water and Global Change (WATCH) special collection: Current knowledge of the terrestrial Global Water Cycle'. *Journal of Hydrometeorology*, **12**, 1149–1156. doi:10.1175/JHM-D-11-024.1

Harding, R.J., Blyth, E.M., Tuinenburg, O.A. *et al.* (2013) Land atmosphere feedbacks and their role in the water resources of the Ganges basin. *The Science of the Total Environment*, **468**, S85–S92.

Harding, R.J., Huntingford, C. & Cox, P.M. (2000) Modelling long-term transpiration from grassland in southern England. *Agricultural and Forest Meteorology*, **100**, 309–322.

Huntingford, C., Jones, P.D., Livina, V.N., Lenton, T.M. & Cox, P.M. (2013) No increase in global temperature variability despite changing regional patterns. *Nature*, **500** (**7462**), 327–330. doi:10.1038/nature12310

IPCC (2007) Parry, M.L., Canziani, O.F., Palutikof, J.P., van der Linden, P.J. & Hanson, C.E. (eds), *Climate Change 2007: Impacts, Adaptation and Vulnerability. Contribution of Working Group II to the Fourth Assessment Report of the Intergovernmental Panel on Climate Change*. Cambridge University Press, Cambridge, UK, pp. 7–22.

IPCC (2012) Field, C.B., Barros, V., Stocker, T.F., Qin, D., Dokken, D.J., Ebi, K.L., Mastrandrea, M.D., Mach, K.J., Plattner, G.-K., Allen, S.K., Tignor, M. & Midgley, P.M. (eds), *Managing the Risks of Extreme Events and Disasters to Advance Climate Change Adaptation. A Special Report of Working Groups I and II of the Intergovernmental Panel on Climate Change*. Cambridge University Press, Cambridge, UK, and New York, NY, USA, pp. 582.

IPCC (2013) *Working group 1 Climate Change 2013: the physical science basis.* Contribution to the IPCC fifth assessment report.

Lafon, T., Dadson, S., Buys, G. & Prudhomme, C. (2013) Bias correction of daily precipitation simulated by a regional climate model: a comparison of methods. *International Journal of Climatology*, **33**, 1367–1381.

Min, S.-K., Zhang, X., Zwiers, F.W. & Hegerl, G.C. (2011) Human contribution to moreintense precipitation extremes. *Nature*, **470 (7334)**, 378–381.

Moore, R.J. (1985) The probability-distributed principle and runoff production at point and basin scales. *Hydrological Sciences Journal*, **30**, 273–297.

Oki, T., Nishimura, T. & Dirmeyer, P.A. (1999) Assessment of annual runoff from land surface models using total runoff integrating pathways (TRIP). *Journal of the Meteorological Society of Japan*, **77**, 135–255.

Prudhomme, C. & Davies, H. (2009a) Assessing uncertainties in climate change impact analyses on the river flow regimes in the UK. Part 1: baseline climate. *Climatic Change*, **93**, 177–195.

Prudhomme, C. & Davies, H. (2009b) Assessing uncertainties in climate change impact analyses on the river flow regimes in the UK. Part 2: future climate. *Climatic Change*, **93**, 197–222.

Prudhomme, C., Jacob, D. & Svensson, C. (2003) Uncertainty and climate change impact on the flood regime of small UK catchments. *Journal of Hydrology*, **277**, 1–23.

Prudhomme, C., Wilby, R.L., Crooks, S., Kay, A.L. & Reynard, N.S. (2010) Scenario-neutral approach to climate change impact studies: application to flood risk. *Journal of Hydrology*, **390**, 198–209.

Prudhomme, C. et al. (2011) How well do large-scale models reproduce regional hydrological extremes in Europe? *Journal of Hydrometeorology*, **12 (6)**, 1181–1204.

Prudhomme, C. et al. (2012a) Future Flows Climate: an ensemble of 1-km climate change projections for hydrological application in Great Britain. *ESSD*, **4 (1)**, 143–148.

Prudhomme, C., Young, A., Watts, G. et al. (2012b) The drying up of Britain? A national estimate of changes in seasonal river flows from 11 Regional Climate Model simulations. *Hydrological Processes*, **26**, 1115–1118.

Prudhomme, C. et al. (2013a) Future Flows Hydrology: an ensemble of daily river flow and monthly groundwater levels for use for climate change impact assessment across Great Britain. *ESSD*, **5 (1)**, 101–107.

Prudhomme, C., Crooks, S., Kay, A.L. & Reynard, N. (2013b) Climate change and river flooding: part 1 classifying the sensitivity of British catchments. *Climatic Change*, **119**, 933–948.

Prudhomme, C., Kay, A.L., Crooks, S. & Reynard, N. (2013c) Climate change and river flooding: Part 2 sensitivity characterisation for British catchments and example vulnerability assessments. *Climatic Change*, **119**, 949–964.

Prudhomme, C., Giuntoli, I., Robinson, E.L. et al. (2014) Hydrological droughts in the 21st century, hotspots and uncertainties from a global multimodel ensemble experiment. *Proceedings of the National Academy of Sciences of the United States of America.* doi:10.1073/pnas.1222473110

Prudhomme, C. et al. (2014). Future Flows: a dataset of climate, river flow and groundwater levels for climate change impact studies in Great Britain. In: T. Daniell (Ed.), Hydrology in a Changing World: Environmental and Human Dimensions. *Proceedings of FRIEND-Water 2014, Hanoi, Vietnam. IAHS.* Publ. **363**, 330–335.

Reynard, N.S., Prudhomme, C. & Crooks, S.M. (2001) The flood characteristics of large UK rivers: potential effects of changing climate and land use. *Climatic Change*, **48**, 343–359.

Schewe, J. et al. (2013) Multi-model assessment of water scarcity under climate change. *Proceedings of the National Academy of Sciences of the United States of America* **111**, 3245–3250.

Sheffield, J., Wood, E.F. & Roderick, M.L. (2012) Little change in global drought over the past 60 years. *Nature*, **491**, 435–43.

Shiklomanov, A.I., Lammers, R.B., Rawlins, M.A., Smith, L.C. & Pavelsky, T.M. (2007) Temporal and spatial variations in maximum river discharge from a new Russian data set. *Journal of Geophysical Research, Biogeosciences*, **112**, G04S53.

Shuttleworth, W.J., Gash, J.H.C., Roberts, J.M., Nobre, C.A., Molion, L.C.B. & Ribeiro, M.N.G. (1991) Post deforestation Amazonian Climate: Anglo-Brazilian research to improve prediction. *Journal of Hydrology*, **129**, 71–86.

Solomon, S.I., Beran, M.A. & Hogg, W. (1987) *The Influence of Climate Change and Climatic Variability on the Hydrological Regime and Water Resources.* International Association of Hydrological Sciences, IAHS Press, Wallingford, UK.

Taylor, C.M. (2010) Feedbacks on convection from an African wetland. *Geophysical Research Letters*, **37**, L05406.

Taylor, C.M., Harding, R.J., Thorpe, A.J. & Bessemoulin, P. (1997) A mesoscale simulation of land surface heterogeneity from HAPEX-Sahel. *Journal of Hydrology*, **189**, 1040–1060.

Taylor, C.M. & Lebel, T. (1998) Observational evidence of persistent convective-scale rainfall patterns. *Monthly Weather Review*, **126 (6)**, 1597–1607.

Taylor, C.M., de Jeu, R.A.M., Guichard, F. et al. (2012) Afternoon rain more likely over drier soils. *Nature*, **489**, 423–426.

Warszawski, L. et al. (2013) The Inter-Sectoral Impact Model Intercomparison Project (ISI-MIP): Project framework. *Proceedings of the National Academy of Sciences*, **111**, 3228–3232.

Weedon, G.P., Prudhomme, C., Crooks, S., Ellis, R.J., Folwell, S.S. & Best, M.J. (2015). Evaluating the performance of hydrological models via cross-spectral analysis: case study of the Thames Basin, UK. *Journal of Hydrometeorology*, **16**, 214–231.

Wetterhall, F., Graham, L.P., Andréasson, J., Rosberg, J. & Yang, W. (2011) Using ensemble climate projections to assess probabilistic hydrological change in the Nordic region. *Natural Hazards and Earth System Sciences*, **11** (8), 2295–2306.

Willett, K.M., Jones, P.D., Gillett, N.P. & Thorne, P.W. (2008) Recent Changes in Surface Humidity: Development of the HadCRUH Dataset. *Journal of Climate*, **21**, 5364–5383.

Young, A.R. (2006) Stream flow simulation within UK ungauged catchments using a daily rainfall-runoff model. *Journal of Hydrology*, **320** (**1–2**), 155–172.

Zhang, X., Zwiers, F.W., Hegerl, G.C. *et al.* (2007) Detection of human influence on twentieth-century precipitation trends. *Nature*, **448**, 461–464.

11 Hydrological Data Acquisition and Exploitation

TERRY MARSH[1], ROGER MOORE[2], HARRY DIXON[1],
JAMIE HANNAFORD[1], ALAN GUSTARD[3], ANDY YOUNG[4],
MELINDA LEWIS[2], COLIN NEAL[1], AND GWYN REES[1]

[1]Centre for Ecology and Hydrology, Wallingford, Oxfordshire, UK
[2]British Geological Survey, Wallingford, Oxfordshire, UK
[3]Water Resource Associates, Henley-on-Thames, Oxfordshire, UK
[4]Wallingford HydroSolutions, Wallingford, Oxfordshire, UK

Progress in Modern Hydrology: Past, Present and Future, First Edition. Edited by John C. Rodda and Mark Robinson.
© 2015 John Wiley & Sons, Ltd. Published 2015 by John Wiley & Sons, Ltd.

11.1 Introduction

Over the last 40–50 years, there was a gradual transition from managing water, and later the environment, as a set of largely independent areas of activity to managing those areas as an integrated whole. Achieving the latter has depended on parallel developments in many technologies, with advances in one unlocking opportunities in others, often far removed from the study of the water cycle, but in the end coming back to aid its progress. Technological advances have been paralleled by a remarkable growth in the volume and variety of data relevant to hydrological science. At the beginning of the

1960s, the amount of nationally collated daily river flow data was less than 3% of the current holdings on the National River Flow Archive (NRFA) and almost all hydrological computation and the derivation of catchment characteristics was carried out by hand. Autographic chart recorders and simple mechanical calculators were among the few forms of automation available – this severely limited the techniques open to hydrologists. Although it was possible to conceive of better methods, the assembly of data on the scale required and their processing were beyond the technology of the time. In common with other sciences, many ideas which are now in everyday use were then considered so farfetched as to be classed absurd. Now, with the breadth of applications for hydrological data burgeoning, many data acquisition programmes are of continental or global extent. System design, data management techniques, standards of data stewardship and access capabilities are having to evolve rapidly into a world of interoperable databases and integrated models, necessary to meet the rapidly expanding information needs of the twenty-first century

This chapter examines the growth of the UK's hydrometric monitoring networks and data acquisition capabilities, together with the development of hydrological archives and digital catchment characterisations. It reviews the design of increasingly sophisticated databases and quality assurance mechanisms together with the use of hydrological data both in support of major research programmes and in addressing complex river and water management issues. There is a necessary focus on the need to identify and quantify hydrological trends as the assumptions, current half a century ago, that runoff variability occurred around a sensibly stable long-term average have been replaced by a priority need to index the impact of man – particularly through global warming – on river flow patterns.

11.2   The Need for Water Data

Hydrological data underpin the science of hydrology and provide the foundation upon which today's water management has been built – providing an objective framework within which to balance society's often conflicting demands on the aquatic environment. Users' requirements for data necessarily change over time. Fifty years ago, the focus was on water resource assessments and the development of improved engineering design procedures to counteract flood risk. Such uses remain important today but form only a part of a much broader spectrum of scientific, environmental and social applications including, for example, enhancing the ecological health of rivers and wetlands and assessing the potential of new sources of renewable energy. Hydrological data also underpin the science behind policy evolution and the formulation of regulatory mechanisms.

To help address the increasingly complex challenges that society faces, those responsible for hydrometric data acquisition and management have capitalised on new data sensing and transmission

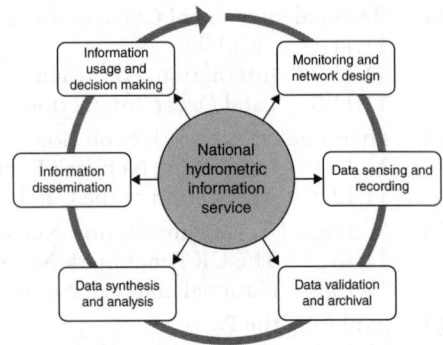

Fig. 11.1 Components of a national hydrometric data management system. (Source: Reproduced with permission of CEH.)

opportunities and rapidly evolving computing, modelling and visualisation capabilities. Such technological advances, together with a continuing dialogue with an expanding user community and the use of rigorous data quality assurance protocols have been the key to ensuring the fitness-for-purpose of modern hydrological datasets (see Fig. 11.1). Expanding user needs and research objectives has driven the need to exploit the synergistic benefits of the integrated management of a range of environmental datasets from individual plot scale to global extent. The latter reflects an increasing appreciation of the degree to which terrestrial, atmospheric and oceanic interactions influence climate and hydrological systems.

11.3   Evolution of the UK Hydrometric Network

The UK is exceptionally diverse in terms of its climate, geology, land use and patterns of water utilisation. In a global context, its rivers are mere streams and most are subject to considerable artificial disturbance (see 11.5.2). Geological and land use contrasts between neighbouring catchments can result in markedly different flow regimes (see Fig. 11.2), and such factors, together with regional and local differences in vulnerability to flood and drought risk, underpin the need for a dense UK gauging station network. However, whilst today it is recognised that for many applications river flow is the most important hydrometeorological variable, for the greater part of the twentieth century the gauging station network was inadequate to meet the needs of water management and to provide real impetus to hydrological research.

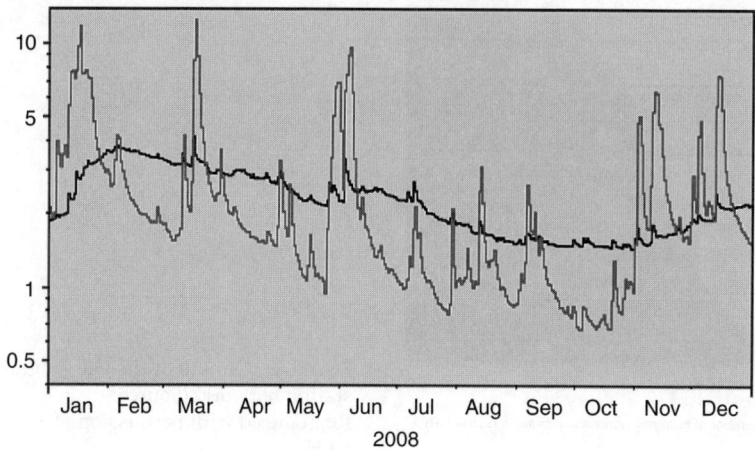

39019 Lambourn at Shaw, Gauged Daily Flow   39081 Ock at Abingdon,
Gauged Daily Flow m³s⁻¹

2008

Fig. 11.2 Contrasting river flow patterns in neighbouring Chalk (black trace) and Clay (grey trace) catchments in central southern England. (Source: Reproduced with permission of CEH.) (Source: Reproduced with permission of CEH.) *(See insert for colour representation of this figure.)*

11.3.1 History

Water levels have been measured since antiquity, but the Enlightenment's focus on an observation-driven understanding of the earth sciences provided a real stimulus to hydrological enquiry and hydrometric measurement. An early scientific concern was the need to better understand the individual components of the water cycle and, from the middle of the seventeenth century, raingauges – of varying design – were increasingly deployed across accessible parts of the UK. Notwithstanding a dearth of upland gauges, John Dalton derived a realistic average rainfall for England and Wales in the 1790s and the raingauge network grew rapidly through the nineteenth century. By the 1880s, there were well over 2000 raingauges in operation across the UK; 30 years later the total reached around 5000 (Lees, 1987).

Somewhat perversely, the globally exceptional density of the UK's raingauge network was a disincentive to the development of a functioning network of river flow gauging stations. Engineers and scientists continued to capitalise on the wealth of available rainfall data, making ad hoc adjustments for evaporation losses in order to assess runoff patterns. Isolated attempts to measure river flows, using floats, were made in the eighteenth century and more routine measurements were instigated in the middle of the nineteenth century, but continuous flow measurement did not begin until 1883 at Teddington Weir on the Thames. In

north-east Scotland just prior to the First World War, Captain W.N. McClean established a pioneering network of gauging stations and developed practical guidelines for hydrometric measurement and data recording procedures (Werritty, 1987). However, in the absence of any statutory requirement to gauge rivers, network growth continued to be very sluggish. By the early 1950s, the number of operational gauging stations remained meagre – less than 100 very unevenly distributed across the country. Real growth awaited the passage of the 1963 Water Resources Act and, to a more moderate degree, the stimulus provided by the International Hydrological Decade (IHD, 1965–74).

Whilst a few pioneering groundwater level monitoring initiatives began in the early nineteenth century (see 11.14.2) sustained growth in the UK's groundwater monitoring network did not occur until the two decades following the Second World War. Over 1000 observation wells were featured in the 1964–1966 Groundwater Year Book, but a substantial proportion of these were monitoring groundwater levels affected by abstractions for water supply purposes. Further subsequent growth through the 1970s, often targeted on aquifer units largely unaffected by artificial influences, provided the foundation for the comprehensive modern network, the data from which constitute the basis on which the Environment Agency (EA) and the privatised water companies now control the development and operation of the major aquifers throughout the UK.

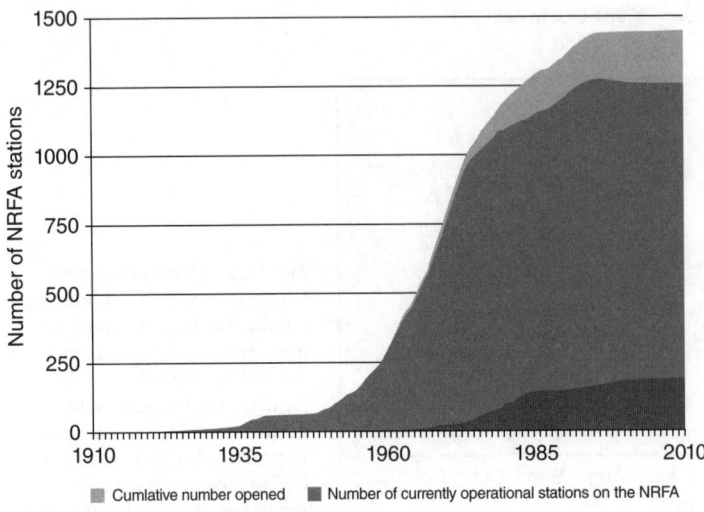

Fig. 11.3 Growth of the UK gauging station network. (Source: Reproduced with permission of CEH.)

11.3.2   The Water Resources Board

A severe drought in 1959 and the dry winters of the early 1960s, together with an expectation that reindustrialisation, increased commercial activity and population growth following World War II would produce a doubling in water demand by the end of the century, were the major stimuli for the 1963 Water Resources Act. This created regional River Authorities and established the Water Resources Board (WRB), the latter with a remit to develop a national plan for water resources in England and Wales. Recognising the urgent need for better a understanding of runoff patterns, the WRB authorised grant aid to fund a rapid expansion in the national monitoring capability. Raingauges and evaporation pans were included in the scheme, but the main focus was on expanding the network of flow measurement stations. In a parallel initiative, the WRB also sponsored, or undertook, research into new gauging techniques and the design of purpose-built weirs, most of the latter being undertaken at the Hydraulics Research Station in Wallingford (Herschy, 1978).

The WRB's foresight and the associated financial incentives enabled a very rapid expansion in the UK gauging station network. More than 600 gauging stations were commissioned in the 15 years from 1965 (Fig. 11.3) and, by the early 1980s, Britain[1] had

a genuine national hydrometric monitoring capability and one of the densest gauging station networks in the world. By contrast, in the mid-1960s meagre record lengths remained a problem; there were only 12 gauging stations, patchily distributed across the UK, with flow records of 30 years or more on the NRFA. Importantly also, later analysis would confirm that the relatively depressed runoff patterns in the 1960s and 1970s were unrepresentative of longer term hydrological variability (see Section 11.17.1). As a consequence, the mean runoff for England and Wales was underestimated. This was reflected in the National Plan (Water Resources Board, 1973), which made provision for a rapid expansion in reservoir provision and a phased introduction of a number of major estuary storage schemes. The latter did not materialise, but many large reservoirs were constructed in the next 20 years and overall water resources provision was exceptional by the late 1980s; by contrast, no reservoirs with capacities greater than $10^7 \, \mathrm{m}^3$ were commissioned in the following two decades. A more comprehensive river flow database in the 1960s would have allowed for a more rational water resources development programme.

11.3.3   Research catchments

Hydrology is necessarily an international science and in the 1960s member countries of UNESCO agreed to support a long-term co-operative

[1] In Northern Ireland, the gauging station network remained sparse until well into the 1980s.

programme (the IHD) to further hydrological research (Lees, 1987). In particular, this helped to extend the existing UK networks into small headwater catchments. Research programmes were identified for about 50 catchments and, for most, hydrometric instrumentation was installed and project-based data archiving capabilities established.

The most important and enduring of the research catchments was established by the Institute of Hydrology (IH) at Plynlimon in central Wales (Chapter 2). Initiated in the 1960s, it incorporated bespoke monitoring networks for a wide range of hydrometeorological variables; these were later extended to embrace water quality. The challenging physical and climatic conditions at Plynlimon proved a valuable test bed for a number of novel gauging structures, as well as standard gauging weirs which often operated successfully beyond their laboratory-authenticated range. Among the innovative approaches to flow measurement needed to address the particular challenges encountered in rugged headwater catchments, a variety of rectangular or V-notch weirs, often incorporated in a rectangular weir box, were deployed with water levels registered on a chart recorder tracking the vertical movement of a float in a small stilling well. Subsequently, similar devices were used to monitor the exceptionally rapid response from largely impermeable areas in IH's pioneering urban hydrology programme (Blyth and Kidd, 1977).

A substantial proportion of the gauging stations installed under the IHD banner were later decommissioned but a minority, including much of the Plynlimon network, were incorporated in the UK national network, whereas others continued to support long-term university or research institute programmes. Generally though, the proportion of the UK network in the latter category declined substantially over the next 20 years. In part, this trend reflected the increasing availability of flow data from the NRFA to support a range of research activities.

11.4 Acquisition and Archiving of UK River Flow Data

The first national collation of river flow data was organised by the Inland Water Survey in the mid-1930s and published in the inaugural Surface Water Yearbook (1935/36), but functioning national archives were not fully established until the 1960s. At that time, departmental responsibilities for water management were split between the Ministry of Agriculture (land drainage and flooding) and the Department of Housing and local Government (water resources). Such divisions, not uncommon around the world, presented a number of data acquisition and processing challenges, especially in relation to hydrological extremes. From its inception, the NRFA was established primarily as an archive of daily mean flows[2] to support water resource investigations; flood data were archived separately – see below. Smaller databases of sub-daily levels and flows were maintained for a number of research programmes, including Plynlimon, but the failure to replicate this at the national level was a notable strategic shortcoming. A fuller integration of floods data with the national daily flow database had to wait until the second decade of the twenty-first century (see Section 11.16.1).

11.4.1 Hydrometric Data Processing and Archiving

The NRFA remained a paper-based archive until well into the 1960s. To accelerate the establishment of a comprehensive digital archive, the WRB encouraged the deployment of digital water level recorders – these rapidly superseded their analogue precursors – and established a national capability to process 16-track and 5-track paper tapes on which 15-minute river levels were registered. The commissioning of an ICL 1902 series mainframe computer in the early 1970s facilitated the rapid derivation of stage–discharge relationships, the introduction of rudimentary quality control procedures and the archiving of daily mean flows on magnetic tape. In a parallel but initially sluggish exercise beginning in the late 1960s, the available historical daily flow data were coded-up from original manuscript records, transferred to punched tape, merged with the digitally derived flow series and archived on magnetic tapes; a very labour intensive exercise. Average monthly catchment rainfall was also derived alongside a rudimentary set of metadata, including the National Grid Reference of the station and its catchment area – the precursors

---

[2]Monthly peaks were also archived but not subject to systematic scrutiny.

of the much more comprehensive reference and spatial information now archived.

At Plynlimon, outputs from logger cassettes were typically converted to 7-track tape thence to 9-track magnetic tape for permanent storage (see below). Data management was a very substantial task embracing the processing, validation and archiving of the range of data types needed to support a major research undertaking (Roberts, 1981). In a parallel development, servicing a number of small urban catchments, bespoke loggers recorded gulley flows at 30-second intervals on magnetic tape cassettes. These, together with short-interval records from tipping bucket raingauges, were then transferred to an early mainframe computer (Univac 1108) allowing, for example significant rainfall–runoff events to be identified and archived for further analysis.

Fifty years ago hydrometric data recording and processing involved rather more perspiration than inspiration, but the late 1970s saw an increasing deployment of electronic solid-state loggers to supersede the old mechanical water levels recorders (Willis, 1999) and a rapid increase in computing capabilities. As a consequence, data acquisition and management became increasingly streamlined. At the same time, there was also a steep growth in the commissioning of ultrasonic gauging stations, the instrumentation of which incorporated substantial field-based computing power and allowed flow measurement to be undertaken at sites where traditional hydrometric techniques were impractical, for example where backwater was a common problem (Herschy, 1978).

Throughout this period of technological innovation, the validation of river levels and flows remained very labour intensive, continuing to rely on the visual appraisals of hydrographs – displayed using first plotters, and later VDU screens – by individuals familiar with behaviour of hydrometric time series. This was, as now, the best method of identifying erroneous patterns, but increasingly more automated procedures were developed to help identify unrealistic sequences of river levels prior to their conversion to flow (Herbertson *et al.*, 1971). Subsequently, the arrival of database management systems in the 1980s provided additional validation opportunities as well as more flexible data access, manipulation and retrieval options (Kirby *et al*, 1991).

11.5   Getting to Grips with Extreme Flows

Floods and droughts have major impacts throughout the world and river flow data, together with a range of other time series and spatial data, are essential to both the understanding of extreme events and the development of effective coping strategies. Unfortunately, the absence of sub-daily flow data[3] on the NRFA, together with the limited record lengths of most river flow time series and only a superficial indexing of the impact of artificial influences on flow regimes, meant that both the pioneering research studies in the 1970s, examining respectively flood and drought flows, required major new data acquisition programmes to be established.

11.5.1   *The flood studies*

The floods of March 1947, the most extensive across England and Wales in the twentieth century, and the Lynmouth Disaster of 1952 underlined the UK's continuing vulnerability to extreme rates of runoff. Subsequently, two major flood episodes in 1968, and the associated report by the Committee on Floods in the UK, were the stimulus for the IH Flood Studies (FS) research programme which began in 1971. The main objective was to improve engineering design procedures.

A primary component of the Flood Studies work programme was the extraction of independent peaks-over-threshold (about 4–6 per water-year) and water-year maxima from autographic river level charts, normally using a digitising table (Fig. 11.4). As part of this exercise half a million weekly river level charts were microfilmed. This provided a comprehensive and unique record of river level behaviour across the UK which now constitutes an important component of the NRFA at Wallingford. Bespoke flood ratings, generally derived using the highest available gaugings, were used to compute annual maxima and peaks-over-threshold for over 1100 gauging stations.

A pioneering aspect of the initial phase of the Flood Studies was the compilation of details of the hydrometric performance of individual gauging stations in the high flow range. Field visits were made to all gauging stations and a standard pro forma was used to record details of the station

---

[3] Apart from 15-minute monthly peaks.

Fig. 11.4 Digitising a river level hydrograph. (Photo NRFA.) (Source: Reproduced with permission of CEH.)

and the measuring section. Relevant upstream influences (e.g. reservoirs) and any known bypassing were also noted. This information, together with the degree of gauging scatter in the high flow range, was captured in a grading scheme which indicated the expected performance of the gauging station around the mean annual flood range. This documentary evidence, including photographic material, was to prove of great hydrological benefit over the following 40 years.

To improve the estimation of flood magnitudes at ungauged sites – a primary objective of the research programme – a range of catchment characteristics were derived to facilitate the development of regionalisation techniques. In the 1970s, this was a very time-consuming task. Many index characteristics were derived manually from Ordnance Survey 1:25,000 maps; the river channel was conventionally assumed to be the blue line on the map and stream frequency was assessed by counting channel junctions and dividing by the basin area – which itself would have been derived either using a planimeter or by counting squares. The lack of digital characterisations of spatial information (e.g. land use) was a significant constraining factor in the Flood Studies programme. Correspondingly, the derivation of more penetrating catchment

descriptors became a focus for considerable future research effort.

11.5.2 The low flow study

In 1974, responding to a need for national procedures for estimating the frequency of low flows at both gauged and ungauged sites (Chapters 1 and 6) the Department of the Environment commissioned IH to undertake a major research programme. It consisted essentially of three main components:
1 Assembling a database of river flows and deriving a number of low flow indices for each gauging station record.
2 Deriving the characteristics of upstream catchment areas.
3 Analysing the relationship between low flows and upstream catchments and presenting the methodology in a format which could be readily used by the UK Water Industry.

At this time, computing facilities for large-scale research were in their infancy. However, the study benefited from a computer-based master list of gauging station information and catchment characteristics developed for the Flood Studies programme, and from the increasing volume of daily flow data held on the NRFA database.

Relationships were sought between the low-flow regime of a river and the upstream catchment to develop design procedures which could be used where there was no observed river flow data. It was therefore essential that the catchments used in the study were relatively natural and had accurate measurements of low flows. To this end, a detailed assessment of the errors of flow measurement was undertaken together with an examination of the extent of upstream artificial influences including reservoir impoundment, groundwater pumping, mine drainage, sewage effluent, river abstractions and the impact of mills, locks and canals. It was important that the net effect of these transfers was established together with an assessment of the redistribution of water in time. The required information existed in hard copy files of a very disparate nature or was established from meetings with experienced water industry staff located in over 40 offices – established (in England and Wales) by the River Boards Act in 1948 and reorganised in the Water Resource Act 1963. About one-third of the overall research effort went into this task. In addition to grading stations for use in the study, it gave the authors a unique understanding of the myriad ways in which UK river flows have been influenced by man. It was essential to look not only at current conditions but also changes over the period for which gauged flows were available.

Over 1400 flow records were considered for the study with a minimum of one year of continuous daily flow data. Of these only 517 were suitable for daily analysis. The remainder were rejected because daily flows had not been derived, the record was shorter than one year, low flows were poorly recorded or there was excessive human impact. Importantly also, only 22 of the 517 had records longer than 20 years. Most of the required flow data were transferred from the Water Data Unit – the means of data transfer was rudimentary: magnetic tapes of daily flows being carried in an Mk 1 Ford Cortina from Reading to Wallingford. A substantial number of record gaps were in-filled using data provided by the River Authorities and River Purification Boards and, in total, over two million daily flow values were loaded on to the IH's UNIVAC 1108 computer.

As was the case with the Flood Studies, many catchment boundaries needed to be drawn manually on 1:50,000 Ordnance Survey maps and a number of catchment characteristics were then manually calculated. These included the average annual rainfall, length of the main stream, stream frequency, slope and the proportion of the catchment covered by a lake or urban area. This was carried out for over 700 catchments and represented another one-third of the research effort. The data were all entered on to a database together with low flow indices analysed in a statistical package. Although the Low Flow Study provided useful design procedures for the UK Water Industry, it was constrained by relatively short flow records, often the result of long delays in processing and archiving flow data. A second significant constraint, as with the Flood Studies, was the absence, at the time, of digitised databases of UK thematic data and the tools to analyse them.

11.6   Towards Integrated Data Management: The Water Data Unit

Whilst the 1963 Act was an important milestone in relation to water resources management and the establishment of the national hydrometric network, the passing of the 1973 Water Act (and the Control, of Pollution Act the following year) was an important step towards integrated water management. The 1973 Act, which established multifunctional Regional Water Authorities (RWAs) across England and Wales, helped to drive forward the development of more coordinated and flexible data management capabilities. The WRB was disbanded and the work of collecting and publishing surface and groundwater water information at the national scale passed to the newly created Water Data Unit (WDU) of the Department of the Environment (DoE). The WDU was instrumental in implementing a more holistic approach to data management and pioneering many initiatives which, in future years, helped deliver the synergistic benefits offered by the integrated management of hydrological and other environmental datasets.

11.6.1   *The Joint Exploitation of Hydrometric and Water Quality Data*

Throughout the 1970s, the routine processing of river levels into flows was devolved to the RWAs and, in Scotland, the River Purification Boards. These organisations continued to submit daily flow data to the WDU which assumed stewardship responsibility for an increasingly wide range of water-related data. In addition to the national surface water and national groundwater level (NGLA) archives, the WDU hosted the Harmonised Monitoring Scheme (HMS), which established a water quality database for Great Britain and provided the information needed to meet a number of national and international obligations (Simpson 1978). The HMS sampling network included 230 sites, mainly located at the tidal limits of major rivers or at the points of confluence of significant tributaries. Recognising the value of the co-location of water quantity and quality monitoring activities, over 80% of the HMS sites were paired with existing flow measurement stations; for most of the others, flows could normally be derived by straightforward modelling based on a suitable upstream gauging station.

The coordinated management of the UK river water quantity and quality archives allowed the associations between runoff rates and contaminant concentrations to be explored at much broader scales than was hitherto possible. It also facilitated the assessment of contaminant loads in rivers and their aggregation to estimate overall loads entering the surrounding seas (Littlewood *et al*, 1998). Through the 1970s, and particularly after the 1976 drought, concern regarding the increasing nitrate concentrations in water supplies stimulated an examination of the impact of increasing fertiliser applications on river water quality across the UK. A comprehensive analysis of the HMS data, augmented by historical data gathered mostly from water abstraction points, confirmed a significant increase in river-borne nitrate since the 1950s and provided an objective framework for developing national policy. A parallel study, capitalising on a recently assembled national database of nitrate concentrations in rainfall, allowed – for the first time – a realistic estimate of the nitrate balance for England and Wales to be derived (Marsh, 1980).

In the context of river water quality more generally, the WDU – charged with preparing the 1975 River Pollution Survey – extended its utility by mapping river stretches according to both their water quality class and their average flow class. This innovation required each river stretch to be uniquely coded in a logical downstream sequence so that quantity and quality could be linked in the underlying database, an important prerequisite for the development of later and more sophisticated digital mapping capabilities (see 11.8.2). Assembling the data and producing the maps was a huge manual undertaking; the average flow was computed for each of thousands of river reaches. The mean flow assessments were based on average catchment rainfall and estimated evaporation losses with the results compared with and, if necessary, adjusted whenever a gauging station was encountered. Although the new maps were a great step forward, there was no way to easily show the change between surveys. However, in a research exercise during the late 1970s, it was shown that by using the digital river network and exploiting its structure, not only could the same maps be produced almost instantaneously on the newly available graphics screens, but also comparisons between successive surveys could also be delivered in visual form.

11.6.2   The Water Archive: linking data to the river network

The legislative changes in the 1970s served to increase both the volume of data and the range of data types needed to deliver more holistic water management. In turn, this required a similarly holistic approach to data management. However, there were few if any, sufficiently versatile archiving systems available at the time. Further, there was no mechanism in place for drawing regional data together to obtain a national picture. Therefore, in 1974 the WDU invited the water authorities to design and implement a system, the Water Quality Archive (later renamed the national Water Archive System (WAS)).

The key system concepts developed very rapidly. Initially, the system was simply concerned with storing the analytical results from river sampling. Quickly, this was extended to include descriptions of the sampling points and drawing on the concepts used in the US Storet system, which incorporated a mechanism to link all sampling point locations to the river network, and allowing upstream causation and downstream impacts to be similarly linked. After significant experimentation, a revised version (the hydrological reference) was adopted for UK application. It required the storage of the locations of all the confluences and sources of rivers. One of the key innovations was that each location was labelled with its geographic location using the UK National Grid, if required, to an accuracy of 1 m. This had the unforeseen benefit of enabling computer drawn schematic maps of the UK river network and its associated features to be produced for the first time. However, the task of setting up hydrological references proved to be a major manual exercise which included measuring with dividers the length of every section of mapped river. Fortunately, digitising was emerging as feasible technology and gradually a programme was introduced to capture digitally the UK river centreline network at 1:50,000 scale. By ensuring that each stretch of digitised centreline could be linked to those upstream and downstream, it became possible to generate hydrological references automatically. It also became possible to replace the schematic river maps by maps that showed the true path of the river.

At this time, the data observed at, or describing, a feature (e.g. a dam or series of daily flows) were considered to be of two types, time-variant (or time series) data and time-invariant data. All feature description data including feature names, initial details, locational references and feature type specific data were considered to be time invariant. Observations of meteorological, hydrological and chemical variables were treated as time series. Collectively these variables were called '*determinands*'. Although the classification of data as time variant or time invariant was quickly realised to be a misjudgement, it was not until the development of the Water Information System's design in the 1980s and implementation in the early 1990s (see 11.10.1) that the issue could be revisited and addressed. It was the reorganisation of the water industry that really highlighted the issue. This reorganisation led the authorities to rename many sewage treatment works to water reclamation works and to assign them new identifiers; tens of thousands of feature descriptions had to be manually edited. In the process, the old names and IDs were lost to the WAS yet remained in everyday use in other systems. This was a sharp and expensive lesson that taught the designers that all data have the potential to change and that their past values can remain of interest for many years; the moral was that everything should be treated as a time series, especially such apparently fixed items as names, IDs and locations.

At the outset of the Water Archive's development, it was believed that the range of chemicals for which analytical results were to be stored was known and fixed – somewhere in the order of twenty. Accordingly, the initial design only made provision for a fixed set of determinands. However, as new feature types were added, it was realised that some provision for future additions had to be made.

After a few false starts, a dictionary approach was adopted. This was indeed fortunate as the number of chemical determinands rose from twenty to two thousand. However, the dictionaries came to have a further role beyond that of allowing users to extend the range of variables that they could store in the system (see 11.16.3).

The Water Archive System was developed just before the emergence into operational use of CODASYL and relational databases and their attendant query languages such as SQL. Nevertheless, its design was influenced by these then new technologies, particularly the facilities for querying the archive. The aim was to give all users as much freedom as could be devised at the time (1975) to query the database and make selections for subsequent input to reports. Its particular innovation was to allow users to search for time series data based on geographic criteria such as 'upstream of', 'downstream of' or within a boundary, where the boundary could be any user-defined geographic area delineated by a string of X, Y UK National Grid coordinates.

The Water Archive System remained in use for 25 years which was remarkable given the rate of change that took place during its lifetime. There were several reasons for its longevity. First and foremost it was simple in concept, easy to use and extremely robust. Very early on, it was appreciated that if the system was to remain relevant, it would have to withstand major administrative changes to the structure of the water industry. In so far, as it was possible nothing to do with the administrative structure of the day was hardwired into the system. The WAS dictionary system made it very flexible. If a new type of feature had to be recorded, the user simply defined the new type in the feature type dictionary and the system could handle it. No system change was required. Similarly, if new variables (determinands) needed to be measured, their descriptions just had to be entered into the determinand dictionary. Again, no system change was required. The last reason for its longevity was that as it grew both in terms of its capability and the volume of data that it held, the cost of replacing it grew even faster.

11.6.3 Publishing and disseminating river flow data

For much of the twentieth century, the primary means of disseminating river flow data was via the

Surface Water Yearbooks (Lees, 1987), but at the start of its custodianship of the NRFA the WDU was confronted with a publication lag of around seven years. This was partially addressed by the introduction of multi-year editions primarily featuring summary flow statistics and gauging station details. In addition, and in response to a growing demand for a broader range of water-related information, the WDU launched an annual 'Water Data' series. Each edition brought together – for the first time in the UK – summary statistics on rainfall, evaporation river flows, groundwater, water quality and water demand. In part, this served to meet an obligation under the 1963 Water Act that data be published upon which assessments of water resources of England and Wales could be made.

Of more enduring significance, and capitalising on the rapidly increasing volume of digitally archived daily flow data, the first UK computer-based national river flow data retrieval service was introduced to meet the needs of a growing user community. Subsequently, the range of available retrieval options increased steadily and, by the end of the 1970s, 15 standard outputs were available with a choice of output to line printer, punch paper tape, magnetic tape (to be provided by the customer) or via a digital plotter. Overall data usage remained modest, but the retrieval service was an important milestone in the progress towards the comprehensive web-based facilities available a quarter of a century later.

11.7 Managing the National Hydrological Archives at Wallingford

With the disbanding of the WDU in 1982, the NRFA and the National Groundwater Level Archive (NGLA) were transferred, respectively, to the Institute of Hydrology and the British Geological Survey, co-located at Wallingford (and both component bodies of the Natural Environment Research Council). The transfer of some of the WDU staff to the IH brought together a critical mass of people working on floods and droughts, archive management, database design, river networks and mapping. This created a very productive dialogue between a wide range of users of hydrometric data and had real benefits for the development of innovative approaches to data management and manipulation. Fortuitously, it also coincided with the emergence of enhanced

computing capabilities – the predecessors of Personal Computers such as the Research Machines 380-Z and the availability of new options for capturing both vector (lines) and raster (gridded) spatial data. The following decade witnessed major advances in hydrological data stewardship and exploitation.

11.7.1 The National Hydrological Monitoring Programme

An initial focus of the coordinated management of the national hydrological databases was the need to address a growing scientific and public requirement for authoritative commentaries on contemporary hydrological conditions across the UK. Prior to the 1980s, monitoring and reporting of major floods and droughts was very spasmodic. Even the extreme drought conditions experienced in 1976 were inadequately documented at the time and scientific appraisals of earlier exceptional events relied primarily on rainfall data. The absence of convincing documentation of extreme events hampered the development of improved water management strategies, national policy evolution and informed public debate. In recognition of this shortcoming, the Department of the Environment requested IH to provide monthly reports on the developing drought conditions through the spring and summer of 1984. Capitalising on the established data gathering arrangements with the regional measuring authorities and the Met Office, monthly summaries of hydrological conditions and water resources status were published until the drought broke in the autumn. These arrangements were resurrected during the very dry winter of 1988/89 and, thereafter, the National Hydrological Monitoring Programme (NHMP) became formally established; its objectives are outlined in Table 11.1.

In addition to the monthly Hydrological Summaries, the NHMP publications portfolio included the Hydrological Data: UK series which superseded the old 'Yearbooks'. The new series incorporated both surface and groundwater data; water quality data was subsequently added for a range of determinands. The amount of hydrological data featured in the new series was markedly greater than in its precursors and commentaries on the main hydrological events in each year were included together with feature articles on a range of hydrometric issues. Beginning with the 1984 drought, comprehensive

Table 11.1 Objectives of the UK National
Hydrological Monitoring Programme

• Provide independent, authoritative documentation of
hydrological conditions across the UK
• Index the extent, severity and impacts of flood and
drought episodes
• Place contemporary hydrological extremes in a
historical context
• Maintain a national capability to identify and interpret
hydrological trends
• Service national and international needs for
hydrological information and advice
• Raise public awareness and understanding of water
issues

reports were also published on major hydrological
episodes, helping to ensure that the NHMP became
established as the primary source of scientific
appraisals of hydrological conditions across the UK.

11.7.2 The Hydrometric Registers

The geographical diversity of the UK, together
with the ubiquity of artificial influences on river
flow patterns (and the behaviour of groundwater
levels), means that hydrological data can rarely
be left to speak for themselves. They need to be
supported by a range of reference and descriptive
material to give analysts the best chance of drawing
valid conclusions from the data. To this end,
the Hydrometric Register and Statistics series
was launched in 1988, complementing the other
publications published under the NHMP banner
(Fig. 11.5). Compiled in close collaboration with
the UK measuring authorities, each edition of the
Register provides a comprehensive catalogue of
gauging station details and hydrological statistics
with particular attention focussed on the hydrometric performance of individual gauging stations.
The range of regime characterisations, summary
statistics and catchment characteristics (see next
section) was substantially extended in the later
editions.

The Registers became indispensible tools for
most users of hydrological data. Although by the
turn of the century, the bulk of featured metadata
was available to download from the NRFA website
(www.ceh.ac.uk/data/nrfa/), many regular users
favoured the retention of the hard copy versions;

they offered a level of convenience that is hard to
duplicate even with the most flexible web navigation tools. Evidence to support this is provided by
the number of the Hydrometric Registers – over
3000 – downloaded in 2012 alone. Importantly also,
the need to publish and update such compendia on
a regular basis imposes a valuable discipline on any
organisation charged with stewardship of national
or regional databases.

11.8   River Networks and Spatial Data

The increasing range of spatial information featured
in the Hydrometric Registers paralleled the remarkable growth in digital mapping and the derivation
of a much wider range of catchment characteristics.
These created exceptional opportunities for the
scientific community; yet, it was only 30 years ago
that a proposal for a digital hydrological terrain
model of the UK at $50 \times 50\,\mathrm{m}$ was ridiculed as
totally infeasible. And if that was absurd, then the
idea of being able to derive and display a watershed
boundary automatically for any point in a fraction
of a second was something out of fantasy land.
This section charts the evolution of ideas for
deriving and exploiting the digital mapping of river
networks. This pivotal capability facilitated the
development of more objective and sophisticated
catchment descriptors, which, in turn, stimulated
more penetrating hydrological analyses and the
development of more effective regionalisation
techniques.

11.8.1   Towards digital mapping

During the 1960s, and early 1970s when computing
facilities were becoming increasingly available to
researchers, the first attempts at computerised mapping were concerned with automating the drawing
of maps; it was about capturing the line work from
existing hand drawn paper maps, correcting for the
stretch and skew in the paper, labelling each line
with the pen width and colour for drawing it and
then sending the information to a pen plotter for
drawing. For many years, the resulting maps were
far inferior to anything that a cartographer could
achieve; problems which today seem inconsequen-
tial were major challenges. Examples were how
to place names on maps automatically without
overwriting; how and when to simplify a motorway

Fig. 11.5 Examples of publications in the National Hydrological Monitoring Programme series. (Source: Reproduced with permission of CEH.) *(See insert for colour representation of this figure.)*

junction when drawing at smaller scales and more generally, how to produce maps by computer that were as artistically pleasing to the eye as those drawn by hand. These and many similar problems were gradually solved in the course of thousands of MSc and PhD projects. During this phase, there was a growing awareness that digitising was the key to unlocking the information that a map contained. If the centreline of a river was recorded as a list of *x*, *y* coordinates, it became a matter of simple geometry to compute its length and other statistics. As other characteristics were captured, for example topography, rainfall, geology, soils and land use, and the coverage became first national and then global, it became possible to recalibrate techniques for flood and low-flow estimation. Now, instead of having to rely on the very sparse sampling of a catchment's characteristics possible by manual means from a printed map, increasingly high-resolution data were becoming available for the whole catchment. The advent of remotely sensed data from satellite and airborne campaigns further extended the information available and was increasingly global in its coverage.

The concept of converting maps to digital form was not new to hydrologists. Flood hydrologists had been rasterising thematic maps of rainfall, geology, land use and other variables by hand since the early 1970s, albeit at very coarse scales by today's standards. However, during the 1980s scanning technologies reduced the cost and increased the resolution and reliability of the data. Remote sensing greatly extended the range of variables that could be monitored and was beginning to make global datasets possible. Although the idea of delineating catchment boundaries automatically from contours had existed since the 1970s, the potential benefits were not realised until very substantial effort had been devoted to overcoming the deficiencies in the available contour and river mapping capabilities, and new techniques for fitting surfaces to contours had been developed.

The first major hydrological application of digital mapping was the capture and exploitation of the UK river network; an exercise begun by the Water Data Unit in 1975 and subsequently completed by the Institute of Hydrology. At the outset, the work was driven by the need to automate the process of creating hydrological references. At the time, the work was highly innovative as the aim was not just to capture the course of the rivers, but to create a fully structured network reflecting the direction and connectivity of the entire river and canal system. Digitising was in its infancy and much of the early work was concerned with achieving a stable base map of the rivers from which to digitise. In conjunction with the OS, blue line maps were prepared. However, contrary to expectation, these did not show continuous blue lines flowing from source to sea, rather they showed a mass of small blue fragments. The rivers were omitted wherever the river flowed under a road, in a culvert, or needed to be omitted to clarify other data. Counter intuitively, it is not all obvious, given a map of river centreline fragments, how they should be joined together. Adding the missing data and flow directions was a huge manual undertaking. In addition, there was no software at the time to check the integrity of the digital data returned by contractors and a whole new suite of programs had to be written to check the connectivity, flow directions and positional accuracy of the data.

However, once the data set was created, it opened up opportunities for computing new catchment characteristics and for developing upstream and downstream operators for database search engines.

11.8.2 *The IH digital terrain model*

The IH digital terrain model (Morris and Flavin, 1996) was a truly pioneering innovation, for the first time giving hydrologists access to detailed height information across the entire surface of a catchment in a form that was easily analysed. However, substantial development work was required to produce a hydrologically realistic DTM.

The early prototype DTMs suffered from many problems. At 50 m resolution, the general shape of the land surface is captured, but important features that can determine the direction of surface flow, such as road embankments and canals, may be missed. Whilst the paper maps, from which the first DTMs were derived, were entirely adequate for their intended purpose, the process by which DTMs were created quickly revealed a range of detailed inconsistencies. The contours at sheet edges were not exactly matched and rivers occasionally crossed the same contour more than once. Most DTMs provide heights at the nodal points of a square grid. If each point is connected to the lowest of its 8 nearest neighbours and so on downhill, you might reasonably expect to reach the sea. However, if the adjacent points in the grid on the downhill side do not pick up the bottom of the valley but by chance lie on either side, the appearance of a sink or pit is created from which there is no way out. These are but a few examples of the hundreds of issues that had to be resolved in the development of a DTM fit for hydrological purposes and from which useful catchment properties could be derived.

The first objective in creating a hydrological DTM was the automation of catchment boundary delineation (Morris and Heerdegen, 1988). Starting around 1983/1984, it took about ten years to develop a DTM and an algorithm which could consistently outperform a person in the computation of a river catchment boundary whatever the terrain in which it lay. But once the ability of the IHDTM to identify drainage directions had been refined, it allowed the automatic computation of catchment boundaries. These, in turn, facilitated the derivation of catchment values from an increasingly wide range of thematic datasets. Of the latter, two of the most important in a hydrological context are those relating to soil type and land cover. The Hydrology of Soil Types (HOST) utilises a 1-km grid classification of the soils of the UK, assigning soils to 29 classes on the basis of their physical properties and their effects on the storage and transmission of soil water both vertically and horizontally (Boorman *et al*, 1995). HOST is used to estimate the Baseflow Index (BFI) and standard percentage runoff, from which low flow and flood statistics can be generated for ungauged catchments. The Land Cover Map of Great Britain 1990 (LCMGB) was the first complete map of the land cover of Great Britain since the 1960s; later versions were released in 2000 and 2011 (http://www.ceh.ac.uk/landcovermapping.html). LCMGB was produced using satellite imagery to produce a 25-m grid map recording 25 land cover types (e.g. developed and arable land) and 18 types of semi-natural vegetation.

The major advances in digital mapping capabilities, together with the development of techniques and software to exploit them (see Fig. 11.6), found very wide application. Amongst the most important was the automated production of the first indicative flood risk maps of England and Wales which defined flood risk to a consistent standard. It showed those areas that would be inundated by floods of the 100-year return period level from non-tidal rivers, in the absence of flood defences (Morris and Flavin

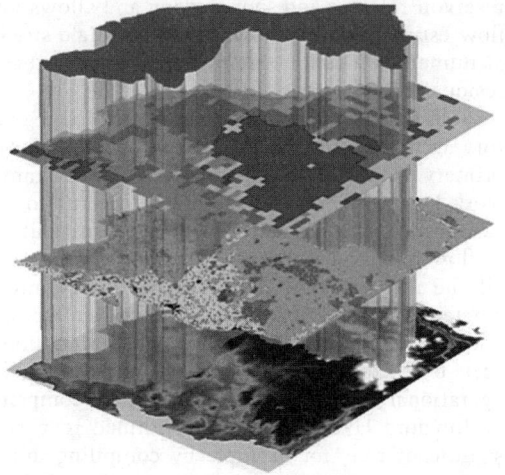

Fig. 11.6 Spatial characterisations for a selected region, from top: target area, soil type, land use and topography. (Source: Reproduced with permission of CEH.)

1996). The regional maps identify the built-up areas that would be at risk and formed the basis of the Environment Agency's 1999 map of indicative flood extent. This very high utility end product, and its later enhancements, testifies to exceptional advances in both environmental mapping and the analytical procedures to capitalise on them.

11.9 Exploiting the New Catchment and Regime Descriptors

The availability of the IHDTM and an associated set of powerful new catchment descriptors, together with a substantial increase in the record lengths of most river flow time series, helped facilitate the updating and extending of the major low flow and high flow research programmes undertaken in the 1970s. These subsequent studies delivered state-of-the-art engineering design procedures to estimate flood and drought flows throughout the river network; these methodologies continue to underpin river and water management across the UK.

11.9.1 Low Flows Studies

Through the 1980s and 1990s information on the magnitude and variability of flow regimes at the river reach scale became a central component of many aspects of water resources and water quality management. Notwithstanding the rapid growth in the gauging station network, directly measured flows remained available for only a small proportion of river reaches across the UK. Correspondingly, the second Low Flow Study (Gustard *et al.*, 1992, see 6.3) refined and extended the procedures developed in the earlier research programme to provide improved estimation procedures for a range of low flow measures at ungauged locations throughout the country. For the first time, data for Northern Ireland were incorporated and, overall, there was an increase of 60% in the gauging station records used relative to the earlier study. Importantly also, the analyses benefited from the exceptional low flows recorded in the droughts of 1990, 1984 and 1976; the latter event redefined low flow regimes across much of the country – minimum gauged flows fell below any previously recorded at most gauging stations, and the majority of these minima have yet to be eclipsed (Rodda and Marsh, 2011).

A more quantitative and digital indexing of the net impact of abstractions and returns on low flow patterns (see Section 11.9.2) enabled the first development of generic methods for estimating the impacts of water use and return on the estimated values of natural low flows (Bullock *et al.*, 1994); the methodology was made available to the UK water industry through the Micro Low Flows package in the early 1990s (see Chapter 6.3.2). Later in the decade, the gaps in hydrological understanding identified by the operational use of these tools led to a root and branch revision of the underpinning low flow estimation science (see 6.4.2). These methods, together with a revision of the work of Bullock *et al*, were deployed through a new software system – Low Flows 2000 – which was developed as a decision-support tool to estimate river flows at ungauged sites throughout the country and to aid the development of water resources at the regional and catchment scales (Young *et al.*, 2003). This software system has been used extensively to aid the implementation of the EA's Catchment Abstraction Management Strategy and contribute to the implementation of the Water Framework Directive. The software has become the standard software system used by the EA and SEPA for estimating river flows, as represented by annual and monthly flow duration statistics, for any river reach in the UK. Many of these innovative developments in data acquisition and exploitation developed at Wallingford were subsequently transferred to new research areas (e.g. hydropower estimation, resource mapping and indexing the impact of artificial influences on low flows).

11.9.2 *Quantifying artificial impacts*

A key problem in estimating low flows in the UK is the impact of a wide range of artificial disturbances. These include surface and groundwater abstractions, effluent discharges and the operating characteristics of impounding reservoirs. Although qualitative guides to the degree of flow regime disturbance have been available since the late 1960s, the first wholesale exercise in the collation of UK-wide information on artificial influences was the 1987 DoE funded Compensation Flow Study undertaken by IH which considered the compensation regimes and other characteristics of all impounding reservoirs with a capacity of greater than 400 Ml.

The need to quantify the influence of water use was initially driven from a research perspective and the desire to improve the estimation of river flows within natural catchments. Over the period 1989–1992, a database of the details for over 43,000 abstraction licences and 96,000 discharge consents was compiled for Great Britain (Gustard *et al.*, 1992). This was no mean feat as this entailed harmonising datasets from nine different regions within England and Wales, all with different data structures, nomenclatures and levels of detail. At that time, there was no abstraction licensing within Scotland and thus the work required the compilation of information from many disparate data sources. An analysis of the available material relating to the artificial influences concluded that less than 20% of the primary gauging station network could be considered to be gauging sensibly natural catchments.

In the early 1990s, as part of a comprehensive study commissioned by the National Rivers Authority (Bullock *et al.*, 1994) an approach was developed to combine monthly natural flow statistics with the monthly variation in artificial influences. The resulting flows could then be accumulated at all points above any specified location. This methodology was implemented in an enhanced version of the Micro Low Flows software system (see Chapter 5) which incorporates details of abstraction licences, discharge consents, reservoir releases and spot gauging and allows low flow estimates to be displayed for a single site or at numerous locations along a river to construct a residual flow diagram.

Subsequently, the formation of national regulatory agencies across the UK and the enactment of primary legislation to implement the Water Framework Directive provided an added impetus to the search for mechanisms to quantify the magnitude of flow regime disturbance. The harmonisation of all the regional and basin level databases into a single national database was a major step forward and, in the early 2000s, the focus on national scale characterisation of water use moved into the operational arena. CEH and a spinout company, Wallingford HydroSolutions, continued to play a significant role, for example by compiling initial characterisations of water use across the whole of England and Wales (on behalf of the EA). A later refinement, focusing on the basin level, yielded a detailed database characterising the impacts of over 40,000 abstraction licences, 10,000 significant discharges and 200 impounding reservoirs upon river flows. In a full circle, this national database has proved invaluable in further refining the understanding of the influence of water use on gauged records thereby enabling research into low flow methods to be further enhanced.

11.9.3 The Flood Estimation Handbook

As with the Low Flow Studies II programme, the IHDTM opened new opportunities to derive a powerful range of innovative catchments descriptors to support floods research, and regionalisation procedures in particular. The Flood Estimation Handbook project (Reed, 1999), which ran from 1994 to 1999, benefited from these novel catchment characterisations and substantial updates to the national holdings of flood data since the publication of the FSR in 1975. In relation to the updating exercise, the extraction of annual maxima and peaks-over-threshold was eased by the availability of digital datasets of river level, thus obviating the time-consuming manual extraction of level data from charts. The original IH flood data holdings were supplemented by post-1973 peaks-over-threshold data for Scottish catchments sourced from researchers at St Andrew's University, and for Northern Ireland where very few flood records existed in the 1970s. Overall, the updating exercise incorporated peak flows for 628 gauging stations, providing an FEH dataset comprising approximately 1000 series of annual maxima and 88,000 flood peaks. The average record length was almost 20 years, and for seven sites the record length exceeded 50 years (Bayliss and Jones, 1993), a substantial improvement on the situation two decades earlier, but still a constraining factor with regard to capturing long-term variations in flood magnitude and frequency.

The digitally derived catchment descriptors utilised in the FEH programme included catchment area, mean drainage path length, the extent of urban/suburban land cover and individual soil class fractions. The Flood Attenuation by Reservoirs and Lakes (FARL) index, developed specifically for the FEH, provided a valuable guide to the degree of flood attenuation attributable to reservoirs and lakes in the catchment above a gauging station. All of the data and catchment descriptors used in the research programme were made available via a CD-ROM which provides a comprehensive suite of descriptors for any river catchment in the UK

with an area $> 0.5 \, \text{km}^2$, together with rainfall Depth Duration Frequency Data, essential for assessments of site-specific surface water runoff and storage (see Chapter 2).

The Flood Estimation Handbook became the authoritative 'standard of practice' in the UK for the design and planning of schemes affected by, or designed to mitigate, flooding. It also bequeathed a rich repository of time series data and catchment characterisations to support future research initiatives extending well beyond those relating to flood issues. In a hydrometric context, the catchment descriptors, like those developed in support of the Low Flow studies, were especially useful in the selection of catchments for particular investigations and in enabling more penetrating reviews of the gauging station network (see 11.12.1).

11.10   A Holistic Approach to Managing Water Information

In addition to the enhancements to engineering design procedures through the 1990s, the decade also witnessed major advances in the design of systems to store and manipulate water data which by then at Wallingford, embraced not only hydrometeorological time series, but also those arising from freshwater and marine water quality monitoring and later biological monitoring.

11.10.1 The Water Information System

Although the Water Archive system's basic design concepts proved extremely durable, it soon became possible to see ways in which they could be improved by simplification, removal of inconsistencies and introducing the presumption that all data have the potential to change over time. Through the 1980s, the new ideas were refined and developed and the first manifestation of a new system was HyGIS which aimed at assembling all the time series and geographic data necessary for flood and low flow estimation within a single system, including not only a database but also all the analysis software. A major IT firm, International Computers Ltd (ICL) (now Fujitsu), saw this system and also saw its potential to create a market for ICL hardware. The Water Authorities had been superseded by the National Rivers Authority which would need to replace its inherited hardware and software systems. Accordingly, around 1989, ICL offered to invest in the development of HyGIS and transform it into a fully operational commercial product under terms that were very generous to IH. The working title for the joint project was the Water Information System (WIS). It was a remarkable and highly fruitful collaboration in which IH provided the ideas and ICL undertook the project management, funded all the software development and marketing.

WIS, like its predecessor WAS, allowed users to record the history of any object, or feature, as it moved through space and time. Descriptions of features and the events observed at them were recorded in terms of attributes (in this context, synonymous with the terms variable, parameter and determinand). The users had complete freedom to define the types of feature of interest to them, and the nature of the attribute data held about or observed at each one.

One of the most significant innovations of WIS was that users could hold most data which were of interest to them, spatial and non-spatial, in a single unified data model. Essentially everything, whether it was a discharge measurement, an animal's weight, the description of a site, the centre line of a river segment or a grid of heights, was treated as a potential time series. Thinking of all variables in the same way enabled data to be linked and correlated very easily within and across features of the same or different types.

At the user level, the world as perceived by WIS was composed of **features**. These were any objects whose description in space and time the user wished to record. It was the user who decided and defined the types of feature for which data were to be held. WIS had no preconceptions about the nature of the particular data which a specific user might wish to hold. The user created new feature types (e.g. boreholes, licences and sites of scientific interest) by defining them in a feature dictionary. The descriptions of features and the events observed at them were recorded in terms of **attributes**. A wide range of spatial and non-spatial data types were supported, allowing WIS to record most types of attribute information likely to be of interest to users, for example, feature identifiers (names, reference codes, serial numbers, etc.), locational information (3-D points, lines and polygons), hydrometeorological variables, chemical and biological sample data, ecological and biological observations and many others.

WIS regarded all values as potentially changeable over time; thus enabling it to handle time series data such as river flow. Because spatial attributes were also treated as time series, the 4-D location of mobile sampling platforms such as ships and aeroplanes could be recorded alongside the observed data. The time series approach also dealt with the reality that key feature identifiers such as IDs and names change in response to administrative change or fashion, for example, the renaming of sewage treatment works to water reclamation works. Hence, old identifiers and names remained available for searching the database.

Users created new attributes by defining them in an attribute dictionary. At the time, the removal of the distinction between graphic (spatial) and non-graphic attributes was a considerable innovation, as was storing both within the same logical framework.

A useful way to visualise how data were stored in WIS was to imagine a large cube, made up of individual cells (Tindall and Moore, 1997) – see Fig. 11.7. The three axes of the cube represented features (WHERE observations were made), attributes (which recorded WHAT the observations were) and occasions (WHEN the observations were made). Thus, each cell in the cube recorded the value of an attribute at a particular feature for a particular point in time. For example, one cell might have recorded the 'concentration of calcium' (WHAT) on 29th June 1995 at 10:15 (GMT) (WHEN) at a sampling point in the river Swale at Catterick (WHERE). The rows, columns and faces of the cube provided different views of the database and an easy way of referencing data for selection, analysis and or display.

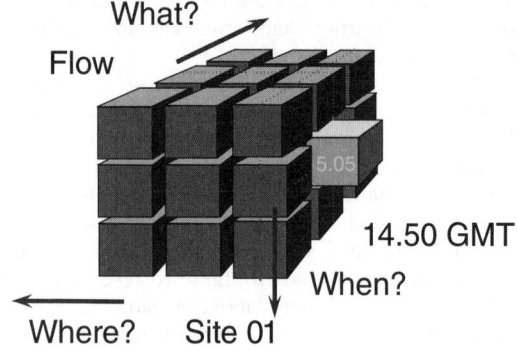

Fig. 11.7 The Water Information System cube. (Source: Reproduced with permission of CEH.)

WIS was a completely generic system and had no inbuilt knowledge of particular data types such as those required for hydrology or biology. Therefore, the first task in using WIS was to describe to the system the nature of the data to be stored and how they should be mapped onto the generic WIS model. This mapping was held in a series of dictionaries whose contents were collectively referred to as *reference data* to distinguish them from *user data*. Of particular interest were the dictionaries of attribute definitions. Originally introduced to make the system flexible and generic, they turned out to be essential user reference documents. They had two roles. The first was to enable the system to recognise and handle the data for each attribute; the second was to hold metadata defining as clearly as possible 'what' the values of the attribute recorded and therefore for what present and future uses the data might be appropriate. In addition to defining the attribute, there was also provision for describing the methods by which the values of the attribute could be determined.

The WIS data model was in operational use for twenty years and supported the LOIS database described later in this chapter. Its great strength, but also its weakness was its generic structure. This allowed it to handle an immense range of data, but the vast majority of users found generic systems cumbersome and extremely difficult to cope with. It is entirely possible that were WIS to have been packaged into a number of different boxes labelled Hydrological Information System, Biological Information System, Meteorological Information System and so on, the problem would have disappeared. The system appeared to have quite a large setting-up cost. If the users had been offered a series of pre-setup dictionaries and datasets such as the river network, it is again possible that this problem could have been reduced to an acceptable level. WIS had a remarkable Graphical User Interface (GUI) for its time but shortly after its release, PC interfaces caught up and the resources were not available for WIS to maintain a comparable level of ease of use.

11.11 Regional and Global Collaborations

A recognition that advances in the understanding of hydrological science would benefit from studies embracing an ever broader geographical canvas has driven a trend towards regional, national and international multidisciplinary research collaborations over the last 40 years. Such collaborative programmes have been facilitated by (a) advances in materials science, digital communications and IT innovations and the associated development of modern data management systems and (b) the manpower resources for the long hard slog of capturing, storing and validating time series data, transforming mapped information into digital form and developing effective protocols to support and maintain datasets as they are brought together to create regional, national or global databases.

11.11.1   FRIEND

As a component in a number of major international collaborations, and building on its coordination of the European Flood Study in the 1980s, IH led the establishment of FRIEND (Flow Regimes from International Experimental and Network Data), a project aimed at developing better understanding of hydrological variability and similarity across time and space, through the exchange of data, knowledge, models and techniques at the regional level (see Chapter 6.5).

FRIEND was launched as a major contributor to the UNESCO International Hydrological Programme, establishing seven regional networks across the world and involving over 100 countries by the beginning of the twenty-first century. At that time, the database of the European FRIEND project comprised river flow data and associated metadata (gauging station details and catchment characteristics) for some 5000 gauging stations in 25 countries. Similar databases were established in other regions including the Hindu Kush/Himalaya, Southern Africa and the Asian Pacific. In 2004, responsibility for the European datasets were transferred to the WMO's Global Runoff Data Centre (GRDC) in Koblenz (see Fig. 11.8). The advances in knowledge of hydrological processes and flow regimes gained through FRIEND made substantial improvements to the methods available for planning and managing water resources.

11.11.2   *International outreach*

Wallingford's expertise in managing both project-based and national hydrological databases was increasingly exploited to support data management and rescue projects around the world. These involved working with state authorities in Russia

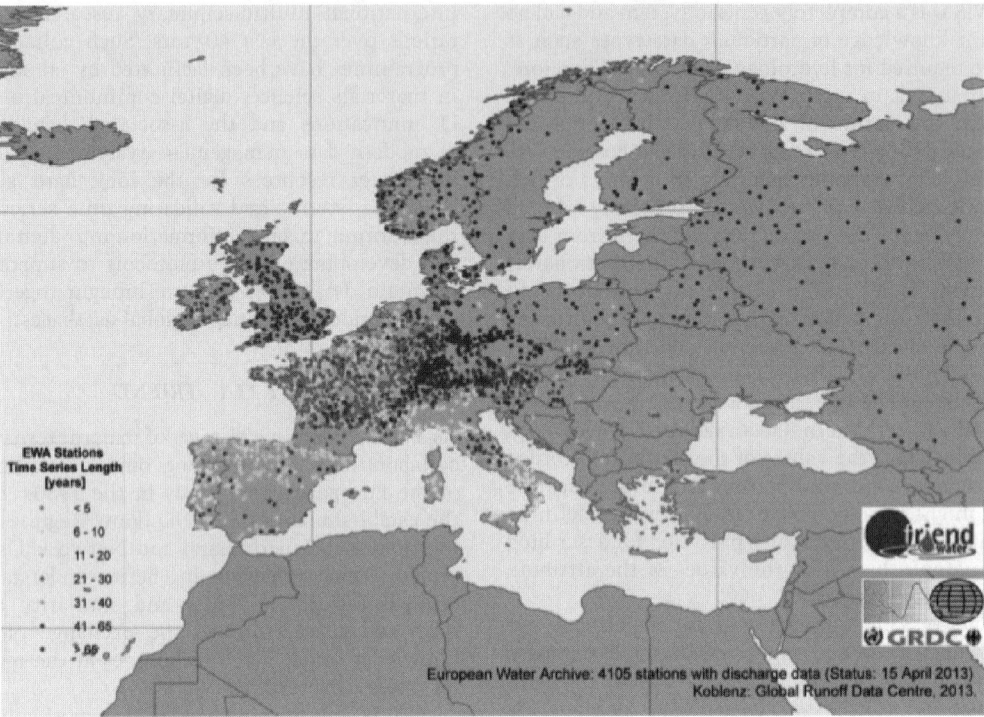

Fig. 11.8 Distribution of gauging stations for which flow records are held in the European Water Archive. (Source: Reproduced by permission of Global Data Runoff Centre.) *(See insert for colour representation of this figure.)*

(and states in the former Soviet Union), Belarus and the Ukraine, Africa, India, Malaysia and in the Lebanon. In addition, capacity building and knowledge transfer was reinforced through the delivery of training courses (e.g. in Bangladesh, Bhutan, India and Nepal) and in contributions to a number of World Meteorological Organisation publications. In addition, adaptations of UK standard hydrological design tools have found wide international application, for example, Micro Low Flows. The latter was a forerunner of the hydropower estimation software, HYDRA, which, in addition to widespread application in the UK, has been adapted and applied in several European countries (Austria, Belgium, Ireland, Italy, Portugal and Spain) and in India and Nepal.

The requirement for robust, powerful and affordable data processing and archiving capabilities is particularly pressing in the developing world. To address this need a purpose-built PC-based archiving and analysis system – HYDATA – was

developed at IH in the late 1980s and was one of the first designed specifically for use by staff with little experience of computers. Crucially, training support was provided as an integral component of the package. HYDATA exploited the same database system used by the NRFA and was designed to provide the hydrometeorological information essential for planning and operating water-related schemes worldwide. It was able to store the types of data most commonly required in water resources studies including river levels and flows, reservoir and lake levels, rainfall and other meteorological data. Included as standard were facilities to develop rating curves and convert reservoir and lake levels into storages. Outputs included flow duration curves, summary hydrological statistics and tabulations in 'Yearbook' style; a range of basic data retrieval options was also incorporated. Within ten years, the system was being used in over 50 countries and was the national database system for surface water data in more than 20 developing countries. It also provided the original archival system for the WMO's World Hydrological Cycle Observing System (WHYCOS) in Southern Africa.

11.11.3 Land Ocean Interaction Study and Data Centres

In the context of the integrated management of data deriving from a range of environmental monitoring programmes, the LOIS (Land Ocean Interaction Study) research programme was a ground-breaking initiative. Launched in 1992, it was NERC's largest multidisciplinary research programme and addressed the measuring and modelling of the nutrient and contaminant fluxes from watersheds in eastern England and Scotland to the North Sea, the continental shelf and ultimately the deep ocean (Huntley *et al*, 2001).

The study brought together data collected by a range of atmospheric, terrestrial and marine monitoring programmes and integrated model outputs such that, for example the impact of rainfall patterns could be traced through runoff responses, contaminant impacts to the North Sea and responses in the marine environment (e.g. the creation of algal blooms). The work included a major geological study of the sedimentary record across the coastal zone to determine how fluxes of material have influenced sea level, climate and land use. At the time, the LOIS programme was unique in its scope and scale – making heavy demands on software, system design and data management.

LOIS utilised the WIS platform for data storage; its generic design being well suited to the demands of a programme for which, at the outset, little was known about the types of data that would be encountered. It was appreciated that data from a wide variety of sources and subject areas would be required and that their assembly into an orderly database would be one of the keys to attaining the LOIS programme's scientific objectives. An important aspect of securing the support of data suppliers had been the agreement that there should be a clearly defined contact hub for all the suppliers. Consequently, five specialised thematic Data Centres were created with responsibility for the acquisition, management and distribution of atmospheric, terrestrial, river, marine and geological data (Moore, 1997); IH hosted the centre responsible for hydrological data. The key roles of the Centres were to (a) bring together and harmonise the data collected by the different agencies and (b) organise the different data types so that the interrelationships between them could be explored easily by the scientific community.

The arguments for having Data Centres compared with leaving researchers to gather their own data were chiefly those of the economies of scale and value for money. Individual researchers could never have assembled and quality controlled such large amounts of data as effectively as a well-staffed and well-equipped Data Centre. As many projects required the same or similar data, much wasteful repetition would have occurred.

The first task of the Data Centres was the identification, assembly and distribution of the data required and produced by researchers. The diversity of the data types, the range of suppliers and the volume of the data made this a challenging undertaking. However, it was not simply a mechanical task of moving data from A to B. The second and greater challenge was to understand and reconcile the different ways of looking at data adopted by scientists working in the atmosphere, river basins, estuaries and the marine components of LOIS. A simple example of these differences is to be found in the different approaches to monitoring marine and freshwater quality. At the time (the 1990s), in the freshwater area, all monitoring was performed at fixed sites though the depth of sampling was of relatively minor importance. However, within marine studies, most data were collected on cruises; fixed sites were rarely used and depth was critical. A consequence of these slightly different perspectives of essentially similar monitoring problems was that different data storage strategies evolved for the same data types. As a result it was difficult to formulate database queries that spanned the marine and freshwater datasets. The third challenge related to matching the terminologies of the different components and determining whether variables with the same name recorded the same quantities. This problem was compounded by the immense variety of data types: river water quality and quantity, land use, biology, geology, soil types, atmospheric chemistry, meteorology, marine water quality and sediment data. Recognising these problems, an early conclusion was that, to be successful the Data Centres would have to combine an understanding of both science and computing; computing capability alone would not be sufficient.

The LOIS Data Centre was an important forerunner of the more comprehensive data centres established through the first decade of the twenty-first century (see 11.16).

11.12   Hydrometric Network Evolution: Capitalising on Regionalisation Techniques

As with most environmental monitoring networks, the density and distribution of UK gauging stations needed to adapt to changing information demands. This fuelled the need for more refined network design procedures. Progress was enabled largely by the fine-resolution characterisations of, for example, rainfall patterns, soil types, geology and land use, which became available from the mid-1990s. They provided the capability to index the representativeness of individual catchments far more effectively than hitherto and underpinned the development of more rigorous network design mechanisms.

11.12.1   *Influencing Network Evolution*

In the early twenty-first century, new gauging stations have been commissioned to monitor, for example, small upland catchments to assess hydropower potential or improve regionalisation techniques, particularly in relation to the design of flood alleviation schemes. However, as the costs of managing the networks increased, particularly through the 1980s, so the rate of decommissioning of gauging stations also began to rise (see Fig. 11.3). Mechanisms for identifying redundant gauging stations varied widely, but a problem characterising many network reviews was the greater weight normally given to the operational value of gauged catchments relative to any assessments of their strategic utility. This is understandable given the compelling operational reasons for measuring flows at particular locations and an historical lack of any convincing mechanisms to index strategic value. This could, for example, be in relation to the improved understanding of hydrological processes or the utility of individual time series in relation to the detection of trends in runoff patterns (see Section 11.17.1). Strategic value also reflects the role of individual catchments in the development and application of improved regionalisation techniques. New methodologies, mostly developed through the 1990s, have helped

to index the contribution individual catchments (and the gauging stations at their outfall) make to the overall information delivery from regional or national hydrometric networks.

The Catchment Utility Index (CUI) (Laizé *et al.*, 2008) is a particular decision-support tool, which exploits the regionalisation techniques developed for the FEH; it provides a more objective framework within which to optimise the number and distributions of gauging stations across the UK (see Section 11.12.2). The CUI approach, combined with other measures of catchment utility (e.g. related to the degree of disturbance to the natural flow regime and assessments of the hydrometric performance of individual gauging stations), was exploited in a new methodology for network optimisation which was used to appraise the England and Wales gauging station network as part of a review undertaken for the EA (Hannaford *et al*, 2013). Such capabilities have assumed greater importance as pressure has increased to remove gauging structures from the UK network to allow easier fish movement and help achieve compliance with the Water Framework Directive.

11.12.2   *Application of the Catchment Utility Index: an example*

The River Dulas drains a wet upland catchment in central Wales and has a sensibly natural flow regime. The gauging station at Rhos-y-Pentref is relatively remote, difficult to maintain, and has only a limited operational role in water management. Gravel accretion can impact adversely on low flow accuracy, but the gauging station, a trapezoidal flume, has the rare ability to contain the 50-year flood (Fig. 11.9). This, together with a continuous flow record exceeding 45 years, emphasises its value both to flood hydrology and in the application of engineering design procedures. To index the catchment's strategic value more quantitatively, the CUI was used to assess the frequency with which all gauging stations in the UK were incorporated into FEH pooling groups. This involved identifying the 20 nearest gauging stations (in Area, Rain and BFIHost space) to all potential target catchments across Great Britain.[4] Scores were then allocated according to their proximity and then accumulated across all target catchments (in this case those with catchment areas between 25 and 100 km^2) to give

---

[4] Some four million catchments.

Fig. 11.9 Large capacity flume at Rhos-y-Pentref in central Wales (Photo: NRFA.) (Source: Reproduced with permission of CEH.)

an overall Utility Score for each catchment. The particular combination of size, wetness and soil type of the Dulas catchment ranks it among the top 25 in the UK – and the top three or four in Wales, thus underlining the particular strategic value of the gauging station at its outlet.

11.13   Strategic UK Gauging Station Networks

The availability of comprehensive hydrometric reference material, spatial characterisations and more effective mechanisms to assess the representativeness of individual catchments helped to define two sub-networks to address priority information demands. Firstly, the need for a viable capability to detect trends – a primary objective of most environmental monitoring programmes – and secondly, the ability to meet international data exchange obligations more effectively than hitherto. Both required bespoke gauging station networks be identified.

11.13.1   The UK Benchmark Network

The expectation that global warming is having a discernible impact on UK flow regimes was the main stimulus for the designation of a network of Benchmark Catchments to provide a focus for the identification of hydrological trends. In most parts of the world, candidate catchments for Benchmark status can be selected from many rivers with regimes little affected by man. That is not the case in the UK.

For the generality of flow records in the UK, separating any climate-driven regime change from the pervasive influence of artificial influences remains a substantial challenge. Less than 20% of UK gauging stations can be considered to monitor sensibly natural flow regimes (Marsh and Hannaford, 2008) and even in the Scottish Highlands hydropower development widely impacts on natural flow patterns. A failure to allow for such impacts – which can vary substantially over time – can of itself introduce compelling long-term trends. For example, abstractions to meet London's water supply needs have seen average flows in the lower Thames decline by around 40% over the last 130 years. The qualifying criteria for Benchmark Catchment status therefore included an absence of any substantial net impact of abstractions and discharges on the natural flow regime. In addition, good hydrometric performance, particularly in the high and low flow ranges, was required together with a flow record of at least 30 years. Catchment selection was undertaken in close collaboration with the regional measuring authorities and a compilation of Representative Basins, compiled by IH in the 1980s, was useful when sifting candidate catchments for the UK Benchmark Network. When initially designated in 1998 the network comprised around 130 catchments (Bradford and Marsh, 2003); see Fig. 11.13. The selection procedure necessarily involved compromises and for a significant minority of the selected catchments final endorsement remains contingent on an upgrading of hydrometric performance or a fuller understanding of the net impact of artificial influences.

The UK Benchmark Network has been very widely exploited in national and international monitoring and research programmes to index changes in runoff over time (see Section 11.20). Subsets of the network also form key components in a number of UK water quality and ecological monitoring programmes.

11.13.2   *National and regional outflow series*

Increasing demands for regional or national assessments of UK runoff (e.g. outflows to the surrounding seas to meet Paris Commission commitments and annual water resources assessments for the European Environment Agency), subsequently reinforced by issues associated with political devolution in the UK, created a pressing need for convincing daily runoff series for the UK as a whole, and its constituent parts.

In the late eighteenth century, John Dalton (1766–1844) was the first to attempt an assessment of mean runoff from England and Wales and several subsequent attempts have used selected, long river flow series to assess overall water resources and index changes over time (Marsh and Littlewood, 1978). However, the inadequate gauging station network coverage prior to the 1980s, together with the difficulties associated with assessing runoff from ungauged areas, limited the precision of these broad-scale runoff assessments. Subsequent improvements in modelling capabilities have allowed considerably greater accuracy to be achieved.

For the quantification of hydrological trends in particular, homogeneity in the national and regional runoff time series is of paramount importance. Accordingly, a stable network of 40 index catchments, which are demonstrably representative

of the total gauged outflow from the component countries, was selected based on gauging stations with lengthy flow records and a limited (or quantifiable) net disturbance to the natural flow regime. Outflows for the ungauged areas – around one-third of the contiguous UK – were assessed using Grid-2-Grid, a spatially distributed hydrological model developed at Wallingford and widely used in Britain for both real-time flood forecasting and the continuous simulation of river flows (Bell *et al.*, 2013). The model was generally configured to a 1-km^2 grid across the UK and is underpinned by digital spatial datasets on topography, soil/geology and land cover. By coupling flows from the index catchments with modelled estimates of runoff from the ungauged areas, convincing outflow series were able to be generated at the regional and national scale.

The national and regional outflow series begin in 1961 and have been used routinely since the late 1990s to identify hydrological trends and to monitor broad-scale hydrological conditions across the country. The time series provide confirmation that, for example, outflows from the English Lowlands in the latter stages of the 1976 drought have not been approached on any other occasion in the 55-year series, whilst, at the other end of the flow spectrum, the January and February runoff in 2014 is the highest for any two-month sequence on record.

11.14   Exploring the Past

As evidence increased that climate variability, as well as anthropogenic climate change, exercised a considerable influence on UK runoff patterns across a range of time spans, the need to place contemporary hydrological variability in a broader historical context became an increasing priority. However, with only four extant gauging stations having sensibly continuous flow records extending back to the 1920s, the paucity of historical river flow information remains a clear barrier to both the indexing of multi-decadal runoff variability and the detection of long-term runoff trends in the UK.

11.14.1   *Historical river flow patterns*

Various approaches have been used to assess runoff patterns prior to the modern era. A rainfall–runoff model, developed originally at the Central Water Planning Unit (Wright, 1978), was used to extend monthly flow series for ten rivers in England and Wales back to the nineteenth century, and further in the case of the Thames (Jones *et al*, 1984). Subsequently, the need for more direct historical assessments of runoff triggered a number of initiatives to uncover, or derive, early river flow series. Such datasets are inevitably of a lower accuracy and often less complete than their modern counterparts. In addition, considerable curatorial effort may be required to assemble a convincing provenance. Nonetheless, such series do substantially extend our understanding of historical river flow variability and runoff patterns.

In Scotland, a monthly series based on loch levels, sluice settings and spillway discharge from Loch Leven was derived back to 1855 (Sargent and Ledger, 1992). This provided important quantitative insights into to early droughts (e.g. 1886/87 and 1892/93) as well as more recent, but poorly documented low flow episodes (e.g. in 1955 and 1959). In southern Britain, a monthly count of lockages on the Grand Junction (now the Grand Union) canal allowed outflows from the Wendover Springs, which issue from the scarp slope of the Chilterns, to be assessed back to 1841 (Bayliss *et al*, 2004). The Wendover Springs record provides a unique insight into runoff variability throughout much of the Victorian era, but it terminates in 1897; flow measurement was, however, recommenced in the 1960s. The cause of the cessation of the record is not known with certainty, but evidence from another spring source in Dorset with discontinuous flow data extending back to the 1850s (Limbrick *et al*, 2002) supports the contention that the flows at Wendover failed following a sequence of dry winter half-years in the early 1890s which heralded a further sustained period of depressed groundwater recharge. Any modern repetition of such climatic patterns would imply considerable water resources and environmental stress.

The inherent rarity of extreme flood events and their tendency to exhibit a degree of clustering through the historical record imply that assessments of return periods based exclusively on data assembled during the last 40 or 50 years may be unrepresentative. In relation to historical flooding, estimates of peak magnitudes based on documentary and epigraphic sources (e.g. flood marks, Macdonald and Black, 2010) provide a more extensive temporal context within which to assess

the rarity of modern extreme events. However, in the absence of contemporary high flow gaugings, stage–discharge relations may only be approximations and, in relation to changes in flood risk, particular care needs to be exercised where, as in the Thames, channel conveyance has changed substantially over time. Valuable contextual material to supplement historical flow series may be accessed in a number of carefully researched catalogues giving details of drought and flood events (e.g. Potter, 1978). The annual British Rainfall series, extending back to 1860, provides important insights into notable hydrological events, but the most comprehensive compendium of historical flood and drought episodes is the British Hydrological Society's Chronology of Hydrological Events (Black and Law, 2004). This was launched as a website in 1998 (http://www.trp.dundee.ac.uk/cbhe/welcome.htm) and is a rich source of documentary evidence of the impact of exceptional weather patterns.

11.14.2 Historical groundwater level data

The first systematic record of fluctuations in groundwater levels in Britain may be considered to be a series of monthly observations made in a well near Sittingbourne in Kent by William Bland from January 1819 to June 1831 (Mather, 2004). Subsequently, the formation of the Geological Survey in 1835, and the rapid growth of groundwater as a source of public supply through the nineteenth century, stimulated a steady increase in the routine monitoring of groundwater levels. Of particular note was the systematic recording of groundwater levels in the Chalk of southern England during the 1840s by the Reverend James Clutterbuck. He was the first Briton to apply observations of groundwater levels in a practical and innovative way to the study of groundwater levels and flow. He recognised the intimate relationship between surface water and groundwater and also, through a very sustained monitoring programme, identified the major decline of groundwater levels below London caused by the increase in abstractions.

Notwithstanding, the growth of groundwater monitoring through the nineteenth century, there remain only six extant observation wells and boreholes in the NGLA (which focuses on natural groundwater level variations) having records extending back to the nineteenth century, and some of these series are incomplete. Correspondingly, assessing historical variability in recharge patterns relies particularly heavily on a few index time series, the most notable of which is that maintained for Chilgrove (in the Chalk of the South Downs). Beginning in 1836, it is one of the longest continuous series of groundwater levels in the world providing an opportunity to examine recent episodes of artesian flow or exceptional summer recharge within a very long historical framework. Long-term trends in groundwater levels are examined in Section 11.17.

11.14.3 *Reappraising historical extremes*

As implied above, a substantial body of evidence indicates that many extreme flood or drought episodes pre-date the modern era but the quality of the evidence and the associated estimates of flow rates can often limit their utility. Improvements in rainfall–runoff modelling capabilities, often used in combination with contemporary documentary evidence, now allow more quantitative reviews of historical extremes to be undertaken. Here, the very lengthy Thames river flow series is used to illustrate the practical utility of such reviews.

Analyses using the IHACRES model and capitalising on a lengthy daily catchment rainfall series for the Thames, strongly suggest that pre-1951[5] low flows in the Teddington record are underestimated to an appreciable degree (Littlewood & Marsh, 1996). The primary cause is almost certainly leakage through the many gates and sluices incorporated in Teddington Weir (the leakage tends to be greatest in both relative and absolute terms during periods of depressed flow). Whilst model uncertainties preclude the substitution of the modelled natural daily series for the dataset held in the NRFA, the analysis did provide confirmation that the minimum flows registered during, for example, the 1921, 1934 and 1944 droughts were significantly higher than those recorded during the extreme drought of 1976. The Station Description associated with the Teddington flow series now alerts all users the to this lack of homogeneity in the low flow record; an important consideration given that the Thames series is one of the most commonly analysed flow records in the world.

Critically reviewing high-magnitude flood events remains a challenging exercise, but models can provide valuable guidance when assessing

the credibility of flows attributed to historically outstanding events. The peak daily flow for the Thames during the extreme flood of November 1894 – 1094 m³/s – remained the highest daily flow on the NRFA for any river in England and Wales throughout the twentieth century. Following persistent doubts about both its magnitude and the exceptional steepness of the flood hydrograph, a joint CEH/EA review was instigated in 2003. It utilised two independent modelling techniques and the results clearly indicated that the archived flows were substantially too high (Marsh *et al*, 2005). A revised maximum daily flow of 800 m³s⁻¹ was adopted; a deliberately rounded figure to avoid any spurious implied level of accuracy. As in many such cases, the utility of the overall flood time series is far better served by the inclusion of a (well-documented) estimate than leaving a blank entry corresponding to an outstanding flood peak.

11.15   Maximising the utility of hydrological data

Efforts to address complex environmental challenges at scales ranging from individual plots to global extent are demanding ever increasing access to interoperable observations of freshwater systems (Dixon *et al*, 2013). As the breadth of application of hydrological data increase and user requirements become more exacting, the development of rigorous data quality assurance frameworks has become crucial components in any data management enterprise.

11.15.1 *Hydrometric data quality assurance*

Over the 50 years that hydrological databases have been maintained at Wallingford, there has been a focus on improving data validation techniques relating both to data collected in experimental basins and acquired through the stewardship of the NRFA. In relation to the latter – for which the quality assurance procedures complement those employed by the regional measuring authorities – a comprehensive Service Level Agreement (SLA) was developed through the early years of the twenty-first century. It aims at maximising the continuity and fitness-for-purpose of river flow time series; there is a particular focus on those gauging station records judged to be of most strategic value (e.g. those in the UK Benchmark Network).

---

[5] Post-1951 flows are based on a downstream ultrasonic gauging station.

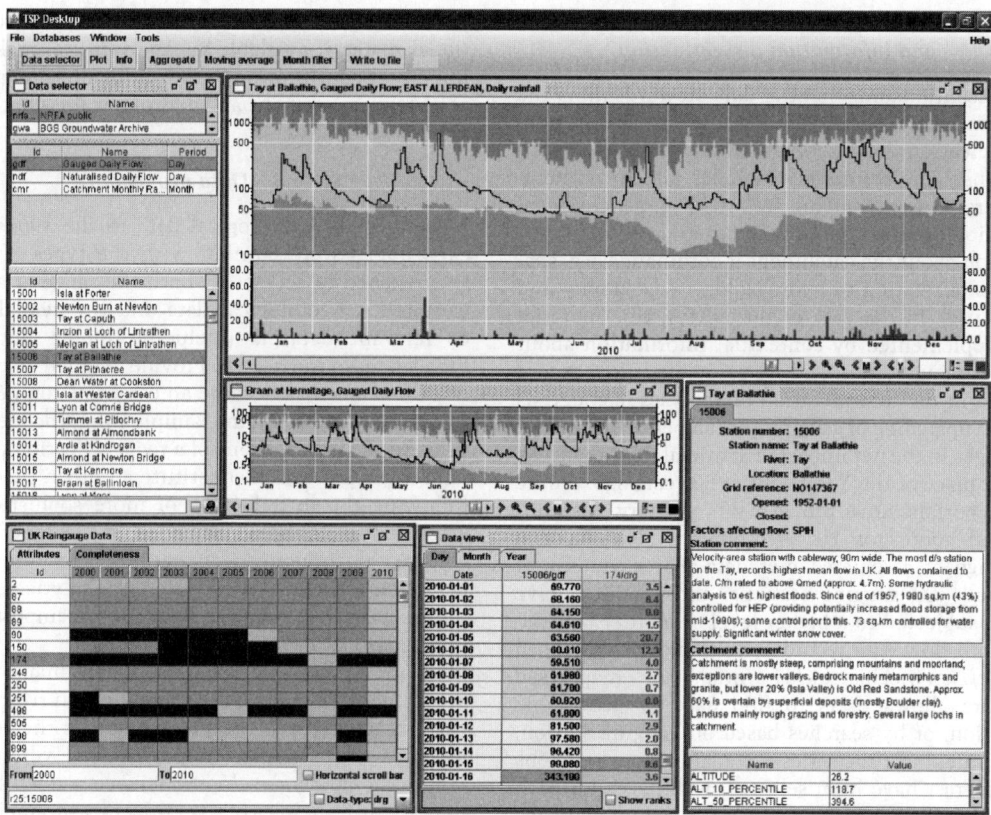

Fig. 11.10 Components used in the Service Level Agreement to identify and rectify errors in river flows. (Source: Reproduced with permission of CEH.) *(See insert for colour representation of this figure.)*

Validation remains a computer-assisted rather than computer-controlled process. Nonetheless, the development of versatile plotting and manipulation software, together with access to a wide range of hydrometeorological time series, analytical tools and catchment information, has facilitated more rigorous examinations of river flow datasets. The NRFA validation suite enables comparisons between different near-neighbour or analogue flow measurement sites, assessments of catchment rainfall hyetographs, automatic baseflow separation and assessment of time series statistics (Fry & Swain 2010). A fast-access web-based GIS system allows consideration of topography, areal rainfall gradients and land use across the catchment under scrutiny, as well as the identification of analogue gauging stations and raingauges for time series comparisons. Additionally, validation procedures (and the user community) benefit from ready access to details of gauging stations histories and their hydrometric performance, along with previous quality control logs (see Fig. 11.10).

Since the introduction of the SLA in 2005, there has been a demonstrable reduction in error frequency and the volume of missing river flow data (Muchan and Dixon, 2014). Data completeness is particularly important as erroneous or missing data can severely impact on the usability of hydrometric time series; a reflection of the fact that missing data are disproportionately concentrated in periods of extreme flow. Correspondingly, where appropriate, the NRFA employs auditable modelling procedures to derive estimates of missing daily mean flows (Harvey *et al.*, 2012); these are routinely flagged to ensure that users can be informed of their provenance.

11.15.2   Hydrological data retrieval and information dissemination

An essential component of data stewardship is the dissemination of user guidance information to aid the selection of datasets suitable for particular applications. This commonly includes discovery metadata relating to gauging stations (e.g. geographical location, station type and catchment area) and information about the catchment it commands (e.g. spatial data relating to hydrogeology, topography and land use). In the case of the NRFA, this material is supplemented by contextual information about the gauging station and its associated flow record to aid in the interpretation of analyses (e.g. details of the upgrading of gauging station performance and changes in the net impact of artificial disturbances).

In place of the Yearbooks of 50 years ago, online web portals now provide the primary dissemination route for UK hydrometric datasets and a range of associated metadata. For a sub-set of around 400 gauging stations, those which have been subjected to a higher level of quality control, the full flow data record is available for download (http://www.ceh.ac.uk/data/nrfa). A map-based feature allows users to find data by geographic location, or by searches based on gauging station, catchment or time series metadata. Once locations of interest have been selected, web pages for each gauging station provide access to comprehensive reference material and other metadata. The surface water data is complemented by groundwater data published on the BGS website – live displays of groundwater time series are available for 28 index wells and boreholes – backed-up by retrieval facilities providing time series records for around 170 sites across the country. As with the NRFA, the migration to web portal access has been used as an opportunity to provide more background material and interpretative notes on individual monitoring sites.

The NRFA portal also provides access to some 40 reports, published under the NHMP umbrella, summarising annual hydrological conditions or documenting notable hydrological events since the early 1980s. For the research community, these are complemented by a similar number of papers published in the scientific literature. In addition, an extensive range of briefing material is made available to government departments and the media. Over the hydrological volatile 2010–2012 period, NHMP material was featured in more than 1000 media articles and reports. These, together with a range of supporting background, and educational material available via the website, make an important contribution to the public understanding of water management and hydrological issues.

11.15.3   Data provenance

Most users of hydrological data in the 1960s and early 1970s were familiar with the types of data they wished to access. As both the range of applications for hydrological data and the availability of relevant datasets (including model outputs) broadened, so the need to provide detailed evidence of provenance has become an essential component of data stewardship. An example might be alerting users to the implications of a reclassification of the land use types which constitute an urban typology following the introduction of more sophisticated satellite surveillance capabilities. In a water quality context, documenting changes in the limits of detection associated with different analytical techniques used over time can be of crucial importance (Neal *et al.*, 2011).

In the hydrometric context, where many users are primarily concerned with changes in river flow patterns over time, the provenance of the record assumes a particular importance. Hydrometry remains a rapidly developing science and while the underlying principles of river flow measurement are broadly the same, the technology and techniques applied have been subject to significant change. For example, the development of time-of-flight ultrasonic gauging stations in the 1970s and, twenty years later, the rapid growth in the use of acoustic Doppler current profilers (ADCPs, see Fig. 11.11) has revolutionised river flow measurement. ADCPs allow rapid and safe river flow gauging, including the measurement of velocities across floodplains, with a greater accuracy than could normally be achieved using current meters. However, considerable further analytical effort may be required to determine whether revisions are required to, for example, earlier extreme flows in order to maintain homogeneity in the overall river flow time series.

While changes in hydrometric technology have undoubtedly delivered significant improvements to the quality and accuracy of flow data, such developments present challenges for users of time series records and those charged with their stewardship. Where changes to gauging stations or hydrometric techniques have impacted upon the uncertainty of the flow record – for example, where gauging

Fig. 11.11 Acoustic Doppler Current Profiler (ADCP) being deployed on the River Thames (Source: Reproduced with permission of CEH.)

weirs have been removed – such information needs to be communicated to the user so that they may consider the possible impact upon their analyses. NRFA practice has been to use textual commentaries for this purpose – reflecting the fact that the routine derivation of quantitative estimates of uncertainly in hydrometry remains elusive.

Considering data provenance generally, the hydrogeochemical datasets assembled as part of a range of Plynlimon research programmes provides an object lesson in good practice (Neal *et al.*, 2011). The final published datasets span over thirty years of record and, as a result, changes in analytical methodologies meant there were issues of accuracy and precision as instrumentation improved. For this reason, detailed metadata was collated and published alongside the time series data giving information on analytical techniques, detection limits, quality protocols and so on. Critically, this relied on the knowledge of long-serving analytical chemistry staff and detailed recording of methodological detail. A huge amount of work

was undertaken to transfer the data and metadata from laboratory datasets, laboratory notebooks and, in some cases, the unwritten knowledge of long-serving members of staff. The resulting metadata provides a exceptionally detailed chronicle of the raw data, containing all the relevant analytical qualifiers, including detailed information on monitoring locations and sample pre-treatment such as filtration and acidification in the field. The resulting 'gold standard' datasets are accessible through CEH's Environmental Information Data Centre (see Section 11.16.2) and are expected to prove of exceptional value for environmental scientists, researchers and modellers alike for many years to come.

11.15.4 The provenance of areal rainfall assessments

Issues of provenance are not limited to data collected in the field or analysed in laboratories.

Outputs from models have long been exploited in hydrological studies (e.g. for the estimation of evaporation losses or soil moisture deficits) but the burgeoning use of model characterisations carries with it a need for comprehensive documentation of their origin and implicit assumptions. Assessments of areal rainfall serve to underline this need. Fifty years ago monthly catchment rainfall totals for the NRFA were derived by averaging standardised monthly rainfall for a fixed raingauge network selected to be representative of the catchment as a whole; the aim being to maximise the homogeneity of the areal rainfall series. Today, significant advances in computing power and data access mean that a variety of models can be used to compute averages at a range of grid scales. However, differences in areal assessments can arise relating to the averaging technique used, grid resolutions and the calculation sequence for producing long-term averages (average then grid, or grid then average). Such differences can be non-trivial, systematic and impact substantially on, for example, the derivation of convincing water balances at the catchment scale.

To further complicate matters, areal rainfall assessments are now available from radar and satellite observations as well as terrestrial monitoring networks – presenting improved spatial and temporal estimates of precipitation but also added issues of compatibility. On the ground, the homogeneity of rainfall records has largely been maintained by the use of standardised gauges, however as demands to adopt (and in some cases replace) such technologies with new gauges grow, so too does the need to carefully document the provenance of the time series.

Considering models more generally, provenance is important for two reasons: the results of different versions may not be compatible and in some situations, especially engineering design, it may be necessary to reproduce a set of results at a later date (e.g. in the event of a dam failure). Correspondingly, there is a continuing need to promote provenance as an essential component of data management and the development of standards both nationally and internationally, in order to harmonise monitoring techniques and promote best practice amongst a disparate and growing hydrometric community.

11.15.5 *Promoting best practice*

Key components in any hydrometric quality assurance scheme are the maintenance of high standards of field practice and the expertise and commitment of those undertaking the data validations procedures. The experience of Wallingford personnel in a wide range of international data acquisition programmes has been captured in a set of practical guidelines for hydrological data collection (Gunston, 1998). The focus in the UK has been on the promotion of consistent operational procedures in line with appropriate national (British Standards Institution – BSi) and international (European Committee for Standardization – CEN) and International Organization for Standardization (ISO) standards.

A continuing dialogue with both the data acquisition and user communities meant that the NRFA team was well placed to lead the development of the first British Standard on Hydrometric Data Management (BSI 1997); an updated version of this document is scheduled for publication in 2015. UK expertise in this field has also aided the development of international guidelines on hydrometric data management, for example, through contributions to the World Meteorological Organization's *Guide to Hydrological Practices* (WMO 2008). With regard to data management, system design and development, Wallingford staff have also contributed to international standards setting organisations, such as the WMO-OGC (Open Geospatial Consortium) technical working groups aiding development of the WaterML2 standard for hydrological data exchange and the OpenMI model integration standard.

11.15.6 *Data utilisation*

The usage and the breadth of application provide convincing measures of the value of environmental databases. Historically, the monitoring of archive usage was sporadic, but the advent of the World Wide Web has made more effective quantification possible.

Statistics for the CEH online Information Gateway (https://gateway.ceh.ac.uk) rank catchment data from Plynlimon as the third most requested dataset whilst, nationally, downloads from the NRFA website since 2010 have averaged around 10–12,000 per year servicing users across a very broad science base, embracing UK and international research communities. In addition, the data provide evidence, methods and tools for government agencies and they support business across a range of sectors: construction, transport, energy, water, finance, agriculture, food and groceries, forestry, fisheries, tourism, sports and recreation. Flood and low flow datasets – and the software developed to aid their exploitation – are used to enable the design of infrastructure at risk of flooding, the setting of license conditions for water abstractions from rivers and the development of environmental flow standards as part of the implementation of the Water Framework Directive. More generally, such detailed quantitative information allows regulatory guidelines to be established that reflect practical restraints and current licensing policies, the latter often balancing sustainable economic growth with the need to maintain and improve the health of the aquatic environment. The last two decades have witnessed a particularly rapid expansion in the use of hydrometric databases to assess the impact of climate change and in the development, calibration, validation of operational hydrological models, in flood forecasting and national water quality models used in identifying waters at risk from pharmaceuticals and agrichemicals.

11.16   Integration of Databases and Models

In response to the increasing complexity of user needs, skills in the linkage of hydrological data with other environmental datasets improved markedly over the last 50 years. By the early twenty-first century, data management systems had evolved to cater effectively with a multiplicity of information streams, address the increasing complexity of the user community's requirements including, crucially, progress in improving medium-term hydrological forecasting capabilities. However, institutional, commercial or technical issues have restricted the degree to which aspirations for 'one-stop' data access capabilities have actually been satisfied.

11.16.1   UK Floods data

The handling of UK floods data provides an illustration of the difficulties that can be encountered when attempting to create a single river flow database capable of addressing user requirements across the full range of flows. The limitations of the NRFA in relation to sub-daily flows (see 11.5) have meant that the bulk of flood data acquisition in the UK has been undertaken as part of time-limited major research programmes. The lack of continuity

in maintaining, updating and validating the UK's floods database is reflected in issues of time series homogeneity and, particularly, the uncertainties associated with the flow magnitudes during extreme events.

To address the hiatus following the end of the FEH data gathering activities in the early 1990s, the EA, SEPA and Rivers Agency Northern Ireland set up the HiFlows-UK project in 2001. This updated and extended the FEH flood time series and incorporated additional data and background information sourced from the NRFA and the University of Dundee. One aim of the HiFlows initiative was to capitalise on recent changes to stage–discharge relations and also to provide more comprehensive hydrometric information to support the flood time series. An additional objective was to make the generality of flood data more readily accessible. Correspondingly, a HiFlows website was launched in 2004; its primary function was to service the needs of experienced flood hydrologists wishing to estimate peak flood flows by FEH methods for locations throughout the UK.

In 2014, responsibility for updating the HiFlows database of peak flows passed to CEH, foreshadowing a closer integration of the NRFA and the national floods database than had been the case for over 30 years (www.ceh.ac.uk/data/nrfa). Reconciling the monthly highest 15-minute flows on the former with the peaks-over-threshold held in the latter – often derived using different stage–discharge relations – represents a considerable challenge, one that needs to be addressed by experienced hydrologists familiar with the hydrometric characteristics of individual gauging stations and measuring reaches. But with access to an extensive range of flood-related metadata and the capability to exploit advanced analytical procedures (allowing, e.g. return periods to be rapidly estimated or utilising fine-resolution assessments of storm rainfall at the catchment scale), there is clear potential to establish a unified flood event database to support a wide range of scientific and societal objectives.

11.16.2   The Environmental Information Data Centre

Considering the exploitation of hydrological data in a wider context, the need to capitalise on an increasingly wide range of information sources to address the environmental challenges of the twenty-first century and support the multidisciplinary aspirations of many research initiatives underpinned the establishment of the UK Environmental Information Data Centre (EIDC). Launched in 2009, the EIDC is hosted at Wallingford and, as well as coordinating CEH data activities, it serves as the NERC Data Centre for the Terrestrial and Freshwater Sciences. It brings together wide-ranging nationally important datasets, including the NRFA, the Biological Records Centre and the Environmental Change Network, and the expertise to manage such a diverse range of environmental data. Access to these resources is normally via the CEH Information Gateway, the tool for finding, viewing and accessing data held by the EIDC and other data providers in the UK and beyond. At the heart of the Gateway is the Data Catalogue – a repository of records that describe the nature and scope of CEH data resources.

The objective of the EIDC is to provide researchers and other data users access to the coordinated data resources and informatics tools needed to find answers to complex, multidisciplinary environmental questions. In addition, the EIDC delivers informatics research to meet CEH information needs and contributes to national and international data sharing and monitoring initiatives (including the Shared Environmental Information System (SEIS), the Global Earth Observation System of Systems (GEOSSs) and the European Earth Observation Programme (GMES)).

11.16.3 Linking models to models: integrated environmental modelling

In parallel to the development of integrated data centres, considerable research effort over the last decade has been devoted to integrated environmental modelling (IEM). Under this broad umbrella, an important initiative, the Open modelling interface (OpenMI), aims to simplify the linkage between water-related simulation models in support of the implementation of the Water Framework Directive (Moore and Tindall, 2005). The objective is to facilitate the simulation of process interactions and make it easier for catchment managers to explore the wider implications of the policy options open to them. The main deliverable of the project is the definition of a standard interface that will allow new and existing models to exchange data at run time. To help users adapt their models to use the interface, and then link and run them with other models, the standard will be supported by software tools for migration, linking, monitoring performance and displaying results.

It is convenient to think of IEM under four headings (Laniak *et al.*, 2013):

• IT science
• Modelling science
• Organisation
• Applications

Conceptually, IEM is simple; an integrated model is created by linking together models of individual or groups of processes (Moore *et al.*, 2012). The links allow data to pass between the models and the technical challenge is how best to achieve the transfer. Essentially, it is a pure IT problem whose solution should be independent of the data being passed. The problem is challenging because the requirement is to be able to link models based on different concepts, working at different scales, using different spatial and temporal resolutions and representations (including none at all) and sourced from different suppliers. In the simple case where the processes being simulated are sequential, the results of the first model can become the input of the next. More sophisticated approaches have been developed to handle situations where the processes are running in parallel and interacting time step by time step. This challenge can become greater if the models are running in different computing environments or in the high performance or cloud environments. Several approaches exist for solving the linkage problem and, in principle, most are very similar; they work by providing each model with a standard interface through which the model can request and receive data or respond to requests.

Having developed a number of viable solutions, the IT science focus is now turning to the problem of transforming integrated modelling from a research tool, which requires a high degree of skill to use, into a tool that anyone can apply with ease and confidence. Can the processes of making models linkable, finding appropriate models, linking and running them be made invisible as the processes that take place when two zip codes are entered into a satellite navigation system and it computes the way from A to B as you drive?

The organisational level is concerned with creating the conditions for IEM to flourish. At present, much of our knowledge about earth system processes and their integrated modelling is spread across the world in islands of excellence, often relating to very specific areas. For some time, there has been a shared belief among some of these groups that if they could be brought together, then the opportunities for innovation that will follow are vast. This belief is based partly on observation of the advances over the last fifty years in other disciplines, especially GIS, and partly on brainstorming what might be possible, given a large library of linkable models spanning diverse processes.

IEM applications have the potential to deliver what modellers have long sort which is plug and play modelling and decision-support system development. In practical terms, this means that the modeller has access to a pool of linkable models and other modelling components such as GUIs. In the future, these components will be located and discoverable from anywhere in the world. Anybody will be able to contribute components and they will be available as open source or commercially. There are already available a number of modelling platforms that allow users to browse available models and link them together. The linking process simply involves dragging the output variable of one model onto the corresponding input of another. When the models compute values of each variable for a number of nodes, a second step maps the nodes of the two models onto each other. Larger models can involve many thousands of connection points. While the process of identifying and joining geographical linkages is largely a manual exercise at present, work is underway to automate it. An example from the water domain might be connecting a hydraulic model of the river network to an underlying groundwater model. The linking process also handles differences in time steps and units of measurement should they arise. The changes that have brought this simplification about are the emergence of object-oriented programming and open standards. The latter are allowing modellers working independently to produce components which they can be confident will link to other independently developed components. This of course has profound implications for the modelling market place, for science and for innovators.

Integrated modelling will make semantics increasingly important, especially as links are made across disciplinary boundaries and between models sourced from non-English speaking countries. Even within disciplines many terms used have variable names, here the names of the model inputs and outputs are very loosely defined. As modelling is opened up to a wider range of scientists and the outside world, it will become increasingly important that clear definitions of what each variable represents are made available, so that modellers are aware of what they are linking to what and can make a judgement as to its validity.

11.16.4 Models and hydrological forecasting

A crucial test of modern data acquisition and modelling capabilities is the degree to which they enhance forecasting skill. The rapid growth of global monitoring and modelling capabilities over the last 15 years has led to very valuable improvements in short-term and seasonal weather forecasting. By linking meteorological and hydrological models, a number of countries, including the United States and Australia, have established long-range hydrological forecasting programmes. In the UK, a series of damaging flood episodes and sustained droughts during the last decade have underlined its continuing vulnerability to extreme weather conditions and emphasised the strategic benefits that indicative seasonal forecasts of hydrological conditions could provide. Correspondingly in 2012, CEH and BGS, in close collaboration with the Met Office and the EA, began to explore the development of a long-range hydrological forecasting capability for the UK.

Short-term flow forecasting, particularly in relation to flood risk, has been applied operationally since the 1980s in the UK, but progress in seasonal forecasting of hydrological variables has been much slower. In part, this reflects the limited skill levels of long-term meteorological forecasts in mid-latitudes, as compared with the tropics where there is a much higher level of inherent predictability due to the slowly varying mechanisms of atmospheric forcing, such as El Nino – Southern Oscillation. While a number of scientific developments in the late 1990s and the first decade of the 2000s pointed towards the possibility of medium-term river flow forecasts for the UK, generally skill levels were modest and the developments remained primarily of scientific interest rather than paving the way for operational systems. At CEH over this period, there was relatively limited research effort focused on seasonal forecasting, although a range of studies highlighted important links between streamflows and large-scale atmospheric circulation indices and other potential predictors (e.g. Svensson and Prudhomme, 2005).

The need for more effective seasonal forecasting in the UK was given particular impetus in 2011 and early 2012 as exceptional rainfall deficiencies developed across large parts of the country. There was an obvious requirement to make judgements about how long the drought would last and how severe it could become; and then, following the record rainfall in April 2012 and the subsequent hydrological transformation (Parry *et al.*, 2013), the likelihood of receiving sufficient rainfall over a given time in order to terminate the drought. This goal prompted the development of a new methodology for appraising soil and groundwater storage deficits based on the Grid-2-Grid model (Bell *et al.* 2013). In turn, this stimulated efforts to combine the Met Office's seasonal climate outlooks, which have been routinely produced since the mid-2000s, with the state-of-the-art hydrological and hydrogeological models available within CEH and BGS. Following successful trialling, an operational version of the UK Hydrological Outlook was launched in October 2013, via a dedicated website: http://www.hydoutuk.net. The Outlook's potential has been enhanced by the parallel development of digital maps identifying areas susceptible to groundwater flooding instances of which occurred very widely during the late winter and spring of 2014 (see Fig. 11.12).

11.17   Capturing Hydrological Change

A primary driver for the establishment of rainfall, river flow and groundwater level monitoring networks throughout the world was to establish baseline conditions and index hydrometeorological variability. In the UK, as network densities and the length of time series increased, these objectives have been largely fulfilled. However, as the possible impacts of anthropogenic global warming became an increasing focus of public and political concern, the need to identify and quantify hydrological trends assumed a clear strategic priority. Whilst models will undoubtedly have a major role in managing future floods and droughts, the uncertainties involved at the current time place major limitations on their utility. At present therefore, river flow and groundwater level time series – notwithstanding their inherent limitations, in the historical series particularly – provide the most reliable basis on which to base hydrological trend analyses.

11.17.1   *Trends in runoff patterns*

The degree of climatic variability, together with the pervasive impact of artificial influences on river flow patterns in the UK, implies that identifying resilient long-term trends in, for example, drought and flood magnitude, remains a considerable scientific challenge. Fortunately, the UK's rich legacy of hydrometeorological data provides direct observational evidence of changes in rainfall, runoff and, less extensively, recharge patterns over a period which has seen a substantial increase in temperature. The challenge is how best to interpret and extrapolate those changes.

A range of trend detection activities undertaken at IH and CEH over the last 20 years have placed Wallingford at the forefront of international efforts to document long-term hydrological change. In the early 1990s, a comprehensive assessment of variability in river flows across the UK, using NRFA datasets, provided a context for an extensive study on future changes using scenario-based modelling (Arnell *et al*, 1990). The study was a notable early example of an attempt to establish past changes based on observed data at a time when such studies were comparatively rare in the international hydrological literature. Subsequently, a major national trend detection study capitalised on the floods data assembled as part of the FSR and FEH studies (Robson, 2002); the trend analyses were unparalleled in scope in the UK, embracing more than 800 flood time series. Little evidence of any increase in flood magnitude or frequency consistent with anthropogenic climate change was found, but the importance of inter-annual and inter-decadal variability in long-term hydrometric records in the UK became firmly established; an outcome endorsed repeatedly by many subsequent studies. One of the issues with the key finding of a lack of significant trends was that the studies primarily adopted a UK-wide focus and aggregated across all sites, including catchments with very different characteristics and incorporating rivers affected by a range of anthropogenic disturbances (urbanisation, reservoir development, etc.) which could modify any climate-driven flood signals.

The identification of the Benchmark Network allowed trend detection studies to focus on near-natural catchments across the UK, following similar programmes in North America where the designation of benchmark networks is much less of a challenge. The NRFA team was thenceforth very active in trend detection work, publishing a string

Susceptibility to groundwater flooding of upper Pang valley, Berkshire

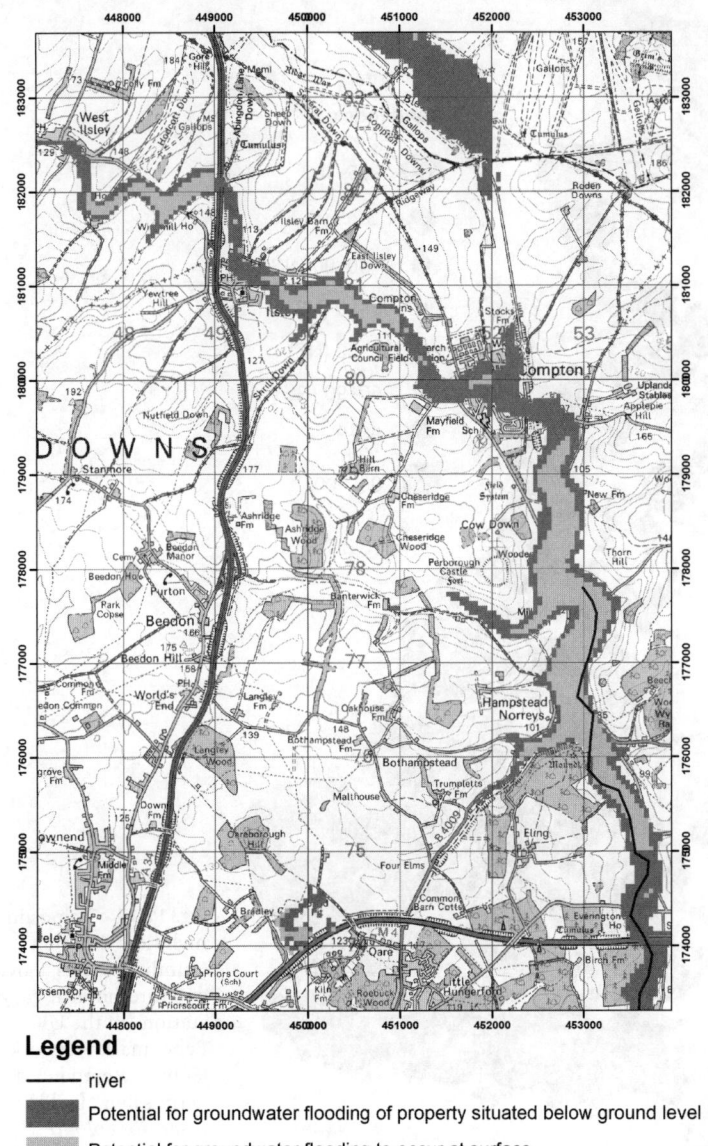

Legend

—— river

Potential for groundwater flooding of property situated below ground level

Potential for groundwater flooding to occur at surface

Fig. 11.12 Susceptibility to groundwater flooding in the upper Pang valley, Berkshire. (Source: Reproduced with permission of British Geological Survey.) *(See insert for colour representation of this figure.)*

of papers on trends in runoff and low flows (Hannaford and Marsh, 2006), high flows (Hannaford and Marsh, 2008) and seasonal runoff (Hannaford and Buys, 2012). As a consequence, the UK Benchmark Network became widely recognised internationally as an exemplar of a 'Reference' network (Whitfield *et al.*, 2012) and forms a key component in a Europe-wide network of near-natural catchments that has been extensively studied from a change detection perspective (Stahl *et al.* 2010).

A 2013 review of the literature on hydrological trends in the UK (Hannaford *et al.* 2013) confirmed that the results of the analyses undertaken at Wallingford chime with a multitude of

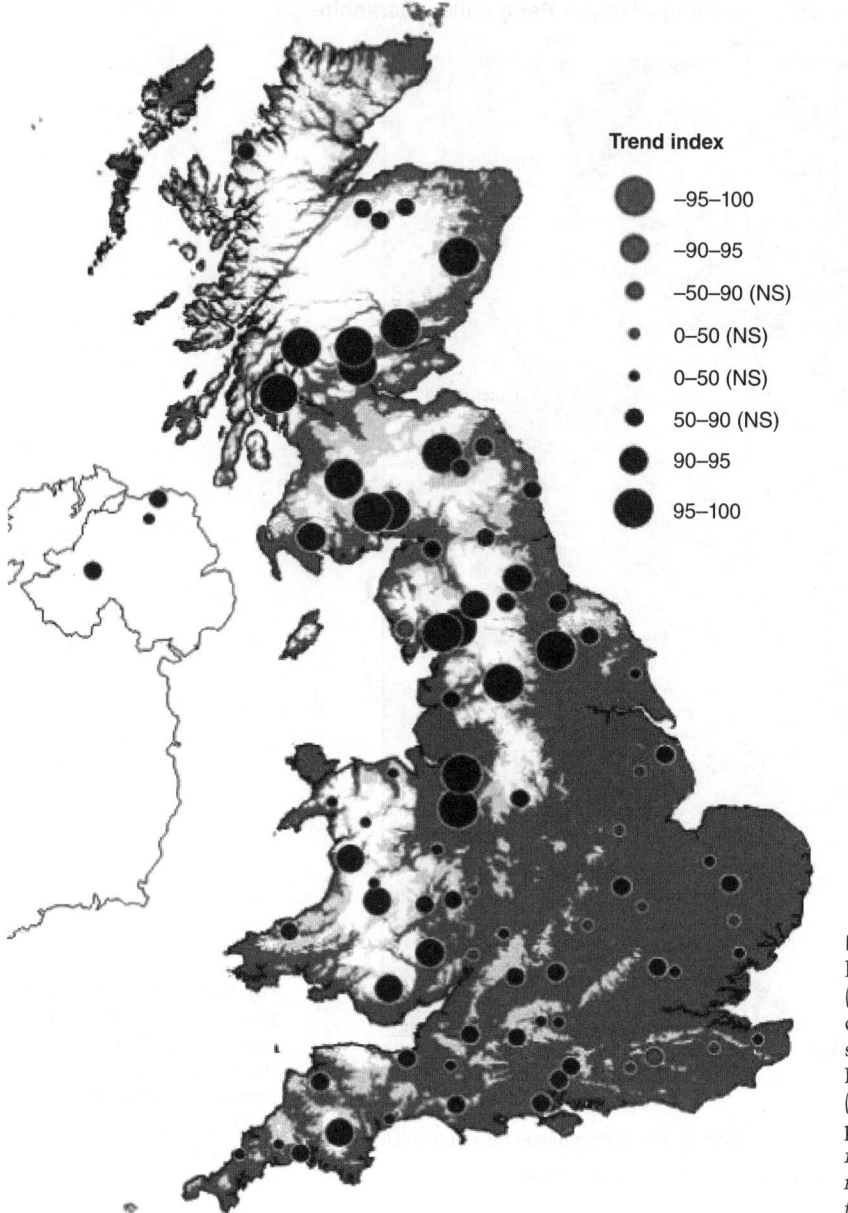

Fig. 11.13 Trends in high flow duration (number of days above q10 threshold) at gauging stations in the UK Benchmark Network. (Source: Reproduced with permission of CEH.) *(See insert for colour representation of this figure.)*

contemporary, but generally more local studies conducted by other research teams. Although constrained by the paucity of long river flow series, a general consensus continues to emphasise the limited compelling evidence for long-term trends in river flows in the UK. The most statistically significant trends are found for periods from the 1960s to the 2000s, predominantly in northern and western upland areas where a tendency towards increased winter runoff and an increasing frequency of high flows may be recognised (Fig. 11.13). Even these trends are difficult to interpret; whilst

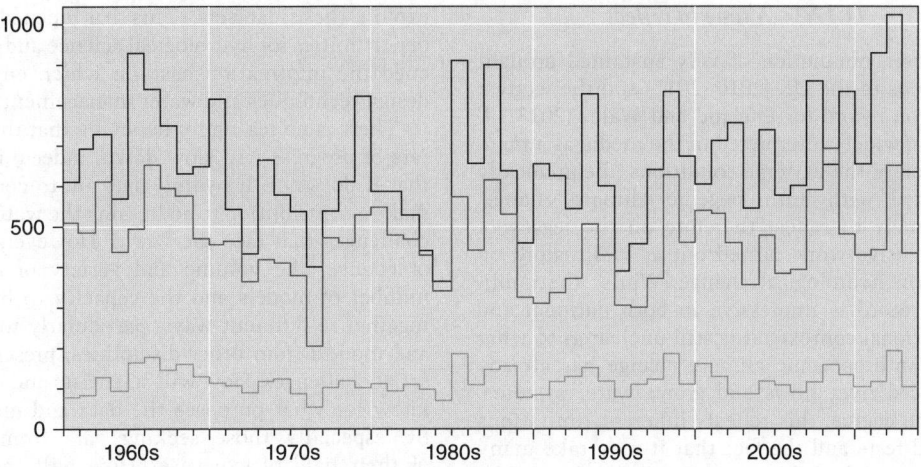

Fig. 11.14 Annual Q_{95} outflows: Scotland (black), England (dark grey) and Wales (light grey). (Source: Reproduced with permission of CEH.) *(See insert for colour representation of this figure.)*

increases in winter rainfall and runoff are entirely consistent with future modelling studies, they are also strongly associated with variations in the North Atlantic Oscillation, the leading mode of atmospheric variability in Northern Europe which may itself be affected by global warming, but also incorporates much internally driven natural variability. These complications hamper the attribution of runoff changes to anthropogenic climate change. In seasons other than the winter, the picture is even more complex; there is no obvious trend towards lower minimum flows (see Fig. 11.14), in contrast to projections based on climate models for the not-too-distant future.

11.17.2 Trends in groundwater levels

Groundwater systems are naturally very variable and may be expected to respond to changes in rainfall, evaporation and temperatures in a complex manner. In addition, across much of the UK groundwater level changes can also be profoundly influenced by abstraction patterns. Compelling trends in groundwater levels can be readily identified in those parts of the country where groundwater abstraction regimes have changed markedly over time. Heavy groundwater pumping from the Chalk below London resulted in a 70-metre decline in groundwater levels at Trafalgar Square over the period from the 1830s to the 1960s; levels recovered subsequently as water supplies were increasingly

drawn from the River Thames. Similar long-term variations in groundwater levels have been noted for other conurbations and in old mining areas (as pumping to dewater the mines was discontinued).

Where abstraction rates are of limited significance, evidence for long-term trends in groundwater levels is, as with runoff patterns, generally weak. The poorer quality of much of the early groundwater level data, and the variations in sampling frequency over time, imply that any apparent trends need to be treated with caution. Nonetheless, many extended time series do share a number of characteristics – for example notably low levels in the 1930s, 1970s and 1990s and exceptionally high levels through the late winter of 2000/01, during the latter half of 2012 and through the winter of 2013/14. Generally, groundwater level time series suggest an enhanced degree of seasonal variability over the last 25 years and the recent past has witnessed a relatively high frequency of groundwater flooding episodes. However, given the relative rarity of such episodes, and the limited availability of groundwater level data prior to the 1960s, further research is needed – some necessarily capitalising on indirect evidence of historical groundwater level behaviour (e.g. indices of winterbourne flows) – to help determine whether this is a reflection of climatic variability or indicative of a growing increase in groundwater flood risk in a warming world.

11.17.3   A time to reflect

The close conjunction of very sustained drought conditions in the UK (2010–2012) and the wettest winter on record for England and Wales (2013/14) has been widely interpreted in the media as a manifestation of the extreme conditions unequivocally associated with anthropogenic climate change. It is timely to reflect therefore on the legacy of the extensive work carried out at Wallingford on long-term hydrological change. Whilst the publication record is impressive in both national and international contexts, it is still unclear as to what extent anthropogenic climate change has already influenced river flows and groundwater systems. To some degree, this reflects inherent limitations of trend tests and the fact that it may take many years for trends to become statistically detectable, given the low signal/noise ratios in hydroclimatic time series. But the apparent disagreement between model-based projections and observational data (Wilby *et al.*, 2008) represents a core challenge for the hydrological community.

The message of 'no compelling change' in runoff patterns arising out of many of the Wallingford studies may not set scientific pulses racing, but it is a vitally important one. Members of the NRFA team in particular have attempted to ensure that the voice of the more sober, empirically informed assessments are widely accessible. It remains possible, of course, that the notable volatility of recent years foreshadows hydrological patterns through the twenty-first century which do differ significantly from those that failed to exhibit any marked trends through the twentieth century. What is certain however is that monitoring programmes will need to be sustained and databases professionally managed if any emerging trends are to be captured and quantified.

11.18   Conclusion

Data acquisition and stewardship are now integral components of nearly all scientific endeavour. Technological advances over the last 50 years have given hydrologists access to a huge range of observations embracing many new variables at high spatial and temporal resolutions, and often with global coverage. At the same time, enhancements in computing power, system design, analytical capabilities, visualisation techniques and information management have provided the means to exploit these datasets. This has opened up rich opportunities for hydrological science and strengthened the information base on which engineering design techniques and water management depend.

There is no reason for believing that the present rate of progress will slow down; indeed, it is vital that it does not, if hydrologists are to continue to make a contribution to addressing the water-related challenges that face the world. However, the rate of change, the volume and variety of data, the number of models and the capacity to link them together in different ways, particularly to datasets and models from other disciplines, present a new set of challenges. How will scientists and managers know for what purposes the data and models are fit, especially those seeking data from outside of their field of expertise? How will they know if an apparent trend observed in the data is real or simply a reflection of more accurate measurement techniques? There are potential dangers here which echo the cautionary computer archive management mantra of the 1970s: 'Garbage In – Garbage Out'. If we cannot effectively address these issues, what will be the value of our science and will we be able to have confidence in our design work? The key to answering these questions must lie in maximising the ease of access to hydrologically relevant information, whilst finding a feasible way of capturing adequate metadata about our datasets and models so that future users understand their provenance.

Through the early years of the twenty-first century, increased data availability has made very important contributions to the understanding of hydrological processes, but the wealth of data now at our command has also underlined the complexity of natural systems. It is a sobering thought that even 340 years after Pierre Perrault's pioneering study of rainfall and runoff in the Paris Basin (Perrault, 1674), achieving convincing water balances at the catchment or national scales remains a considerable challenge. More importantly in relation to the threats posed by too little or too much water in a warming world is the limited agreement between many model-based projections of changes in the magnitude and frequency of extreme flows and the generality of observational evidence. This remains a focus of considerable research effort and continuing model enhancement, often exploiting advanced data acquisition and integrated modelling capabilities, offer real potential for a more complete reconciliation. Nonetheless, there is a pressing need to maintain and upgrade existing hydrometric monitoring networks and to focus on the accuracy and homogeneity of the data they provide.

Today it requires no more skill than that needed to rub your finger across the glass screen of your mobile to find the nearest restaurant for lunch. All the complexity involved in making that possible is hidden. This last point highlights a very important lesson – one that has clear resonance in relation to the acquisition and exploitation of hydrological data. Most of the initial funding that provided the UK's comprehensive hydrological monitoring capabilities and the development expertise to transform visionary ideals in the 1960s and 1970s into today's integrated archives, engineering design capabilities and geographical information systems came from the public purse. The current unprecedented ease of access to an extraordinary array of hydrological data is testimony to the importance of that investment.

11.19   References

Arnell, N.W., Brown, R.P.C & Reynard, N.S. (1990) *Impact of climatic variability and change on river flow regimes in the UK*. Institute of Hydrology Report 107.

Bayliss, A.C. & Jones, R.C. (1993) *Peaks-over-threshold flood database: summary statistics and seasonality*. Institute of Hydrology Report No. 121, Wallingford.

Bayliss, A.C., Norris, J. & Marsh, T.J. (2004) The Wendover Springs Record: an insight into the past and a benchmark for the future. *Weather*, **59**, 267–271.

Bell, V.A., Davies, H.N., Kay, A.L., Marsh, T.J., Brookshaw, A. & Jenkins, A. (2013) Developing a large-scale water-balance approach to seasonal forecasting: application to the 2012 drought in Britain. *Hydrological Processes*, **27**, 3003–3012.

Black, A.R. & Law, F.M. (2004) Development and utilisation of a national web-based chronology of hydrological events. *Hydrological Sciences Journal*, **49**, 237–246.

Blyth, K. & Kidd, C.H.R. (1977) The development of a meter for the measurement of discharge through a road gulley. *Chart. Munic. Engrs*, **104**, 24–27.

Boorman, D.B. Hollis, J.M. & Lilly, A. (1995). *Hydrology of Soil Types: a hydrologically-based classification of the soils of the United Kingdom*. IH Report No. 126. Institute of Hydrology, Wallingford.

Bradford, R.B. & Marsh, T.J. (2003) Defining a network of natural benchmark catchments for the UK. *Journal of ICE Water, Maritime and Energy*, **156**, 109–116.

Bullock, A., Gustard, A., Irving, K. & Young, A.(1994). *Low flow estimation in artificially influenced catchments*. NRA R&D Project 257.

British Standards Institution (1997) *Guide to hydrometric data management*. BS 7898:1997. 28p.

Dixon, H., Hannaford, J. & Fry, M. (2013) The effective management of national hydrometric data: experiences from the United Kingdom. *Hydrological Sciences Journal*, **58**, 1383–1399.

Fry, M.J. & Swain, O. (2010) Hydrological data management systems within a national river flow archive. In: Kirby, C. (ed.), *Role of Hydrology in Managing Consequences of a Changing Global Environment. Proceeding of the BHS Third International Symposium*. British Hydrological Society, pp. 808–815.

Gunston, H. (1998) *Field Hydrology in Tropical Countries – A Practical Introduction*. Intermediate Technology Publications, pp. 107.

Gustard, A., Bullock, A. & Dixon, J.M. (1992) *Low flow estimation in the United Kingdom*. Institute of Hydrology Report No. 108, Wallingford, pp 88 with 6 Appendices.

Hannaford, J., Holmes, M.G.R., Laizé, C.L.R., Marsh, T.J. & Young, A.R. (2013) Evaluating hydrometric networks for prediction in ungauged basins: a new methodology and its application to England and Wales. *Hydrology Research*, **44**, 401–408.

Hannaford, J. & Marsh, T.J. (2006) An assessment of trends in UK runoff and low flows using a network of undisturbed catchments. *International Journal of Climatology*, **26**, 1237–1253.

Hannaford, J. & Marsh, T.J. (2008) An assessment of trends in UK high flow magnitude and frequency using a network of undisturbed catchments. *International Journal of Climatology*, **28**, 1325–1338.

Hannaford, J. & Buys, G. (2012) Trends in seasonal river flow regimes in the UK. *Journal of Hydrology*, **475**, 158–174.

Harvey, C.L., Dixon, H. & Hannaford, J. (2012) An appraisal of the performance of data infilling methods for application to daily mean river flow records in the UK. *Hydrology Research*, **43**, 618–636. doi:10.2166/nh.2012.110

Herbertson, P.W., Douglas, J.W. & Hill, A. (1971) *River level sampling periods*, NERC IH Report No. 9.

Herschy, R.W. (1978) *Hydrometry – Principles and Practices*. John Wiley & Sons, London, pp. 478.

Huntley, D., Leeks, G. & Walling, D. (eds) (2001) *Land-Ocean Interaction*. IWA Publishing, pp. 286.

Kirby, C. Newson, M.D. & Gilman K. (1991). *Plynlimon research: the first two decades*. Institute of Hydrology Report 109. 188p.

Jones, P.D., Olgilvie, A.E.J. & Wigley, T.M.L. (1984) *River Flow Data for the United Kingdom: Reconstructed Data Back to 1844 and Historical Data Back to 1556*. Climatic Research Unit CRURP8, pp. 166.

Laize, C.L.R., Marsh, T.J. & Morris, D.G. (2008) Catchment descriptors to optimise hydrometric networks. *Proceedings of ICE, Water Management*, **161**, 117–125.

Laniak, G.F., Olchin, G., Goodall, J. *et al.* (2013) Integrated environmental modeling : a vision and roadmap for the future. *Environmental Modelling & Software*, **39**, 3–23.

Lees, M.L. (1987) *Inland Water Surveying in the United Kingdom*. 1985 Yearbook – Hydrological data UK series., Institute of Hydrology, Wallingford pp. 35–48.

Limbrick, K., Willows, J. & Marsh, T.J. (2002) Springflow variability since 1858 in south Dorset, UK. *Weather*, **57**, 413–421.

Littlewood, I.G. & Marsh, T.J. (1996) Re-assessment of the monthly naturalised flow record for the Thames at Kingston since 1883, and the implications for the relative severity of historical droughts. *Regulated Rivers: Research and Management*, **12**, 13–26.

Littlewood, I.G., Watts, C.D. & Custance, J.M. (1998) Systematic application of United Kingdom river flow and quality databases for estimating annual river mass loads. *The Science of the Total Environment*, **210 (11)**, 21–40.

Macdonald, N. & Black, A.R. (2010) Reassessment of flood frequency using historical information for the River Ouse at York, UK (1200–2000). *Hydrological Sciences Journal*, **55 (7)**, 1152–1162.

Marsh, T.J. & Littlewood, I.G. (1978) An estimate of annual runoff from England and Wales, 1728–1976. *Hydrological Sciences Bulletin*, **23 (1)**, 3.

Marsh, T.J. (1980) Towards a nitrate balance for England and Wales. *Water Services*, **84**, 601–606.

Marsh, T.J., Greenfield, B.J. & Hannaford, J.A. (2005) The River Thames flood of November 1894 – a reappraisal of the maximum flow. *ICE Water Management*, **158**, 103–110.

Marsh, T.J. & Hannaford, J. (eds) (2008) *UK Hydrometric Register*. Hydrological data UK Series. Centre for Ecology and Hydrology, pp. 212.

Mather, J.D. (ed) (2004) *200 years of British Hydrogeology*. Geological Society, London, pp. 394. Special Publication 225.

Morris, D.G. & Heerdegen, R.G. (1988) Automatically derived catchment boundaries and channel networks and their hydrological applications. *Geomorphology*, **1**, 131–141.

Moore, R.V. (1997) The logical and physical design of The Land Ocean Interaction Study database. *Science Of The Total Environment*, **194**, 137–146.

Moore, R., Hughes, A., Gaber, N. *et al.* (2012) *International Summit on Integrated Environmental Modeling*. U.S. Environmental Protection Agency, Washington, DC, EPA/600/R/12/728, http://cfpub.epa.gov/si/si_public_file_download.cfm?p_download_id=509013.

Moore, R.V. & Tindall, I.C. (2005) An overview of the open modelling interface and environment (the OpenMI). *Environmental Science and Policy*, **8 (3)**, 279–286.

Morris D.G. & Flavin R.W. (1996). *Flood risk map for England and Wales*. Institute of Hydrology Report No. 130. 87p.

Muchan. K. & Dixon. H. (2014) Ensuring hydrometric data are fit-for-purpose through a national Service Level Agreement. Hydrology in a Changing World: Environmental and Human Dimensions. *Proceedings of FRIEND-Water 2014*, Montpellier, France, October 2014. IAHS Publ. 363: 323–329.

Neal, C., Reynolds, B., Norris, D. *et al.* (2011) Three decades of water quality measurements from the Upper Severn experimental catchments at Plynlimon, Wales: an openly accessible data resource for research, modelling, environmental management and education. *Hydrological Processes*, **25**, 3818–3830.

Parry, S., Kendon, M. & Marsh, T.J. (2013) 2012: from drought to floods in England and Wales. *Weather*, **68 (10)**, 268–274.

Perrault, P. (1674). *On the origins of springs (Trans: Aurele LaRoque)*. Hafner. ASIN B0026MBIHE.

Potter, H.R. (1978). *The use of historic records for the augmentation of hydrological data*. Institute of Hydrology Report No.46, 58p.

Reed, D. (1999) *Flood Estimation Handbook – Vol 1, Overview*. Institute of Hydrology, Wallingford, 108 p.

Roberts, G. (1981). *The processing of hydrological data*. IH report No. 70. 113p.

Robson, A.J. (2002) Evidence of trends in UK flooding. *Philosophical Transactions of the Royal Society*, **360**, 1327–1343.

Rodda, J.C. & Marsh, T.J. (2011). *The 1976 drought – a contemporary and retrospective review*. Centre for Ecology & Hydrology, 58p.

Sargent, R.J. & Ledger, D.C. (1992) Derivation of a 130-year runoff record from sluice records for the Loch Leven catchment, South East Scotland. *Proceedings of the Institution of Civil Engineers–Water and Maritime Engineering*, **96**, 71–80.

Simpson, E.A. (1978) The harmonisation of the monitoring of the quality of inland fresh water. *Journal of the Institutional of Water Engineers and Scientists*, **32**, 45–56.

Stahl, K., Hisdal, H., Hannaford, J. *et al.* (2010) Streamflow trends in Europe: evidence from a dataset of near-natural catchments. *Hydrology and Earth System Sciences*, **14**, 2367–2382.

Svensson, C. & Prudhomme, C. (2005) Prediction of British summer river flows using winter predictors. *Theoretical and Applied Climatology*, **82**, 1–15. doi:10.1007/s00704-005-0124-5

Tindall, C.I. & Moore, R.V. (1997) The Rivers Database and the overall data management for the Land Ocean Interaction Study programme. *Science of the Total Environment*, **194**, 129–135.

Water Resources Board (1973) *Water Resources of England and Wales*. HMSO.

Werritty, A. (1987) *The McClean Hydrometric Data Collection*. 1985 Yearbook – Hydrological data UK series, Institute of Hydrology, Wallingford, pp. 49–54.

Whitfield, P.H., Burn, D.H., Hannaford, J. *et al.* (2012) Reference hydrologic networks I. The status and potential future directions of national reference hydrologic networks for detecting trends. *Hydrological Sciences Journal*, **57**, 1562–1579.

Wilby, R.L., Beven, K.J. & Reynard, N.S. (2008) Climate change and fluvial flood risk in the UK: more of the same? *Hydrological Processes*, **22**, 2511–2523.

Willis, A. (1999) Hydrometric data processing. In: Herschy, R.W. (ed), *Hydrometry Principles and Practices*. Wiley, pp. 317–354.

World Meteorological Organisation (2008) *Guide to Hydrological Practices: Volume 1*, 6 edn, WMO-No.168,. World Meteorological Organisation, Geneva.

Wright, C.E. (1978) *Synthesis of river flows from weather data*. Technical Note No. 26,. Central Water Planning Unit, Reading, UK.

Young, A.R., Grew, R. & Holmes, M.G.R. (2003) Low Flows 2000: a national water resources assessment and decision support tool. *Water Science and Technology*, **48**, 119–126.

12 Looking Towards the Future

JOHN C. RODDA[1], MARK ROBINSON[1], ALAN JENKINS[1],
KEITH BEVEN[2], MAX BERAN[3], AND GRAHAM LEEKS[1]

[1]Centre for Ecology and Hydrology, Wallingford, Oxfordshire, UK
[2]Lancaster Environment Centre, University of Lancaster, Lancaster, UK
[3]Ex-Institute of Hydrology, Wallingford, UK

12.1 Introduction

The future has always been of great concern to humanity. Prophets, astrologers, oracles and the like have had important roles in trying to establish what was going to happen. Battles have been won, fortunes made and countless enterprises launched when the outcome was foretold to be favourable. History does not chronicle the failures. But now, looking into the future has an established scientific methodology in many endeavours, not least in hydrology, where forecasting, particularly flood forecasting, is a well-developed activity. However, forecasting the future of hydrology itself is considered a risky business (Yevjevitch and Harmancioglu, 1987), whereas trying to predict the directions of hydrological research can be even more hazardous. There are so many factors to consider, many far from the control of the hydrologist: economic growth, political pressures, advances in technology, the availability of funding, progress in allied sciences and the ability to write project proposals and present them, to name but a few. Will the days ever return when an individual's innovative idea will spark a new line of research and attract the funds to promote it?

One view of the future from the mid-1970s (Rodda *et al.*, 1976) was that population growth, water pricing and the retreat from the 'climatic optimum' of the 1940s were the pertinent influences on UK research in hydrology. In the 1980s, the IAHS Hydrology 2000 Working Group reported on a range of topics likely to be important for the future (Kundzewicz *et al.*, 1987), but underestimated the magnitude of the impact of progress in information technology, remote sensing and GIS. Later, the Hydrology 2020 Group (Oki *et al.*, 2006) recognised trends, eras and fashions in hydrological research, identifying coming world water problems as an important driver. Wallace *et al.* (2000), in considering future UK research, also noted how the focus changed with time. In the 1960s, it was on the hydrological effects of upland afforestation; in the 1980s, it was acidification and then drought in the 1990s after some particularly dry years. Future climate change and its effects on the hydrological cycle, deforestation and diminishing water resources were considered by Arnell (1996) to be the important factors. McCulloch (2007) saw future research being built on vastly improved instrumentation for sensing, logging and retrieving data from all corners of the earth, manipulation of these data, together with the modelling of systems using computers with undreamed of capabilities. However Szollosi-Nagy (2014), amongst others, cautions that 'Forecasting is a difficult thing, particularly if it concerns the future'.

Probably, the programmes of future research mapped out by the successive phases of the IHP have had a certain amount of influence on directions. The IHP is currently in its eighth phase and is addressing Water Security under 6 themes. Similarly, the discharging of Acts of Parliament concerned with water, most recently the Flood and Water Management Act, 2010 and the Water Act, 2014, together with the implementation of the European Water Framework Directive (EU 2000) and Floods Directive (CEC, 2007) have guided certain aspects of research. However, the cooperative research 'Framework' programmes emanating from Brussels have driven a greater body of research in the past and can be expected to do so in the future. Likewise, the different initiatives coming from the international science community, ICSU for example, will have a bearing on what direction science will take initiatives such as Future Earth and the World Climate Research Programme. The space missions planned by NASA, the European Space Agency and similar bodies will also be important factors. Then in 2015, the UN General Assembly has to agree a fresh set of sustainable development goals to pilot the global development agenda, whereas a new global climate deal is expected to be reached at COP 21, the 21st Conference to the Parties to the UN Convention on Climate Change. The outcome of these sessions and that of the 3rd World Conference on Disasters will be important for the water community and for the research that supports it.

As important are the NERC collaborative funded programmes, programmes which in the past have explored different areas, such as LOIS (Land Ocean Interaction Study), LOCAR (Lowland Catchment Research) and FREE (Flood Risk from Extreme Events), each placing emphases on different aspects of the hydrological cycle. And, of course, there have

been other relevant programmes underway within NERC and in the other research councils, such as the Engineering and Physical Sciences Research Council's Flood Risk Management Research Consortium. Now, the NERC focus for water is on Drought and Water Scarcity and on the Changing Water Cycle (CWC). The CEH leaflet 'Meeting the Challenges of Environmental Change' – the Science Strategy 2014–2019 maps out a framework for future research by addressing areas such as: assessing the risks from environmental hazards, soil diversity and dynamics, improving the understanding of the quantity and quality of water resources and their variability, together with the risks presented by chemicals in the environment. The 10-year IAHS programme entitled *Panta Rhei* which started in 2014, as the follow-up to PUB (Prediction in Ungauged Basins), is likely to be influential for international and national hydrological research.

It is worth noting that the boundaries of hydrology and hydrological research appear to be widening continually. Evidence for the expanding hydrological universe is captured by the 5 volume *Encyclopedia of Hydrological Sciences* (Anderson, 2005) which embraces not less than 203 topics from: 'Fractals and Similarity Approaches in Hydrology' to 'Role and Importance of Paleohydrology in the Study of Climate Change and Variability'. Another example is Herschy and Fairbridge (1998) who list 261 entries from 'a' to 'y' in the *Encyclopedia of Hydrology and Water Resources*. The increasing number of journals devoted to hydrology also points to this expansion.

There is a contrast between contemporary topics for research and those underway prior to the widespread use of computers and mathematical modelling. Now, modelling is central to most research, but there are comparatively few studies where advances in methods of measurement and novel instrumentation are discussed. How these instruments are deployed – the design of networks – is another rarely addressed topic. Nevertheless, the need continues for improved physical measurements of the variables and parameters included in catchment models and their fluctuations in space and time. High-resolution temporal flux measurements are important for assimilation in weather forecasting models. Snow is one variable where there is concern, particularly its irregular distribution. Infiltration is another in terms of crusts, hydrophobicity and frozen soil which bring about heterogeneity. Soil moisture at relevant spectral scales is an elusive component of the continuity equation. Groundwater research must be encompassed along with that embracing the other components of the hydrological cycle. Groundwater is one of two lines of research for the future suggested by Uhlenbrook (2006), namely the physical and biogeochemical processes of groundwater systems, and the second is impact of climate variability and climate change for hydrological processes. Indeed, the implications of climate change, including sea level rise for hydrology and hydrological research, require more rigorous examination – alterations to the spatial and temporal characteristics of the hydrological cycle will have a greater impact on humanity than the rising global temperature.

12.2   Concern for the State of the Environment

Over the last 50 years or so, concern has been growing for the state of the environment and the impact of human activities on it. This period has been punctuated by numerous expressions of this concern both individual, such as Rachael Carson's 'Silent Spring' (Carson, 1962) and corporate ones, the 'Brundtland Report' (Brundtland, 1987), for example. Brundtland coined the term 'Sustainable Development' – an oxymoron which became a sort of **philosopher's stone** for the ambitions of projects and programmes promoted by many intergovernmental and non-governmental bodies alike. Then at the intergovernmental level, the UN Stockholm Conference on the Human Environment in 1972 stirred many governments and NGOs into action on environmental problems. Its successor the UN 'Earth Summit' in Rio de Janeiro in 1992 highlighted many more areas of concern and spawned a number of initiatives such as the Framework Convention on Climate Change, the Convention on Biological Diversity and Agenda 21, the, so-called, blueprint for the next century. National and local governments were urged to produce their own versions of Agenda 21; in the UK, Agenda 21 was promoted at the four levels of government (one action in the aquatic environment was the designation of 43 nitrate protection zones across the country). The subsequent successions of UN and NGO conferences, forums, water decades and water years have highlighted water's multitude of problems. Various initiatives have been influenced by the Report of the Dublin Conference on

Water and the Environment (WMO, 1992) which announced the four Dublin Principles and advocated a holistic approach to the management of water resources. Successive UN conferences in the wake of Rio have placed providing water supply and sanitation for those without them high amongst the goals, the Millennium Development Goals for example, but the environmental consequences of reaching these goals have not always been recognised. Amongst the exceptions is the Report of the World Commission on Dams (UNEP/IUCN 2000) which attempts to reach a compromise between promoters and opponents of dams. Where will the UN focus on water after 2015?

Since the first session of the WMO/UNEP Intergovernmental Panel on Climate Change in 1988 and that of the Conference of the Parties to the UN Framework Convention on Climate Change in 1994, it has become increasingly clear that the global climate is changing and changing at a rate matched in few other periods in geological history. What was not recognised in the initial sessions of IPCC was that the most serious impacts of climate change are those that affect the hydrological cycle and water resources. However, the recent IPCC Technical Paper on Climate Change and Water (Bates *et al.*, 2008) highlighted these impacts making clear the vital nature of the relationship between climate change and water resources. This points to the need for an intergovernmental water equivalent of IPCC to consider and act on world water problems before it is too late and the world water crisis materialises. The IPCC Fifth Assessment Report (2013) notes that since 1901 precipitation amounts averaged over mid latitudes in the northern hemisphere have increased and that there were increases in the frequency or intensity of heavy rain in Europe and North America over the same period. But for droughts, it is projected that increases are likely in some regions. Coming changes in the hydrological cycle were not expected to be uniform during the twenty first century. Contrasts between wet and dry regions and between wet and dry seasons will increase.

12.3   Interactive Water

Evidently almost every human activity alters some facet of the water environment and can result in radical effects over a range of time scales (Defra, 2013). Lovelock (1982), in advancing his new theory of life, the **Gaia Hypothesis**, stressed that the

Earth is a self-regulating system seeking a physical and chemical environment influenced by biota. Can Gaia be applied to river basins where fluvial processes shape the channels, flood plains and estuaries, whereas biological ones determine the life forms and how they develop? Indeed, a river basin can be considered as an ecosystem, where energy and materials interact and the living and non-living components are interdependent. Human actions, such as constructing a dam, can change the nature of the ecosystem. The *ecosystem approach* to river basin management parallels the *catchment management concept* (Ferrier and Jenkins, 2010) with its attention to the physical pathways of water, the movement of nutrients and pollutants and its consideration of the social, political and economic forces operating in the basin. Calder's (1999) **'Blue revolution'** espouses the same concept of water resources and land use being managed together, based on the new understanding of how land use influences hydrological processes. Together, these approaches offer a basis for *integrated water resources management,* the widely recognised term which developmental bodies and others advocate, but rarely seem to achieve. They provide opportunities for the holistic rather than the present largely *'a la carte'* approach to land and water management. The EU Water Framework Directive (EU 2000) makes considerable progress in this direction, offering a holistic view for the protection of surface and groundwater and their management based on the river basin – the natural geographical and hydrological unit. The Directive aims to achieve good status for all waters by a set deadline through a series of measures leading to the development of river basin management plans, including public participation in their preparation and execution. Efforts supporting the Directive in this country hinge largely on Defra's (2013) encouragement for an integrated catchment-based approach to improving the quality of the water environment (Caba) with activities in 80 or so basins. Earlier, Chorley (1969) in 'Water, Earth and Man' illustrated the advantages of an integrated view of earth and social sciences through the link that water and hydrology provide, to the extent that Newson (2009) felt able to use the title, 'Land, Water and Development' because the new integrated 'beast' had developed so much flesh. He considered that not only has hydrology helped to define anthropogenic impacts on planetary systems far beyond 'water' alone, but also it has changed almost every aspect of the broad range of governmental responsibilities known as 'integrated water resource management'.

12.4   Outreach

When the idea spread that a Hydrological Research Unit (HRU) should be set up, and later when the Unit was established at Wallingford, there could have been little awareness that the 'new science' of hydrology would extend its tentacles to so many academic disciplines and become such a vital component of the interdisciplinary effort needed to 'save the planet'. Of course, hydrology, as understood by the original members of HRU and others at that time, has expanded far beyond the then encompassing bounds of science and engineering and the limits imposed by the tacit agreements with other research groups about trespassing too far into their research territories. It is different today: meteorologists, climate scientists and others have become engaged in hydrology but often without much reference to the hydrological literature.

Judging the extent of the outreach of Wallingford hydrology may be difficult, but IH/CEH annual reports and science reviews illustrate what has happened over the years. Take the 1993–1994 report as an example (IH, 1995). Committee membership provides one guide – there were 22 international committees containing IH staff as members or chairmen and 36 national committees. Collaboration in research is another – IH was collaborating with 10 UK universities, with 19 research organisations in Europe and with 23 over the remainder of the world. These were in addition to the universities collaborating in TIGER (Terrestrial Initiative in Global Environmental Research), an NERC community programme coordinated by IH. The same annual report lists 13 UK government and private sector as clients for IH research along with 9 foreign governments and international bodies such as UNESCO. Outreach also includes education and training. The HOMS technology transfer system developed by WMO operates through about 100 national centres around the world, one being IH. Through this system, some 400 packages of information are available covering a wide range of topics, techniques and software, most contributed by national centres in the developed world and transferred on request to centres in the developing world. Some include training courses; in 1993, IH provided two in East Africa. IH also offered a number of training courses in the UK, on flood and low flow estimation and on the use of software packages, HYDATA, for example. Doctoral students at 4 universities were supervised by IH staff members along with 13 registered postgraduate students. Students from 10 local schools were given work experience and, unusually, 7 daughters of staff members came with their fathers to spend a day at IH. In addition, the Institute hosted visits by a wide range of business people, scientists and engineers and officials from government and industry. They came from the UK, the European Union and countries beyond.

Chapter 1 lists, from time to time, the numbers of peer-reviewed scientific publications and reports produced in different years. Members of staff and former members have written and continue to write scientific texts published commercially. The British media, at times of serious flooding and extensive drought, seek comments from CEH as a basis for articles in the press and on radio and television. Members of CEH appear on national and local radio and television during such events. Some have contributed to major of television and radio series. During 2013–2014, the National River Flow Archive received over 500 requests for data, many sparked by the extensive winter flooding. The intellectual capital originally invested in IH/CEH has been spread by movement of staff to chairs and senior positions in a number of universities in the UK and Ireland and to the USA, Australia, Brazil and several other countries. Staff members have also moved from Wallingford to posts in consulting engineers, environmental bodies, and abroad to WMO in Geneva, to the UN Department of Technical Cooperation in New York, to the Economic Commission for Africa in Addis Ababa and to NASA and IBM in the USA. Many members of staff have resided in developing countries for periods of up to 3 years to manage projects. It may be noted that in common with hydrologists in other countries, few have progressed to positions in central government where they might influence water policy.

The International Association of Hydrological Sciences' Press has been located at IH and CEH since 1972, adding another dimension to outreach from Wallingford. The Press produces the *Hydrological Sciences Journal*, now a monthly production shared with Taylor and Francis Ltd., and the IAHS Publications, known as the famous 'Red Books', the proceedings of symposia the Association organises, up to 10 per year. Several other publications come from the Press, such as the 'Blue Books' (special publications) and the thematic Benchmark Series in Hydrology which reprints the papers which provide the scientific foundations for hydrology. Together, these different publications have a global coverage. Members of IH and CEH have been the editors of the Journal (one is currently) and of many of the IAHS book publications. They have also edited a number of other hydrological journals. Increasingly research is being conducted with, or led by, universities. But that is a story for perhaps another book!

12.5   Community Research Programmes

NERC developed the concept of the Community Programme through the 1980s as a complementary funding model to the standard one of studentships and post docs where the initiative lies with the Principal Investigator. The result was that major new NERC-funded research programmes emerged, which integrated large research communities across universities and research institutes to achieve thematically focused research objectives. IH took the leading role in the development, science and management of several of these programmes. Research themes were pre-determined by NERC through a committee structure and managed by convenors who invited proposals that matched the stated themes. The first of these were the £20m TIGER (Oliver *et al.*, 1999) and the £25m LOIS (Huntley *et al.*, 2001). Later examples include Urban Regeneration and the Environment (URGENT), LOCAR and, at the time of writing, the CWC and the Drought and Water Scarcity Initiative programmes. In 2013, NERC launched the Initiative as a 5-year programme focussed on the UK with a budget of up to £12 million to feature research on drought and water scarcity. With shifting frequencies and patterns of rainfall and an accelerating demand for water, the impact of water shortages is becoming more critical for domestic consumers, food production, industry, energy production and the environment. The aim is to characterise the drivers of drought, to examine the multiple impacts of drought and water scarcity and to develop methods to support decision making. CEH has been awarded over £2.5 million from the total to fund the four areas which it will research including: a system-based study of impacts and interactions, decision making and risk, drought prediction and the impacts of uncertainties. The programme includes teams from a number of UK universities.

Of course, the organisation of science research and funding has changed considerably since the early days of hydrological research at Wallingford. That was a time when research was conducted in a formal top-down way. Individual scientists had considerable freedom, but science funding was very much controlled by centrally by departments, and then to research centres. This structure made it very difficult to conduct cross-institute and interdisciplinary research. Now, the science budget is no longer the main source of funds, and funding may come from a number of other sources. Individual scientists may have less freedom, however, to explore promising areas. Conversely, data are now far more readily available, such as from the CEH Information Gateway (https://gateway.ceh.ac.uk).

12.6   Contributions from Space Science

Since the advent of the *Space Age,* hydrological research has benefitted from the ability of satellites to present areal appreciations of hydrological variables and associated features of the surface of the globe. Each new technological advance in remote sensing has revealed more information about the hydrological cycle and added to understanding of the processes operating (Christopher *et al.*, 2012). Perhaps as important is the ability to access data worldwide, because of satellite-based communication systems and regionally from radar networks, compared to the long period in history when the mail was the only organised means for collecting data. Enormous volumes of data flow from online sources, the internet, mobile telephony and many others. Even greater volumes of data are anticipated for the future with the need for computational algorithms to process these streams of data, manipulate them and extract knowledge.

In many past missions, hydrology has often been neglected, or has been an add-on, rather than the main thrust. However in recent years, interest has grown in the global water and energy cycles. One example is the Tropical Rainfall Measuring Mission (TRMM), launched in 1997. Another example of the value added by space is the Global Precipitation Measurement (GPM) international satellite mission launched in February 2014 to provide next-generation observations of rain and

snow worldwide every three hours. Now, the focus of a number of earth observing missions has changed and this change appears to be continuing into the future. For instance, the 2017 mission named GRACE FO (Gravity Recovery and Climate Experiment Follow On) has the objective of providing high temporal resolution gravity field measurement for the purpose of tracking large-scale water movements. This is to improve on GRACE which was launched in 2002 (Hafeez *et al.*, 2011). That pair of satellites map gravity anomalies and have been employed to follow the vicissitudes of the hydrological cycle in the Amazon Basin. Will GRACE FO offer hydrologists a means of tracking soil moisture changes across basins? Surface soil moisture is a variable which ESA's SMOS satellite, launched in 2009, is able to map from its microwave imaging radiometer and this same facility will be provided by the SMAP satellite's synthetic aperture radar (https://smap.jpl,nasa.gov) which is to be launched in 2014. The SWOT mission (https//.jpl.nasa.gov./mission/hydrology), which is due in 2020, is being designed to resolve rivers 100 m wide and elevations to 10 cm. Future missions can be expected to improve resolutions, include more variables and cover more of the globe with shorter revisit times.

Looking outwards from Earth, currently a large amount of attention is directed at the presence of water on the Moon, Mars and some of the satellites of Saturn and Uranus, such as Enceladus and Triton. Past fluvial processes on the Red Planet seem to have shaped the surface. Is there a role for hydrologists in discovering more about the processes that were active? Experience in IH and CEH with automatic weather stations (AWS) suggest that their deployment on the Martian surface would give a valuable appreciation of the planet's weather and climate (Strangeways, 2014).

12.7   Future Modelling and Future Management

The future is uncertain, but making good investment decisions to ensure a sustainable supply of good quality rivers, lakes and groundwaters *and* ensuring their continuing good ecological status in the face of increasing human pressures requires some basis for prioritising options. One way of assessing the relative merits of potential options is by attempting to model future outcomes. The climate change predictions as used by IPCC are of this type, but models are also used to assess flood frequencies, to estimate future water demand, to manage water supplies through projected droughts, for the short-term future in flood forecasting and for many other purposes.

In all of these activities, ways need to be found for coping with the uncertainty that might arise from a plethora of causes. In the short term, it is the uncertainties in rainfall amounts and the effect of antecedent conditions on runoff generation that are the most important in flood forecasting. In some situations, it may be the greater uncertainty in rainfall forecasts, from radar patterns and numerical weather prediction and the provision of an adequate lead time. Further into the future, uncertainties about the economy, population changes and lifestyles that may affect emissions scenarios, as an input to predictions of climate change, as well as the uncertainties in global circulation models (and their hydrological components that control latent and sensible heat fluxes) that will be important.

Many of these sources of uncertainty are epistemic in nature, that is they result from a lack of knowledge rather than from simple random variability. For example, a statistical distribution might be fitted to historical floods, as a way of estimating flood frequencies for making decisions about investments in flood defences. In the Flood Estimation Handbook (IH, 1999), for example the Generalised Logistic distribution is recommended for the analysis of annual maximum floods. This allows future occurrences to be treated in terms of probabilities, so that risk-based decision theories can be used in assessing different investment options. But it is not known if the chosen distribution has the correct behaviour, or whether the actual occurrences are random, or if they follow a stationary distribution in the long term. The extremes of any underlying distribution with Poisson occurrences should follow a Generalised Extreme Value distribution given enough data, but there are never enough data to check either the form or stationarity of the distribution. With climate change, it is *expected* that the distribution will be non-stationary, though it has, as yet, proven difficult to show this in the data series (Robson, 2002; Kundzewicz and Robson, 2004), whereas current climate models do not actually do that good a job of reproducing historical rainfall statistics.

There are more epistemic uncertainties to be considered, not only in the analysis of past data,

but also in making some quantitative statements about future conditions. In many such cases, it is simply not possible to make probabilistic predictions and *scenario* modelling is resorted to. In this way, different potential scenarios of the future can be examined without suggesting that they have any associated probability of turning out that way. Climate change projections are of this type. Many different scenarios are run, but without any estimate of probability that any one scenario will be a correct scenario of the future.[1] But if probabilities cannot be associated with future scenarios, then, even though the best projections of the future might provide some evidence on which to act, risk-based decision theory cannot be used in assessing investment options. Some other mechanism for assessing different options is required, see for example the conditional loss exceedance probability approach in Rougier and Beven (2013), or the precautionary principle approach to adapting to future climate changes advocated by Wilby and Dessai (2010) and Beven (2011).

In one sense, being precautionary in making decisions in this way is only a way of not being too reliant on model predictions in an uncertain future. What seems a long time ago, Konikow and Bredehoeft (1992) reported on a post-audit analysis of a range of past groundwater modelling studies, evaluating how their predictions had actually compared with reality. In nearly every case, the models had not done well, but in some cases that was only because the projections of future boundary conditions used had been quite wrong. Running the model with the actual boundary conditions, once they were known, could then produce much improved results. However, some argue that there are alternatives to being precautionary, such as in adaptation planning and management where adaptation actions should be robust, flexible and/or reversible offering soft rather than hard options

The lessons learned from comparing predictions with reality seem to have been largely forgotten. Much of the modelling efforts in hydrology today

---

[1] UK Climate Projections 2009 (UKCP09) provide 10, 50 and 90% quantiles for its projection that have been sometimes misinterpreted as probabilities of future outcomes. In fact, they are quantiles derived from multiple runs of the same model with parameter perturbations. They are only relevant as a measure of uncertainty associated with the projections of that model scenario (and even then do not represent a full range of uncertainty). The quantiles differ for different emissions scenarios.

appear to be focussed on improving the representation of past data as information about past reality. There is perhaps an implicit assumption that if improved performance can be demonstrated when evaluated against historical data, then improved projections of future conditions should be expected. To have models that give the right results for the right reason is desirable (Kirchner, 2006). There is, however, a problem in representing the past. Not all of the data may be informative. For example, it can be shown that some catchment input–output might be inconsistent with water balance constraints and might therefore be disinformative in evaluating model performances (Beven *et al.*, 2011; Beven and Smith, 2014). In predicting the next event, even if there is some information about the inputs for that event, there is then no way of telling whether *a priori* if that event will be consistent or inconsistent with water balance constraints. In part, this comes from the limitations of the techniques of hydrological measurement, particularly for rain and snow inputs to catchments, and for measuring catchment discharges. Even after the decades of hydrological research recorded in this volume, more effort is needed to improve the assessment of certain of these very basic hydrological quantities.

But there may also be others ways that models can be evaluated that will change the way that hydrological modelling is carried out in the future. One of the advantages of improved computing and graphics capabilities is that the results of hydrological models can be enhanced. There is the possibility of hyperresolution simulations, systems which operate at much higher resolutions, even at global scales (Wood *et al.*, 2011); most earth system simulation centres are working towards providing simulations at finer and finer resolution. Having high-resolution model outputs, of course, does not mean that the resulting simulations are realistic, because in catchment science high resolution does not overcome the basic problem of epistemic uncertainty about inputs and water flow pathways (Beven and Cloke, 2012; Beven *et al.*, 2015). However, high-resolution visualisations allow local people and stakeholders to evaluate local predictions and, in particular, to suggest that those predictions might not be accurate, which requires a response. In such a future, modelling becomes much more of an interactive learning process that allows an interaction between modeller and user (see Beven, 2007; Lane *et al.*, 2011; Whatmore *et al.*, 2011; Beven and Alcock, 2012). This will change the activity of modelling and the way model results feed into decisions about catchment management in quite a fundamental way.

12.8 Opportunities Ahead

Advances in weather and climate forecasting will have an increasing impact on hydrology. Over recent decades, there has been impressive progress in WMO's World Weather Watch (WWW), particularly in numerical modelling, in observational capabilities and their standardisation, as well as in computer and communication technologies. This progress has facilitated more reliable forecasts and forecasts which look further ahead. More detail in space and time, in coupled meteorological and hydrological models has meant improved flood forecasts. In the UK, the products from the weather radar network used in distributed hydrological models have created a powerful tool for forecasting floods. However, errors remain in rainfall assessment and in the model parameters and these need to be resolved, as do scale issues. Improvements come from using the distributed radar rainfall data with distributed hydrological models. Matching the radar rainfall with gauge-measured totals still presents problems. And are there ways in which forecasting models can be improved? One difficulty is coping with the large amounts of data that are being generated. Another is the fact that these data exhibit trends, breaks and changes which render them non-homogeneous and non-stationary, forgetting the challenges posed by missing data, errors and outliers. These problems need to be solved, but perhaps more intriguing will be the opportunities that coming climatological forecasts will offer.

The recent development of novel ground-based and satellite soil moisture measurement technology offers the possibility for huge advances in flood forecasting, weather prediction and real-time soil moisture mapping. In particular, the ongoing development of a UK scale network of cosmic ray sensors (COSMOS-UK) will provide ground-based measures of soil moisture which can be blended with satellite observations and hydrometeorological models (such as JULES) run in near real time to provide soil moisture maps on a daily basis. The near real-time data from the UK raingauge network, flow monitoring network and soil moisture network offer the opportunity for development of methods for data assimilation to provide a step change in the utility of hydrometeorological, weather forecast and flood prediction models.

A major step forward in water resources assessment, flood forecasting and drought forecasting is the issuing of a Hydrological Outlook for the UK in the monthly Hydrological Summary (http://www.hydoutuk.net/). CEH carries out this work in cooperation with BGS, the Met Office and the gauging agencies. The science still has some way to go to achieve the necessary skill in the forecasts, but the science is advancing rapidly and will benefit from future investment in water cycle monitoring, telemetry and computational expansion. This will be a major focus for CEH in the immediate future in conjunction with the existing partners.

Pollution of surface and groundwater is an increasing problem globally and to a lesser extent in the UK. 'Fracking' to produce shale gas is being advocated by some authorities, including the government, as a palliative for Britain' s rising demand for energy and over-reliance on imported coal, gas and oil. However, fracking demands the deep injection of water under high pressure mixed with sand and chemicals to prompt the release of the gas, whereas this spent mixture may have potential risks to water supplies and the aquatic environment. Is there sufficient water available in the South East of England, as opposed to the north and west, to power the release of the gas? Can the EA ensure that its regulations will avoid any potentially harmful effects from exploration and production wells? There seems scope for research in these areas. The construction of onshore wind farms and solar farms may also present environmental problems which require hydrological research.

Of course, hundreds of new pharmaceuticals, radionuclides, macronutrients and other potential pollutants are being released into the environment every year which may have hazardous properties. They may enter potable water resources, particularly in basins like the Thames where river water is recycled several times (see Chapter 8). The novel pollutants which are emerging demand the attention of scientists and the pushing forward of the limits of detection, notably oestrogenic compounds, volatile organic compounds and human and animal pharmaceuticals. CEH is already in the vanguard of the new technology required and the scientific understanding needed in interpreting the major databases that will ultimately be assembled. Increasing emphasis is being placed on characterising the water quality of agricultural and urban/industrially impacted river systems as part of the environmental management schemes of regulatory agencies. This points to more intensive monitoring of a wider suite of pollutants at an increasing number of locations and the better management of these data. Future developments in water quality research will undoubtedly focus on the development of smarter technologies for monitoring chemicals and pollutants in waters to provide a higher resolution understanding of the sources of pollution, its fate and potential toxicity.

Recent water legislation has been driven by the adoption of targets that define the structural characteristics of the freshwater ecology of water bodies (populations and species present). New science needs to forge ahead with the concept of ecosystem services through the development of functional indicators of quality described by targets that reflect the services that water provides in different geographical settings. A further focus will be the natural capital concept whereby our freshwater systems need to be valued for the services they provide for humans and the environment. This demands that the definition of environmental flows is moved forward, the concept that will underpin new water abstraction policy but as yet lacks the necessary science for appropriate implementation.

CEH (2014) proposes that understanding today's and tomorrow's environmental challenges requires the marshalling of high-quality data and the tools for their analysis – *environmental informatics.* Discipline-leading data management capabilities have been developed through the Environmental Information Data Centre increasing the potential to exploit information and data in the terrestrial and freshwater sciences. Neal and Clarke (2007) considered such capacities as vital to the progress of research.

Ultimately, the future focus in CEH science must be to address the issues of environmental change on the water environment. This spans the research now underway to introduce the non-stationarity caused by climate change to methods for design flood estimation, the impact of climate change and population changes on water resource availability together with the work initiated on the impacts of climate and land use change on lake ecosystems, to name but a few. The concept of climate services is central to the future ambition at CEH. Water science is both a user and a provider of climate services and services are already being developed to describe variation in the hydrological cycle at a range of spatial (from individual catchments to the whole globe) and temporal (from days to centuries) scales. There is very much science still to do and it is always appropriate to consider where we have come from and what has been achieved to help guide future direction and priorities.

Human life in all its aspects will face mounting uncertainties as the future unfolds. In the current Anthropocene Era, among these uncertainties and the risks attached to them, those concerning water will be a large component at every scale from the local to the global. Hence, a rational assumption would be that it is vital to foster capabilities, particularly for governments to foster capabilities, which can anticipate problems before they arise to impact society and to offer solutions. This is the role of research. What this research might encompass in the next 20–50 years is a matter of considerable speculation and some imagination. With so many and rapid changes taking place currently, perhaps lessons from the past offer less plausible guidance to the future than previously.

Starting more than 50 years ago with small, seemingly pristine, headwater basins, lysimeters and plots, Wallingford's hydrological research has extended to the largest river basins in the UK and worldwide to the Amazon. Both natural and human-modified basins have been the subjects of study. The depth of this research has developed from simply solving a catchment's water balance to encompass investigations of the governing processes, physical, chemical and biological. Forecasting and prediction have been essential elements of this research. Atmospheric processes and those operating below the ground surface, as well as processes active in the stream and river, have been included to give a greater environmental dimension. At the same time, the purview of hydrological research incorporates aspects of climate change and touches on various facets of society and economy. Principle No.1 of the Dublin Conference's Guiding Principles (WMO, 1992) calls for a **holistic approach** to the effective management of water resources. The research described in this volume must be close to achieving this ambition.

In his Foreword to the HESS Special Issue dedicated to Jim McCulloch, Jim Dooge (2007) reflected on the mission of the Institute of Hydrology. He saw it as 'the transformation of hydrology over the past five decades from a compendium of unplanned measurements and practical manipulations to systematic measurements and procedures based on scientific principles. This has advanced the acceptance of hydrology as a branch of science capable of offering key solutions to important social problems'. What transformations to hydrology can CEH and the wider research community accomplish in the future? This is the challenge.

12.9   References

Anderson, M. (2005) *Encyclopedia of Hydrological Sciences*. John Wiley & Sons Ltd., pp. 3145.

Arnell, N. (1996) *Global warming, river flows and water resources*. Institute of Hydrology/John Wiley.

Bates, B.C., Kundzewicz, Z.W., Wu, S. & Palutikof, J.P. (2008) *Climate Change and Water*. Technical Paper of the Intergovernmental Panel on Climate Change, . IPCC Secretariat, Geneva, pp. 210.

Beven, K.J. (2007) Working towards integrated environmental models of everywhere: uncertainty, data, and modelling as a learning process. *Hydrology and Earth System Science*, 11 (1), 460–467.

Beven, K.J. (2011) I believe in climate change but how precautionary do we need to be in planning for the future? *Hydrological Processes*, 25, 1517–1520. doi:10.1002/hyp.7939

Beven, K., Smith, P.J. & Wood, A. (2011) On the colour and spin of epistemic error (and what we might do about it). *Hydrol. Earth Syst. Sci.*, 15, 3123–3133. doi:10.5194/hess-15-3123-2011

Beven, K.J. & Alcock, R. (2012) Modelling everything everywhere: a new approach to decision making for water management under uncertainty. *Freshwater Biology*, 56. doi:10.1111/j.1365-2427.2011.02592.x

Beven, K.J. & Cloke, H.L. (2012, Comment on Wood *et al.* (2011),) Hyperresolution global land surface modeling: Meeting a grand challenge for monitoring Earth's terrestrial water. *Water Resources Research*, 48, W01801. doi:10.1029/2011WR010982

Beven, K.J., Cloke, H., Pappenberger, F., Lamb, R. & Hunter, N. (2015) Hyperresolution information and hyperresolution ignorance in modelling the hydrology of the land surface. *Science China Earth Sciences*, 58 (1): 25–35.

Beven, K.J. & Smith, P.J. (2014) Concepts of information content and likelihood in parameter calibration for hydrological simulation models. *ASCE J., Hydrol. Eng.*, doi: 10.1061/(ASCE)HE.1943-5584.0000991

Brundtland, G.H. (1987) *Our Common Future, The World Commission on Environment and Development*. Oxford University Press, pp. 400.

Calder, I.R. (1999) *The Blue Revolution: Land Use and Integrated Water resources Management.* Earthscan Publications Ltd, pp. 193.

Carson, R. (1962) *Silent Spring.* Houghton Mifflin.

CEC (Commission of European Communities) (2007) *Directive of the European Parliament and of the Council on the Assessment and Management of Floods.* CEC, Brussels.

CEH (2014) *Meeting the Challenges of Environmental Change, Science Strategy 2014–2019,* Centre for Ecology and Hydrology, 9.

Christopher, M., Neale, U. & Cosh, M.H. (eds) (2012) *Remote sensing and hydrology.* IAHS Press Pub No. 352, pp. 499.

Chorley, R.J. (ed) (1969) *Water, Earth and Man.* Methuen & Co., Ltd, pp. 588.

Defra (2013) *Catchment Based Approach to Improving Water Quality.* Department for the Environment, Food and Rural Affairs, pp. 36.

Dooge, J.C.I. (2007) Foreword. In: Neal, C., Clarke, R.T. & McCulloch, J.S.G. (eds), *A view from the watershed revisited,* Special Issue, *Hydrology and Earth System Sciences,* 11, 1, 1.

EU (2000) Directive 2000/60/EC of the European Parliament and the Council of 23 October establishing a framework for community action in the field of water policy. *Official Journal of the European Communities,* **L327,** 1–32.

Ferrier, R.C. & Jenkins, A. (eds) (2010) *Handbook of Catchment Management.* Wiley-Blackwell, pp. 540pp.

Flood and Water Management Act (2010), Chapter 29, 81.

Hafeez, M., van de Giesen, N., Bardsley, E., Seyler, F., Pail, R. & Taniguchi, M. (eds) (2011) *GRACE, Remote sensing and ground-based methods in multi-scale hydrology.* IAHS Press Pub No. 343, pp. 206.

Herschy, R.W. & Fairbridge, R.W. (eds) (1998) *Encyclopedia of Hydrology and Water Resources.* Kluwer Academic Publishers, pp. 803.

Huntley, D., Leeks, G.J.L. & Walling, D. (eds) (2001) *Land Ocean Interaction - Measuring and Modelling Fluxes from River Basins to Coastal Seas.* International Water Association, London, pp. 286.

IH (1995) *Annual Report 1993–1994,* Institute of Hydrology, 67.

IH (1999) *Flood Estimation Handbook.* Vol. **5.** Institute of Hydrology.

IPCC (WMO/UNEP Intergovernmental Panel on Climate Change) (2013) Summary for Policymakers. In: Stocker, T.F., Qin, D., Plattner, G.-K., Tignor, M., Allen, S.K., Boschung, J., Nauels, A., Xia, Y., Bex, V. & Midgley, P.M. (eds), *Climate Change 2013: The Physical Science Basis. Contribution of Working Group I to the Fifth Assessment Report of the WMO/UNEP Intergovernmental Panel on Climate Change.* Cambridge University Press, Cambridge, United Kingdom and New York, NY, USA.

Kirchner, J.W. (2006) Getting the right answers for the right reasons: Linking measurements, analyses, and models to advance the science of hydrology. *Water Resources Research,* **42,** W03S04. doi:10.1029/2005WR004362

Konikow, L.F. & Bredehoeft, J.D. (1992) Groundwater models cannot be validated. *Advances in Water Resources,* **15,** 75–83.

Kundzewicz, Z.W., Gottschalk, L. & Webb, B. (eds) (1987) *Hydrology 2000* IAHS Publication No. 171. IAHS Press, pp. 100.

Kundzewicz, Z.W. & Robson, A.J. (2004) Change detection in hydrological records—a review of the methodology. *Hydrological Sciences Journal,* **49** (**1**), 7–19. doi:10.1623/hysj.49.1.7.53993

Lane, S.N., Odoni, N., Landström, C., Whatmore, S.J., Ward, N. & Bradley, S. (2011) Doing flood risk science differently: an experiment in radical scientific method. *Transactions of the Institute of British Geographers,* **36** (**1**), 15–36.

Lovelock, J. (1982) *Gaia: A New Look at Life on Earth.* Oxford University Press.

Neal, C. & Clarke, R.T. (2007) Contemporary and future perspectives for hydrological and catchment sciences. In: Neal, C., Clarke, R.T. & McCulloch, J.S.G. (eds), *A view from the watershed revisited.* Special Issue, *Hydrology and Earth System Sciences,* 11(1), 663.

Newson, M. (2009) *Land, Water and Development,* 3rd edn. Routledge, pp. 441.

McCulloch, J.S.G. (2007) All our yesterdays: a hydrological perspective. In: Neal, C., Clarke, R.T. & McCulloch, J.S.G. (eds), *A view from the watershed revisited,* Special Issue, *Hydrology and Earth System Sciences,* **11**(1), 3–11.

Oki, T., Valeo, C. & Heal, K. (eds) (2006) *Hydrology 2020, An Integrating Science to Meet World Water Challenges* IAHS Publication No. 300. IAHS Press, pp. 190.

Oliver, H.R., Gash, J.H.C. & Gurney, R.J. (eds) (1999) *The TIGER Programme.* Special Issue, *Hydrology and Earth Systems Science Journal.* 3, 1–149.

Robson, A.J. (2002) Evidence for trends in UK flooding. *Philosophical Transactions of the Royal Society of London, Series A: Mathematical., Physical and Engineering Sciences,* **360** (**1796**), 1327–1343.

Rodda, J.C., Downing, R. & Law, F.M. (1976) *Systematic Hydrology.* Newnes Butterworth, pp. 399.

Rougier, J. & Beven, K.J. (2013) Model limitations: the sources and implications of epistemic uncertainty. In: Rougier, J., Sparks, S. & Hill, L. (eds), *Risk and uncertainty assessment for natural hazards.* Cambridge University Press, Cambridge, UK, pp. 40–63.

Strangeways, I. (2014) Automatic weather stations for Mars. *Weather,* **69,** 8–13.

Szollosi-Nagy, A. (2014) *Valedictory Address on his retirement as Rector from UNESCO – IHE Delft,* September 2014.

Wallace, J., O'Connell, E. & Whitehead, P. (2000) Future UK Hydrological research. In: Acreman, M. (ed), *The Hydrology of the UK*. Routledge.

Water Act (2014) *Reform of the water industry to increase competition and protect the environment*. Chapter

Whatmore, S.J., Lane, S.N., Odoni, N.A., Ward, N. & Bradley, S. (2011) Coproducing flood risk knowledge: redistributing expertise in critical participatory modelling'. *Environment and Planning A*, **43** (7), 1617–1633.

Wilby, R.L. & Dessai, S. (2010) Robust adaptation to climate change. *Weather*, **65** (7), 180–185.21, 246

WMO (1992) The Dublin Statement and Report of the Conference, *Proc. International Conference on Water and the Environment*, 26–31 January 1992, Dublin, Ireland. 55.

Wood, E.F., Roundy, J.K., Troy, T.J. *et al.* (2011) Hyper-resolution global land surface modeling: Meeting a grand challenge for monitoring Earth's terrestrial water. *Water Resources Research*, **47** (5), W05301. doi:10.1029/2010WR010090

UNEP/IUCN (2000) *Dams and development: a new framework for decision-making* Report of the World Commission on Dams. Earthscan Publications Ltd, pp. 404.

Uhlenbrook, S. (2006) *Inaugural Address: Catchment hydrology with satellites, models and rubber boots*. UNESCO-IHE Delft, pp. 22.

Yevjevitch, V. & Harmancioglu, N.B. (1987) Some reflections on the future of hydrology. In: Rodda, J.C. & Matalas, N.C. (eds), *Water for the Future*. IAHS IAHS Publication No.164, pp. 405–414.

River current meter gauging from a 'bosun's chair' – Invergarry 1913: Some of the earliest river gauging in the UK was carried out by Captain William Newsam McClean in 1913, when he installed a water level gauge on the River Garry, close to the lodge of Invergarry House, Inverness-shire. He subsequently set up gauging stations and raingauges across the River Ness basin. With a strong desire to see large-scale hydro power established in the Scottish Highlands as a means of stimulating economic and social development, McClean used his own resources to establish a hydrological survey that was truly ahead of its time. The appointment of an Inland Water Survey Committee in 1935 was likely, in no small part, to have been due to his efforts. (Source: Reproduced with permission of Dundee University.)

Index